S-Heterocycles
Retrospect, Prospects, and Biological Applications

S-Heterocycles
Retrospect, Prospects, and Biological Applications

Edited by

Shrikaant Kulkarni
Victorian Institute of Technology, Australia
Email: srkulkarni21@gmail.com

Hemantkumar Akolkar
Abasaheb Marathe Arts & New Commerce Science College, India
Email: hemantakolkar@gmail.com

Vijay M. Khedkar
Vishwakarma University, India
Email: vijaycpn@gmail.com

and

A. K. Haghi
University of Coimbra, Portugal
Email: akhaghi@gmail.com

ROYAL SOCIETY OF CHEMISTRY

Print ISBN: 978-1-83767-401-5
PDF ISBN: 978-1-83767-485-5
EPUB ISBN: 978-1-83767-486-2

A catalogue record for this book is available from the British Library

The Royal Society of Chemistry is a charity, registered in England and Wales, Number 207890, and a company incorporated in England by Royal Charter (Registered No. RC000524), registered office: Burlington House, Piccadilly, London W1J 0BA, UK, Telephone: +44 (0) 20 7437 8656.

Visit our website at books.rsc.org

Printed in the United Kingdom by CPI Group (UK) Ltd, Croydon, CR0 4YY, UK

Preface

Heterocycles are a key class of vital importance in organic chemistry. The heterocyclic compounds present frameworks that constitute the structural building blocks of numerous bioactive naturally occurring products. The formation of heterocycles has been the mainstay for a long time and dragged tremendous attention toward the formulation of better, milder, cleaner, and highly productive approaches. Numerous classic initiatives demand the formation of bonds between carbon and heteroatoms, resulting in reactions such as dipolar cycloaddition leading to the closure of rings, reductive aminations, or nucleophilic substitutions for synthesizing heterocycles. This edited volume aims to focus on synthesizing many S-heterocycles and their derivatization, property profile, and pharmaceutical and therapeutic applications.

The heterocycles serve as the core structure for innumerable bioactive compounds exhibiting biological and pharmacological behaviors. Different classes of S-heterocycles exhibit biological activity. It is most sought-after to evolve novel green methods of synthesis for deriving biological activity.

S-enabled heterocycles have emerged as one of the key classes of heterocycles and have drawn tremendous attention attributed to their identity as 'privileged pharmacological structures' across natural products and highly active pharmaceutical agents. A host of methods for the synthesis of such S-heterocycles have been reported. This edited volume brings to the fore the latest strategies used for the synthesis of S-heterocycles from many precursors, including β-nitrostyrenes and their relevant counterparts. An emphasis has been laid upon the furtherance and limitations in synthesizing such S-heterocycles.

Benzo-fused five-membered S-heterocycles, such as indole, benzofuran, and benzothiophene, containing a solitary heteroatom, possess vital applications in the fields of medicine, agrochemistry, and materials science.

S-Heterocycles: Retrospect, Prospects, and Biological Applications
Edited by Shrikaant Kulkarni, Hemantkumar Akolkar, Vijay M. Khedkar and A. K. Haghi

Published by the Royal Society of Chemistry, www.rsc.org

Metal-catalyzed reactions provide well-formulated synthesis routes for the formation of C–S bonds, resulting in the direct synthesis of heterocyclic compounds. The approaches adopted have a comparative advantage over traditional ones in terms of fewer steps, improved atom economy, low catalyst loading, and regio- and stereoselectivity. Due to their widespread applications across the pharmaceutical industry, the formation of the C–S bond has gained tremendous attention. This book focuses on various mechanistic approaches, including the metal-catalyzed synthesis of S-heterocycles, particularly indole, benzofuran, and benzothiophene, reviewing the developments in the past and deliberating upon the unexplored future prospects.

Heterocyclic compounds are very vital in synthetic chemistry, the pharmaceutical industry, agrochemistry, and materials science. This edited volume reports on the latest developments in practical and selective reactions and methods involving carbenes and nitrenes to fulfill the growing needs of structural diversity in heterocycles. Carbenes and nitrenes are highly reactive, but controllable, intermediates instrumental in a range of catalytic transformations. Together, these chapters provide a comprehensive view of the diversity of methods, properties, and the variety of applications in the construction of S-heterocycles that are made available through carbene, nitrene, and thiazine chemistries. We are sincerely grateful to the authors for their contributions to the developments described in this volume that have opened new vistas for the synthesis of heterocycles.

This edited volume is divided into four parts. Part 1 is devoted to chemistry, synthesis pathways, property profile, and applications of thiazoles, and covers five chapters. Chapter 1 discusses the methods of synthesis and biological applications of 1,3-thiazoles due to their properties. Chapter 2 elaborates on the synthesis and therapeutic applications of dithiazoles attributed to their unique properties. Chapter 3 throws light on the methods of synthesis and pharmacological applications of isothiazole because of its characteristic properties. Chapter 4, on the other hand, delves into the synthesis and biological applications of benzothiazole, while Chapter 5 gives an account of synthesis routes and biological applications of 2,4-thiazolidinediones due to their distinct properties.

Part 2 is dedicated to the chemistry, synthesis routes, property profile, and applications of thiadiazoles, and covers two topics. Chapter 6 discusses the synthesis pathways and biological applications of derivatives of 1,2,4-thiadiazole, while Chapter 7 sheds light on the synthesis protocols and biological applications of derivatives of 1,3,4-thiadiazole.

Part 3 is assigned to the chemistry, synthesis protocols, property profile, and applications of thiopyrans, thiazine, and thiazepine. Chapter 8 discusses synthesis methods and virological applications of thiopyrans. Chapter 9 summarizes the synthesis and biological applications of thiazine. Chapter 10 sheds light on schemes of syntheses and biological applications of benzothiazepine due to its distinct properties.

Last, part 4 focuses on the chemistry, synthesis methods, property profile, and applications of thiophenes, and contains two chapters. Chapter 11 deals with synthesis schemes and biological applications of thiophene attributed to its biological activity, while concluding Chapter 12 covers synthesis methods and the biological behavior of benzothiophene.

In general, S-heterocyclic compounds are quite a useful class of organic compounds due to their characteristic structural features, functional groups and substituents, and five- and six-membered cyclic structures.

This volume will be of immense help to academicians, researchers, and students in pursuing their academic and research initiatives. It will pave the way for future research directions and in enriching the knowledge base of heterocyclic chemistry.

Shrikaant Kulkarni
Hemantkumar Akolkar
Vijay Khedkar
A. K. Haghi

Editorial: S-Heterocycles: Navigating the Past and the Future

Introduction

The world of organic chemistry is punctuated by molecules that not only define eras but also shape the trajectory of scientific inquiry. S-heterocycles, characterized by the incorporation of a sulfur atom into cyclic structures, stand as enigmatic compounds with a rich history and promising future. This editorial embarks on a journey through the retrospect and prospects of S-heterocycles, shedding light on their historical significance, current applications, and the boundless potential they hold for the future of chemistry.

The narrative of S-heterocycles in the vast tapestry of organic chemistry weaves together a rich history with promising prospects. As we reflect on the retrospect of their journey, we find a story of pivotal discoveries, diverse applications, and contributions to multiple scientific domains. Simultaneously, the future beckons with exciting prospects and promising advancements in green chemistry, nanotechnology, and biomedical applications. In this editorial, we delve into the retrospect and prospects of S-heterocycles, celebrating their legacy and anticipating the trail they are set to blaze in the years to come.

S-heterocycles stand as remarkable motifs, weaving together a story of innovation, discovery, and future potential. This editorial embarks on a reflective journey, examining the historical contributions of S-heterocycles and envisioning their promising prospects in shaping the future landscape of chemistry and beyond.

S-Heterocycles: Retrospect, Prospects, and Biological Applications
Edited by Shrikaant Kulkarni, Hemantkumar Akolkar, Vijay M. Khedkar and A. K. Haghi

Published by the Royal Society of Chemistry, www.rsc.org

Sulfur-containing heterocycles, or S-heterocycles, occupy a distinctive place in the landscape of organic chemistry, a realm where the fusion of sulfur and cyclic structures has yielded compounds of profound significance. As we look back on their historical journey and gaze forward to the promising prospects that lie ahead, S-heterocycles stand as a testament to the enduring spirit of discovery and innovation within the scientific community.

Retrospective Glance

Discovery and Early Recognition: The tale of S-heterocycles begins with the discovery of sulfur-containing compounds, such as thiophene, in the late 19th century. These molecules sparked curiosity, eventually leading to the systematic exploration of their properties and reactivities. The early recognition of their unique structures laid the foundation for subsequent breakthroughs.

Role in Pharmaceuticals: The retrospection of S-heterocycles reveals their pivotal role in the development of pharmaceuticals. From the penicillin structure featuring a thiazolidine ring to modern antivirals and anticancer agents, sulfur-containing heterocycles have been indispensable in the medicinal chemistry landscape.

Catalysis and Coordination Chemistry: Over the years, S-heterocycles have emerged as key players in catalysis. Their participation in coordination chemistry resulted in the discovery of novel catalysts, affecting the efficiency and selectivity of various chemical transformations.

Pioneering Discoveries: The roots of S-heterocycles trace back to pioneering discoveries, where early chemists identified sulfur-containing rings as fundamental components. Thiophene, a cornerstone in this journey, emerged from coal tar and paved the way for the exploration of various sulfur-containing structures.

Medicinal Chemistry Milestones: Looking back, the contributions of S-heterocycles to medicinal chemistry stand out prominently. From the penicillin breakthrough featuring a thiazolidine ring to the synthesis of powerful antibiotics and anticancer agents, S-heterocycles have played a crucial role in advancing pharmaceutical sciences.

Materials Science Marvels: The retrospective lens reveals S-heterocycles as key players in materials science. Their incorporation into conductive polymers, liquid crystals, and other materials with unique electronic properties has fueled advancements in electronic devices, sensors, and displays.

Foundations in Natural Products: The story of S-heterocycles begins with the pages of nature itself. Thiophene, a simple five-membered ring containing a sulfur atom, was among the first sulfur-containing heterocycles discovered in coal tar. Over the years, derivatives of thiophene have emerged as key players in natural products, laying the foundations for further exploration.

Medicinal Marvels: Retrospectively, S-heterocycles have been pivotal in the realm of medicine. The discovery of antibiotics, such as penicillin, with its thiazolidine ring, marked a turning point in combating bacterial infections. The journey continues with modern anticancer agents and antivirals, where S-heterocycles play a central role in the development of life-saving pharmaceuticals.

Foundations in Natural Products: The roots of S-heterocycles stretch deep into the historical soil of organic chemistry. Early discoveries in coal tar and natural products, such as the ubiquitously present thiophene, laid the foundations for a class of compounds that would later find applications across diverse scientific disciplines.

Current State of Affairs

Catalytic Innovations: In the present era, S-heterocycles continue to shine in catalysis, influencing the development of novel synthetic methodologies. The catalytic prowess of thiourea derivatives in asymmetric synthesis exemplifies the ongoing contributions of S-heterocycles to the toolkit of organic chemists.

Biomedical Breakthroughs: S-heterocycles are at the forefront of biomedical research. The exploration of their potential in drug discovery, particularly in the synthesis of compounds with potent anticancer and antiviral properties, underscores their importance in addressing pressing global health challenges.

Sustainable Synthesis: The contemporary era witnesses a growing emphasis on sustainable chemistry. S-heterocycles are actively contributing to this paradigm shift, with researchers exploring green and eco-friendly synthetic routes, utilizing bio-based feedstocks, and adopting environmentally benign reaction conditions.

Current Applications

Medicinal Chemistry Advancements: In the present era, S-heterocycles continue to shine in medicinal chemistry. Their structural diversity and reactivity have positioned them as crucial components in the synthesis of pharmaceuticals, demonstrating efficacy against various diseases.

Catalytic Innovations: The contemporary landscape witnesses S-heterocycles spearheading catalytic innovations. Thiourea derivatives, for instance, have become versatile organocatalysts, playing a pivotal role in asymmetric synthesis and green chemistry practices.

Materials Science Prowess: S-heterocycles contribute significantly to the materials science domain. Their involvement in the design of conductive polymers, liquid crystals, and advanced materials for electronics showcases their adaptability and versatility in creating materials with tailored properties.

Catalytic Innovations: In the current landscape, S-heterocycles have become instrumental in catalysis. Their role as ligands in metal coordination complexes has paved the way for catalytic systems, influencing the efficiency and selectivity of various chemical reactions.

Green Chemistry Initiatives: S-heterocycles are finding a renewed purpose in the realm of green chemistry. Researchers are exploring sustainable synthesis methods, leveraging S-heterocycles to develop environmentally friendly routes and eco-friendly reaction conditions.

Biomedical Frontiers: Current applications extend to the biomedical field, where S-heterocycles are being explored for their potential in drug discovery and therapeutics. The versatile reactivity of sulfur-containing heterocycles provides a platform for designing compounds with enhanced pharmacological properties.

Catalytic Marvels: In the present era, S-heterocycles have found prominence in catalysis. Acting as ligands in metal coordination complexes, they influence reaction pathways and selectivity. This catalytic prowess extends to both academic research and industrial processes, underscoring their importance in modern synthetic methodologies.

Medicinal Chemistry Renaissance: The journey of S-heterocycles in medicinal chemistry continues with the discovery of novel drug candidates. Their presence in antibiotics, antivirals, and anticancer agents highlights their therapeutic potential. As researchers delve deeper into the intricacies of molecular interactions, new vistas for drug development are unveiled.

Prospects for the Future

Green Chemistry Initiatives: The future of S-heterocycles is intricately woven with the fabric of green chemistry. Researchers are exploring sustainable synthesis routes and eco-friendly methodologies, aiming to minimize the environmental impact while maximizing the efficiency of synthetic processes.

Advanced Materials and Nanotechnology: Prospects for S-heterocycles extend into the realm of nanotechnology. Their unique electronic properties make them promising candidates for applications in nanomaterials, sensors, and other cutting-edge technologies, contributing to the ongoing revolution in materials science.

Biomedical Discoveries: As the understanding of S-heterocycles deepens, the future holds the promise of unlocking new biomedical discoveries. Continued research aims to harness the potential of these compounds for drug development, diagnostics, and therapeutics, paving the way for breakthroughs in medical science.

Sustainability in Synthesis: The prospects for S-heterocycles shine brightly in the domain of green and sustainable synthesis. The focus on environmentally friendly processes, bio-based feedstocks, and reduced environmental impact positions S-heterocycles as key players in the pursuit of sustainable chemistry.

Nanotechnology and Advanced Materials: The future envisions S-heterocycles at the forefront of nanotechnology and advanced materials. Their unique electronic properties make them promising candidates for applications in electronic devices, sensors, and materials with tailored properties for specific technological needs.

Biomedical Frontiers: Anticipating the future, S-heterocycles are poised to make groundbreaking contributions to biomedical sciences. Continued research holds the promise of discovering novel drug candidates, targeting specific biological pathways, and addressing emerging health challenges of these compounds in shaping the materials that define our technological landscape.

Advancements in Nanotechnology: The future holds the promise of S-heterocycles as key components in nanotechnology. Their unique electronic properties position them as potential candidates for the development of nanomaterials and advanced devices, fueling innovations in electronics, sensors, and beyond.

Biomedical Frontiers: As we peer into the future, S-heterocycles are poised to continue their impact on biomedical research. The exploration of novel derivatives for targeted drug delivery, imaging agents, and theranostics unveils exciting possibilities for addressing complex medical challenges with precision and efficacy.

Materials Beyond Conventional Boundaries: The prospects for S-heterocycles extend to unconventional materials. Their integration into emerging fields, such as flexible electronics, organic photovoltaics, and energy storage devices, holds the potential to revolutionize the landscape of materials science.

Green Chemistry Revolution: The future promises a green chemistry revolution, with S-heterocycles playing a pivotal role in sustainable synthesis. Researchers are actively seeking innovative approaches to reduce the environmental impact, making S-heterocycles key players in the development of cleaner and more sustainable chemical processes.

Biomedical Marvels: The prospect of S-heterocycles in biomedical applications continues to expand. Researchers envision the development of next-generation drugs, diagnostic agents, and theranostic platforms that leverage the unique properties of these molecules for personalized and targeted healthcare.

Sustainable Synthesis and Green Chemistry: The future promises an even more pronounced role for S-heterocycles in the realm of sustainable synthesis. The quest for greener chemical processes aligns seamlessly with the versatility of sulfur-containing structures, offering a pathway to minimize environmental impact and reduce the carbon footprint of synthetic procedures.

Advanced Materials and Nanotechnology: The dawn of nanotechnology brings with it exciting prospects for S-heterocycles. Their unique electronic properties position them as candidates of interest in the development of advanced materials for electronic devices, sensors, and other nanotechnological applications.

Conclusion

In retracing the footsteps of S-heterocycles from historical milestones to contemporary applications, one cannot help but marvel at their enduring impact on diverse scientific disciplines. As we stand at the crossroads of retrospect and prospects, S-heterocycles beckon researchers and innovators to explore the untapped realms of their chemistry. The future promises a continued journey of discovery, with S-heterocycles poised to play a central role in shaping the landscape of organic chemistry, materials science, and medicine. As we navigate the intricate pathways of these compounds, we are not merely uncovering the secrets of the past but also charting a course toward a future where S-heterocycles will undoubtedly continue to captivate and inspire generations of scientific minds.

In the symphony of organic chemistry, S-heterocycles play a melody that resonates through time – a melody that encapsulates the retrospect of their historical significance and harmonizes with the prospects of future advancements. As scientists and researchers continue to unravel the complexities of these sulfur-containing structures, we witness a narrative that extends beyond the laboratory, influencing diverse fields and shaping the trajectory of scientific progress. The journey of S-heterocycles is far from complete, and as we navigate the uncharted terrain ahead, the promise of innovation and discovery stands as a testament to their enduring legacy. The journey of S-heterocycles from their historical roots to their current applications exemplifies their enduring significance in the vast tapestry of chemistry. As we stand at the threshold of the future, the prospects for S-heterocycles are as dynamic and promising as ever. Whether catalyzing chemical transformations, serving as therapeutic agents, or contributing to sustainable synthesis, S-heterocycles are set to play pivotal roles in shaping the scientific landscape. The unfolding chapters in their story promise not just retrospection but a forward-looking narrative that holds the key to unlocking new dimensions in science and technology. The journey of S-heterocycles is far from over, and the best, it seems, is yet to come.

In conclusion, the retrospect and prospects of S-heterocycles unveil a captivating narrative of scientific progress and future possibilities. From their historical significance in shaping the foundations of organic chemistry to their current roles in catalysis, medicine, and sustainable synthesis, S-heterocycles have left an indelible mark on the scientific landscape.

As we look ahead, the future promises an exciting chapter where S-heterocycles will lead the way in forging a more sustainable, innovative, and interconnected world. The journey continues, with researchers, chemists, and visionaries collaboratively authoring the next chapters in the evolving saga of S-heterocycles – a saga that holds the promise of addressing global challenges and reshaping the frontiers of scientific discovery.

In retracing the evolutionary path of S-heterocycles and peering into the horizon of their future applications, it becomes clear that the journey of these compounds is far from concluded. Their molecular intricacies, catalytic prowess, and therapeutic potential continue to beckon researchers toward uncharted territories of scientific exploration. As we navigate the retrospect and prospects of S-heterocycles, we find ourselves at the intersection of tradition and innovation, where each discovery paves the way for the next chapter in the captivating narrative of sulfur-containing heterocycles. The story is ongoing, and the pages yet to be written hold the promise of unveiling new dimensions in chemistry, materials science, and medicine, ensuring that the legacy of S-heterocycles endures and evolves for generations to come.

Shrikaant Kulkarni

Contents

S-Heterocycles: Retrospect, Prospects, and Biological Applications
Edited by Shrikaant Kulkarni, Hemantkumar Akolkar, Vijay M. Khedkar and A. K. Haghi

Published by the Royal Society of Chemistry, www.rsc.org

Part I: Thiazoles

CHAPTER 1

Synthesis, Properties, and Biological Applications of 1,3-Thiazoles

N. R. DAREKAR[a], V. M. KHEDKAR[b], P. K. GADEKAR[c], S. D. MHASKE[d], S. J. TAKATE[e] AND H. N. AKOLKAR*[f]

[a]P.G. and Research, Department of Chemistry, Radhabai Kale Mahila Mahavidyalaya, Ahmednagar, Maharashtra, 414001, India; [b]School of Pharmacy, Vishwakarma University, Pune, Maharashtra, 411048, India; [c]Department of Pharmaceutical Sciences, Bouvé College of Health Sciences, Northeastern University, Boston, Massachusetts 02115, USA; [d]Department of Chemistry, Dadapatil Rajale Arts, Science and Commerce College, Adinathnagar, Ahmednagar, Maharashtra, 414505, India; [e]P.G. and Research, Department of Chemistry, New Arts, Science and Commerce College, Ahmednagar, Maharashtra, 414001, India; [f]Department of Chemistry, Abasaheb Marathe Arts and New Commerce, Science College, Rajapur, Ratnagiri, Maharashtra, 416702, India
*E-mail: hemantakolkar@gmail.com

1.1 Introduction

The largest and most significant subfield in organic chemistry is heterocycles. Heterocyclic compounds are important in the biological, agricultural, pharmaceutical, and drug development industries. The majority of the one-, two-, or three-heteroatom, five- and six-membered heterocyclic

S-Heterocycles: Retrospect, Prospects, and Biological Applications
Edited by Shrikaant Kulkarni, Hemantkumar Akolkar, Vijay M. Khedkar and A. K. Haghi

Published by the Royal Society of Chemistry, www.rsc.org

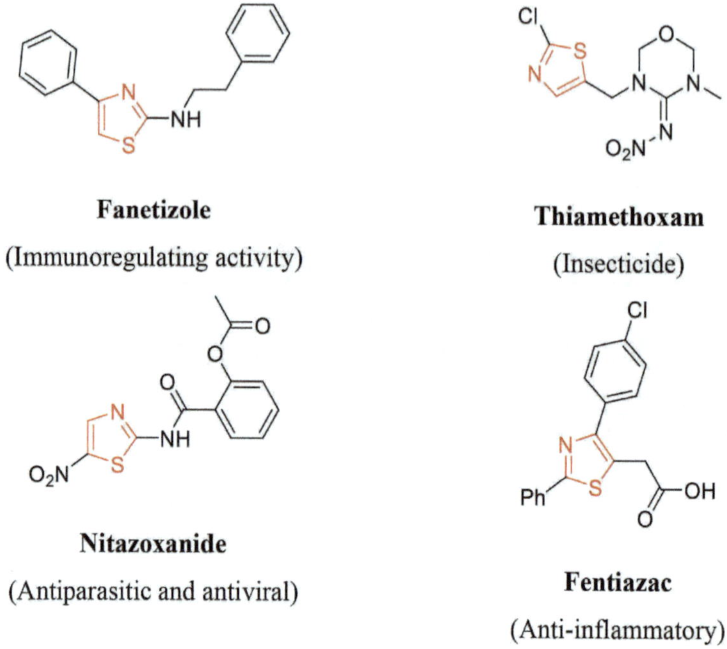

Figure 1.1 Structure of 1,3-thiazole-containing drugs.

compounds play an active role in the metabolism of living things. Over the last few decades, researchers have continued to be interested in and drawn to heterocycles that contain both nitrogen and sulfur atoms. 1,3-Thiazole is a well-known sulfur- and nitrogen-containing five-membered heterocyclic compound found in numerous clinically used drugs, such as nizatidine, meloxicam, ritonavir, tiazofurin, bleomycin, and nitazoxanide (see Figure 1.1).

Thiazole-containing molecules are desirable candidates for medicinal chemistry due to their broad range of biological functions, including anti-inflammatory,[1] antimicrobial,[2,3] and antiproliferative activities,[4] and their roles as adenosine receptor antagonists,[5] *C*-aryl glucoside sodium–glucose cotransporter-2 (SGLT2) inhibitors,[6] histone deacetylase inhibitors,[7] and checkpoint kinase-1 (CHK1) inhibitors.[8]

The thiazole moiety has drawn a lot of attention lately. Numerous review publications have emphasized the importance of the thiazole scaffold in the development and synthesis of more current bioactive pharmaceutical candidates.[9–13] The production of thiazole-containing compounds and their most recent anticancer, antitubercular, antioxidant, and antibacterial characteristics are covered in the current study.

1.2 Recent Advances in the Synthesis of Thiazole

An ecologically friendly technique involving solvent-free multicomponent reactions of primary amines, isothiocyanates, and alkyl bromides with nanorod zinc oxide (ZnO) as a catalyst produces 1,3-thiazole as reported by Hajinasiri *et al.*[14] (see Scheme 1.1).

Wu *et al.* described the one-pot synthesis of 2,4-disubstituted thiazole.[15] In this reaction, Au(I) was used as a catalyst at room temperature, and the terminal alkyne, Mor-DalPhosAuOMs (as a catalyst), ammonium sulfonate, 8-methylquinoline *N*-oxide, and substituted thioamides were also used (see Scheme 1.2).

The synthesis of hydrazinyl thiazole utilizing tosylates, aryl aldehydes, and thiosemicarbazide was described by Gurav *et al.*[16] using Fe_2O_3 nanoparticles (NPs) as the catalyst (see Scheme 1.3).

Dhas *et al.*[17] have synthesized a series of 4-substituted-2-amino thiazoles with an excellent yield using various substituted phenacyl bromides and

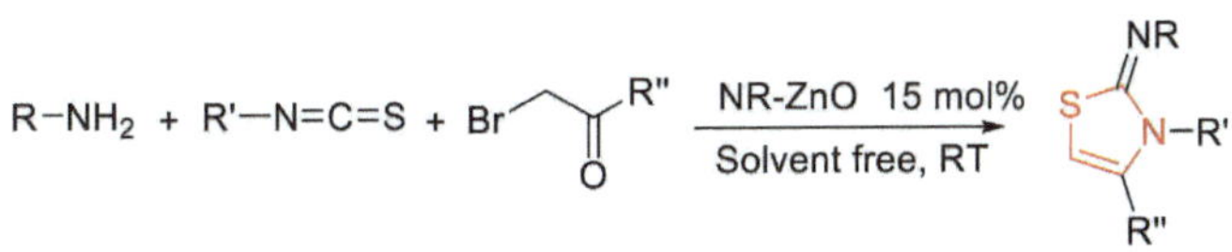

Scheme 1.1 NR-ZnO catalysed synthesis of 1,3-thiazole derivatives. Reproduced from ref. 14 with permission from Bentham Science, Copyright 2023.

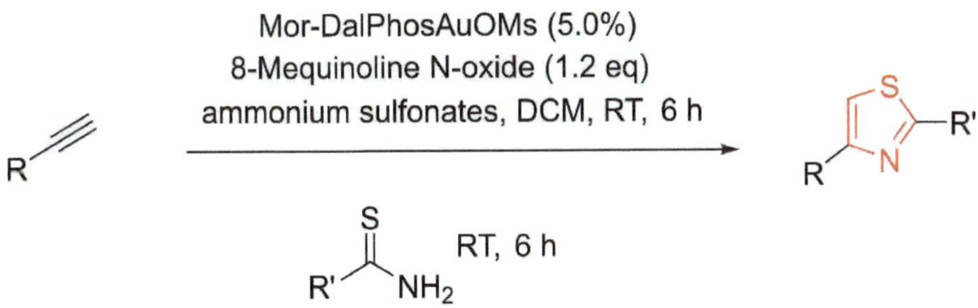

Scheme 1.2 Mor-DalPhosAuOM-catalyzed synthesis of 1,3-thiazole derivatives.

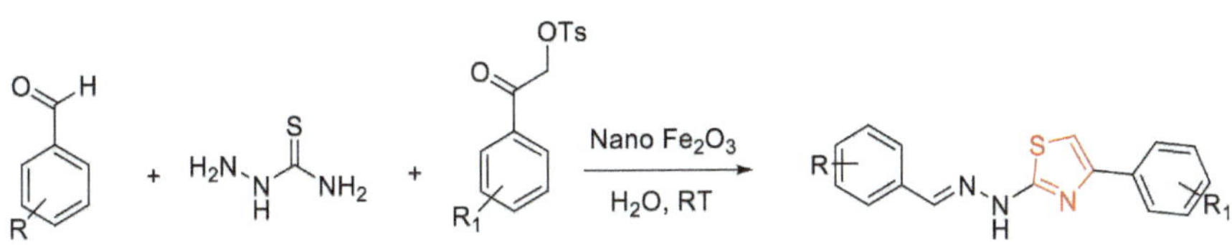

Scheme 1.3 Nano-Fe_2O_3-catalyzed synthesis of 1,3-thiazole derivatives. Reproduced from Ref. 16 with permission from the Royal Society of Chemistry.

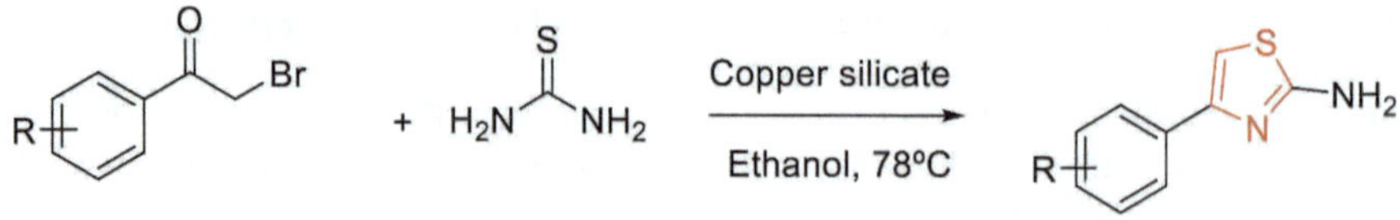

Scheme 1.4 Copper silicate-catalyzed synthesis of 1,3-thiazole derivatives.

thiourea in ethanol as a solvent at a temperature of 78 °C and copper silicate as a heterogeneous and reusable catalyst (see Scheme 1.4).

Masquelin and Obrecht[18] have reported a process for synthesizing amino-1,3-thiazole in the solution phase. Amido-thioureas and thioureido-thioureas were produced by condensing amidines and thiouronium salts with isothiocyanates. The *S*-alkylated intermediate was produced by further treatment with -bromo ketones, which was followed by a base-catalyzed ring-closure reaction to produce 1,3-thiazoles (see Scheme 1.5).

Safaei-Ghomi *et al.*[19] have reported a straightforward, gentle, and practical approach for the synthesis of 3,4,5-trisubstituted-1,3-thiazole-2(3*H*)-thiones involving a primary amine, carbon disulfide (CS_2), and α-haloketone under room temperature utilizing potassium iodide (see Scheme 1.6).

Shahbazi-Alavi *et al.*[20] used a superior catalyst called crosslinked sulfonated polyacrylamide (Cross-PAA-SO_3H) attached to nano-Fe_3O_4 to synthesize substituted thiazole derivatives through three-component reactions involving 4-methoxyphenacyl bromide or phenacyl bromide, CS_2, and a primary amine in ethanol (see Scheme 1.7).

The synthesis of thiazole derivatives has been described by Nakhaei,[21] utilizing Fe_3O_4@ZrO_2-SO_3H as an effective and recyclable catalyst in water

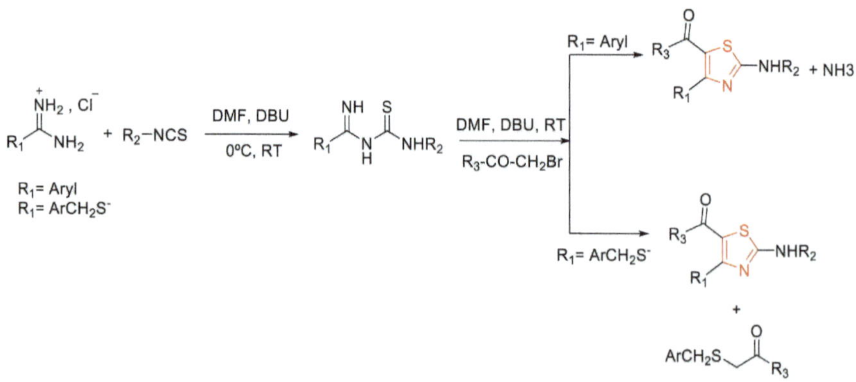

Scheme 1.5 Base-catalyzed ring-closure reaction to synthesize 1,3-thiazoles. Reproduced from Ref. 18 with permission from Elsevier, Copyright 2001.

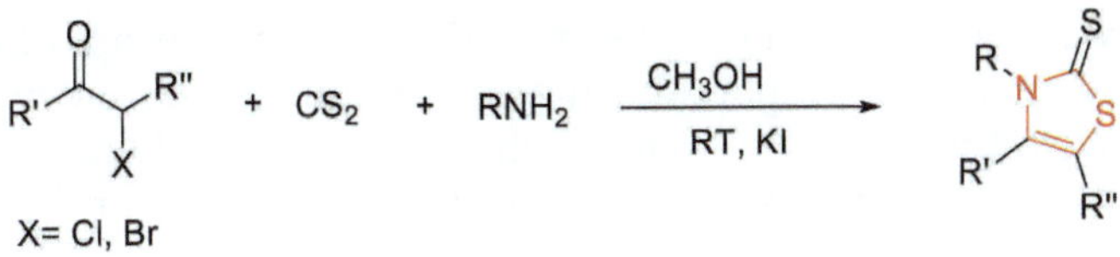

Scheme 1.6 Synthesis of 3,4,5-trisubstituted-1,3-thiazole-2(3*H*)-thiones. Reproduced from Ref. 19 with permission from Springer Nature, Copyright 2013.

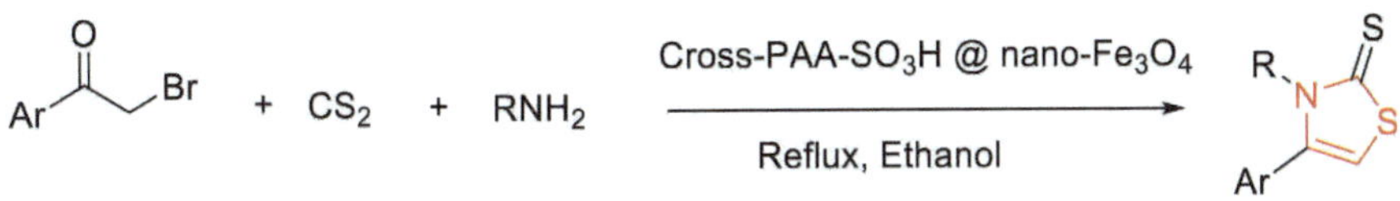

Scheme 1.7 Cross-PAA-SO_3H @ nano-Fe_3O_4-catalyzed synthesis of thiazoles.

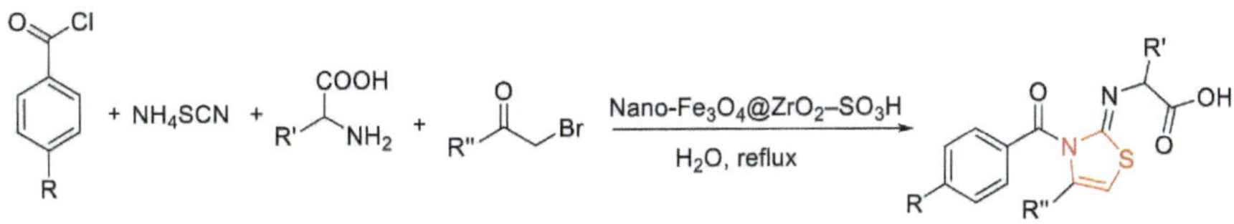

Scheme 1.8 Fe_3O_4@ZrO_2-SO_3H-catalyzed synthesis of thiazoles. Reproduced from Ref. 21 with permission from Springer Nature, Copyright 2017.

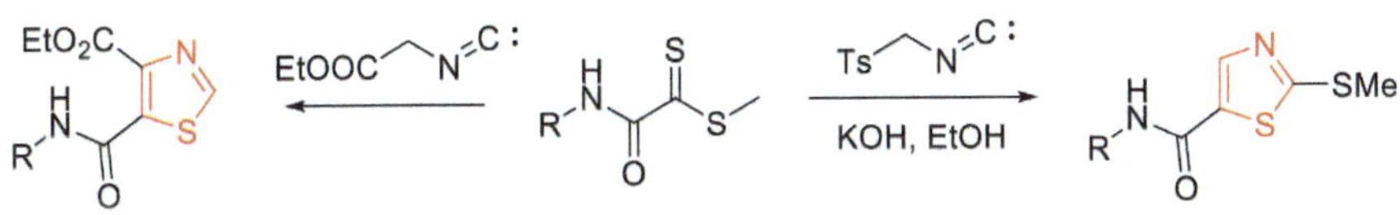

Scheme 1.9 Synthesis of thiazole derivatives using methyl-2-oxo-2-(amino)ethanedithioates. Reproduced from Ref. 22 with permission from the Royal Society of Chemistry.

to facilitate the reaction of acyl chloride with ammonium thiocyanate, amino acids, and alkyl bromides (see Scheme 1.8).

Singh *et al.*[22] have synthesized different thiazole derivatives through the cyclization of methyl-2-oxo-2-(amino)ethanedithioates with ethyl isocyanoacetate and toluenesulfonylmethyl isocyanide (TosMIC), respectively, using bases (see Scheme 1.9).

Mahata *et al.*[23] used two catalyst-free approaches to synthesize a variety of new diphenyl-1,3-thiazole-linked barbituric acid hybrids from easily accessible starting materials. Within 30 minutes, equivalent barbituric acid moiety-tethered trisubstituted thiazoles were produced by the reaction between barbituric acid and arylglyoxal and substituted thioamides. The reaction was carried out in the presence of three to four drops of water. An alternative method was reported using ammonium sulfide and aryl nitriles instead of aryl thioamide in a water medium (see Scheme 1.10).

Balti *et al.*[24] reported two methods for the preparation of thiazole-2(3*H*)-thiones, where aryl groups were substituted at the 4th and 5th positions (see Schemes 1.11 and 1.12). Supported acids and bases are utilized in flow processing as a tool to make the synthesis approach and product separation easier.

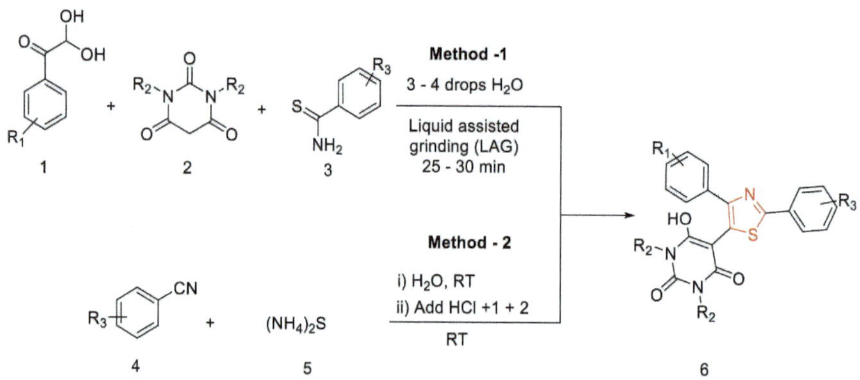

Scheme 1.10 Synthesis of diphenyl-1,3-thiazole-linked barbituric acid hybrids. Reproduced from Ref. 23 with permission from the Royal Society of Chemistry.

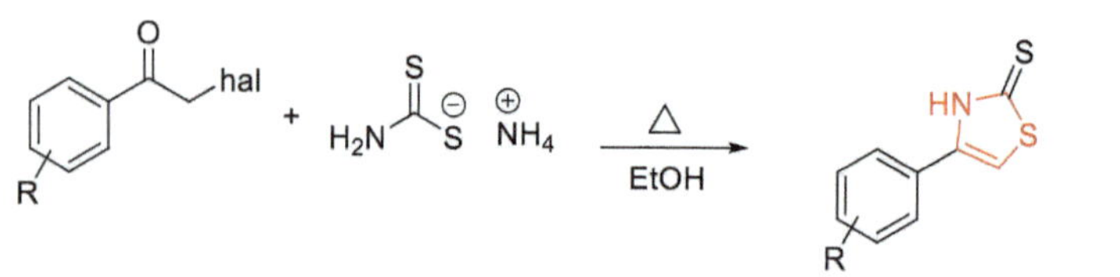

Scheme 1.11 Synthesis of thiazole-2(3*H*)-thiones. Reproduced from Ref. 24 with permission from the Royal Society of Chemistry.

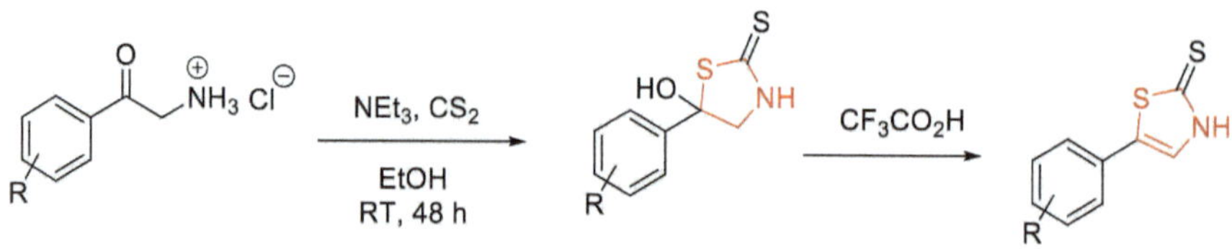

Scheme 1.12 Synthesis of thiazole-2(3*H*)-thiones. Reproduced from Ref. 24 with permission from the Royal Society of Chemistry.

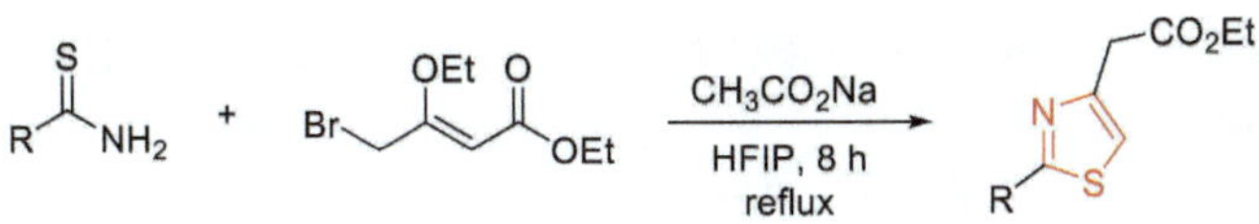

Scheme 1.13 Synthesis of thiazole derivatives.

Thiazole compounds were produced as a result of the interaction of thiourea derivatives with the electrophile 4-bromo-3-ethoxycrotonate, according to Alsharif *et al.*[25] (see Scheme 1.13).

Silver(I) trifluoromethanethiolate ($AgSCF_3$) and *N*-propargylamines have been used successfully to synthesize substituted thiazol-2(3*H*)-ones in good yields by Behera *et al.*[26] *N*-Propargylamines undergo [3,3]-sigmatropic rearrangement and 5-*exo*-dig cyclization as the process progresses (see Scheme 1.14).

In the study by Hassell-hart *et al.*,[27] a sulfur ylide insertion processcatalyzed by iridium was used to synthesize a thiazole library (see Scheme 1.15).

Adole *et al.*[28] described a unique one-pot method for producing thiazole derivatives from 1,2,6,7-tetrahydro-8*H*-indeno[5,4-*b*]furan-8-one, thiosemicarbazide, and a variety of acetophenones in the presence of polyethylene glycol (PEG)-400 (see Scheme 1.16).

Zhou *et al.*[29] reported a direct method for synthesizing thiazole by catalyst-free intramolecular cyclization reaction of *N*-allylbenzothioamide. The reaction proceeded smoothly with *N*-bromosuccinimide (NBS) as an oxidant in dichloroethane at 100 °C (see Scheme 1.17).

According to Nguyen and Retailleau,[30] the reaction involving chalcones, isothiocyanates, and sulfur can be used to synthesize thiazole using bases as a catalyst (see Scheme 1.18).

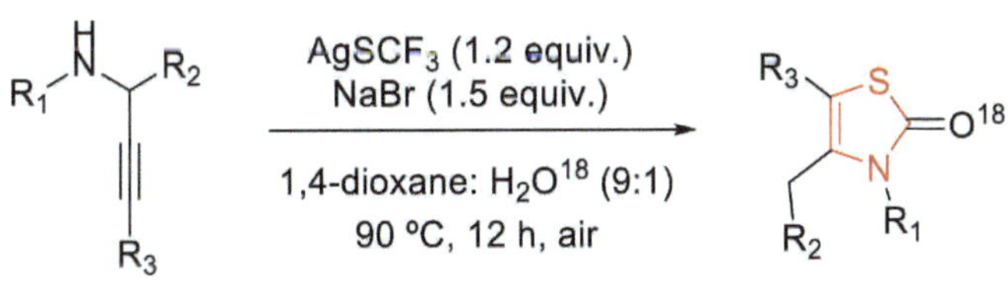

Scheme 1.14 Synthesis of thiazol-2(3*H*)-ones using *N*-propargylamines and $AgSCF_3$. Reproduced from Ref. 26 with permission from the American Chemical Society, Copyright 2022.

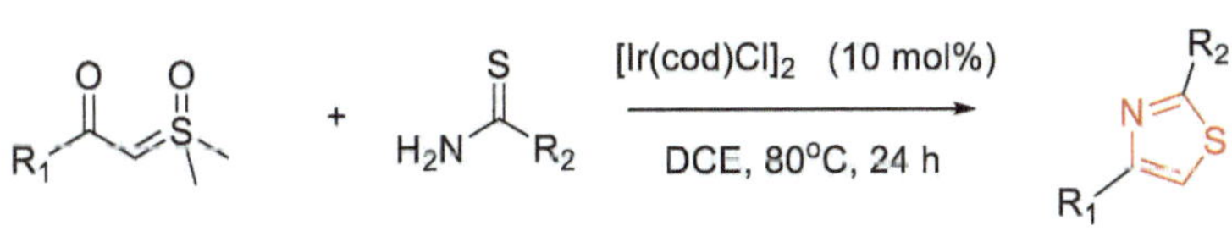

Scheme 1.15 Synthesis of 2,4-disubstituted thiazoles using $[Ir(cod)Cl]_2$.

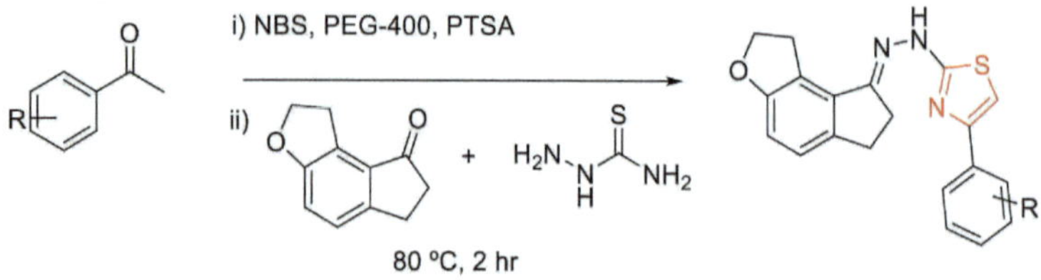

Scheme 1.16 Synthesis of substituted thiazole using PEG-400. Reproduced from Ref. 28 with permission from John Wiley & Sons, Copyright 2020 Wiley-VCH Verlag GmbH & Co. KGaA, Weinheim.

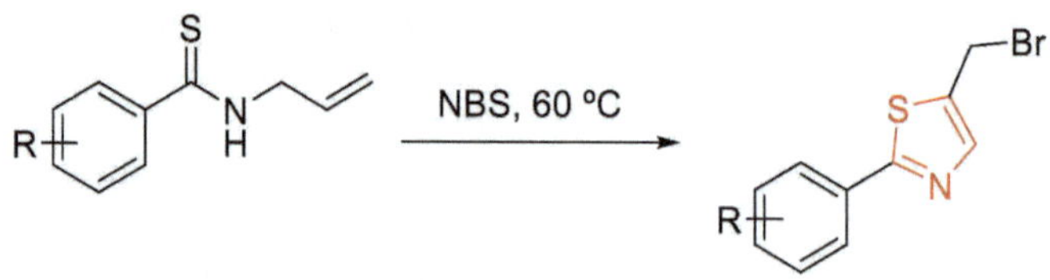

Scheme 1.17 Direct method for synthesizing thiazole using catalyst-free intramolecular cyclization. Reproduced from Ref. 29 with permission from Elsevier, Copyright 2023.

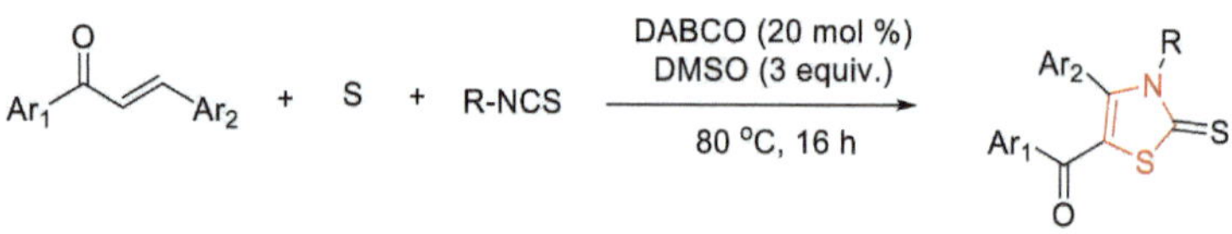

Scheme 1.18 Base-catalyzed synthesis of thiazoles. Reproduced from Ref. 30 with permission from the American Chemical Society, Copyright 2021.

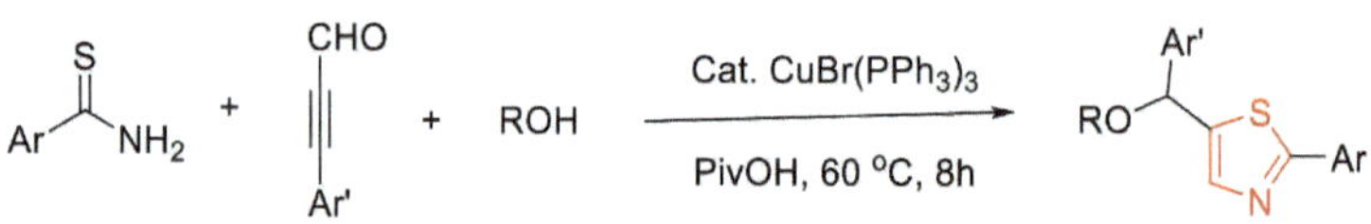

Scheme 1.19 $CuBr(PPh_3)_3$-catalyzed synthesis of thiazole. Reproduced from Ref. 31 with permission from the American Chemical Society, Copyright 2020.

Using a three-component reaction catalyzed by copper(I), Wang *et al.*[31] have developed a unique and simple method for producing functionalized thiazoles from thioamides, ynals, and alcohols (see Scheme 1.19).

Nikpassand *et al.*[32] utilized an aldehyde, a thiosemicarbazide, and a bromoacetophenone in a three-component reaction to synthesize thiazole, which was catalyzed by the ionic liquid 3,3′-(pentane-1,5-diyl)bis(1,2-dimethyl-1*H*-imidazol-3-ium) bromide ([PDBMDIm]Br) (see Scheme 1.20).

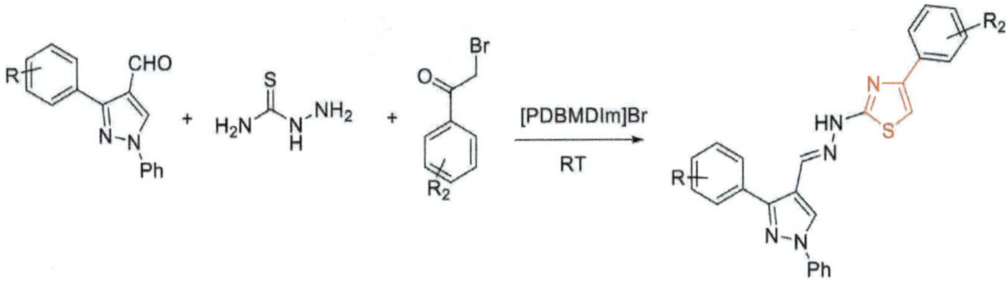

Scheme 1.20 Synthesis of thiazole using ionic liquid [PDBMDIm]Br.

1.3 Biological Applications

1.3.1 Thiazole as an Anticancer Agent

Singh *et al.*[33] have synthesized novel sulfonamide-thiazole-derived acetamide derivatives and evaluated for their anticancer potential. Using the sulforhodamine B (SRB) test and paclitaxel as the reference medication, all the synthesized compounds were evaluated for their ability to inhibit the NCI-H226 lung cancer cell lines. The presence of sulfonamide-carrying thiazole in addition to an electronegative atom, as well as Cl, Br, NO_2, and F compounds, which are connected to the 4th position (R-group) of the thiazole ring, boosts the anticancer activity of compounds (see Figure 1.2).

Using a multicomponent process and microwave assistance, Mamidala *et al.*[34] have designed and produced thiazole derivatives. All the synthesized substances were tested against MCF-7, MDA-MB-231/ATCC, HS 578 T, BT-549, T47D, and MDA-MB-468 to determine their *in vitro* cytotoxic activities. One of the compounds has strong anti-MDA-MB-231/ATCC cell line activity. The results of the anticancer screening revealed that the groups that donate electrons at the locations R, R_1, and R_2 will be advantageous for activity. Based on molecular docking studies, the inclusion of –OH groups in thiazole derivatives may also be helpful for the treatment of breast cancer (see Figure 1.3).

Bandaru *et al.*[35] have designed a novel series of amide-based thiazole–pyrimidines. Using the 3-(4,5-dimethylthiazol-2-yl)-2,5-diphenyl-2*H*-tetrazolium bromide (MTT) assay method, all the newly reported compounds were tested

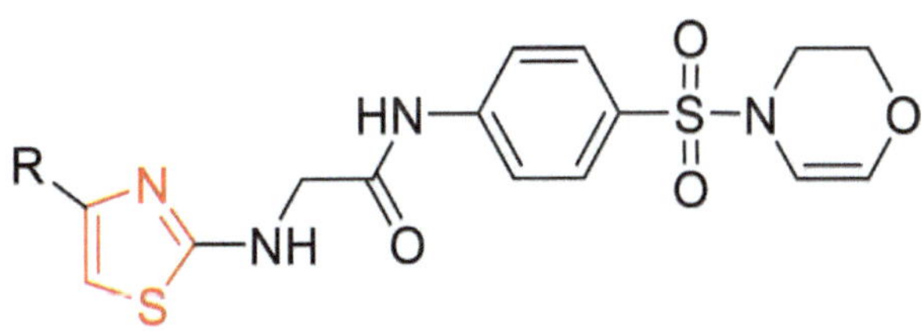

Figure 1.2 Structure of sulfonamide-thiazole-derived acetamide derivatives used as anticancer agents.

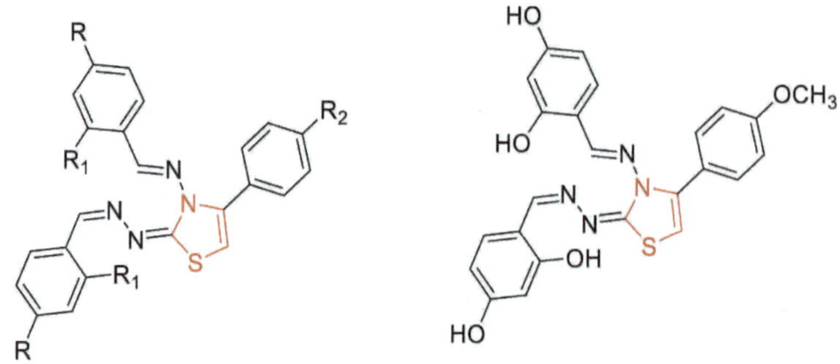

Figure 1.3 Structures of compounds showing strong anti-MDA-MB-231/ATCC cell line activity.

against A549 (lung), MCF-7 (breast), Colo-205 (colon), and A2780 (ovarian) human cancer cell lines. The results of the anticancer screening indicated that tri- and di-substituted analogs were discovered to have greater potential than mono-substituted analogs (see Figure 1.4).

Zhang *et al.*[36] have synthesized a number of novel morpholine-containing thiazole amide derivatives. Using the MTT assay method, all the newly reported compounds were tested against the A549 (human non-small-cell lung cancer cell), HeLa (human cervical cancer cell), HT29 (human colon cancer cell), and Karpas 299 (human lymphoma cell line). The substituent R of the target compounds was changed to examine if the cellular activity could be increased. The results showed that the addition of a meta-halogen (particularly chloro)-substituted phenyl ring at R had a substantial impact on the activity. A comparison of molecules with chloro-substitution at various sites on the phenyl ring has shown their order of activity as m-Cl > 3,4-Cl_2 > 2,4-Cl_2 (see Figure 1.5).

4-(Dimethylamino)-2-(*p*-tolylamino) thiazole-5-carbonitrile, a new thiazole compound synthesized by Biswas *et al.*,[37] demonstrated strong *in vitro* anticancer activity against the targeted Bcl-2 Jurkat cell line as measured by the MTT test (see Figure 1.6).

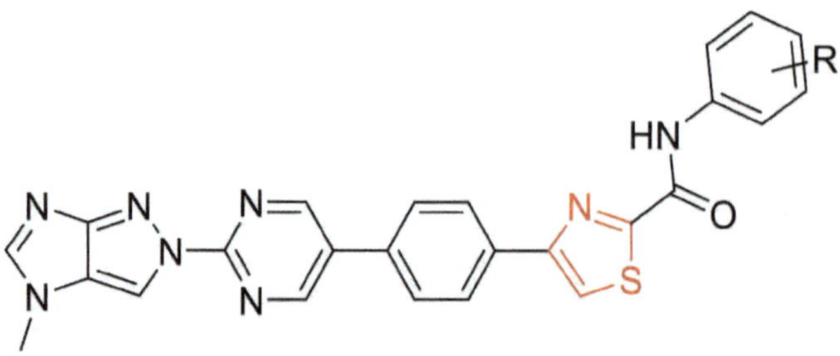

Figure 1.4 Structure of amide-based thiazole–pyrimidines used as anticancer agents.

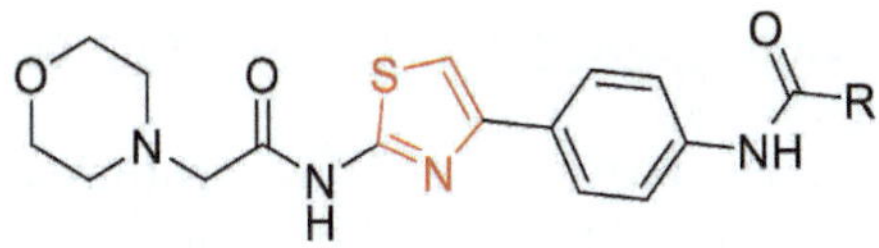

Figure 1.5 Structure of morpholine-containing thiazole amides used as an anticancer agent.

Oliveira *et al.*[38] have synthesized novel phthalimido-bis-1,3-thiazoles. All the synthesized compounds were evaluated for their cytotoxicity against various tumor cell lines. At least eight tumor cell lines showed considerable cytotoxic action for substances containing a phthalimido-bis-1,3-thiazole moiety. Their half maximal inhibitory concentration (IC_{50}) values were calculated, and tests were also performed to observe how they affected peripheral blood mononuclear cells (PBMCs) and red blood cells. The phthalimido-thiazole compounds were found to have IC_{50} values under 12 μM and to be nontoxic to untransformed cells (see Figure 1.7).

Thiazole-naphthalene derivatives were reported by Wang *et al.*,[39] and their ability to inhibit tubulin polymerization and antiproliferative properties were tested. The majority of the compounds under investigation significantly reduced the proliferation of MCF-7 and A549 cancer cell lines. When the antiproliferative properties of the reported compounds were compared, it became clear that the inhibitory action is affected by the substitution. The molecules exhibit the highest action when the ethoxy group is situated at the para position of the phenyl ring, while the

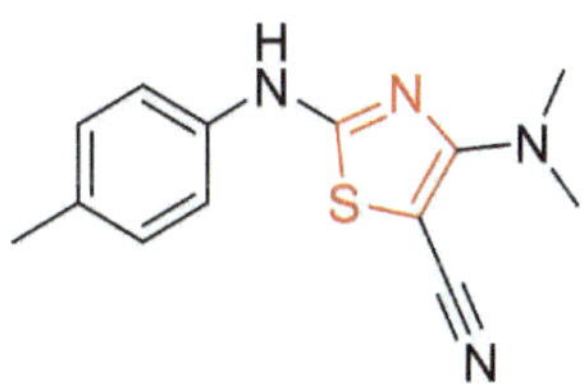

Figure 1.6 Structure of 4-(dimethylamino)-2-(*p*-tolylamino) thiazole-5-carbonitrile used as an anticancer agent.

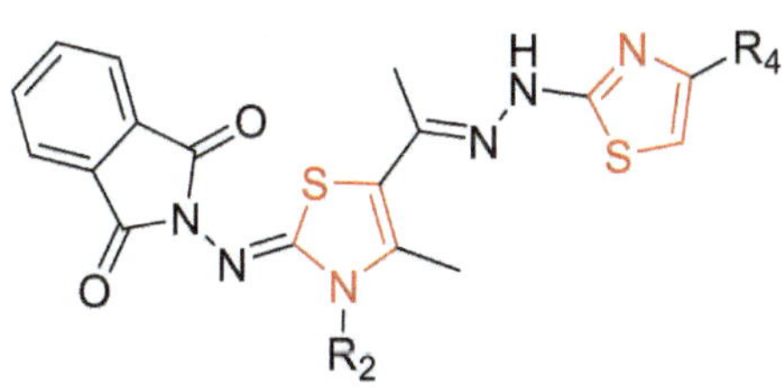

Figure 1.7 Structure of novel phthalimido-bis-1,3-thiazoles used as an anticancer agent.

replacement of the ethoxy group by any other group decreases the activity significantly (see Figure 1.8).

Radwan and Khalid[40] have reported the synthesis of novel thiazole containing thiophene, pyridine, or chromene moieties. The HepG2, HCT-116, and MCF-7 cell lines were used to investigate the anticancer efficacy of each substance. The most potent cytotoxic action was displayed by the thiazolyl-pyridine hybrid **A**. The chemical structure of **A**, where two functional groups (CN and C=O) are present on the one side of the pyridine ring without any steric hindrances, may be responsible for its reactivity. Even though compound **B** includes three different functional groups attached to the pyridine ring (2 CN, NH2, and C=O), its cytotoxicity was lower than that of compound **A**. This may be because position 4 of the pyridine ring contains *p*-methoxybenzene. In comparison with compounds **A** and **B**, compound **C** with chromene was less active (see Figure 1.9).

Liao *et al.*[41] have synthesized a novel 2-amino-4-(aminomethyl)thiazoles and investigated their various anticancer activities. Some of the target compounds showed similar activities compared with 5-fluorouracil (see Figure 1.10).

Raveesha *et al.* have synthesized new thiazole compounds and assessed their anticancer properties.[42] The newly reported compounds were tested against MDA-MB-231 (mammary carcinomas), HeLa (cervical cancer), HCT-116 (colon cancer), HT-29, MCF-7 (human breast cancer), and normal Chang liver cancer cell lines and utilized cisplatin as a positive control (see Figure 1.11).

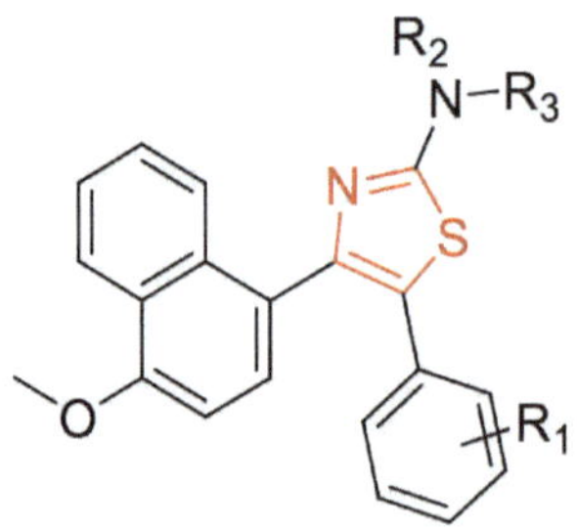

Figure 1.8 Structure of thiazole-naphthalene derivatives used as an antiproliferative agent.

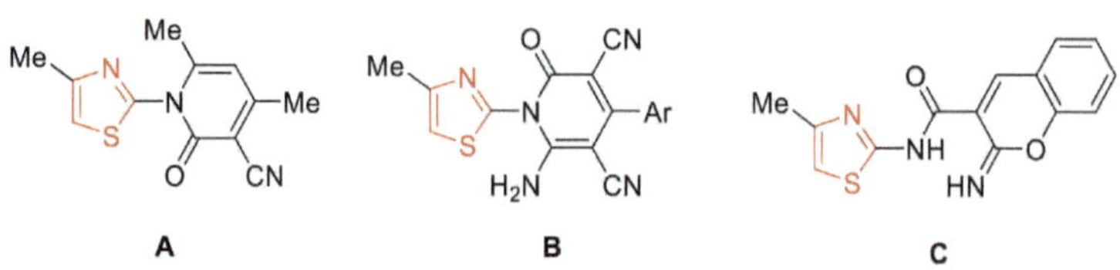

Figure 1.9 Structures of thiazole derivatives used as an anticancer agent.

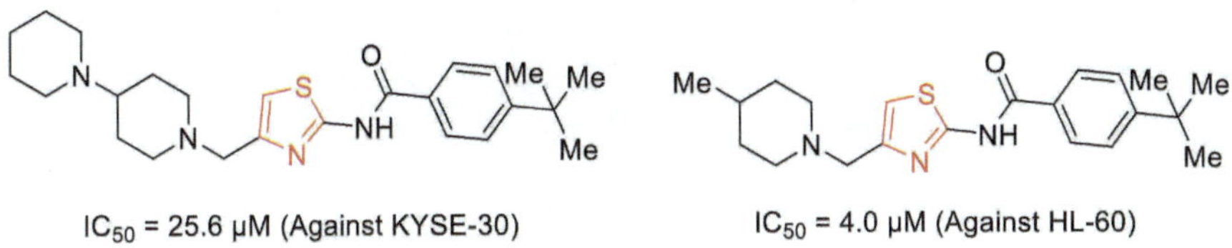

Figure 1.10 Structures of novel 2-amino-4-(aminomethyl)thiazoles used as an anticancer agent.

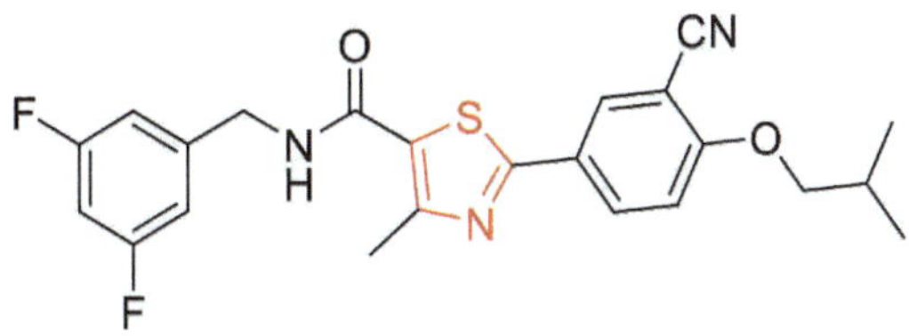

Figure 1.11 Structure of thiazole compounds used as an anticancer agent.

1.3.2 Thiazole as an Antioxidant Agent

Abu-Melha[43] has designed and synthesized thiazole derivatives from 2-chloroacetamido-5-(4-chlorophenylazo)thiazole. The antioxidant activity of each newly reported molecule was assessed. When compared to the standard, the synthesized substances (see Figure 1.12) demonstrated good antioxidant activity. Due to the presence of a 1,3-thiazole ring, the synthesized thiazolyl acetamide compounds demonstrated good antioxidant properties. Because they also include a sulfide moiety, which has considerable antioxidant efficacies, thiazolyl acetamide derivatives, in particular, were shown to have the greatest activity. In addition, as they contain the hydroxy/aminopyrimidine ring, which has good antioxidant reactivity, the pyrimidine-containing thiazolyl acetamide derivative recorded higher activity than the other derivatives.

Asma *et al.*[44] have studied the synthesis of a novel series of pyrazoline-based thiazoles. The *in vitro* antioxidant activity of each produced molecule was evaluated. The tendency of the molecule to experience amine/imine tautomerization may be caused by the highly pronounced activity of the carbothioamide derivative **D.** The 2,2-diphenyl-1-picrylhydrazyl (DPPH) activity decreased quickly when carbothioamide was cyclized into thiazole derivatives. The absence of electron/proton donor capacity in the produced compounds is largely responsible for the less strong radical inhibitory action of the thiazole derivative **E** (see Figure 1.13).

Shah *et al.* have synthesized thiazole Schiff base derivatives.[45] All the synthesized substances were tested for their antioxidant and antibacterial properties. One of the compounds showed very strong antioxidant activity compared to standard ascorbic acid in the DPPH free radical scavenging assay (see Figure 1.14).

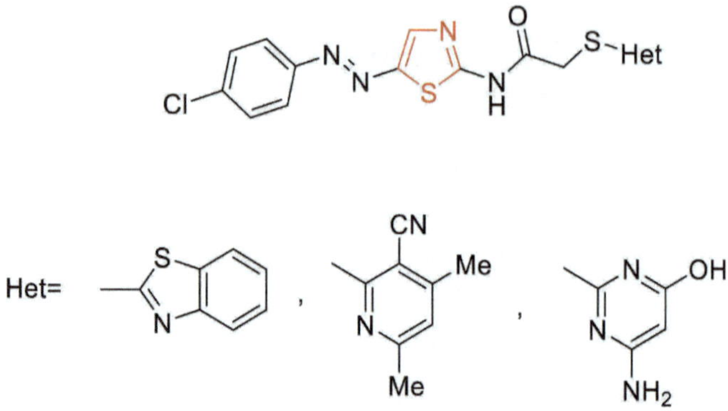

Figure 1.12 Structures of antioxidant agents.

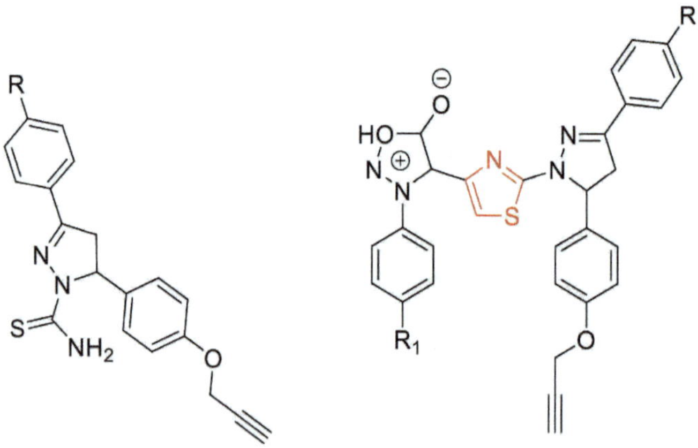

D E

Figure 1.13 Structures of pyrazoline-containing thiazoles used as antioxidants.

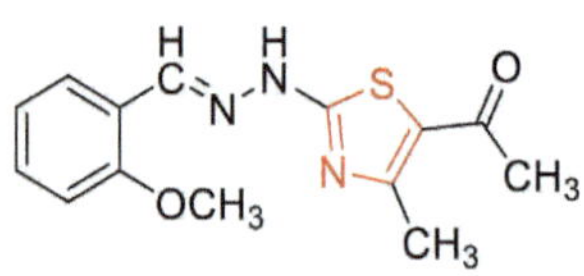

Figure 1.14 Structure of thiazolyl Schiff base used as antioxidants.

Muluk *et al.*[46] have reported the synthesis of novel substituted pyridinyl thiazole derivatives. The *in vitro* antibacterial and antioxidant properties of the synthesized compounds were assessed. Some of the compounds displayed good antioxidant activities (see Figure 1.15).

Jaishree *et al.*[47] have reported the synthesis of novel azetidinyl thiazoles. All the newly reported compounds were tested for their *in vitro* antioxidant properties. Some of the compounds showed good antioxidant properties. In all of the investigated procedures, the replacement derivatives of the acetyl and carboxylic groups have shown strong antioxidant activity. *Ortho-* and *para-*nitro-substituted compounds, on the other hand, exhibited decreased antioxidant activity. A comparison of all thiazole derivatives with ascorbic acid as a reference showed improved reducing ability for these compounds. It verifies that all the substances are capable of reducing ferric to ferrous ions (see Figure 1.16).

New thiazole-clubbed thiophene derivatives were synthesized, their molecular docking was documented, and their antioxidant activity was reported by Abumelha *et al.*[48] The thiazolyl–thiophene hybrids displayed rational antioxidant activity (DPPH assay) (see Figure 1.17).

Jenisha *et al.*[49] have reported the synthesis, density functional theory (DFT) studies, and antioxidant activities of 2-(alkylamino)-4-(naphth-2-yl)thiazole derivatives. In antioxidant studies, 2-(butylamino)-4-(naphth-2-yl)thiazole exhibited a significant antioxidant capacity (see Figure 1.18).

Piperidinone oximes with substituted thiazole groups have been synthesized by Harini *et al.*[50] The compound shown in Figure 1.19 exhibited excellent antioxidant activity among the synthesized compounds.

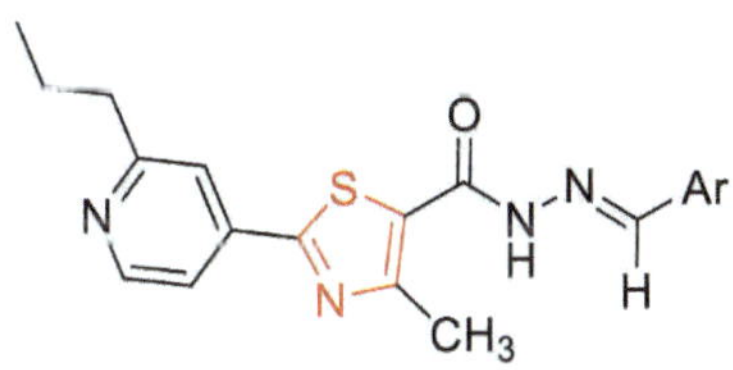

Figure 1.15 Structure of pyridine-containing thiazole used as an antioxidant.

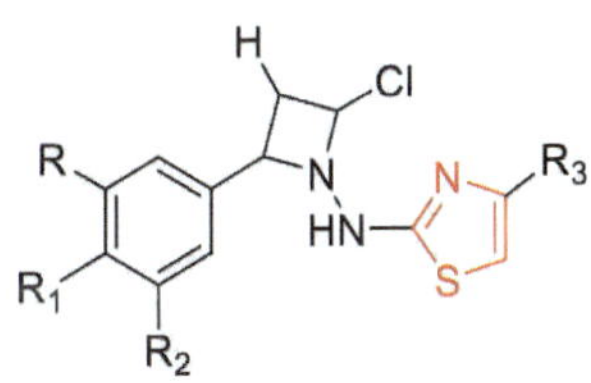

Figure 1.16 Structure of novel azetidinyl thiazoles showing antioxidant activity.

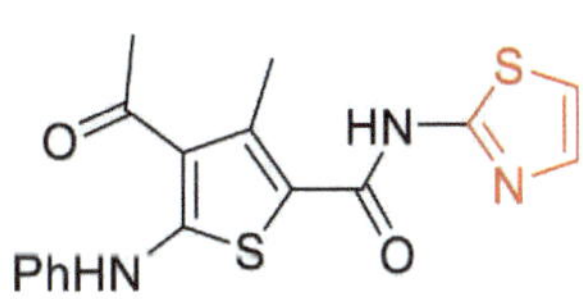

Figure 1.17 Structure of new thiazole-clubbed thiophene derivatives showing antioxidant activity.

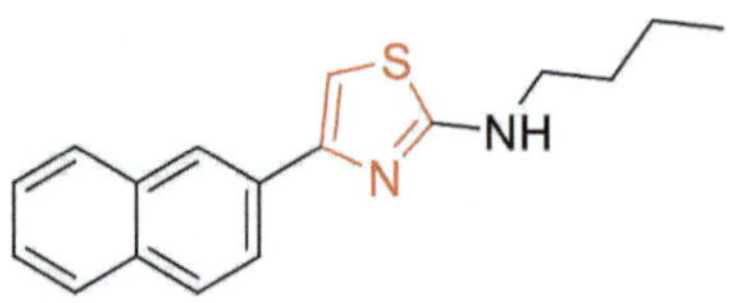

Figure 1.18 Structure of 2-(alkylamino)-4-(naphth-2-yl)thiazole derivatives showing antioxidant activity.

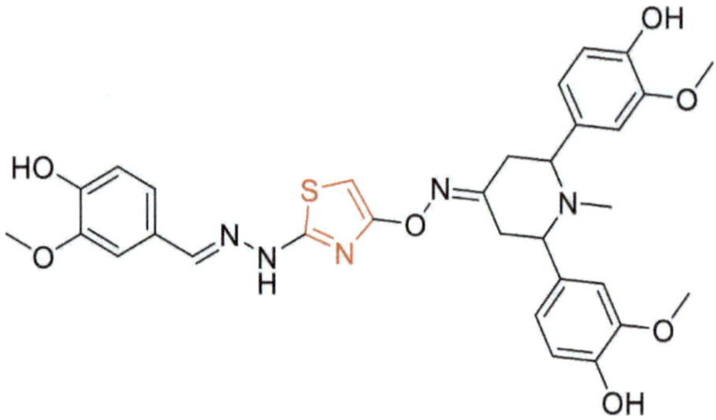

Figure 1.19 Structure of piperidinone oximes-thiazoles exhibiting significant antioxidant activity.

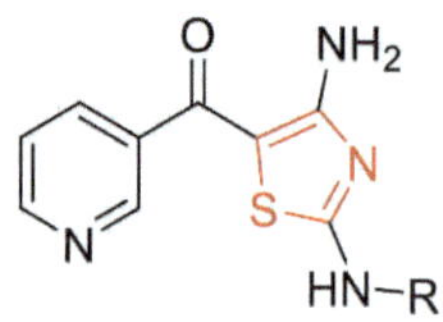

Figure 1.20 Structure of thiazololylpyridines possessing radical scavenging activity.

The investigated compounds showed a spectrum of different antioxidant potencies. In fact, the presence of hydroxyl groups at the *para*-position of the phenyl rings and the particular shape of the compounds are linked to their effective lipid peroxyl (LPO) radical scavenging activities. Contrarily, substances containing electron-withdrawing chloro/nitro groups exhibited significant antioxidant activity (see Figure 1.19).

The synthesis and antioxidant activity of a novel series of thiazololylpyridine derivatives have been described by Rani *et al.*[51] Some of the compounds possess excellent radical scavenging activity (see Figure 1.20).

A study on the synthesis of phenolic thiazole derivatives was published by Marc *et al.*[52] The antioxidant properties of each synthetic molecule were examined (see Figure 1.21).

The synthesis and antibacterial and antioxidant properties of Schiff base containing thiazole derivatives have been described by Lemilemu *et al.*[53]

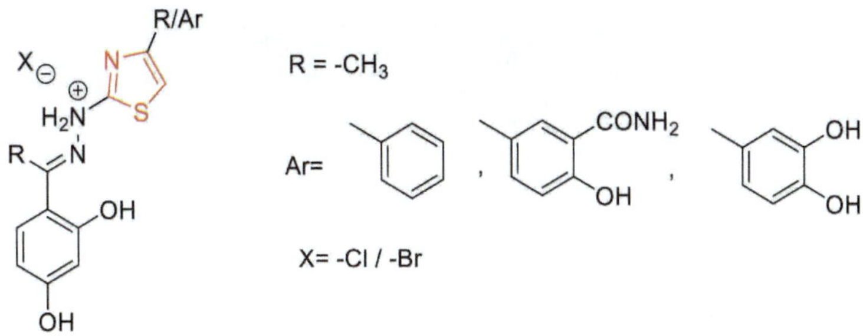

Figure 1.21 Structures of phenolic thiazole compounds used as antioxidants.

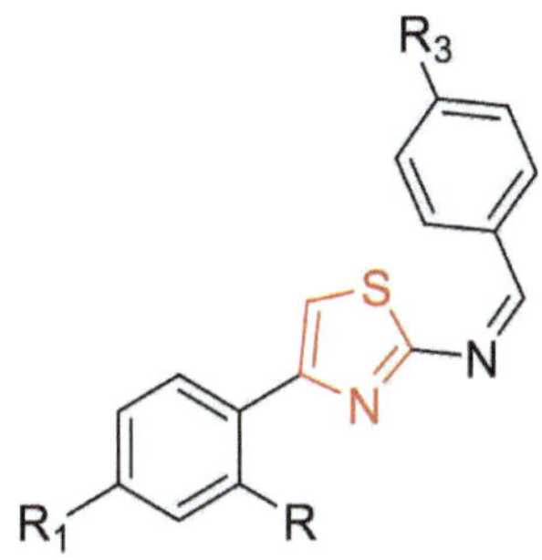

Figure 1.22 Structure of thiazolyl Schiff bases showing antioxidant activity.

Among the synthesized compounds, some of the compounds displayed better antioxidant properties (DPPH radical scavenging potency) compared to ascorbic acid as a standard (see Figure 1.22).

1.3.3 Thiazole as an Antitubercular Agent

Hebade *et al.*[54] have reported silica-supported dodecatungstophosphoric acid (DTP/SiO_2)-assisted synthesis of benzimidazole–thiazole hybrids. All the synthesized compounds were screened for their *in vitro* antitubercular activity against the *Mycobacterium tuberculosis* (*MTB*) (MTCC 300) strain. Some of the compounds displayed promising antitubercular activity (see Figure 1.23).

Novel 2-amino-4-aryl thiazoles have been synthesized by Gundlewad and Patil.[55] By using *MTB* H37Rv as the reference strain, the macro-broth dilution technique was used to assess the antitubercular activity of each compound. The synthesized substance has considerable antitubercular action (see Figure 1.24).

Maganti *et al.*[56] developed and produced new 2-(2-hydrazinyl)thiazole derivatives that include acetylene. The efficacy of each produced compound against *MTB* was assessed *in vitro* and *in silico* (see Figure 1.25).

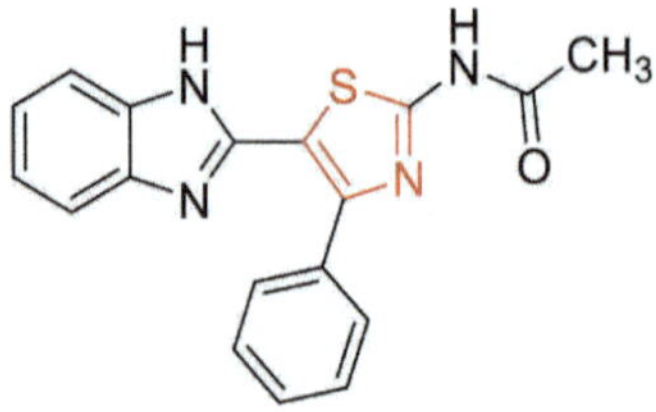

Figure 1.23 Structure of benzimidazole–thiazole hybrids used as antitubercular agents.

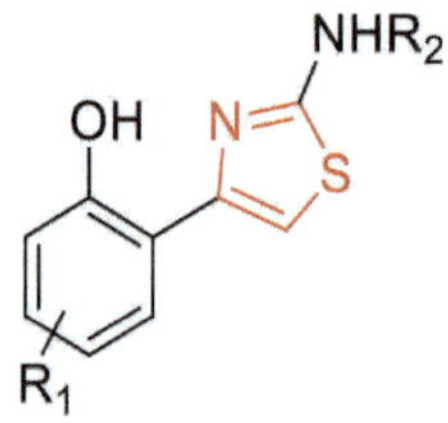

Figure 1.24 Structure of novel 2-amino-4-aryl thiazoles used as antitubercular agents.

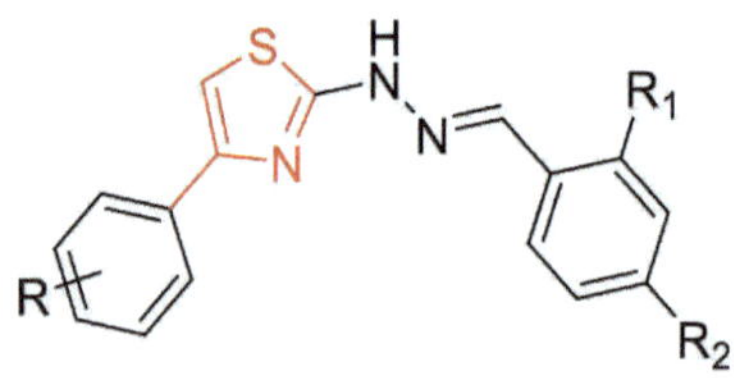

Figure 1.25 Structure of new 2-(2-hydrazinyl)thiazole derivatives used as antitubercular agents.

Hublikar *et al.*[57] have synthesized (*E*)-2-(2-allylidenehydrazinyl)thiazole derivatives. All the new compounds were evaluated for their *in vitro* antimycobacterial properties. Among the synthesized compounds, one compound has been shown to have the highest inhibitory potential, with a minimum inhibitory concentration (MIC) value of 6.5 μg ml^{-1}. However, other compounds with different combinations of substituents have been shown to be moderately active (see Figure 1.26).

The synthesis of Schiff bases of 2-amino thiazole derivatives has been documented by Rachel and Monica.[58] All of the compounds were examined for their antitubercular potency using the microplate alamar blue assay (MABA) method against the *MTB* H37Rv strain. All the reported compounds have shown excellent-to-moderate activity (see Figure 1.27).

Veena *et al.*[59] have designed and synthesized novel benzimidazole-linked thiazole derivatives. The antitubercular activity of every synthesized compound was assessed. There are several substances that are effective against drug-sensitive and drug-resistant *MTB* strains (see Figure 1.28).

A unique sequence of α-amino acids with thiazole linkers has been synthesized by Nalawade *et al.*[60] The inhibitory activity of the synthesized compounds against *Mycobacterium bovis* (BCG) and *MTB* H37Ra strains was tested. One of the compounds showed good antitubercular activity against *M. bovis* (see Figure 1.29).

The synthesis of thiazole-containing pyrazole derivatives was described by Takate *et al.*[61] To test the antitubercular activity, all the produced compounds were examined. Some of the compounds displayed good

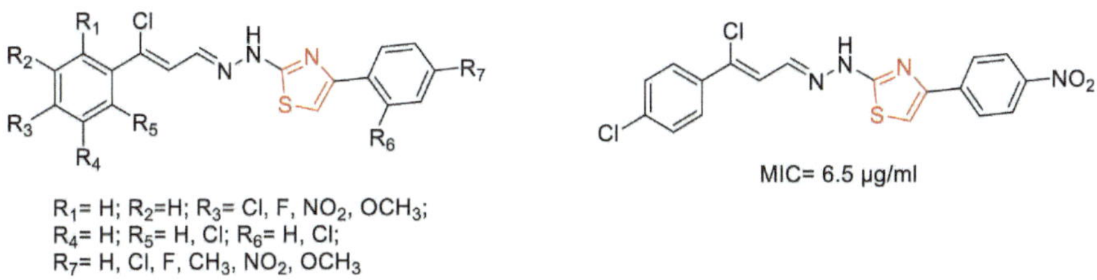

Figure 1.26 Structures of (*E*)-2-(2-allylidenehydrazinyl)thiazole derivatives used as antimycobacterials.

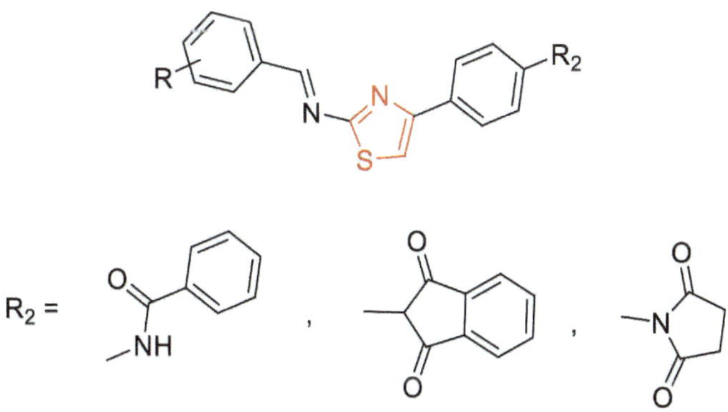

Figure 1.27 Structures of compounds having antitubercular potency.

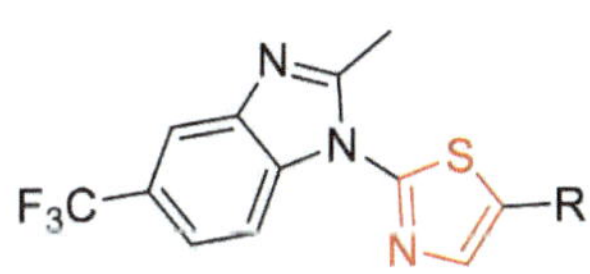

Figure 1.28 Structure of benzimidazole-linked thiazoles used as an antitubercular agent.

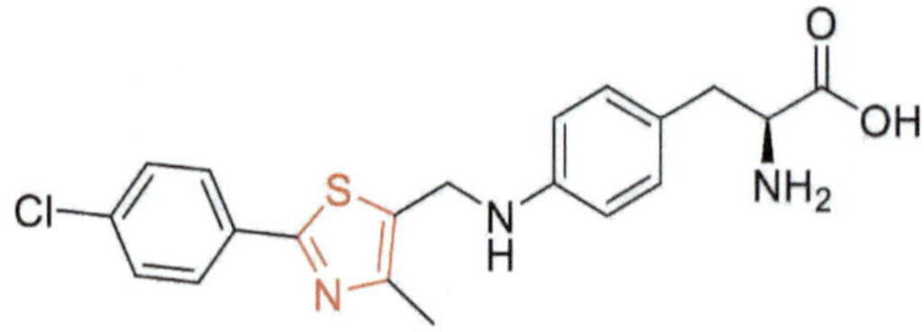

Figure 1.29 Structure of 1,3-thiazoles showing good antitubercular activity.

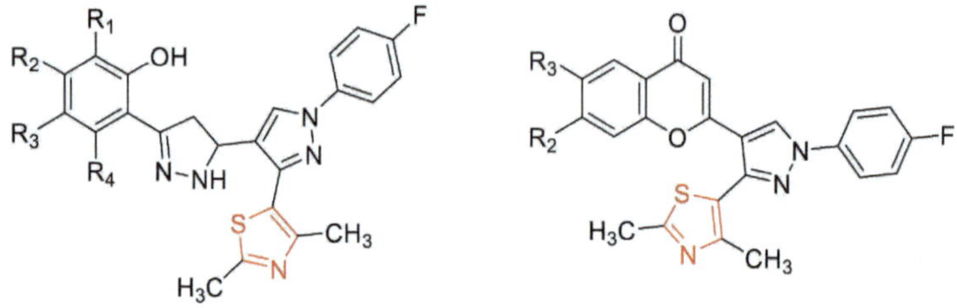

Figure 1.30 Structures of thiazole-containing pyrazole derivatives used as an antitubercular agent.

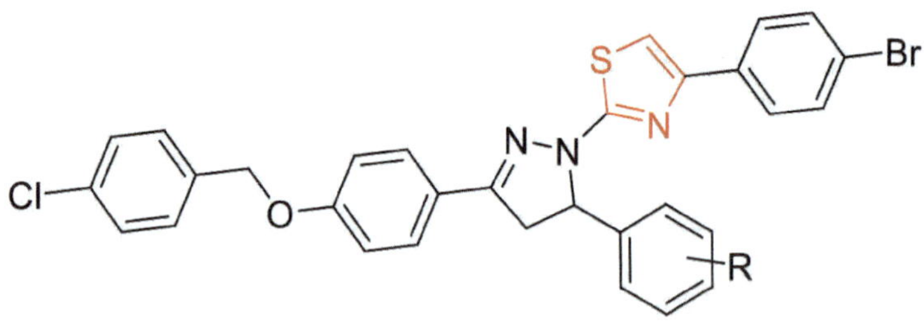

Figure 1.31 Structure of novel thiazole–pyrazoline hybrids showing antimicrobial properties.

antimycobacterial activity against *MTB* and *M. bovis* (BCG) strains (see Figure 1.30).

1.3.4 Thiazole as an Antimicrobial Agent

The efficient synthesis of novel thiazole–pyrazoline hybrids was reported by Salih *et al.*[62] Using fluconazole and ciprofloxacin as positive controls, all the compounds were tested against one fungal strain and four species of harmful bacteria using the well-diffusion and MIC tests. One of them, with an MIC value of 16 μg mL^{-1}, had strong antibacterial activity against *Acinetobacter baumannii*, while ciprofloxacin was ineffective (see Figure 1.31).

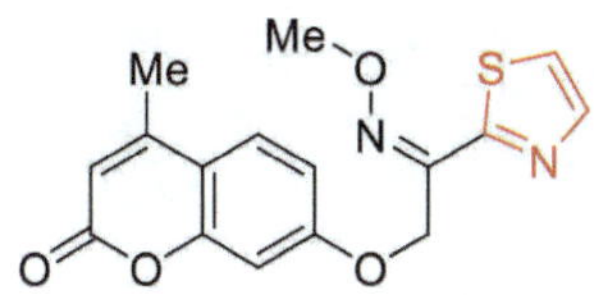

Figure 1.32 Structure of coumarin thiazoles used as antimicrobial agents.

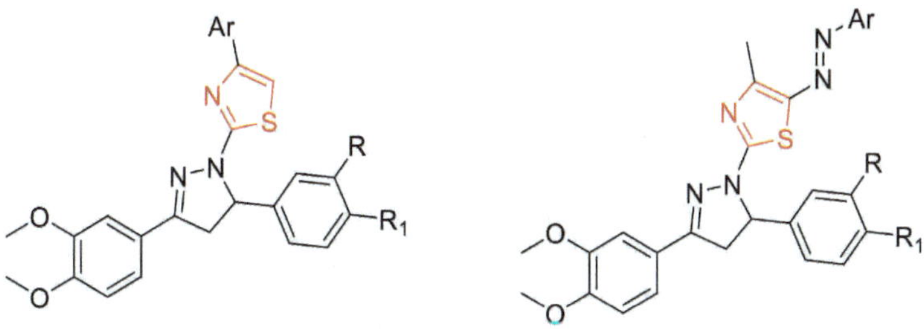

Figure 1.33 Structures of new thiazolyl pyrazolines.

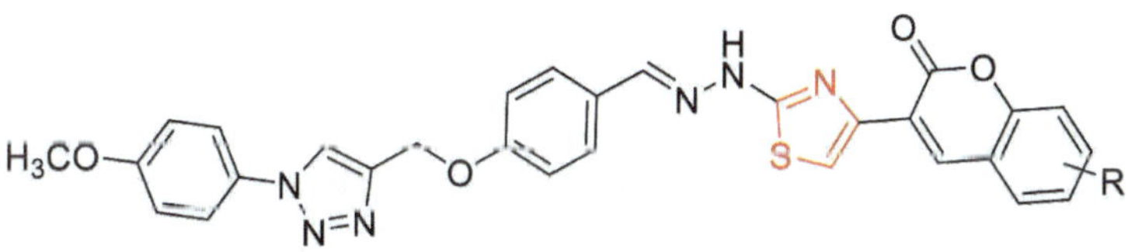

Figure 1.34 Structure of 1,2,3-triazole containing thiazoles used as antimicrobial agents.

Starting with commercial resorcin, Yang *et al.*[63] produced a novel kind of coumarin thiazoles as distinct multi-targeting antibacterial agents by the processes, including condensation, nucleophilic substitution, and cyclization. The antibacterial efficacy of every synthesized derivative was examined. One of them, methyl oxime, demonstrated a broad antibacterial spectrum and significant inhibitory effectiveness against methicillin-resistant *Staphylococcus aureus* (MRSA) at a low dose (1 μg mL^{-1}) (see Figure 1.32).

By reacting thioamide derivatives with 1-aryl-2-bromoethanones, chloroacetones, and hydrazonoyl chlorides, Masoud and co-workers[64] described the synthesis and antimicrobial testing of new thiazolyl pyrazolines. The antimicrobial efficacy of all the freshly synthesized compounds was tested against four bacterial and five fungal strains (see Figure 1.33).

1,2,3-Triazole containing thiazoles was reported by Gondru *et al.*[65] by Hantzsch thiazole synthesis. Testing was performed on all the synthesized substances to determine their *in vitro* antimicrobial and anti-biofilm properties. Some of the compounds displayed a broad spectrum of antibacterial activities (see Figure 1.34).

Figure 1.35 Structures of thiazoles comprising chromones and pyrazoles.

The synthesis of thiazoles comprising chromones and pyrazoles has been described by Darekar *et al.*[66] All the produced compounds were evaluated for their *in vitro* antimicrobial properties. Some of the compounds displayed good activities against bacterial and fungal strains (see Figure 1.35).

Jayachitra *et al.*[67] have reported the synthesis of sulfonamide-clubbed thiazole compounds. Studies on antimicrobial activity show that, when compared with positive controls and other microbes, the compound has strong antibacterial and antifungal action against *Klebsiella pneumonia* and *Candida albicans* (see Figure 1.36).

Bakthavatchala Reddy *et al.*[68] have reported the synthesis of new benzimidazole containing pyrazoles and pyrazolyl thiazoles. Antimicrobial activity was examined in all of the compounds. One of these, a molecule with an aromatic *para*-nitro substituent, had stronger antibacterial action, notably against *Penicillium chrysogenum* and *Pseudomonas aeruginosa* (see Figure 1.37).

The synthesis of thiazole-containing oxadiazole derivatives has been reported by Darekar *et al.*[69] All the produced compounds were tested for their antimicrobial properties. The findings of the studies performed *in vitro* and *in silico* indicate that heterocycles integrated with thiazole and oxadiazole derivatives can meet the structural specifications for the further development of innovative medicines (see Figure 1.38).

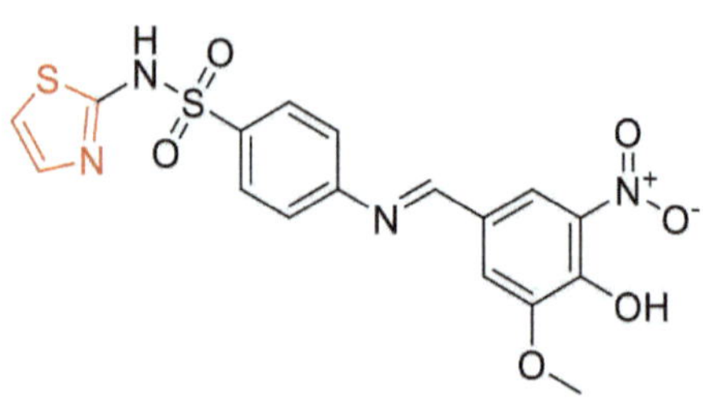

Figure 1.36 Structure of sulfonamide-clubbed thiazole compounds.

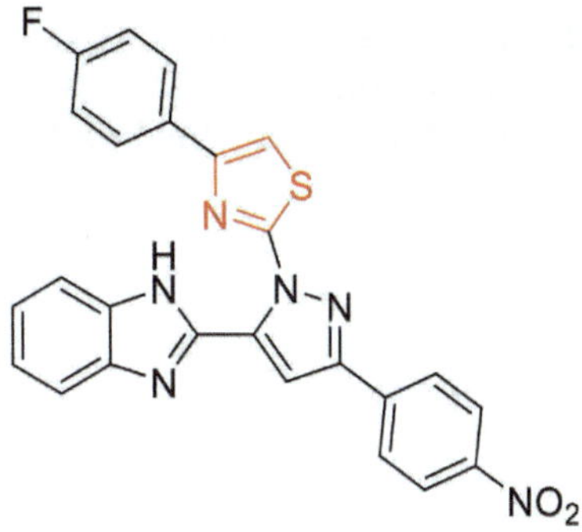

Figure 1.37 Structure of new benzimidazole containing pyrazoles and pyrazolyl thiazoles.

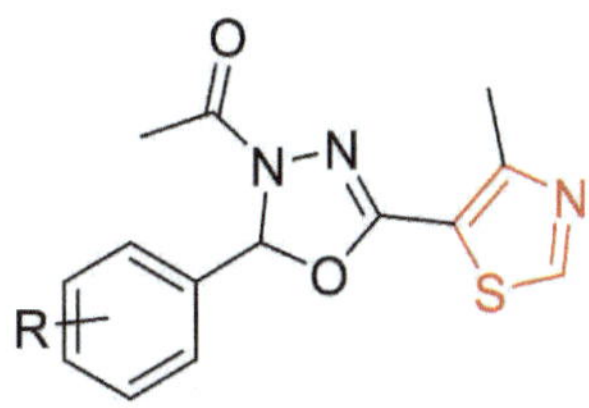

Figure 1.38 Structure of thiazole-containing oxadiazole derivatives.

1.4 Conclusion

In conclusion, the synthesis of thiazole derivatives has been recorded using a number of methods. It is possible to synthesize thiazole compounds with a high yield by applying new practical methods. By replacing the substituent of the thiazole ring with suitable substituent groups, a number of new compounds with improved biological activity were developed and synthesized for use in biological applications. Studies on thiazole derivatives may, thus, be the key area of exploration in the future due to their special properties and promise.

Abbreviations

CHK1 inhibitors	Checkpoint kinase-1 inhibitors
CS_2	Carbon disulfide
DABCO	1,4-Diazabicyclo[2.2. 2]octane
DBU	1,8-Diazabicyclo[5.4.0]undec-7-ene
DCE	1,2-Dichloroethane
DCM	Dichloromethane
DFT	Density functional theory

DMF	Dimethylformamide
DMSO	Dimethyl sulfoxide
DPPH	2,2-Diphenyl-1-picrylhydrazyl
DTP/SiO_2	Silica-supported dodecatungstophosphoric acid
eq./equiv.	Equivalent
h	Hour
HFIP	1,1,1,3,3,3-Hexafluoroisopropanol
LPO	Lipid peroxyl
MIC	Minimum inhibitory concentration
MTB	*Mycobacterium tuberculosis*
MTT	3-(4,5-Dimethylthiazol-2-yl)-2,5-diphenyl-2*H*-tetrazolium bromide
NBS	*N*-bromosuccinimide
NEt_3	Triethyl amine
NPs	Nanoparticles
NR	Nanorod
PBMC	Peripheral blood mononuclear cells
PEG	Polyethylene glycols
PTSA	*p*-Toluenesulfonic acid
RT	Room temperature
SGLT2 inhibitors	Sodium–glucose cotransporter-2 inhibitors
SRB Assay	The sulforhodamine B (SRB) protein stain assay
TosMIC	Toluenesulfonylmethyl isocyanide
ZnO	Zinc oxide

References

1. M. H. M. Helal, M. A. Salem, M. S. A. El-Gaby and M. Aljahdali, *Eur. J. Med. Chem.*, 2013, **65**, 517.
2. S. G. Dengale, B. K. Karale, H. N. Akolkar, N. R. Darekar and K. K. Deshmukh, *Russ. J. Gen. Chem.*, 2019, **89**, 1535.
3. N. R. Darekar, B. K. Karale, H. N. Akolkar and A. S. Burungale, *Russ. J. Gen. Chem.*, 2020, **90**, 1721.
4. A. Grozav, L. I. Gaina, V. Pileczki, O. Crisan, L. Silaghi-Dumitrescu, B. Therrien, V. Zaharia and I. Berindan-Neagoe, *Int. J. Mol. Sci.*, 2014, **15**(12), 22059.
5. J. E. Muijlwijk-Koezen, H. Timmerman, R. C. Vollinga, J. F. Drabbe Künzel, M. de Groote, S. Visser and A. P. IJzerman, *J. Med. Chem.*, 2001, **44**(5), 749.
6. K. Song, S. H. Lee, M. J. Kim, H. J. Seo, J. Lee, S. Lee, M. E. Jung, E. Son, M. W. Lee, J. Kim and J. Lee, *ACS Med. Chem. Lett.*, 2011, **2**(2), 182.
7. M. Zhou, C. Ning, R. Liu, Y. He and N. Yu, *Bioorg. Med. Chem. Lett.*, 2013, **23**(11), 3200.
8. X. Huang, C. C. Cheng, T. O. Fischmann, J. S. Duca, X. Yang, M. Richards and G. W. Shipps Jr, *ACS Med. Chem. Lett.*, 2012, **3**(2), 123.

9. S. H. Ali and A. R. Sayed, *Synth. Commun.*, 2020, **51**(5), 670.
10. A. Petrou, M. Fesatidou and A. Geronikaki, *Molecules*, 2021, **26**(11), 3166.
11. S. Pathania and R. K. Rawal, *Chem. Heterocycl. Compd.*, 2020, **56**(4), 445.
12. T. H. Namitha, S. S. Nair, A. Kumar, B. Vinod and P. A. Daisy, *Int. J. Pharm. Sci. Rev. Res.*, 2021, **70**(1), 189.
13. M. F. Arshad, A. Alam, A. A. Alshammari, M. B. Alhazza, I. M. Alzimam, M. A. Alam, G. Mustafa, M. S. Ansari, A. M. Alotaibi, A. A. Alotaibi, S. Kumar, S. M. B. Asdaq, M. Imran, P. K. Deb, K. N. Venugopala and S. Jomah, *Molecules*, 2022, **27**(13), 3994.
14. R. Hajinasiri, Z. Hossaini and F. Sheikholeslami-Farahani, *Comb. Chem. High Throughput Screen.*, 2015, **18**(1), 42.
15. G. Wu, X. Wang and H. Liu, *Catalysts*, 2016, **6**(8), 126.
16. R. Gurav, S. K. Surve, S. Babar, P. Choudhari, D. Patil, V. More, S. Sankpal and S. Hangirgekar, *Org. Biomol. Chem.*, 2020, **18**(24), 4575.
17. A. Dhas, A. Kanagare, M. Kanetkar and D. Pansare, *Lett. Appl. NanoBio-Science*, 2023, **12**(4), 1.
18. T. Masquelin and D. Obrecht, *Tetrahedron*, 2001, **57**(1), 153.
19. J. Safaei-Ghomi, F. Salimi and A. Ramazani, *J. Chem. Sci.*, 2013, **125**(5), 1087.
20. H. Shahbazi-Alavi, S. Khojasteh-Khosro, J. Safaei-Ghomi and M. Tavazo, *BMC Chem.*, 2019, **13**(1), 120.
21. A. Nakhaei, *Russ. J. Gen. Chem.*, 2017, **87**(8), 1850.
22. K. R. Singh, C. Santhosh, T. R. Swaroop and M. P. Sadashiva, *Org. Biomol. Chem.*, 2022, **20**, 5771.
23. A. Mahata, B. Prabhas, A. K. Panday, R. Yadav, T. Parvin and L. H. Choudhury, *New J. Chem.*, 2020, **44**(12), 4798.
24. M. Balti, S. A. Miller, M. L. Efrit and N. E. Leadbeater, *RSC Adv.*, 2016, **6**, 72165.
25. Z. A. Alsharif and M. A. Alam, *RSC Adv.*, 2017, **7**(52), 32647.
26. B. K. Behera, S. Shit, S. Biswas and A. K. Saikia, *J. Org. Chem.*, 2022, **87**(14), 9259.
27. S. Hassell-hart, E. Speranzini, S. Srikwanjai, E. Hossack, S. M. Roe, D. Fearon, D. Akinbosede, S. Hare and J. Spencer, *Org. Lett.*, 2022, **24**(43), 7924.
28. V. A. Adole, R. A. More, B. S. Jagdale and T. B. Pawar, *ChemistrySelect*, 2020, **5**(9), 2778.
29. W. Zhou, S. Ni, H. Mei, J. Han and Y. Pan, *Tetrahedron Lett.*, 2015, **56**(27), 4128.
30. T. B. Nguyen and P. Retailleau, *Org. Lett.*, 2021, **23**, 5344.
31. Y. Wang, X. Liu, B. Zhu, P. Guo, Y. Pei, Q. He and H. Cao, *J. Org. Chem.*, 2020, **85**(15), 10118.
32. M. Nikpassand, Z. F. Leila and S. Sina, *Heterocycl. Commun.*, 2016, **22**(4), 243.
33. S. Singh, A. P. Jain and A. Nayak, *Indo Am. J. Pharm. Sci.*, 2022, **9**(7), 48.

34. S. Mamidala, V. S. Mudigunda, S. R. Peddi, K. K. Bokara, V. Manga and R. R. Vedula, *Synth. Commun.*, 2020, **50**(16), 2488.
35. C. M. Bandaru, N. Poojith, S. S. Jadav, M. V. B. Rao, K. S. Babu, R. Sreenivasulu and R. Alluri, *Polycycl. Aromat. Compd.*, 2022, **42**(8), 5368.
36. Z. H. Zhang, H. M. Wu, S. N. Deng, R. X. Chai, M. C. Mwenda, Y. Y. Peng, D. Cai and Y. Chen, *Chem. Pap.*, 2019, **73**, 355.
37. N. M. Biswas, A. Shard, S. Patel and P. Sengupta, *Drug Dev. Res.*, 2018, **79**(8), 391.
38. A. R. Oliveira, F. A. Dos Santos, L. P. de L. Ferreira, M. G. da R. Pitta, M. V. de O. Silva, M. V. de O. Cardoso, A. F. Pinto, P. Marchand, M. J. B. de Melo Rego and A. C. L. Leite, *Chem.-Biol. Interact.*, 2021, **347**, 109597.
39. G. Wang, W. Liu, M. Fan, M. He, Y. Li and Z. Peng, *J. Enzyme Inhib. Med. Chem.*, 2021, **36**(1), 1694.
40. A. S. Radwan and M. A. A. Khalid, *J. Heterocycl. Chem.*, 2019, **56**(3), 1063.
41. M. Liao, D. Huang and Y. X. Cheng, *Mendeleev Commun.*, 2023, **33**, 73.
42. R. Raveesha, A. M. Anusuya, A. V. Raghu, K. Yogesh Kumar, M. G. Dileep Kumar, S. B. Benaka Prasad and M. K. Prashanth, *Comput. Toxicol.*, 2022, **21**, 100202.
43. S. Abu-Melha, *J. Saudi Chem. Soc.*, 2022, **26**(2), 101431.
44. B. Asma Kalluraya, N. Manju and C. L. Sharath, *J. Heterocycl. Chem.*, 2020, **57**(8), 3105.
45. M. S. Shah, M. M. Rahman, M. D. Islam, A. Al-Macktuf, J. U. Ahmed, H. Nishino and M. A. Haque, *J. Mol. Struct.*, 2022, **1248**, 131465.
46. M. B. Muluk, P. S. Phatak, S. B. Pawar, S. T. Dhumal, N. N. M. A. Rehman, P. P. Dixit, P. B. Choudhari and K. P. Haval, *J. Chin. Chem. Soc.*, 2019, **66**(11), 1507.
47. V. Jaishree, N. Ramdas, J. Sachin and B. Ramesh, *J. Saudi Chem. Soc.*, 2012, **16**, 371.
48. H. M. Abumelha, A. Bayazeed, A. Alsoliemy, A. Hossan, A. Alharbi, F. A. Saad and N. M. El-Metwaly, *J. Mol. Struct.*, 2022, **1262**, 133112.
49. J. Jenisha, J. Jeba Lenet and T. F. Abbs Fen Reji, *Indian J. Chem. Technol..*, 2023, **30**, 231.
50. S. T. Harini, H. V. Kumar, J. Rangaswamy and N. Naik, *Russ. J. Bioorg. Chem.*, 2017, **43**(2), 186.
51. S. M. Rani, J. Brindha and T. F. Abbs Fen Reji, *Indian J. Chem.*, 2021, **60B**, 605.
52. G. Marc, A. Stana, A. H. Franchini, D. C. Vodnar, G. Barta, S. M. Tertiş, I. Santa, C. Cristea, A. Pîrnau, A. Ciorîta, B. Dume, V. A. Toma, L. Vlase, I. Oniga and O. Oniga, *Antioxidants*, 2021, **10**, 1707.
53. F. Lemilemu, M. Bitew, T. B. Demissie, R. Eswaramoorthy and M. Endale, *BMC Chem.*, 2021, **15**, 67.
54. M. J. Hebade, S. T. Dhumal, S. S. Kamble, T. R. Deshmukh, V. M. Khedkar, R. N. Gacche and B. S. Dawane, *Polycycl. Aromat. Compd.*, 2022, **43**(3), 2866.
55. G. B. Gundlewad and B. R. Patil, *J. Heterocycl. Chem.*, 2018, **55**(3), 769.

56. L. H. B. Maganti, D. Ramesh, B. G. Vijayakumar, M. I. K. Khan, A. Dhayalan, J. Kamalraja and T. Kannan, *RSC Adv.*, 2022, **12**, 8771.
57. M. Hublikar, V. Kadu, J. K. Dublad, D. Raut, S. Shirame, P. Makam and R. Bhosale, *Arch. Pharm.*, 2020, **353**(7), 2000003.
58. C. Rachel and K. Monica, *Bioorg. Med. Chem. Lett.*, 2020, **30**(24), 127655.
59. K. Veena, M. S. Raghu, K. Yogesh Kumar, C. B. Pradeep Kumar, F. A. Alharti, M. K. Prashanth and B. H. Jeon, *J. Mol. Struct.*, 2022, **1269**, 133822.
60. J. L. Nalawade, P. C. Mhaske, A. D. Shinde, A. P. Chavan, Y. K. Abhale, D. Sarkar and V. D. Bobade, *ARKIVOC*, 2022, **ix**, 64.
61. S. J. Takate, A. D. Shinde, B. K. Karale, H. Akolkar, L. Nawale, D. Sarkar and P. C. Mhaske, *Bioorg. Med. Chem. Lett.*, 2019, **29**(10), 1199.
62. R. H. H. Salih, A. H. Hasan, N. H. Hussen, F. E. Hawaiz, T. B. Hadda, J. Jamalis, F. A. Almalki, A. S. Adeyinka, L.-C. C. Coetzee and A. K. Oyebamiji, *J. Mol. Struct.*, 2023, **1282**, 135191.
63. X.-C. Yang, C.-F. Hu, P.-L. Zhang, S. Li, C.-S. Hu, R.-X. Geng and C.-H. Zhou, *Bioorg. Chem.*, 2022, **124**, 105855.
64. D. M. Masoud, R. A. Azzam, F. Hamdy, A. A. I. Mekawey and H. A. Abdel - Aziz, *J. Heterocycl. Chem.*, 2019, **56**(11), 3030.
65. R. Gondru, S. Kanugala, S. Raj, C. G. Kumar, M. Pasupuleti, J. Banothu and R. Bavantula, *Bioorg. Med. Chem. Lett.*, 2020, **33**, 127746.
66. N. R. Darekar, S. J. Takate, H. N. Akolkar, S. P. Salve, B. K. Karale and S. D. Mhaske, *Indian J. Heterocycl. Chem.*, 2022, **32**(4), 433.
67. R. Jayachitra, M. Padmavathy, A. Kanagavalli, G. Thilagavathi, N. Elangovan, S. Sowrirajan and R. Thomas, *J. Indian Chem. Soc.*, 2023, **100**, 100824.
68. N. Bakthavatchala Reddy, G. V. Zyryanov, G. Mallikarjuna Reddy, A. Balakrishna, A. Padmaja, V. Padmavathi, C. Suresh Reddy, J. R. Garcia and G. Sravya, *J. Heterocycl. Chem.*, 2019, **56**, 589.
69. N. R. Darekar, S. J. Takate, H. N. Akolkar, M. H. Shaikh, V. M. Khedkar, B. K. Karale and S. D. Mhaske, *Russ. J. Gen. Chem.*, 2022, **92**, 1511.

CHAPTER 2

Synthesis, Properties, and Therapeutic Applications of Dithiazoles

SUBRATA BARICK, TITIKSHA KUMAR SAGAR AND SANDEEP CHANDRASHEKHARAPPA*

Department of Medicinal Chemistry, National Institute of Pharmaceutical Education and Research, Raebareli (NIPER-R), Transit Campus, Bijnor-Sisendi Road, Sarojini Nagar, Near CRPF Base Camp, Lucknow, Uttar Pradesh, 226002, India
*E-mail: c.sandeep@niperraebareli.edu.in; c.sandeep@niperrbl.ac.in

2.1 Introduction

Heterocycles have been the subject of extensive organic[1–8] and medicinal chemistry[9–16] research for almost a century. They have contributed to the biological[17–19] and industrial advancement of society, as well as to the study of life processes and initiatives aimed at improving the standard of living.[20] Over decades of historical growth in organic synthesis, chemists have remained interested in heterocyclic molecules containing sulfur and nitrogen atoms.[3,21–26] Their biological activities and distinctive structures served as the basis for this interest and led to numerous applications in the pharmacological, agrochemical,[27] and, more recently, materials science fields of study.[28,29] Heterocyclic compounds having nitrogen and sulfur atoms are very important in the field of medicinal chemistry.[30]

S-Heterocycles: Retrospect, Prospects, and Biological Applications
Edited by Shrikaant Kulkarni, Hemantkumar Akolkar, Vijay M. Khedkar and A. K. Haghi

Published by the Royal Society of Chemistry, www.rsc.org

1 2 3 4

1,2,3-dithiazole 1,2,4-dithiazole 1,3,2-dithiazole 1,3,4-dithiazole

Figure 2.1 Possible structures of dithiazoles.

1,2,3-Dithiazoles are one of the four possible dithiazole systems (Figure 2.1). These are five-membered aromatic (6π system) and planar heterocyclic compounds containing one nitrogen and two sulfur atoms.[31] They can be neutral or positively charged (Appel's salt) molecules.[32] This 1,2,3-dithiazole moiety was first synthesized in 1957 by Wannagat and Schindler[33]; however, its first biological activity as an antifungal and herbicidal agent was reported two decades later by Moore.[34] Previously, it was synthesized from cyanothioformamides.[33] Finally, the discovery of Appel's salt (4,5-dichloro-1,2,3-dithiazolium chloride) accelerated the research in 1,2,3-dithiazole chemistry, enabling the formation of this core structure in a single-step reaction.[35] Most of the reactions of Appel's salt involve the C-5 position of the scaffold, and less reactivity in terms of transformation is found at the C-4 position.[36] Nevertheless, some recent advancements (including reactions with oximes and intramolecular cyclization) have aimed at targeting the C-4 position of Appel's salt.[37] Additionally, the benzo-fused dithiazolium salt (Herz's salt) is prepared from aromatic amines and sulfur monochloride (S_2Cl_2).[38]

In medicinal chemistry, heteroatom-rich 1,2,3-dithiazoles are comparatively understudied, despite several reports on their intriguing biological applications, such as antiviral,[39] anticancer,[30] antibacterial,[40] antimicrobial,[41] antifungal,[42,43] and antifibrotic activities.[44] 1,2,3-Dithiazoles also act as precursors for various other medicinally potent moieties that are used in drugs, such as prazosin, ketanserin, and trimetrexate.[45]

2.2 Synthesis and Chemistry

In 1957, 1,2,3-dithiazole was synthesized for the first time by Wannagat and Schindler,[33] and later, in 1976, Joseph E. Moore filed a patent on the preparation and biological application of 4-halo-5-aryl-1,2,3-dithiazole. They used cyanothioformamides as the starting material for the synthesis of 1,2,3-dithiazole.[34]

Various *N*-aryl cyanothioformamides (**5**) were reacted with dihalides (**6**) in the presence of a catalytic amount of formamide compounds or a quaternary ammonium salt to prepare dithiazoles (**9**).[16] *N*,*N*-Dialkylformamides and *N*-alkylformamides were used as suitable formamide catalysts. In the case of quaternary salts, tetramethylammonium chloride

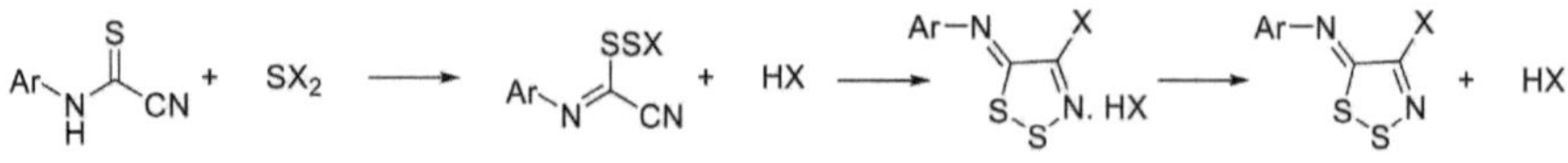

5 6 7 8 9

Scheme 2.1 General synthesis process of dithiazoles as reported by Moore.

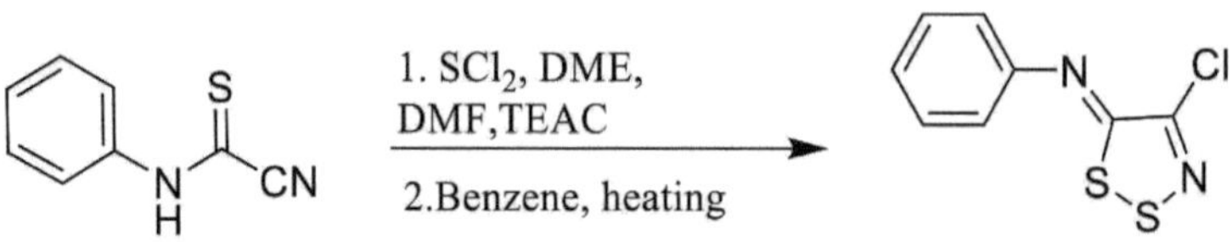

10 11

Scheme 2.2 Preparation of 4-chloro-5-phenylimino-1,2,3-dithiazole.

or tetramethylammonium bromide was used for the reaction. Various suitable inert diluents, such as hexane, isooctane, dichloromethane (DCM), benzene, toluene, chlorobenzene, dioxane, and tetrahydrofuran (THF) were used for the synthesis of dithiazoles (Scheme 2.1).[34]

As an example, the preparation of 4-chloro-5-phenylimino-1,2,3-dithiazole (**11**) involves the reaction of *N*-phenyl cyanothioformamide (**10**) (prepared from phenyl isothiocyanate and potassium cyanide) with sulfur dichloride in dimethoxyethane (DME) in the presence of dimethylformamide (DMF) and tetraethylammonium chloride (TEAC). The reaction mixture was stirred for 24 h, and the solid was filtered and heated in benzene until a homogeneous solution was obtained (Scheme 2.2). Further evaporation of benzene gave a crude product as an oily mass, which was then crystallized in hexane to produce the final dithiazole compound.

In 1985, the synthesis procedure took a great turn with the discovery of Appel's salt (**14**), which catalyzed the formation of various monocyclic dithiazoles in a single-step reaction.[35] The discovery of Appel's salt, chemically known as 4,5-dichloro-1,2,3-dithiazolium chloride (**14**), provided access to various neutral 4-chloro-5*H*-1,2,3-dithiazoles.[47]

Appel's salt was synthesized using chloroacetonitrile (**12**) and disulfur dichloride (**13**) in DCM (Scheme 2.3).[35]

2.2.1 Reaction at the More Reactive C-5 Position of Appel's Salt

Appel's salt (**14**) can condense to produce 4-chloro-5*H*-1,2,3-dithiazolylidenes (**15**) when combined with active methylenes, such as acetonitrile derivatives, diketones, and ketoesters (Scheme 2.4).[48]

When Appel's salt (**14**) reacts with hydrogen sulfide in acetonitrile at room temperature, it produces dithiazole-5-thione (**16**) with a yield of

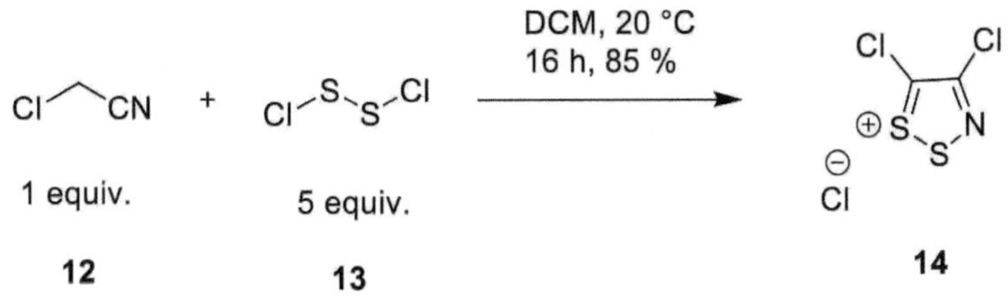

Scheme 2.3 Preparation of Appel's salt.

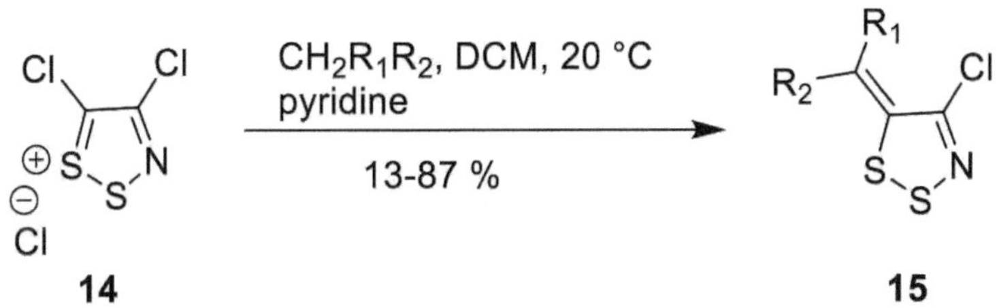

Scheme 2.4 Formation of 4-chloro-5*H*-1,2,3-dithiazolylidenes (**15**) from Appel's salt (**14**).

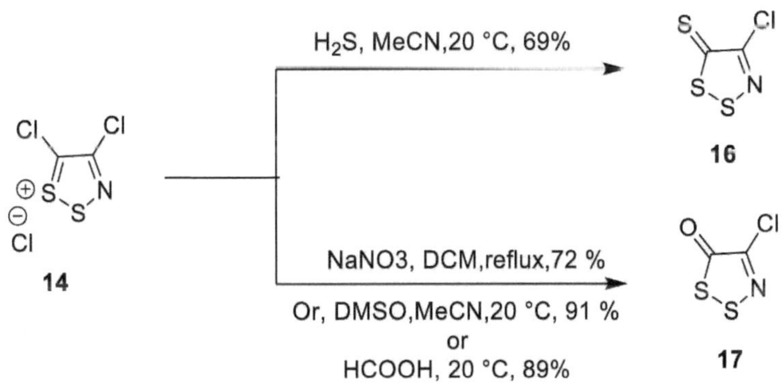

Scheme 2.5 Reaction of Appel's salt with hydrogen sulfide and oxygen nucleophiles.

69%.[35] The reaction of Appel's salt with a nucleophile bearing an oxygen source, such as $NaNO_3$, sulfoxide, or formic acid,[45] produces dithiazole-5-one (**17**) (Scheme 2.5).

Appel *et al.* also showed that in the presence of pyridine, Appel's salt (**14**) reacted with primary aniline (**18**) to produce stable pale yellow- to orange-colored *N*-arylimines (**19**) (Scheme 2.6).[35]

However, in the case of the reaction with aminopyridines (**20**), dithiazoles show less reactivity due to the less nucleophilic character of aminopyridines than primary anilines toward dithiazoles, resulting in the

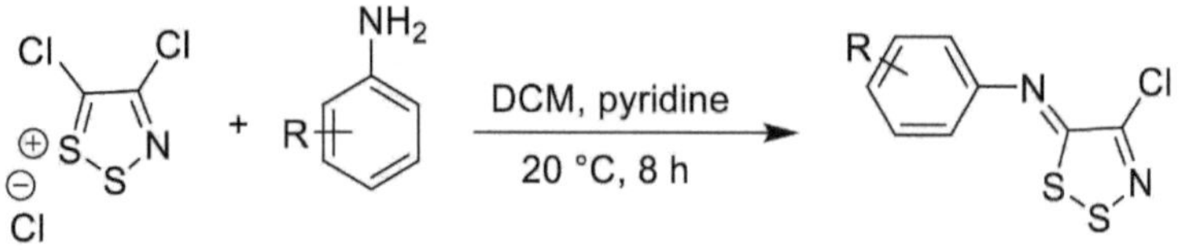

14 18 19

Scheme 2.6 Synthesis of *N*-arylimines from Appel's salt.

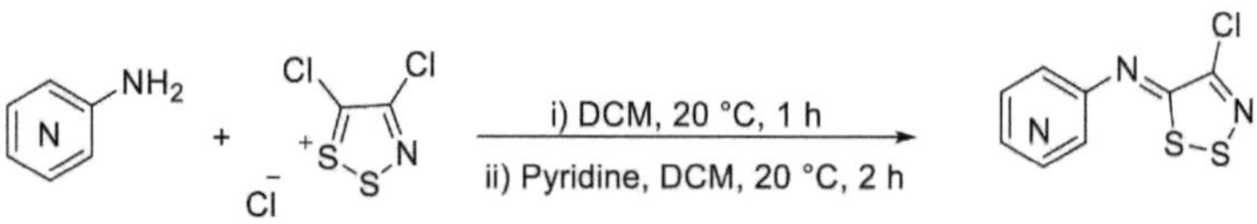

20 14 **21** pyrid-2-yl, pyrid-3-yl, pyrid-4-yl

Scheme 2.7 Synthesis of *N*-(4-chloro-5*H*-1,2,3-dithiazol-5-ylidene)pyridinamines.

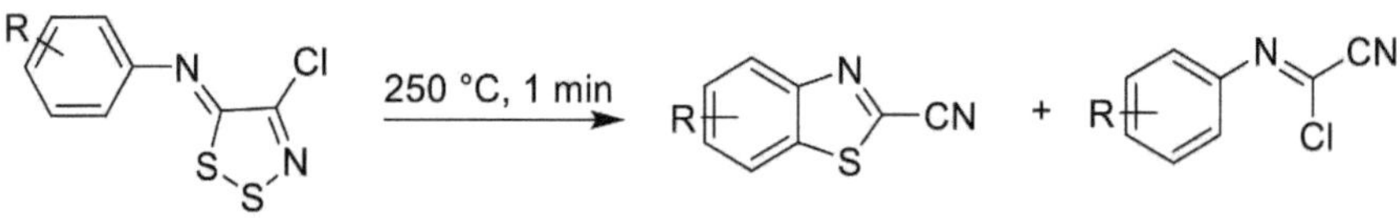

19 22 23

Scheme 2.8 Cyclization reaction of *N*-arylimines.

production of the desired product with low yield and formation of side products (Scheme 2.7). The yield of the desired compound was found to vary over a wide range. The yields of products from 2-aminopyridine, 3-aminopyridine, and 4-aminopyridine were 69%, 24%, and 1%, respectively.[49]

Cyclization of this imino compound (**19**) under vigorous heating conditions has been reported in previous research work.

This cyclization process afforded 2-cyanobenzothiazoles (**22**) (Scheme 2.8) along with sulfur and hydrochloric acid. However, in the presence of electron-withdrawing groups, such as *m*- or *p*-NO_2 in aniline, it favored the formation of cyanoimidoyl chloride (**23**) as the major product.[46]

2.2.2 Reaction at the Less Reactive C-4 Position of Appel's Salt

From the previous research work, it was found that the C-5 position of dithiazole is mostly active in chemical transformations. However, in

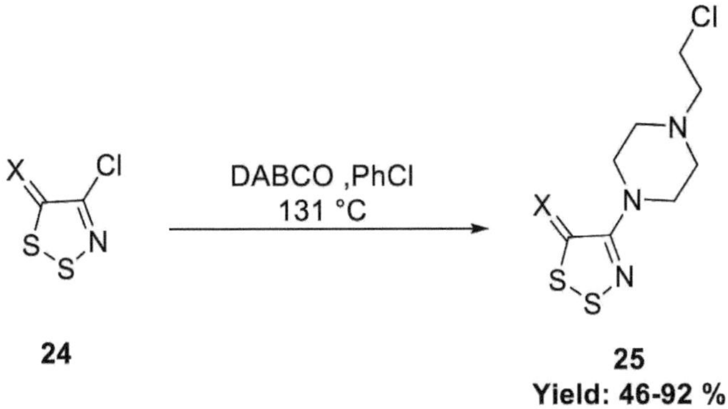

Scheme 2.9 Reaction of 4-chloro-5*H*-1,2,3-dithiazoles (**2**) with DABCO.

some recent advancements in the chemistry of 1,2,3-dithiazole, researchers have tried to functionalize the C-4 position. This reaction involves ring transformation of dithiazoles into pyrazolo[3,4-*d*]thiazoles, pyridothiazoles, pyrido[2,3-*d*]pyrimidines, and the rare 1,2,4-dithiazine system (pyrazolo-[3,4-*e*][1,2,4]dithiazines and benzo[*e*][1,2,4]dithiazines).[50]

Recently, a group of researchers has functionalized the C-4 position by optimizing the reaction of 5-substituted-1,2,3-dithiazole (**24**) with 1,4-diazabicyclo[2.2.2]octane (DABCO) in chlorobenzene at 131 °C to produce *N*-(2-chloroethyl)piperazines (**25**) in good yields (Scheme 2.9). This reaction proceeded faster when reacted with *ortho*-substituted dithiazole than *meta*- or *para*-substituted ones. Again, the rate of this reaction also depends on the groups attached to it. Electron-withdrawing groups showed faster reactions (*e.g.* Me < MeO < Br ≈ Cl < NO_2).[50]

A plausible mechanism of this reaction involves an addition of nucleophile, ring opening, and ring closure (ANRORC)-style mechanism (Figure 2.2), where a nucleophilic attack by DABCO causes the opening of the ring. Therefore, in order to produce amidine, a second DABCO molecule can competitively add to the more accessible nitrile. After that, amidine undergoes intramolecular cyclization to form a quaternized DABCO molecule at the C-4 position. Lastly, the attack of the chloride ion opens the DABCO molecule and produces the desired product.[50]

Rees *et al.* reported the first instance of the intramolecular nucleophilic displacement of the chlorine atom at C-4 by a phenoxide ion and cyclization, which resulted in the formation of dithiazolobenzoxazine (**27**) from the reaction of 4-chloro-5-[(*o*-hydroxyphenyl)imino]-5*H*-1,2,3-dithiazole (**26**) with NaH in THF (Scheme 2.10).[37,51]

Another ring transformation reaction of dithiazoles is the conversion of the dithiazole ring to quinazolines, which is present as a scaffold in the antihypertensive drug prazosin, antitumor drug trimetrexate, and 5-hydroxytryptamine (5-HT_2) antagonist ketanserin. This reaction involves the formation of 4-alkoxyquinazoline-2-carbonitriles (**29**) from

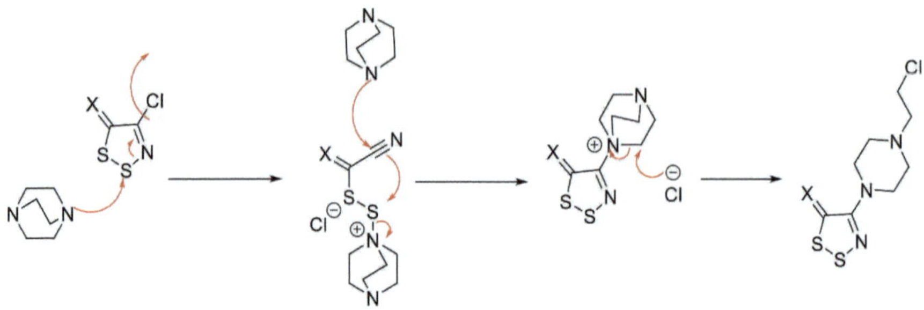

Figure 2.2 Plausible mechanism of 4-[*N*-(2-chloroethyl)piperazin-1-yl]-5*H*-1,2,3-dithiazole synthesis.

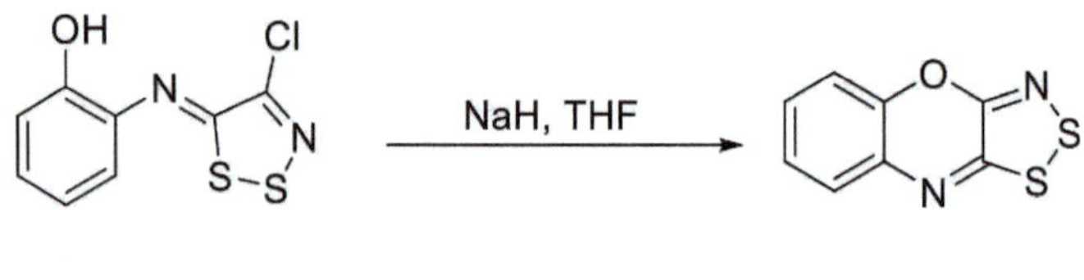

Scheme 2.10 Preparation of 4-chloro-5-[(*o*-hydroxyphenyl)imino]-5*H*-1,2,3-dithiazole.

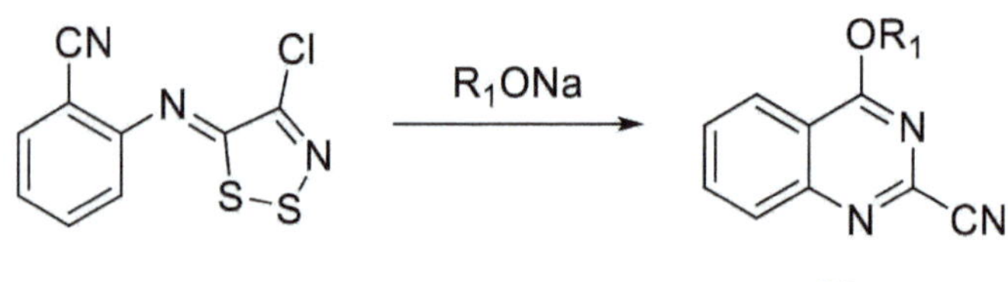

Scheme 2.11 Preparation of quinazoline scaffolds from the dithiazole precursor.

(*Z*)-2-(4-chloro-5*H*-1,2,3-dithiazol-5-ylideneamino)benzonitrile (**28**) in the presence of an alkoxide base with a good yield (Scheme 2.11).[45,52]

Formation of these quinazoline compounds from the dithiazoles has also been reported in some protocols. One protocol demonstrates the reaction of dithiazoles (**28**) with triphenyl phosphine (**30**) to form cyanoformanilide (**31**). Subsequently, the cyclization proceeded in the presence of alkoxide or EtSH to produce quinazoline (**32**).[45]

Moreover, quinazoline *N*-oxides were produced by treating 2-carboxy-substituted 2-(dithiazolylideneamino)benzenes (**33**) first with hydroxylamine (**34**), forming a 2-cyano-4-methylquinazoline 3-oxide ion (**35**). Quinazolines (**36**) were obtained by reduction with $TiCl_3$ (Scheme 2.12).[45]

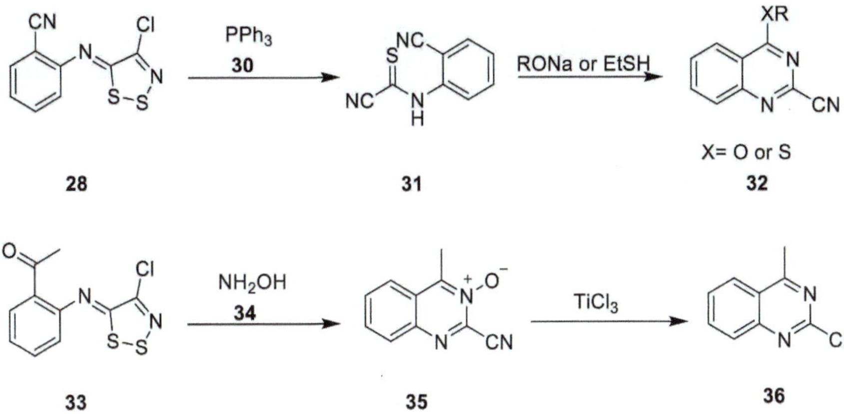

Scheme 2.12 Alternate method for the synthesis of quinazolines from dithiazoles.

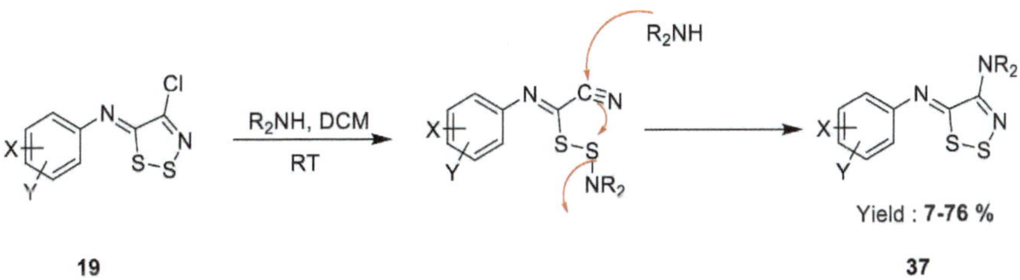

Scheme 2.13 Preparation of *N*-(4-dialkylamino-5*H*-1,2,3-dithiazol-5-ylidene)anilines.

The most general example of C-4 functionalization of dithiazole includes the reaction of 5-(arylimino)-4-chloro-5*H*-l,2,3-dithiazoles (**19**) with an excess amount of dialkyl amines in DCM at room temperature to produce *N*-(4-dialkylamino-5*H*-1,2,3-dithiazol-5-ylidene)anilines (**37**) in variable yields following the ANRORC mechanism.[53] The reaction mechanism follows the addition of nucleophiles, ring opening, and ring closing (Scheme 2.13).

2.2.3 Synthesis of Fused Dithiazoles

Therefore, we have seen that monocyclic 1,2,3-dithiazoles are mainly prepared using Appel's salt. Similarly, benzo-fused 1,2,3-dithiazoles are readily prepared from Herz's salt (benzo-fused 1,2,3-dithiazolium chloride)[54] (**39**) using aromatic naphthalene-2,3-diamines (**38**) and sulfur monochloride (S_2Cl_2) (**13**) *via* a cyclocondensation ring-closure reaction at the reactive 1-position (Scheme 2.14).[55]

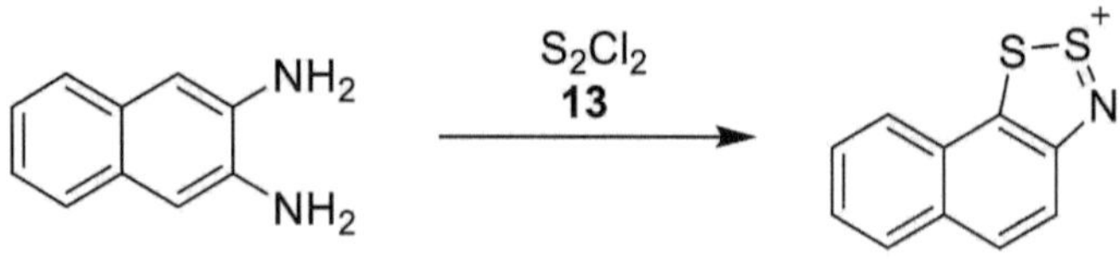

38 39

Scheme 2.14 Formation of fused dithiazole.

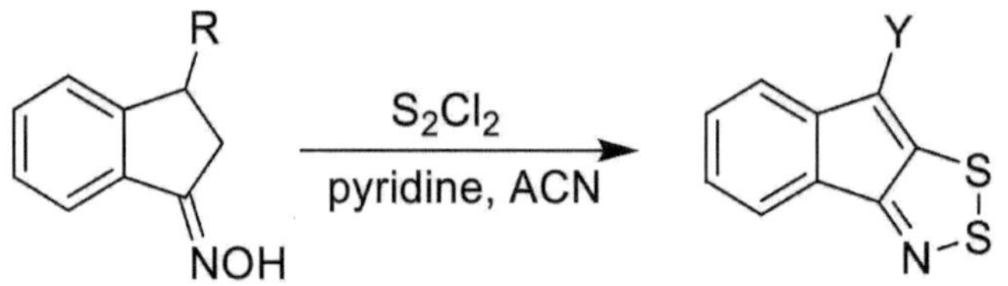

40a R= H
b R= Ph

41a Y=Cl, 81 %
b Y=Ph, 90%

Scheme 2.15 Synthesis of fused 1,2,3-dithiazole from cyclic oximes.

Another possible method for preparing fused 1,2,3-dithiazole involves the use of Hunig's base (*N*-ethyldiisopropylamine) and triisobutylamine. In this procedure, cyclic oximes are reacted with sulfur monochloride (S_2Cl_2) (**13**).[56]

Rakitin *et al.* have found that the yield of this cyclocondensation reaction of 1,2,3-dithiazole depends on the reaction conditions and the organic base used in the reaction. Hence, they demonstrated the selective synthesis of 1,2,3-dithiazole from cyclooximes (**40**) and S_2Cl_2 (**13**) using pyridine as a base. The corresponding 1,2,3-dithiazole (**41**) was selectively produced in 81% yield as the best output when cyclooxime was treated with a threefold excess of S_2Cl_2 or a fourfold excess of pyridine in acetonitrile (ACN) at 5 °C for 1 h. The reaction of S_2Cl_2 with 3-phenylindanone oxime (**40b**) under the same reaction conditions avoids chlorination and increases the yield (Scheme 2.15).[57]

When S_2Cl_2 (**13**) (6 equivalents) and pyridine (8 equivalents) were added to boiling ACN, the classic cyclopentanone oxime (**42**a) produced 4,5,6-trichlorocyclopenta-1,2,3-dithiazole (**43**a) with a moderate yield. Since only **43** was isolated in lower yields, attempts to get less chlorinated products using less S_2Cl_2 or/and pyridine were unsuccessful. Dichlorocyclopentadithiazole (**43b**) was successfully produced from a 4-carbethoxy-substituted derivative (**42b**) using a similar process (Scheme 2.16).[57]

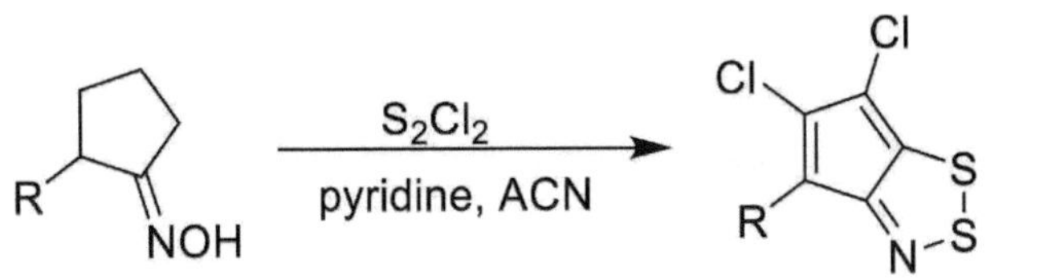

42a, R=H
b, R=COOEt

43a, R=H
b, R=COOEt

Scheme 2.16 Synthesis of dithiazole from cyclopentanone oxime.

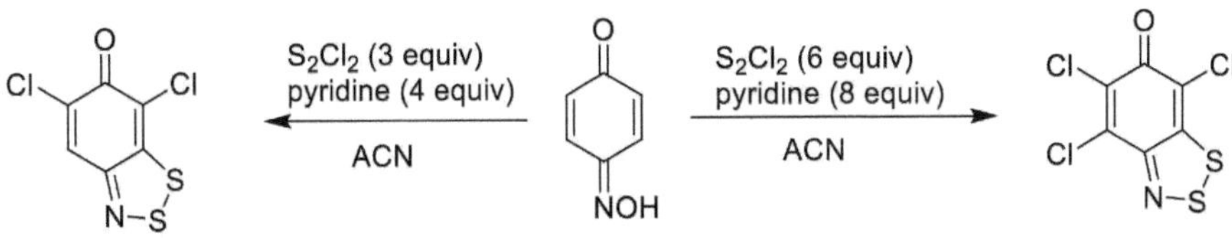

46
15 %

44

45
13 %

Scheme 2.17 Preparation of 1,2,3-dithiazole from *p*-benzoquinone monooxime.

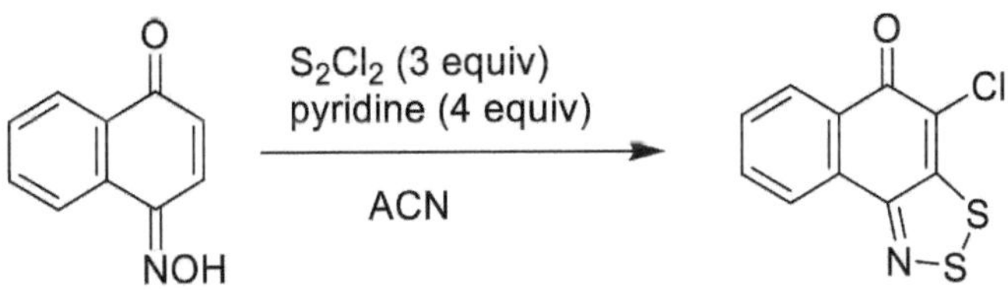

47

48

Scheme 2.18 Synthesis of fused dithiazole from naphthaquinone oxime.

In the case of the reaction of six-membered cyclic oximes, such as *p*-benzoquinone monooxime (**44**) with S_2Cl_2, the expected dithiazole was formed in a very low yield.[58] In addition, the formed product depends on the molar excess of the reagent. A higher excess of reagents in boiling ACN produced trichloro dithiazole (*e.g.* 4,5,7-trichloro-6*H*-benzo[*d*][1,2,3]dithiazol-6-one) (**45**), whereas a lower excess of reagents gave dichloro dithiazole (5,7-dichloro-6*H*-benzo[*d*][1,2,3]dithiazol-6-one) (**46**) (Scheme 2.17).

However, 1,4-naphthoquinone oxime (**47**) afforded the desired dithiazole (**48**) in a good yield (74%) when reacted with a lower excess of the reagents (Scheme 2.18).[57]

Besides all this chemistry of dithiazole, Appel's salt is arguably the most unstable 1,2,3-dithiazole because, although it is quite stable at about 20 °C

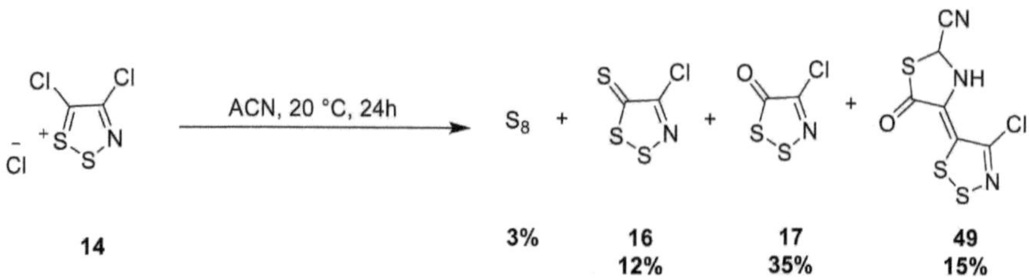

Scheme 2.19 Degradation of Appel's salt in ACN.

when protected by a desiccant, it reacts readily with moisture. According to Kalogirou and Koutentis, it formed elemental sulfur, dithiazole-5-thione, dithiazol-5-one, and thiazol-5-one when stirred in wet ACN for a short period (Scheme 2.19).[59]

2.3 Biological Activity of 1,2,3-Dithiazoles

1,2,3-Dithiazole has many medicinal activities, such as antimicrobial,[36] anticancer,[47] anti-inflammatory,[60] antiviral,[39] and antifibrotic[61]. The medicinal activities of 1,2,3-dithiazoles are diverse and hold promise for the development of new therapeutic agents across various medical fields. However, it is important to note that further research is needed to fully understand the mechanisms of action and optimize the pharmacological properties of these compounds for clinical use.

2.3.1 Antimicrobial Activities of 1,2,3-Dithiazoles

In 1977, J. E. Moore, representing Chevron Research Co., filed a patent that marked the initial documentation of biological activity associated with the 1,2,3-dithiazole scaffold. This patent revealed the synthesis of innovative 1,2,3-dithiazoles involving *N*-aryl cyanothioformamide and sulfur dichloride. The primary purpose of these compounds, as outlined in the patent, was found to be effective in combating fungi, especially *Botrytis cinerea*-caused plant fungal infections and leaf blights caused by organisms such as *Pythium ultimum*.[34]

The ability of the invented compounds to control the conidia of the organism *Alternaria solani*, which causes tomato early blight, was examined by utilizing tomato seedlings that were six to seven weeks old. A 250 ppm solution of the test compound in acetone, water, and a small quantity of a non-ionic emulsifier was sprayed on tomato plants. Based on the percentage of disease decrease in comparison to untreated check plants, the percentage of disease control of the test compound was calculated. Notably, (*Z*)-4-((4-chloro-5*H*-1,2,3-dithiazol-5-ylidene)amino)benzonitrile (**50**) emerged as a promising compound, demonstrating a substantial 90% reduction in infection compared to untreated plants. Conversely, the 2,4-dichloro analogue **51** exhibited weaker activity, resulting in just over half the infection reduction. Subsequent investigations targeted the impact of *Phytophthora infestans* conidia on five- to six-week-old tomato seedlings using the same experimental procedure. The 4-cyano analogue **50** showed a remarkable efficacy of 97% activity against the infection. In comparison, the 2-(4-nitrophenoxy) analogue **52** exhibited an 80% reduction in infection. Further experimentation involved testing the effects on a celery late blight organism *Septoria apiicola* using eleven-week-old plants. The 4-cyano analogue **50** demonstrated approximately 60% protection, while notable improvements were observed with other analogues. Specifically, both the 2-fluoro analogue **53** and 3-(4-trifluoro,2-cyanopenoxy) analogue **54** exhibited an 80% protection rate.

They extended their study to find the activity of testing compounds against *Erysiphe polygoni*, the causing organism of powdery mildew of bean leaves, by spraying a solution of 250 ppm of the test compounds and compared it with untreated check plants. The test compounds containing halogens (**55–59**) showed potent activity against the organism. Among these analogues, the (*Z*)-4-chloro-*N*-(4-chloro-2-methylphenyl)-5*H*-1,2,3-dithiazol-5-imine (**55**) and its corresponding 3-chloro analogue **56** demonstrated 100% inhibitory activity against the causing organism at the above-mentioned concentration. Conversely, the 5-chloro analogue **57** and 4-bromo analogue **58** exhibited marginal reductions in efficacy, with reductions of 10% and 1%, respectively. The 3,5-dichloro analogue **59** was moderately effective, achieving a net protection rate of 76% (Figure 2.3).

The *in vitro* antifungal properties of some synthesized dithiazole compounds (**60–68**) against different yeast strains of *Candida albicans*, *Candida utilis*, *Candida lipolytica*, *Saccharomyces cerevisiae*, *Pichia stipitis*, and molds (*Aspergillus niger* and *Penicillium* sp.) were investigated in 2002 by Baraldi *et al.* Notably, all derivatives exhibited significant antifungal activity against the tested yeasts. The compounds demonstrated the highest efficacy, exhibiting favorable minimum inhibitory concentration (MIC) values within the range of 10–50 mg ml^{-1} (Figure 2.4). Particularly noteworthy was the comparable effectiveness of these products against molds when compared with amphotericin B, a reference compound for fungal inhibitory activity.[36]

Compounds **60**, **61**, and **63** exhibited limited activity, primarily against specific microorganisms, such as *Penicillium* and *C. utilis*. Conversely, out of all the fungi that were examined, products **62** and **68** showed the least amount of antifungal activity. Interestingly, compounds **60–65**, which were generated from *o*-amino nitriles with identical heterocyclic rings, showed that substituents at the *N*–1 position of the pyrazole ring affect the biological activity.

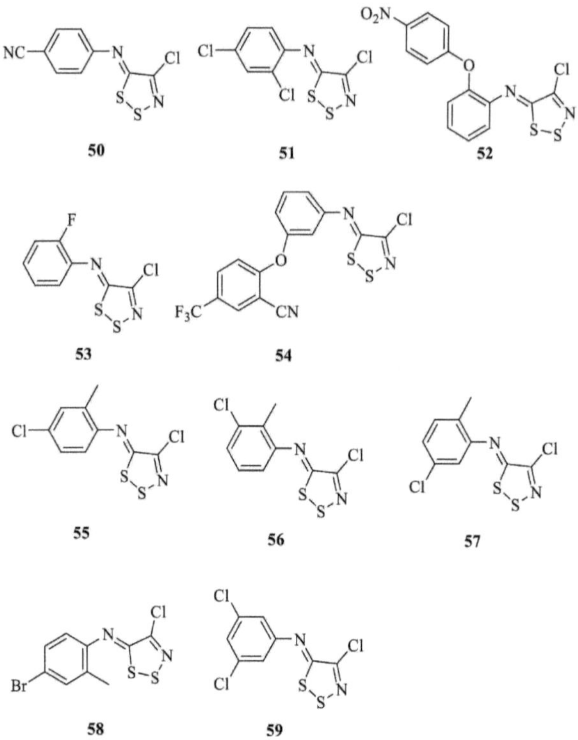

Compound no	% of Tomato early blight control
50	**90**
51	**56**

Compound no	% of Tomato late blight control
50	**97**
52	**80**

Compound no	% of Celery late blight control
50	**63**
53	**80**
54	**80**

Compound no	% of Powdery mildow control
55	**100**
56	**100**
57	**90**
58	**99**
59	**76**

Figure 2.3 Antifungicidal activities of early 1,2,3-dithiazole derivatives.

2.3.2 Antiviral Activities of 1,2,3-Dithiazoles

The antiviral characteristics of compounds based on the 1,2,3-dithazole scaffold were initially described by Hilton *et al.* in 2016. This work concentrated on a group of 1,2,3-dithiazole-5-thiene derivatives with the goal of targeting the nucleocapsid protein potentially using the disulfide bond to function as a zinc ejector. Using the *Feline immunodeficiency virus* (FIV) as a model, antiviral screening was done. Crandell-Rees Feline Kidney (CrFK) cells were used to test the toxicity of compounds and assess their antiviral properties in a feline lymphoid cell line (FL-4) (Figure 2.5). The substances included 5-oxo- and 5-imino-1,2,3-dithiazole derivatives.[62] Among the compounds tested, four were highlighted for their potency and favorable toxicity profile.[63] Specifically, compounds **69–72** demonstrated the most potent antiviral effects with a substantial toxicity window, indicated by the ratio of FL-4/CrFK. Notably, 4-phenyl-5*H*-1,2,3-dithiazole-5-thione (**69**) exhibited an excellent activity-to-toxicity ratio (>4000) and a potency of half maximal effective concentration (EC_{50}) = 23 nM. Another promising compound was 4-(4-fluorophenyl)-5*H*-1,2,3-dithiazol-5-one (**70**), which showed comparable potency to **69** with a slight increase in toxicity at higher

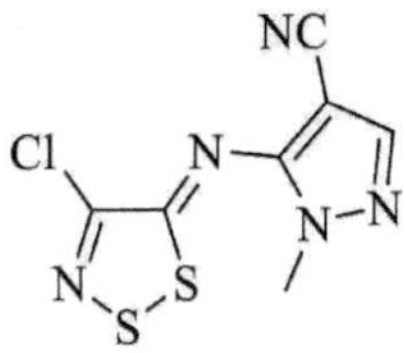

60

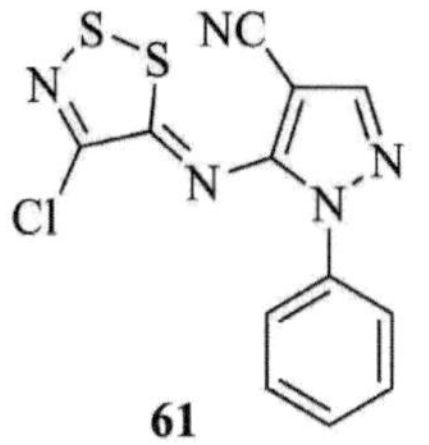

61

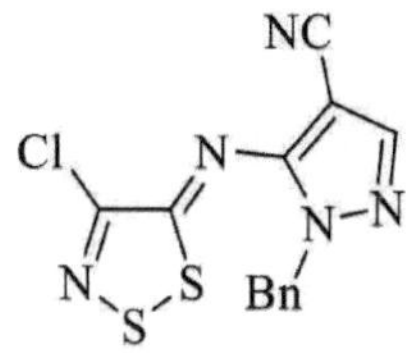

62

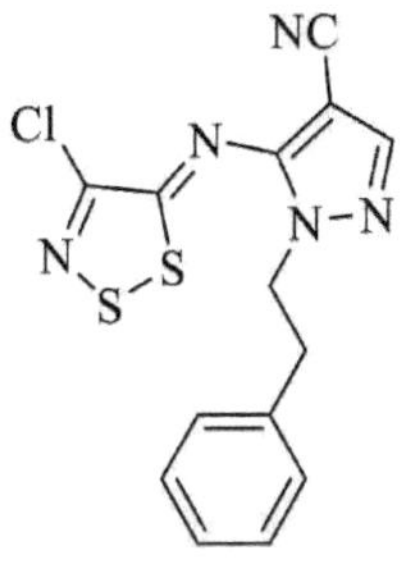

63

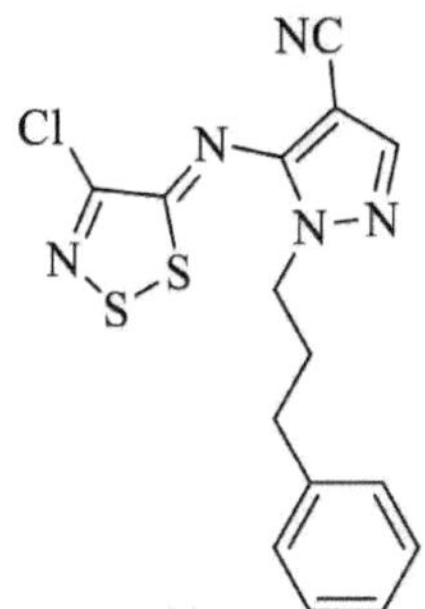

64

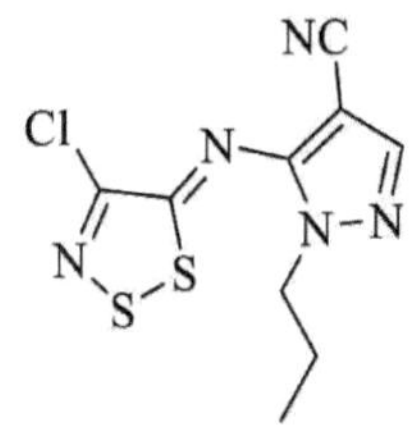

65

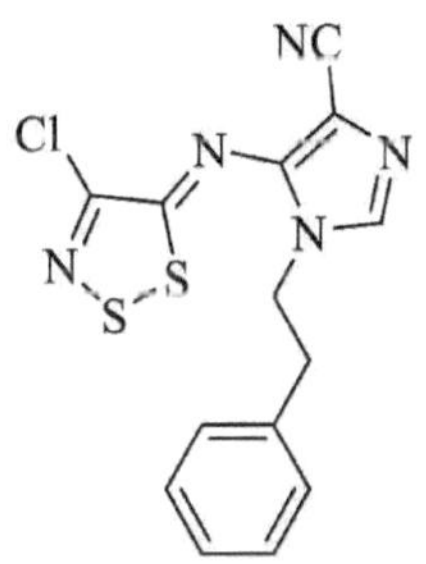

66

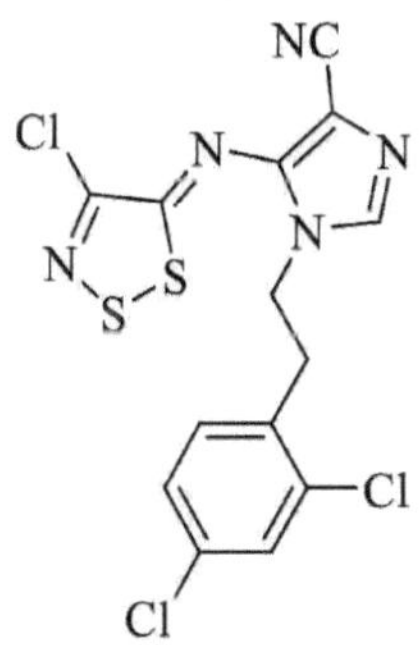

67

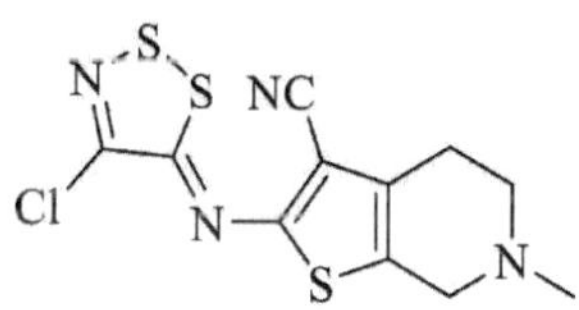

68

Figure 2.4 Antifungal activity of the compound against various yeast strains.

concentrations (concentration of cytotoxicity [CC_{50}] = 64 μM). However, the 4-methoxy analogue **71** exhibited an approximately 8-fold reduction in potency, while the ethyl (*Z*)-5-(phenylimino)-5*H*-1,2,3-dithiazole-4-carboxylate (**72**) analogue exhibited an almost 3-fold decrease in potency, both maintaining a comparable toxicity profile. These findings highlight the potential of 1,2,3-dithiazole derivatives, particularly compound **69**, as effective antiviral agents with a promising activity-to-toxicity ratio, offering a novel approach for combating viral infections, especially the FIV.[39]

Compound **73** featured a disrupted disulfide bridge, and its creation served as indirect evidence confirming the necessity of the disulfide for antiviral activity. This experimental approach strengthens the validity of

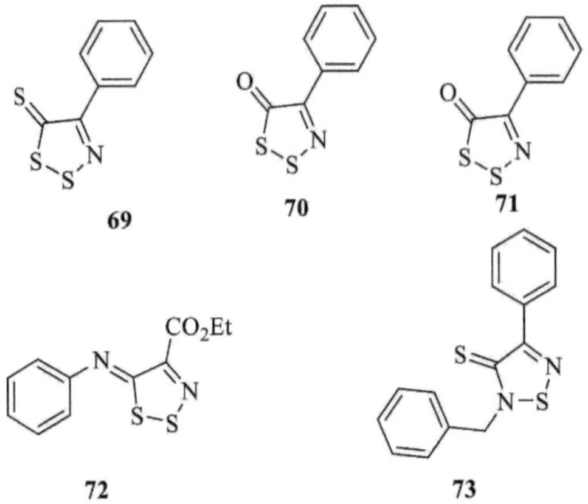

Compound no.	CC_{50}(μM)	EC_{50} (μM)
69	>100	0.023
70	64	0.025
71	41	0.19
72	64	0.067
73	0.68	0.68

Figure 2.5 Antiviral activity of the 1,2,3-dithiazole scaffold on CrFK cells against the FIV.

the proposed mechanism, emphasizing the critical role of the disulfide bridge in the observed antiviral effects, as demonstrated by the synthesized analogue **73**.[51]

2.3.3 Anticancer Activities of 1,2,3-Dithiazoles

In 2002, Baraldi *et al.* first documented potential anticancer properties associated with the 1,2,3-dithiazole scaffold.[36] The study involved the synthesis of ten 1,2,3-dithiazoles, which were subsequently assessed against various targets related to both antimicrobial and anticancer therapeutic applications. Notably, compounds **74–76** exhibited promising low micromolar potency against leukemia cell lines L1210 and K562 (Figure 2.6). Despite a generally flat structure–activity relationship (SAR) within the series, this initial phenotypic analysis revealed encouraging activity across both cell lines. While the antibacterial screening showed limited activity, the antifungal screening revealed that compounds **75** and **76** exhibit some efficacy against *Aspergillus niger*, with an MIC_{50} value of 10 μM. This finding further underscores the potential suitability of the scaffold as an antifungal agent. Overall, the observed activity against the leukemia cell lines and the identified antifungal properties highlight the promising therapeutic prospects of the 1,2,3-dithiazole scaffold in the realm of anticancer and antimicrobial research.

Rakitin *et al.* in 2009 synthesized and screened some compounds for anticancer activity. A series of C-4-substituted dithiazoles underwent screening against two breast cancer cell lines, namely MCF7 and MDA-MB-231.[30] There was minimal activity on the MDA-MB-231 cell line

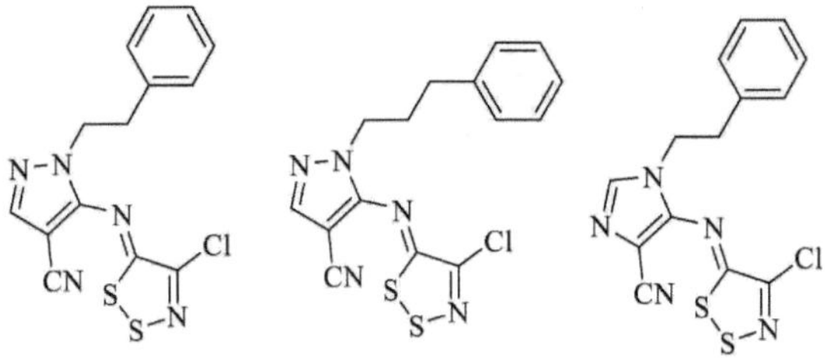

74 75 76

Compound no.	L1210 IC_{50} (μM)	K562 IC_{50} (μM)
74	5	3
75	3	5
76	3	4

Figure 2.6 1,2,3-Dithiazoles with anticancer activity initially reported in 2002 by Baraldi *et al.*

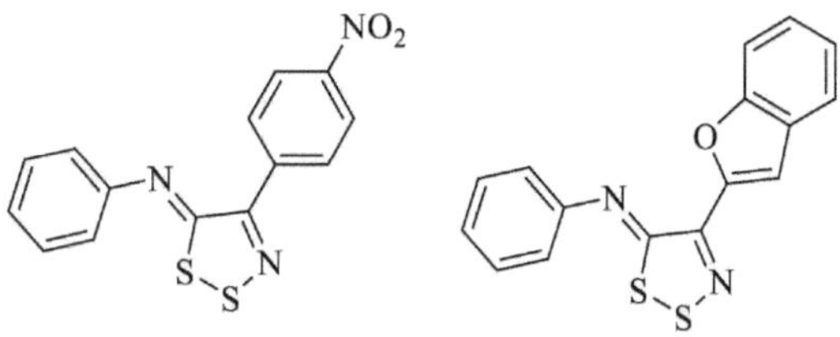

77 78

Breast cancer (MCF7) -GI_{50} = Approx. 10 μM

Figure 2.7 Anticancer activity of 1,2,3-dithiazoles reported in 2009 by Rakitin *et al.*

throughout the series. Nevertheless, two particular compounds—(*Z*)-4-(4-nitrophenyl)-*N*-phenyl-5*H*-1,2,3-dithiazol-5-imine (**77**) and its benzofuran-2-yl analogue (**78**)—showed remarkable outcomes, exhibiting 50% growth inhibition at about 10 μM after 72 h (Figure 2.7). These findings serve as a foundation for expanding the chemical space and conducting further investigations into the SAR of 1,2,3-dithiazoles in the context of anticancer properties.

After 1,2,3-dithiazoles were first shown to have anticancer properties in different cell lines, Indiveri *et al.* screened a library of more than forty compounds in 2012.[64] The glutamine-amino acid transporter ASCT2, which is known to be overexpressed in a number of malignancies, was the focus of the screening. Post-translational changes and interactions with scaffold proteins control the stability, trafficking, and transport function of ASCT2. Rapidly reproducing cells, such as stem cells and inflammatory cells, have a tendency to produce more ASCT2, which makes it easier to deliver the higher glutamine requirements. By overexpressing ASCT2, cancer-promoting pathways can take advantage of this process to satisfy their glutamine need and promote fast development. In addition to being known as an anticancer target, ASCT2 can also transfer viruses that infect human cells.[65] Cancer-promoting pathways can exploit this mechanism to meet their

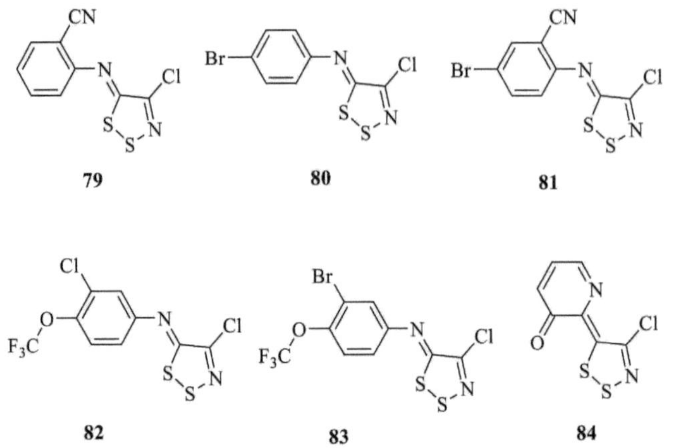

Compound Code	IC_{50} (μM)
79	9.7
80	8.3
81	3.7
82	5.6
83	8.1
84	10.2

Figure 2.8 Anticancer activity of 1,2,3-dithiazoles as ASCT2 transport inhibitors.

glutamine demand and support rapid growth by overexpressing ASCT2. ASCT2, besides being recognized as an anticancer target, also possesses the capability to traffic virions for infecting human cells. Six compounds (**79–84**) from the library showed significant activity in the evaluation of a series of 1,2,3-dithiazoles as transporter inhibitors, with IC_{50} values approximately or less than 10 μM (Figure 2.8). These substances showed strong inhibition of the ASCT2-catalyzed glutamine/glutamine transfer. Finding these active ingredients raises the possibility that 1,2,3-dithiazoles can inhibit ASCT2, opening up a new line of inquiry for both antiviral and anticancer research.[66]

The observed inhibition was characterized as non-competitive, and the reversibility of inhibition upon the addition of dithiothreitol (DTE) suggested the formation of adducts with protein cysteine residues. This indicated that the inhibitory reaction likely proceeded through an ANRORC-style rearrangement. Computational modeling, encompassing both molecular and quantum mechanical studies (MM and QM, respectively), along with Frontier Orbital Theory (FOT) applied to 1,2,3-dithiazole models, provided insights. Following the ASCT2 investigation, Indiveri *et al.* conducted a screening of slightly over fifty 1,2,3-dithiazoles against the LAT1 transporter in 2017. Both ASCT2 and LAT1 are amino acid transporters overexpressed in cancer, and inhibitors against these transporters have been explored for anticancer applications, with JPH203, an LAT1 inhibitor, progressing to a phase 1 clinical trial.

In the library screening against LAT1, eight compounds demonstrated inhibition exceeding 90% at 100 μM. With an IC_{50} value of less than 1 μM, compounds **85** and **86** stood out as the most potent (Figure 2.9). A mixed type of inhibition with regard to the substrate was found by kinetic analysis of inhibition on the above-mentioned inhibitors. Even after the compounds

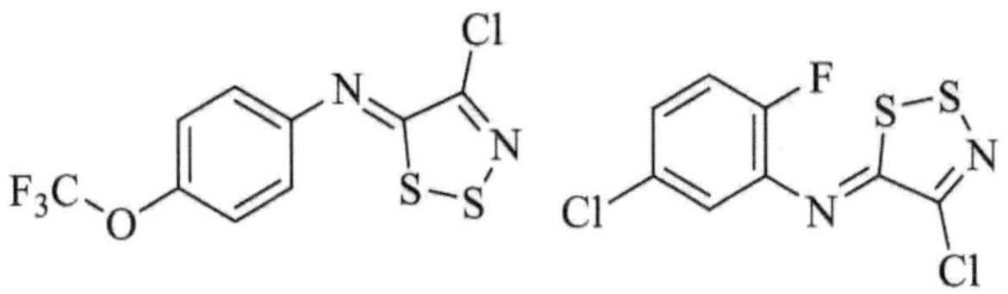

85 **86**

Compound no.	IC_{50} (μM)
85	0.98
86	0.89

Figure 2.9 Anticancer activity of 1,2,3-dithiazoles targeting transporter protein LAT1 (SLC7A5).

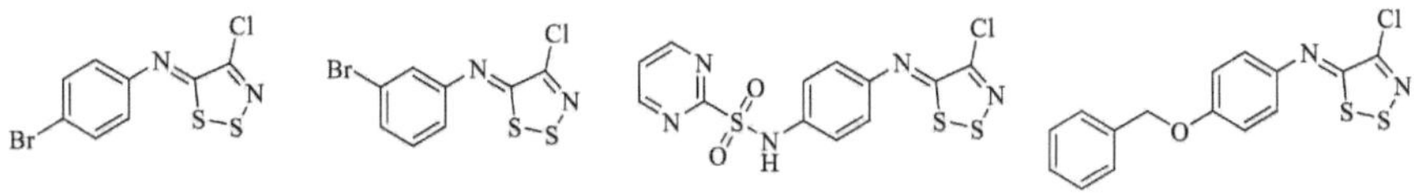

87 **88** **89** **90**

Figure 2.10 The initial screening, highlighting results.

were taken out of the reaction mixture, the inhibition remained, indicating irreversible binding. Nevertheless, dithioerythritol, an S–S reducing agent, has the ability to reverse this irreversible impact, hence bolstering the notion that the chemicals and LAT1 create disulfide bonds. Using an LAT1 homology model, molecular modeling of residues **85** and **86** showed interactions with the substrate-binding site and the creation of a covalent bond with C407. A more thorough investigation conducted in 2021 by Marino *et al.* supported this conclusion by emphasizing the requirement for a water molecule in the reactive pathway.[67]

A study published in 2021 expanded on the phenotypic reports of 1,2,3-dithiazoles as anticancer agents. It evaluated a library of less than forty 1,2,3-dithiazole analogues across a variety of cancer cell lines, including lung, pancreatic, bladder, prostate, breast, and chordoma. As a non-specific toxicity control, a skin fibroblast cell line was utilized (Figure 2.10).[47]

Encouraging initial results were observed with compounds (*Z*)-4-(4-bromophenyl)-*N*-phenyl-5*H*-1,2,3-dithiazol-5-imine (**87**) and its 3-position bromo analogue **88**, exhibiting potency against the breast cancer cell line MCF7 (IC_{50} = 11 and 6.7 μM). Subsequent investigations identified (*Z*)-*N*-(4-((4-chloro-5*H*-1,2,3-dithiazol-5-ylidene)amino)phenyl)pyrimidine-2-sulfonamide (**89**) with good activity against breast cancer (MCF7-IC_{50} = 3.0 μM) and no observed toxicity. The analogue (*Z*)-*N*-(4-(benzyloxy)phenyl)-4-chloro-5*H*-1,2,3-dithiazol-5-imine (**90**) demonstrated the preference for inhibiting bladder cancer (5637-IC_{50} = 13 μM) without toxicity.

A focused screening of 5-membered heteroatomic compounds revealed potency against prostate cancer, with (*Z*)-4-chloro-*N*-(thiazol-2-yl)-5*H*-1,2,3-dithiazol-5-imine (**91**) and two pyrazoles (**92** and **93**) showing activity against prostate cancer (DU145-IC_{50} = 8–11 μM). Thiazole analogue **91** also exhibited activity against the chordoma cell line U-CH1 (IC_{50} = 10 μM) with limited toxicity. Compound (*Z*)-4-chloro-*N*-(3-methyl-1-phenyl-1*H*-pyrazol-5-yl)-5*H*-1,2,3-dithiazol-5-imine (**93**) showed low micromolar activity against bladder cancer (5637-IC_{50} = 2.1 μM) but with associated toxicity (Figure 2.11).[47]

Further evaluation of C-4-substituted analogues revealed a trend of activity against breast cancer (MCF7).[50] Compounds **94–96** (IC_{50} = 2–10 μM) (Figure 2.12) showed potency but no distinguishable SAR features. (*Z*)-5-Bromo-2-((4-(4-(2-chloroethyl)piperazin-1-yl)-5*H*-1,2,3-dithiazol-5-ylidene)amino)benzonitrile is particularly noteworthy. (*Z*)-5-Bromo-2-((4-(4-(2-chloroethyl)piperazin-1-yl)-5*H*-1,2,3-dithiazol-5-ylidene)amino)benzonitrile (**95**) showed efficacy against bladder cancer (5637-IC_{50} = 8.0 μM). Notably, the 4-chloro-1,2,3-dithiazole moiety of previous analogues was not essential for anticancer activity, indicating the possibility of several different modes of action.[47]

2.3.4 Other Biological Activities

2.3.4.1 Antibacterial Activity of 1,2,3-Dithiazoles

The *in vitro* antibacterial activity of *N*-arylimines and *N*-arylcyanothioformamides was evaluated against various bacterial strains, encompassing Gram-negative bacteria (*Escherichia coli*, *Pseudomonas aeruginosa*, *Klebsiella pneumoniae*, *Proteus mirabilis*, and *Salmonella choleraesuis* ser.)

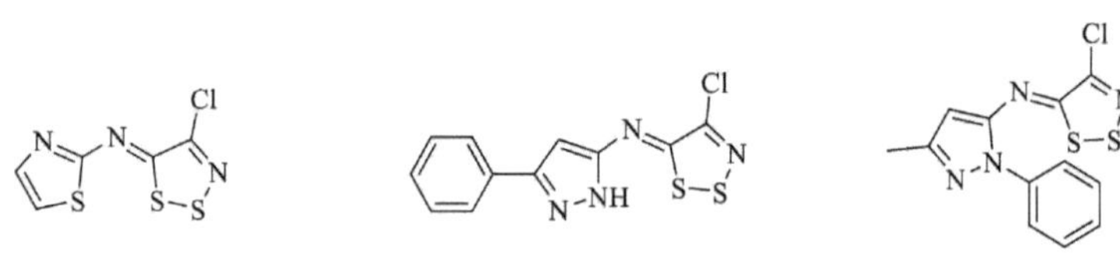

91 92 93

Figure 2.11 Activity of 1,2,3-dithiazoles against prostate cancer.

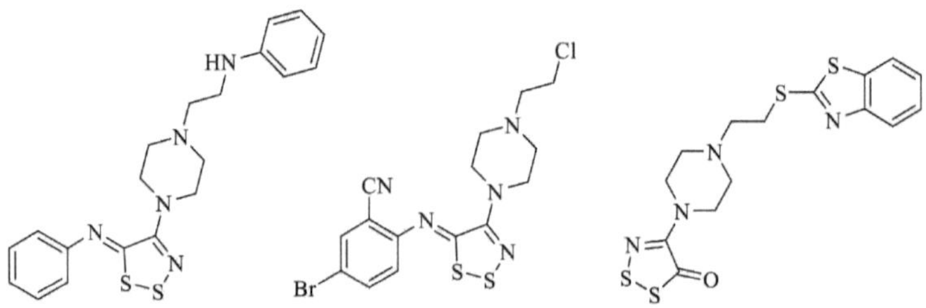

94 95 96

Figure 2.12 C4-substituted 1,2,3-dithazole analogues having anticancer activity.

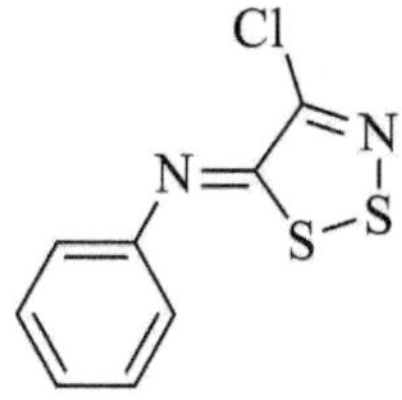
97

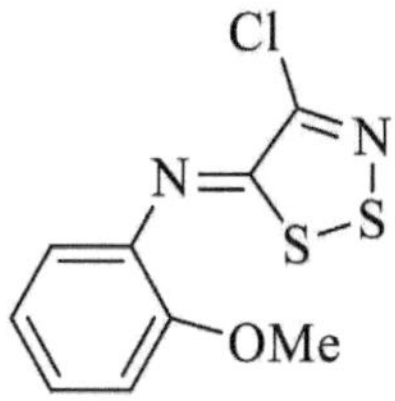
98

Figure 2.13 The initial compound with antimicrobial activity.

and Gram-positive bacteria (*Staphylococcus aureus*, *Streptococcus pyogenes*, *Listeria monocytogenes*, and *Enterococcus faecalis*).[36]

N-Arylimines and *N*-cyanothioformamides did not have any effect on the growth of Gram-negative bacteria in solid media when the disc diffusion method was used for antimicrobial testing. On the one hand, *N*-arylcyanothioformamide was not active against Gram-positive bacteria. On the other hand, *N*-arylimines had a significant antibacterial action that was targeted toward Gram-positive bacteria (Figure 2.13).

To get further information about the effectiveness of *N*-arylimines against Gram-positive bacteria, the broth dilution method was used to determine the MICs. The results of numerous tests consistently showed that the activity of the molecule was unrelated to the functionality of cyanothioformamide. The 1,2,3-dithiazole ring on its own is obviously adequate to exhibit significant inhibitory effects against Gram-positive bacteria. Among the studied series, the unsubstituted aromatic compound **97** and its *o*-methoxy derivative **98** were found to be the most active.

2.3.4.2 *Antifibrotic Collagen-specific Chaperone Hsp47 Inhibitor*

The hit compound methyl 6-chloro-3*H*-benzo[*d*]dithiazole-4-carboxylate 2-oxide (**99**) has been highlighted in two separate reports, first in 2005 and subsequently in 2010.[44] Both reports involved high-throughput screening of compound libraries.

A Maybridge compound library was screened in 2005 by Ananthanarayanan *et al.* against Heat shock protein 47 (Hsp47), an enzyme for which there were no known inhibitors at that time. The pathophysiology of fibrotic illnesses has been linked to the activity of Hsp47, a molecular chaperone that is unique to collagen. The regulation of both Hsp47 and collagen expression is associated with various diseases, including fibrotic conditions in the liver,[68] kidneys,[69] lungs,[70] and skin,[71] and atherosclerosis and cancer. After the initial screening, 4 out of 2080 compounds were found to be Hsp47 inhibitors, yielding a 0.2% hit rate. Compound **99** (Figure 2.14) was found to be the most potent after secondary screening; its IC_{50} value was found to be 3.1 μM.

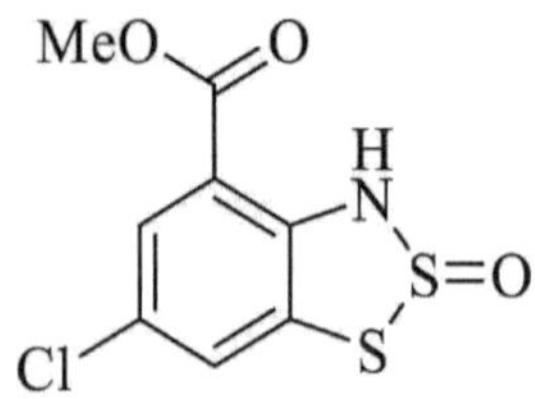

99

Hsp47 - IC_{50} = 3.1 μM

Figure 2.14 Most potent compound screened by Ananthanarayanan *et al.* for antifibrotic activity.

2.4 Conclusion

Overall, the research on the biological activities of 1,2,3-dithiazoles has revealed their diverse potential in medicinal chemistry. The compounds have shown promising antimicrobial, antiviral, and anticancer properties, making them interesting candidates for further drug development. The mechanisms of action and structure–activity relationships are areas of ongoing research to optimize the pharmacological properties of these molecules for clinical use. It is worth noting that the synthesis and functionalization of 1,2,3-dithiazoles play a crucial role in expanding the chemical space and exploring new derivatives with enhanced biological activities. The introduction of Appel's salt and Herz's salt has significantly simplified the synthesis of these compounds, enabling the formation of core structures in a single step. Additionally, recent advancements in the functionalization of the C-4 position of 1,2,3-dithiazoles, traditionally less reactive than the C-5 position, demonstrate the continuous efforts to explore and expand the synthetic toolbox for these molecules. The ring transformation reactions leading to the formation of fused dithiazoles, such as quinazolines, further showcase the versatility of these compounds and their potential as precursors for structurally diverse medicinal scaffolds. As the research in this field progresses, it is likely that more insights into the biological activities and synthetic methodologies of 1,2,3-dithiazoles will emerge, paving the way for the development of novel therapeutic agents across various medical disciplines.

Despite the progress made in understanding the biological activities of 1,2,3-dithiazoles, there is still much to explore. Future research could focus on elucidating the precise mechanisms of action, optimizing the pharmacological properties, and conducting in-depth structure–activity relationship studies to guide the design of more potent and selective derivatives. Additionally, exploring the potential applications of 1,2,3-dithiazoles in other therapeutic areas and uncovering novel synthetic methodologies will contribute to the continued growth of this field. The multifaceted nature of 1,2,3-dithiazoles makes them promising candidates for the development of new therapeutic agents with diverse applications in medicine and beyond.

References

1. S. T. Dorai, K. Lakshmikanth, P. Tiwari, S. M. Saini and S. Chandrashekharappa, One-pot construction of novel trifluoromethyl dihydroimidazo[1, 2-a]pyridine: A greener approach, *Tetrahedron*, 2023, **148**, 133691.
2. P. Arora, V. Arora, H. Lamba and D. Wadhwa, Importance of heterocyclic chemistry: a review, *Int. J. Pharm. Sci. Res.*, 2012, **3**(9), 2947.
3. M. García-Valverde and T. Torroba, Sulfur-nitrogen heterocycles, *Molecules*, 2005, **10**(2), 318–320.
4. K. N. Venugopala, S. Chandrashekharappa, S. Bhandary, D. Chopra, M. A. Khedr and B. E. Aldhubiab, *et al.*, Efficient Synthesis and Characterization of Novel Substituted 3-Benzoylindolizine Analogues via the Cyclization of Aromatic Cycloimmoniumylides with Electrondeficient Alkenes, *Curr. Org. Synth.*, 2018, **15**(3), 388–395.
5. C. Sandeep, B. Padmashali, R. S. Kulkarni, S. M. Mallikarjuna, M. B. Siddesh, H. K. Nagesh and K. S. Thriveni, Synthesis of Substituted 5-acetyl-3-benzoylindolizine-1-carboxylate from substituted 2-acetyl pyridinium bromides, *Heterocycl. Lett.*, 2014, **4**(3), 371–376.
6. C. Sandeep, P. Basavaraj and S. K. Kulkarni, Synthesis of Isomeric Subtituted 6-acetyl-3-benzoylindolizine-1-carboxylate and 8-acetyl-3-benzoylindolizine-1-carboxylate from subtituteded 3-acetyl pyridinium bromides and their antimicrobial activity, *J. Appl. Chem.*, 2013, **2**(5), 1049.
7. G. Sumanth, K. Lakshmikanth, S. M. Saini, P. Mundhe, K. Shivaprasad and S. Chandrashekharappa, Phenyl pyrrolo [1,2-a] quinolines- finding of a key by-product during quinolinium salt preparation, *J. Mol. Struct.*, 2023, **1273**, 134350.
8. G. Sumanth, S. M. Saini, K. Lakshmikanth, G. S. Mangubhai, K. Shivaprasad and S. Chandrashekharappa, Microwave-Assisted Improved Regioselective Synthesis of 3-Benzoyl Indolizine Derivatives, *J. Mol. Struct.*, 2023, **1286**, 135561.
9. A. Al-Mulla, A review: biological importance of heterocyclic compounds, *Der Pharma Chem.*, 2017, **9**(13), 141–147.
10. E. Kabir and M. Uzzaman, A review on biological and medicinal impact of heterocyclic compounds, *Results Chem.*, 2022, 100606.
11. T. Qadir, A. Amin, P. K. Sharma, I. Jeelani and H. Abe, A review on medicinally important heterocyclic compounds, *Open Med. Chem. J.*, 2022, **16**(1), 1–34.
12. P. Mundhe, S. Kidwai, S. M. Saini, H. R. Singh, R. Singh and S. Chandrashekharappa, Design, synthesis, characterization, and anti-tubercular

activity of novel ethyl-3-benzoyl-6, 8-difluoroindolizine-1-carboxylate analogues: Molecular target identification and molecular docking studies, *J. Mol. Struct.*, 2023, **1284**, 135359.
13. K. Shivaprasad, S. Kidwai, S. Gopavaram, S. M. Saini, K. Reddy and S. Chugh, *et al.*, Design, synthesis and in vitro antitubercular evaluation of novel 7-methoxy pyrrolo[1,2-a]quinoline analogues as CYP 121 inhibitors, *J. Mol. Struct.*, 2023, **1284**, 135439.
14. K. N. Venugopala, O. H. A. Al-Attraqchi, C. Tratrat, S. K. Nayak, M. A. Morsy and B. E. Aldhubiab, *et al.*, Novel Series of Methyl 3-(Substituted Benzoyl)-7-Substituted-2-Phenylindolizine-1-Carboxylates as Promising Anti-Inflammatory Agents: Molecular Modeling Studies, *Biomolecules*, 2019, **9**(11), 1–14.
15. S. Rashmi, T. S. Kumara, C. Sandeep, G. Nagendrappa, H. Sowmya and P. S. Ganapathy, Synthesis, antimicrobial, antioxidant and docking study of (3-alkoxy-5-nitrobenzofuran-2-yl)(phenyl) methanone derivatives, *Int. J. Pharm. Pharm. Sci.*, 2015, **7**(2), 493–497.
16. R. Ningegowda, S. Chandrashekharappa, V. Singh, V. Mohanlall and K. N. Venugopala, Design, synthesis and characterization of novel 2-(2, 3-dichlorophenyl)-5-aryl-1, 3, 4-oxadiazole derivatives for their anti-tubercular activity against Mycobacterium tuberculosis, *Chem. Data Collect.*, 2020, **28**, 100431.
17. M. Kidwai, R. Venkataramanan, R. Mohan and P. Sapra, Cancer chemotherapy and heterocyclic compounds, *Curr. Med. Chem.*, 2002, **9**(12), 1209–1228.
18. M. Hossain and A. K. Nanda, A review on heterocyclic: synthesis and their application in medicinal chemistry of imidazole moiety, *Science*, 2018, **6**(5), 83–94.
19. S. Pathania, R. K. Narang and R. K. Rawal, Role of sulphur-heterocycles in medicinal chemistry: An update, *Eur. J. Med. Chem.*, 2019, **180**, 486–508.
20. A. P. Taylor, R. P. Robinson, Y. M. Fobian, D. C. Blakemore, L. H. Jones and O. Fadeyi, Modern advances in heterocyclic chemistry in drug discovery, *Org. Biomol. Chem.*, 2016, **14**(28), 6611–6637.
21. K. N. Venugopala, C. Tratrat, S. Chandrashekharappa, M. Attimarad, N. Sreeharsha and A. B. Nair, *et al.*, Anti-tubercular Potency and Computationally assessed Drug-likeness and Toxicology of Diversely Substituted Indolizines, *Indian J. Pharm. Educ. Res.*, 2019, **53**(2), 545–552.
22. C. Sandeep, K. N. Venugopala, M. A. Khedr, B. Padmashali, R. S. Kulkarni and R. Venugopala, *et al.*, Design and Synthesis of Novel Indolizine Analogues as COX-2 Inhibitors: Computational Perspective and in vitro Screening, *Indian J. Pharm. Educ. Res.*, 2017, **51**(3), 452–460.
23. K. Lakshmikanth, S. M. Saini, S. T. Dorai and S. Chandrashekharappa, Tandem-Michael-cyclization cascade to make pyridines: Use of electron-deficient acetylenes for the synthesis of indolizines in aqueous media, *Tetrahedron*, 2023, **142**, 133516.

24. S. Chandrashekharappa, K. N. Venugopala, R. Venugopala and B. Padmashali, Qualitative anti-tubercular activity of synthetic ethyl 7-acetyl2-substituted-3-(4-substituted benzoyl) indolizine-1-carboxylate analogues, *J. App. Pharm. Sci.*, 2019, **9**(2), 124–128.
25. V. Uppar, S. Chandrashekharappa, K. N. Venugopala, P. K. Deb, S. Kar and O. I. Alwassil, *et al.*, Synthesis and characterization of pyrrolo[1,2-a]quinoline derivatives for their larvicidal activity against Anopheles arabiensis, *Struct. Chem.*, 2020, **31**(4), 1533–1543.
26. V. Uppar, S. Chandrashekharappa, C. Shivamallu, P. Sushma, S. P. Kollur and J. Ortega-Castro, *et al.*, Investigation of Antifungal Properties of Synthetic Dimethyl-4-Bromo-1-(Substituted Benzoyl) Pyrrolo[1,2-a] Quinoline-2,3-Dicarboxylates Analogues: Molecular Docking Studies and Conceptual DFT-Based Chemical Reactivity Descriptors and Pharmacokinetics Evaluation, *Molecules*, 2021, **26**(9), 2722.
27. Y. Sanemitsu and S. Kawamura, Studies on the synthetic development for the discovery of novel heterocyclic agrochemicals, *J. Pestic. Sci.*, 2008, **33**(2), 175–177.
28. T. Torroba, Poly-sulfur-nitrogen heterocycles via sulfur chlorides and nitrogen reagents, *Journal für praktische Chemie.*, 1999, **341**(2), 99–113.
29. H. Mutlu, E. B. Ceper, X. Li, J. Yang, W. Dong and M. M. Ozmen, *et al.*, Sulfur chemistry in polymer and materials science, *Macromol. Rapid Commun.*, 2019, **40**(1), e1800650.
30. L. S. Konstantinova, O. I. Bol'shakov, N. V. Obruchnikova, H. Laborie, A. Tanga and V. Sopéna, *et al.*, One-pot synthesis of 5-phenylimino, 5-thieno or 5-oxo-1, 2, 3-dithiazoles and evaluation of their antimicrobial and antitumor activity, *Bioorg. Med. Chem. Lett.*, 2009, **19**(1), 136–141.
31. I. C. Christoforou, P. A. Koutentis and S. S. Michaelidou, 1,2,3-Dithiazole chemistry in heterocyclic synthesis, *Arkivoc.*, 2006, **7**, 207–223.
32. L. S. Konstantinova, I. V. Baranovsky, E. A. Pritchina, M. S. Mikhailov, I. Y. Bagryanskaya and N. A. Semenov, *et al.*, Fused 1, 2, 3 - Thiaselenazoles Synthesized from 1, 2, 3 - Dithiazoles through Selective Chalcogen Exchange, *Chem. Eur. J.*, 2017, **23**(67), 17037–17047.
33. U. Wannagat and G. Schindler, Reaktionen des Schwefeldichlorids mit Pyridin und verwandten Verbindungen, *Angew. Chem.*, 1957, **69**(24), 784.
34. J. E. Moore, Certain 4-Halo-5-aryl-1, 2, 3-dithiazole Compounds and their Preparation, *Google Pat.*, 1977.
35. R. Appel, H. Janssen, M. Siray and F. Knoch, Synthese und Reaktionen des 4, 5-Dichlor-1, 2, 3-dithiazolium-chlorids, *Chem. Ber.*, 1985, **118**(4), 1632–1643.
36. P. G. Baraldi, M. G. Pavani, M. del Carmen Nunez, P. Brigidi, B. Vitali and R. Gambari, *et al.*, Antimicrobial and antitumor activity of N-heteroimmine-1, 2, 3-dithiazoles and their transformation in triazolo-, imidazo-, and pyrazolopirimidines, *Bioorg. Med. Chem.*, 2002, **10**(2), 449–456.

37. C. W. Rees, Polysulfur-nitrogen heterocyclic chemistry, *J. Heterocycl. Chem.*, 1992, **29**(3), 639–651.
38. A. G. Neo, R. M. Carrillo and C. F. Marcos, A straightforward synthesis of 2-aminobenzothiazoles from Herz compounds, *Org. Biomol. Chem.*, 2011, **9**(13), 4850–4855.
39. C. R. Asquith, T. Meili, T. Laitinen, I. V. Baranovsky, L. S. Konstantinova and A. Poso, *et al.*, Synthesis and comparison of substituted 1, 2, 3-dithiazole and 1, 2, 3-thiaselenazole as inhibitors of the feline immunodeficiency virus (FIV) nucleocapsid protein as a model for HIV infection, *Bioorg. Med. Chem. Lett.*, 2019, **29**(14), 1765–1768.
40. G. Cottenceau, T. Besson, V. Gautier, C. W. Rees and A.-M. Pons, Antibacterial evaluation of novel N-arylimino-1, 2, 3-dithiazoles and N-arylcyanothioformamides, *Bioorg. Med. Chem. Lett.*, 1996, **6**(5), 529–532.
41. V. Thiéry, C. W. Rees, T. Besson, G. Cottenceau and A.-M. Pons, Antimicrobial activity of novel N-quinolinyl and N-naphthylimino-1,2,3-dithiazoles, *Eur. J. Med. Chem.*, 1998, **33**(2), 149–153.
42. S.-L. Yan, M.-Y. Yang, Z.-H. Sun, L.-J. Min, C.-X. Tan and J.-Q. Weng, *et al.*, Synthesis and antifungal activity of 1, 2, 3-thiadiazole derivatives containing 1, 3, 4-thiadiazole moiety, *Lett. Drug Des. Discovery*, 2014, **11**(7), 940–943.
43. J. E. Moore, Method for control of fungi using 4-halo-5-aryl-1, 2, 3,-dithiazoles, *Google Pat.*, 1978.
44. C. A. Thomson, H. M. Atkinson and V. S. Ananthanarayanan, Identification of small molecule chemical inhibitors of the collagen-specific chaperone Hsp47, *J. Med. Chem.*, 2005, **48**(5), 1680–1684.
45. A. S. Kalogirou, S. S. Michaelidou, A. J. White and P. A. Koutentis, Transformation of 2-(4-chloro-5H-1, 2, 3-dithiazol-5-ylideneamino)-6-ethoxy-4-phenylpyridine-3, 5-dicarbonitrile into 4-aminopyrido [2, 3-d] pyrimidines and 2-(pyrid-2-yl) guanidines, *Tetrahedron*, 2015, **71**(12), 1799–1807.
46. T. Besson and C. W. Rees, Some chemistry of 4, 5-dichloro-1, 2, 3-dithiazolium chloride and its derivatives, *J. Chem. Soc., Perkin Trans. 1*, **1995**(13), 1659–1662.
47. K. A. Maffuid, M. Koyioni, C. D. Torrice, W. A. Murphy, H. K. Mewada and P. A. Koutentis, *et al.*, Design and evaluation of 1, 2, 3-dithiazoles and fused 1, 2, 4-dithiazines as anti-cancer agents, *Bioorg. Med. Chem. Lett.*, 2021, **43**, 128078.
48. L. S. Konstantinova and O. A. Rakitin, Synthesis and properties of 1, 2, 3-dithiazoles, *Russ. Chem. Rev.*, 2008, **77**(6), 521.
49. P. A. Koutentis, M. Koyioni and S. S. Michaelidou, Synthesis of [(4-chloro-5H-1, 2, 3-dithiazol-5-ylidene) amino] azines, *Molecules*, 2011, **16**(11), 8992–9002.

50. M. Koyioni, M. Manoli and P. A. Koutentis, The Reaction of DABCO with 4-Chloro-5H-1, 2, 3-dithiazoles: Synthesis and Chemistry of 4-[N-(2-Chloroethyl)piperazin-1-yl]-5H-1, 2, 3-dithiazoles, *J. Org. Chem.*, 2016, **81**(2), 615–631.
51. H. Lee, K. Kim, D. Whang and K. Kim, Novel Synthesis of 5-(Arylimino)-4-(dialkylamino)-5H-1, 2, 3-dithiazoles and the Mechanism of Their Formation, *J. Org. Chem.*, 1994, **59**(21), 6179–6183.
52. T. Besson, C. W. Rees, G. Cottenceau and A.-M. Pons, Antimicrobial evaluation of 3, 1-benzoxazin-4-ones, 3, 1-benzothiazin-4-ones, 4-alkoxyquinazolin-2-carbonitriles and N-arylimino-1, 2, 3-dithiazoles, *Bioorg. Med. Chem. Lett.*, 1996, **6**(19), 2343–2348.
53. H. C. Van der Plas, The SN (ANRORC) mechanism: a new mechanism for nucleophilic substitution, *Acc. Chem. Res.*, 1978, **11**(12), 462–468.
54. A. J. Nicholls and I. R. Baxendale, Benzo [1, 2, 3] dithiazole Compounds: A History of Synthesis and Their Renewed Applicability in Materials and Synthetic Chemistry, Originating from the Herz Reaction, *Reactions*, 2021, **2**(3), 175–208.
55. R. T. Oakley, R. W. Reed, C. M. Robertson and J. F. Richardson, Naphthalene-1, 2, 3-dithiazolyl and its selenium-containing variants, *Inorg. Chem.*, 2005, **444**(6), 1837–1845.
56. S. Macho, D. Miguel, T. Gómez, T. Rodríguez and T. Torroba, From cyclopentanone oximes to bis [1, 2, 3] dithiazolo-s-indacenes, cyclopenta [c][1, 2] thiazine, pentathiepino-, tetrathiino-, and thienocyclopenta [1, 2, 3] dithiazoles as a rich source of new materials, *J. Org. Chem.*, 2005, **70**(23), 9314–9325.
57. L. S. Konstantinova, I. V. Baranovsky, I. G. Irtegova, I. Y. Bagryanskaya, L. A. Shundrin and A. V. Zibarev, *et al.*, Fused 1, 2, 3-dithiazoles: Convenient synthesis, structural characterization, and electrochemical properties, *Molecules*, 2016, **21**(5), 596.
58. C. Polo, V. Ramos, T. Torroba, O. A. Rakitin and C. W. Rees, One pot synthesis of 1, 2, 3-benzodithiazol-6-ones, *Tetrahedron*, 1998, **54**(1–2), 223–232.
59. A. S. Kalogirou and P. A. Koutentis, The degradation of 4, 5-dichloro-1, 2, 3-dithiazolium chloride in wet solvents, *Tetrahedron*, 2009, **65**(34), 6859–6862.
60. K. A. H. M. El-Bayouki, W. M. Basyouni, Y. A. F. Mohamed, M. M. Aly and S. Y. Abbas, Novel 4 (3H)-quinazolinones containing biologically active thiazole, pyridinone and chromene of expected antitumor and antifungal activities, *Eur. J. Chem.*, 2011, **2**(4), 455–462.
61. A. S. Kalogirou, H. J. Oh and C. R. Asquith, The Synthesis and Biological Applications of the 1, 2, 3-Dithiazole Scaffold, *Molecules*, 2023, **28**(7), 3193.
62. C. R. Asquith, L. S. Konstantinova, T. Laitinen, M. L. Meli, A. Poso and O. A. Rakitin, *et al.*, Evaluation of Substituted 1, 2, 3-Dithiazoles as Inhibitors of the Feline Immunodeficiency Virus (FIV) Nucleocapsid

Protein via a Proposed Zinc Ejection Mechanism, *ChemMedChem*, 2016, **11**(19), 2119–2126.
63. K. M. Cook, S. T. Hilton, J. Mecinović, W. B. Motherwell, W. D. Figg and C. J. Schofield, Epidithiodiketopiperazines block the interaction between hypoxia-inducible factor-1α (HIF-1α) and p300 by a zinc ejection mechanism, *J. Biol. Chem.*, 2009, **284**(39), 26831–26838.
64. F. Oppedisano, M. Catto, P. A. Koutentis, O. Nicolotti, L. Pochini and M. Koyioni, *et al.*, Inactivation of the glutamine/amino acid transporter ASCT2 by 1, 2, 3-dithiazoles: proteoliposomes as a tool to gain insights in the molecular mechanism of action and of antitumor activity, *Toxicol. Appl. Pharmacol.*, 2012, **265**(1), 93–102.
65. M. Scalise, L. Pochini, L. Console, M. A. Losso and C. Indiveri, The human SLC1A5 (ASCT2) amino acid transporter: from function to structure and role in cell biology, *Front. Cell Dev. Biol.*, 2018, **6**, 96.
66. S. Bröer, Amino acid transporters as disease modifiers and drug targets, *SLAS DISCOVERY: Advancing Life Sciences R&D*, 2018, **23**(4), 303–320.
67. L. Napolitano, M. Scalise, M. Koyioni, P. Koutentis, M. Catto and I. Eberini, *et al.*, Potent inhibitors of human LAT1 (SLC7A5) transporter based on dithiazole and dithiazine compounds for development of anticancer drugs, *Biochem. Pharmacol.*, 2017, **143**, 39–52.
68. H. Masuda, M. Fukumoto, K. Hirayoshi and K. Nagata, Coexpression of the collagen-binding stress protein HSP47 gene and the alpha 1 (I) and alpha 1 (III) collagen genes in carbon tetrachloride-induced rat liver fibrosis, *J. Clin. Invest.*, 1994, **94**(6), 2481–2488.
69. M. S. Razzaque and T. Taguchi, Collagen-binding heat shock protein (HSP) 47 expression in anti-thymocyte serum (ATS)-induced glomerulonephritis, *J. Pathol.*, 1997, **183**(1), 24–29.
70. M. S. Razzaque, A. Nazneen and T. Taguchi, Immunolocalization of collagen and collagen-binding heat shock protein 47 in fibrotic lung diseases, *Mod. Pathol.*, 1998, **11**, 1183–1188.
71. M. S. Razzaque and A. Ahmed, Collagens, collagen-binding heat shock protein 47 and transforming growth factor-β1 are induced in cicatricial pemphigoid: possible role(s) in dermal fibrosis, *Cytokine*, 2002, **17**(6), 311–316.

CHAPTER 3

Isothiazoles: Synthetic Strategies and Pharmacological Applications

NITIN A. SASANE[a], HEMANTKUMAR N. AKOLKAR[b], NIRMALA R. DAREKAR[c], VIJAY M. KHEDKAR[d], ANAND S. BURANGE[e] AND KULBHUSHAN A. SASANE*[b]

[a]Department of Chemistry, Mahatma Phule Arts, Science and Commerce College, Panvel, Navi Mumbai, 410206, India; [b]Department of Chemistry, Abasaheb Marathe Arts & New Commerce, Science College, Rajapur, Ratnagiri, Maharashtra, 416702, India; [c]P.G. and Research, Department of Chemistry, Radhabai Kale Mahila Mahavidyalaya, Ahmednagar, Maharashtra, 414001, India; [d]School of Pharmacy, Vishwakarma University, Pune, Maharashtra, 411048, India; [e]Department of Chemistry, Wilson College (Autonomous), Chowpatty Seaface Road, India
*E-mail: kulsasane007@gmail.com

3.1 Introduction

In the contemporary organic synthesis, aromatic heterocycles have been considered invaluable compounds for the production of drugs, pharmaceutical intermediates, organic compounds, and agricultural chemicals in both academic and commercial sectors.[1,2] Apart from heterocyclic compounds, isothiazoles and their derivatives are notable for being five-membered aromatic heterocycles containing nitrogen and sulfur atoms, which

S-Heterocycles: Retrospect, Prospects, and Biological Applications
Edited by Shrikaant Kulkarni, Hemantkumar Akolkar, Vijay M. Khedkar and A. K. Haghi

Published by the Royal Society of Chemistry, www.rsc.org

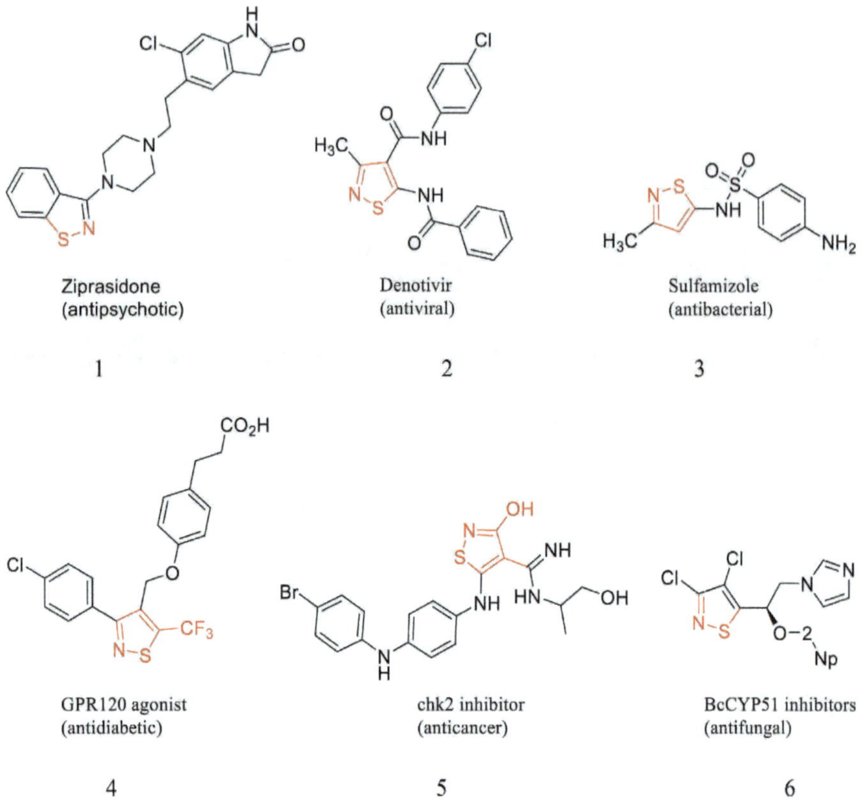

Figure 3.1 Examples of isothiazole derivatives that exhibit pharmacological activities.

are widely used in organic synthesis and medicinal chemistry because of their unique characteristics.[3] A few selected roles of isothiazoles are as antifungals, antipsychotic drugs (ziprasidone **1**, denotivir **2**, and sulfamizole **3**), antiviral agents, antidiabetic agents, and numerous types of anticancer, antibacterial, and neuroprotective agents, wherein the presence of lipophilic sulfur and its arrangement in a five-membered ring structure are responsible for their high activity[4] (see Figure 3.1).

Furthermore, in the realm of medication discovery, isothiazole compounds are widely recognized[5] due to their substantial biological activities, including analgesic, antipyretic, fungicidal, and herbicidal activities. To be more precise, the fundamental structures of pharmaceutically relevant drugs, such as G-protein-coupled receptor (GPR) agonists,[6] mGluR1 antagonists,[7] and the strong inhibitors of aurora kinase,[8] are composed of the skeleton of 3,5-disubstituted isothiazoles. Preclinical research indicates that mGluR1 antagonists are important for modifying nociception, while GPR agonists elevate the levels of triglycerides in the cell membranes and regulate insulin (see Figure 3.2).

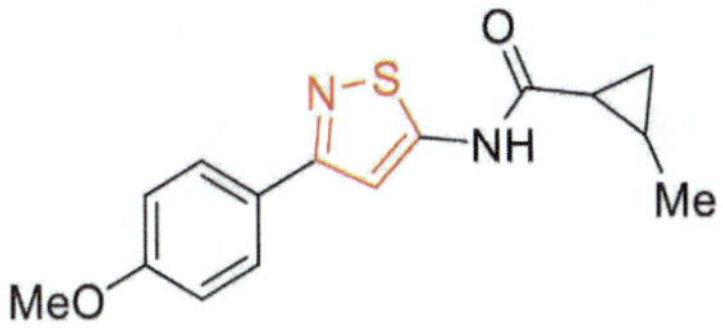

mGluR1 antagonists

7

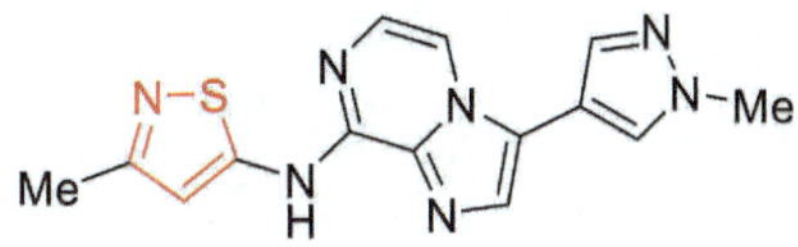

Aurora kinase inhibitor

8

GPR-120 agonist

9

Figure 3.2 An assortment of pharmacologically active isothiazole derivatives.

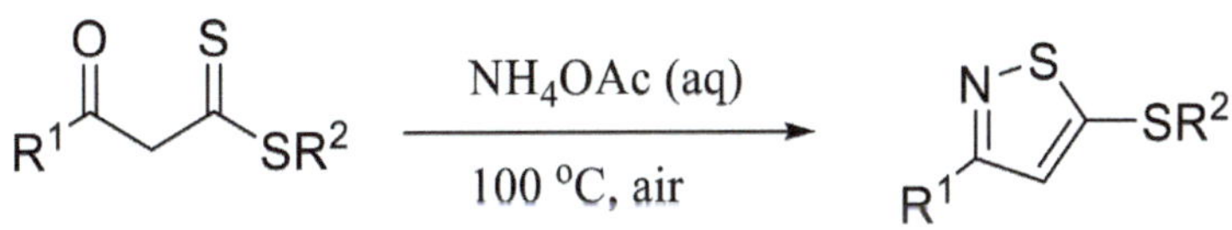

10

11

R^1= Ar, heteroaryl

R^2= alkyl

Scheme 3.1 Ammonium acetate is used in the synthesis of 3,5-disubstituted isothiazoles.

3.2 Recent Advances in the Synthesis of Isothiazoles

The synthesis of isothiazole derivative *via* a one-pot, transition metal-free reaction in the presence of ammonium acetate and air under aqueous conditions was reported by Singh *et al.* β-Ketodithioesters are attacked intramolecularly by sulfur, which is then further aromatized into isothiazole **11** in the presence of air oxygen (see Scheme 3.1).[9]

Yadav *et al.* have reported that a hypervalent iodine reagent, 2-amino-*N*′-arylbenzohydrazides **12**, can be converted into benzoisothiazoles **13** in a single-pot reaction. This tandem approach is used for the synthesis of **13** from simple 2-amino-*N*′-arylbenzohydrazides (see **Scheme 3.2**).[10]

Direct condensation for the synthesis of the desired isothiazoles in their total synthesis of pronqodine A and aulosirazole has been reported by Moody *et al.* Quinones that occur naturally and contain isothiazoles have anticancer properties (see Scheme 3.3).[11]

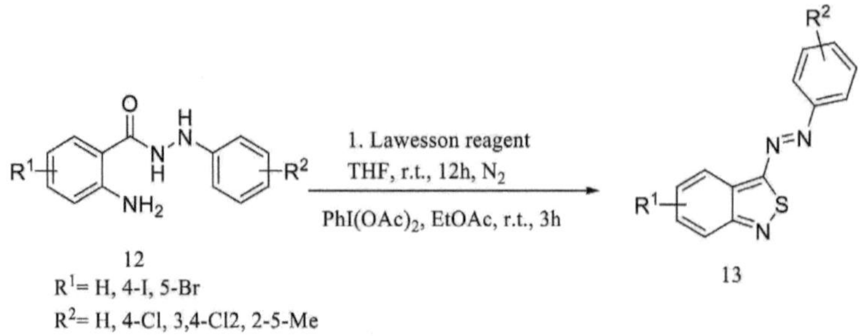

Scheme 3.2 Conversion of 2-amino-*N'*-arylbenzohydrazides **12** into benzoisothiazoles **13**.

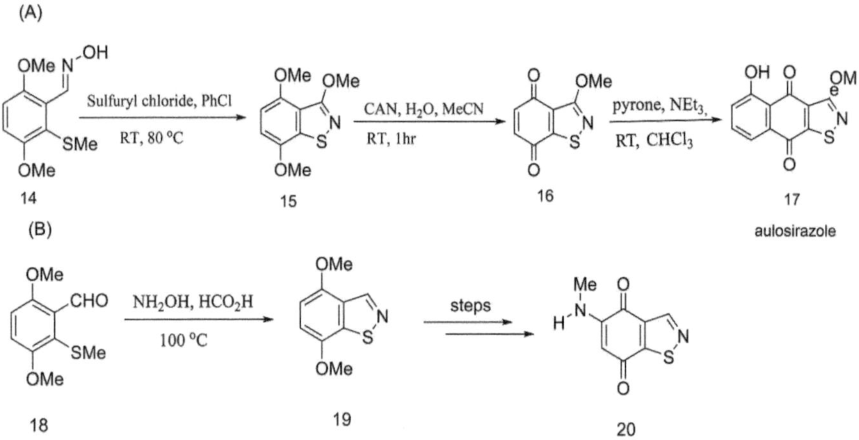

Scheme 3.3 (A) Aulosirazole and its analogues and (B) pronqodine A and its analogues are produced during the synthesis of isothiazoles.

Deng *et al.* proposed the conversion of elemental sulfur and *o*-halo-substituted aryl amidines **21** into 3-aryl-1,2-benzoisothiazoles **22** under aerobic circumstances using K_3PO_4 (see Scheme 3.4).[12]

Condensation reactions that are initiated by conjugate additions have drawn attention to certain current processes in the synthesis of isothiazoles. These are as follows:

Barton has disclosed the synthesis of novel derivatives of amino arylisothiazoles (see **Scheme 3.5**).[13]

The reaction shown in Scheme 3.5 is operationally simple and requires optimized reaction conditions to give **24** and is also applicable with 3-chloropropenenitriles **25** in place of propynenitriles for the synthesis of **26** (see Scheme 3.6).

Lee *et al.* have reported the synthesis of isothiazole by the rhodium-catalyzed transannulation process over a Rh based catalyst. A range of

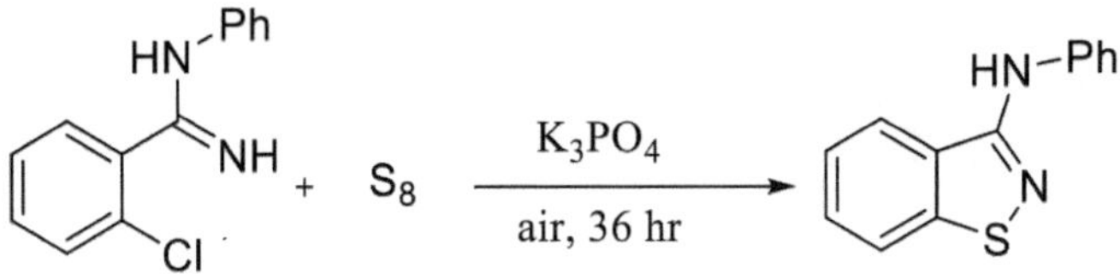

21 22

Scheme 3.4 Using sulfur and NS/CS bond formation, benzoisothiazoles **22** can be synthesized efficiently from *o*-halo-substituted aryl amidines **21** without the need for transition metals.

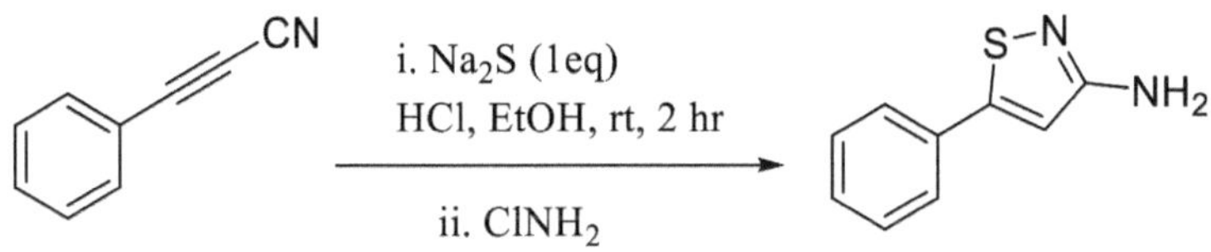

23 24

Scheme 3.5 Synthesis of arylisothiazole derivatives **24** from propynenitriles **23**.

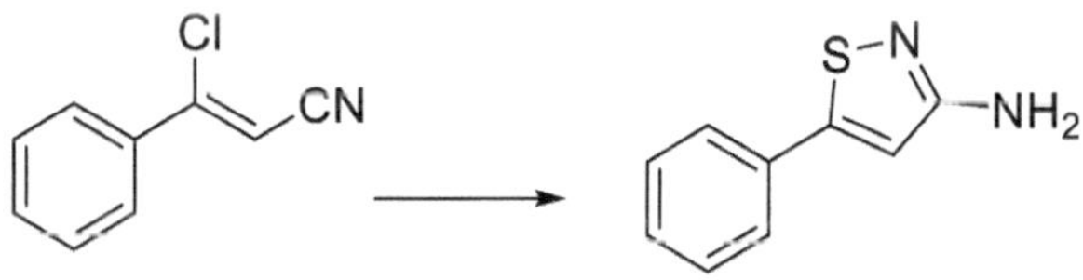

25 26

Scheme 3.6 3-Amino-5-arylisothiazole **26** synthesis using sodium sulfide and chloramine in 3-chloropropenenitriles **25**.

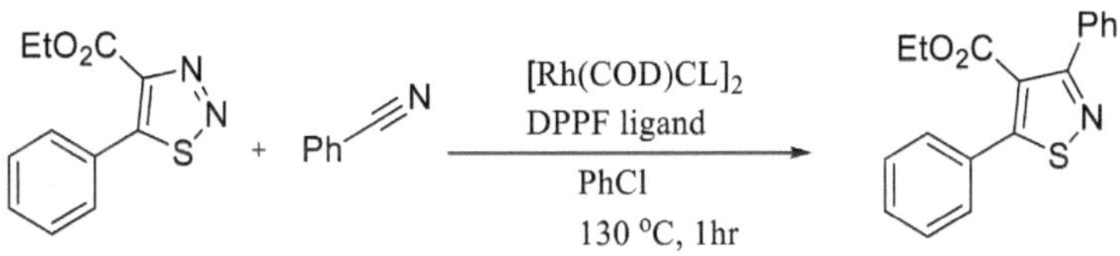

27 28

Scheme 3.7 Conversion of 1,2,3-thiadiazoles **27** with nitriles into isothiazoles **28** through a transannulation reaction catalyzed by rhodium.

isothiazoles **28** can be produced synthetically *via* 1,2,3-thiadiazole transannulation with aryl, heteroaryl, and alkyl nitriles (see Scheme 3.7).[14]

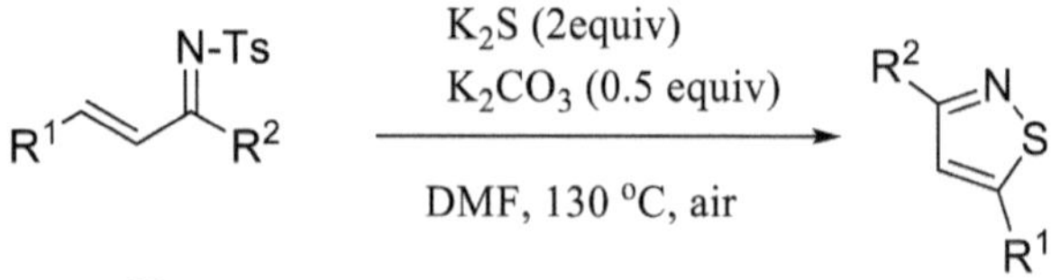

29 30

Scheme 3.8 Trisulfur radical anion-facilitated construction of isothiazoles **30** has been developed.

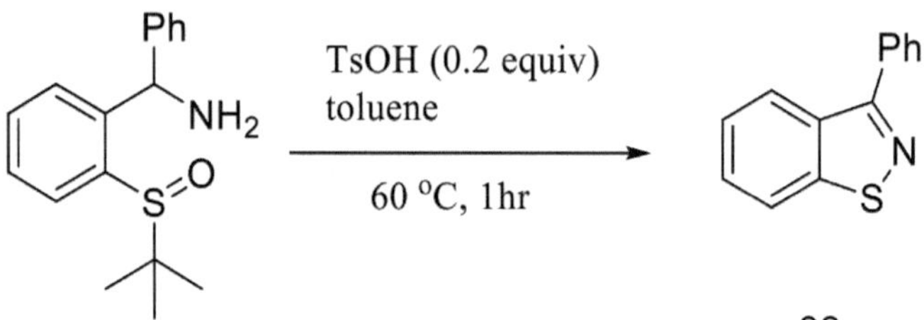

31 32

Scheme 3.9 Synthesis of aryl isothiazole **32** starting from **31**.

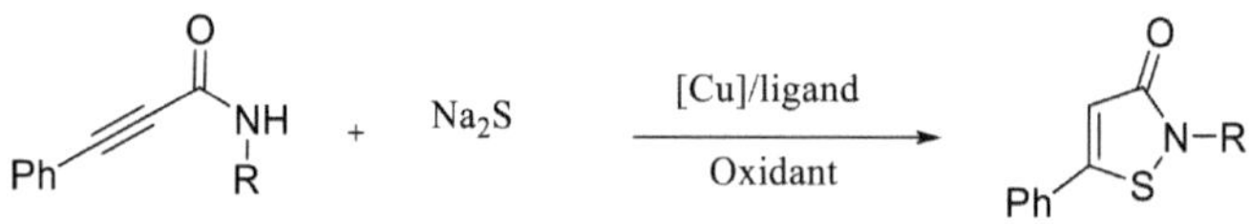

33 34

Scheme 3.10 Thiocyclization of propionamide **33** to obtain isothiazol-3-ones **34**.

According to Ji *et al.*, isothiazole derivatives can be synthesized by detosylation of α,β-unsaturated N-Ts imines and the addition of a cascade of trisulfur radical anion (see Scheme 3.8).[15]

Isothiazole synthesis *via* an acid-catalyzed sulfenic acid technique has been reported by Sun and Yuan (see Scheme 3.9).[16]

The synthesis of isothiazol-3-ones using a copper catalyst and thioannulation of propynamides with sodium sulfide was reported by Zhang *et al.* (see Scheme 3.10).[17]

The synthesis of *N*-iminoisothiazolium ylides using [4+1] cycloaddition of KSCN with α,β-alkynic hydrazones has been reported by Ji *et al.* (see Scheme 3.11).[18]

Huang *et al.* have documented the synthesis of isothiazole derivatives **39** by a three-component annulation of alkynones, xanthate, and NH_4I

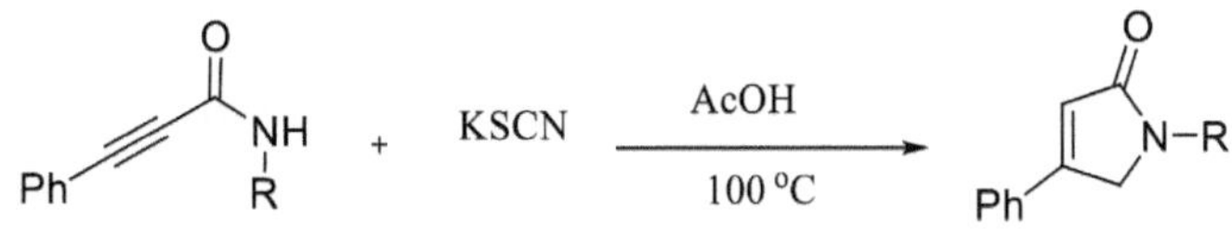

R= Ar, Bn

35 36

Scheme 3.11 Synthesis of isothiazol-3-ones **36** starting from α,β-alkynic hydrazones **35**.

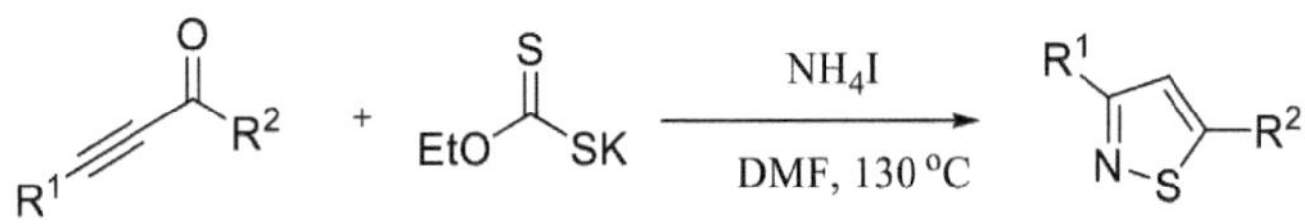

37 38 39

Scheme 3.12 Transition metal-free synthesis of isothiazole derivatives **3**.

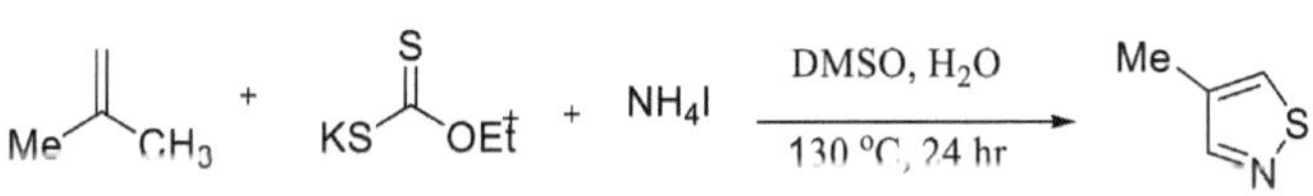

40 41 42

Scheme 3.13 Synthesis of isothiazole derivatives **42** through a cascade annulation reaction.

reagent, which eliminates the need for transition metals (see Scheme 3.12).[19]

Li *et al.* used a three-component cascade annulation process to synthesize 4-substituted isothiazoles **42** (see Scheme 3.13).[20]

Reddy *et al.* synthesized isothiazoles from α,β-alkynylamides using thiocyanate, resulting in an intriguing outcome (see Scheme 3.14).[21]

Matović *et al.* synthesized isothiazole derivatives from **43** and **44** derivatives, which reacted further with bromine in chloroform to yield **46** (**see Scheme 3.14**).[22]

According to Mantelingu *et al.*, acetophenones **46** and dithioester **47** are used in a unique and simple metal-free synthesis of 3,5-disubstituted isothiazole **49**. Using the aforementioned technique, C–O and C–S are functionalized by successive imine production and intramolecular cyclization (see **Scheme 3.15**).[23]

Metal-free visible light-promoted synthesis of isothiazoles was reported by Alemán *et al.* (see Scheme 3.16).[24]

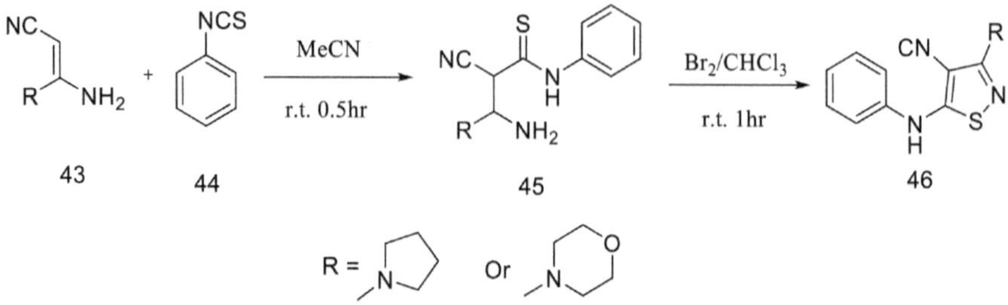

Scheme 3.14 Synthesis of isothiazole derivatives using bromine in the presence of a chloroform reaction system.

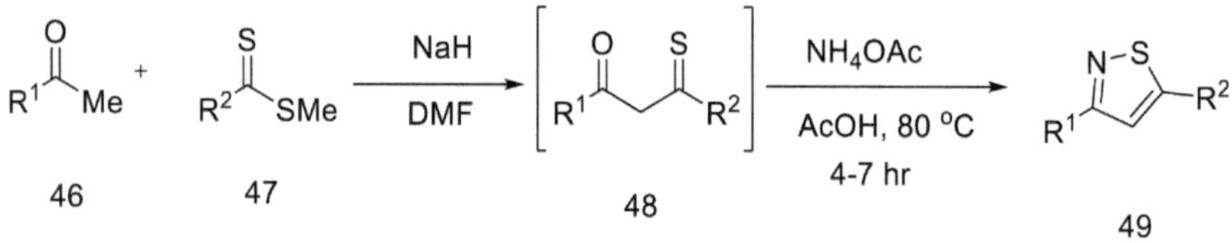

Scheme 3.15 3,5-Disubstituted isothiazole derivative synthesis under metal-free conditions.

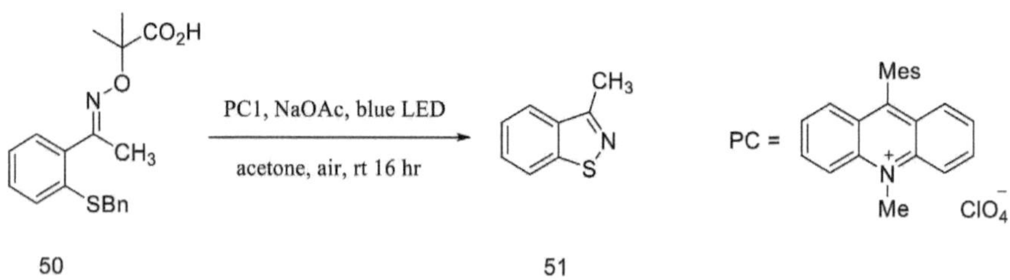

Scheme 3.16 Synthesis of isothiazoles **51**, which is scalable and effective and just requires visible light and easily accessible oxime derivatives **50**.

Isothiazole derivatives can also be synthesized using metal catalysts and transition metal complex-based catalytic systems. For instance, Miura *et al.* synthesized isothiazoles **53** through rhodium-catalyzed oxidative annulation with elemental sulfur (see Scheme 3.17).[25]

Reddy *et al.* reported an interesting synthesis of isothiazoles **56** by a stereoselective conjugate addition of thiocyanate to α,β-alkynylamides **54**. In this study, the proposed mechanism involves 2 steps: (1) alkyne thiocyanation and (2) decyanative intramolecular cyclization induced by a nitrogen atom in the amide (see Scheme 3.18).[26]

The synthesis of 4-amino-isothiazoles **59** by a Thorpe–Ziegler-type cyclization was reported by Tor *et al.* (see Scheme 3.19).[27]

Isothiazole syntheses have been reported over metal and transition metal complexes (Schemes 3.7 and 3.17). On the other hand, side-chain-based

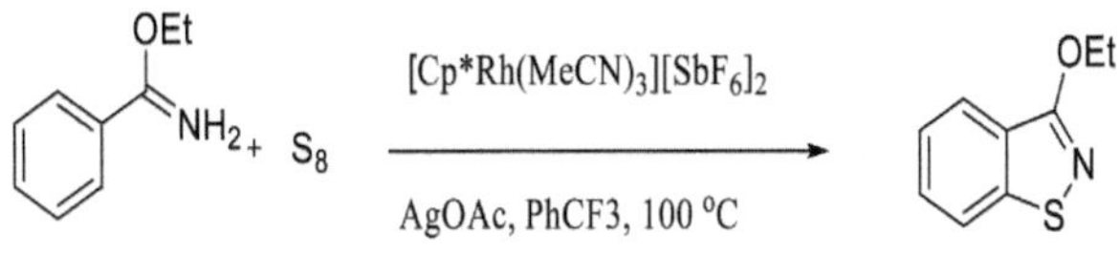

52 53

Scheme 3.17 The direct synthesis of isothiazole rings **53** is achieved using an oxidative annulation of benzimidates with elemental sulfur catalyzed by rhodium.

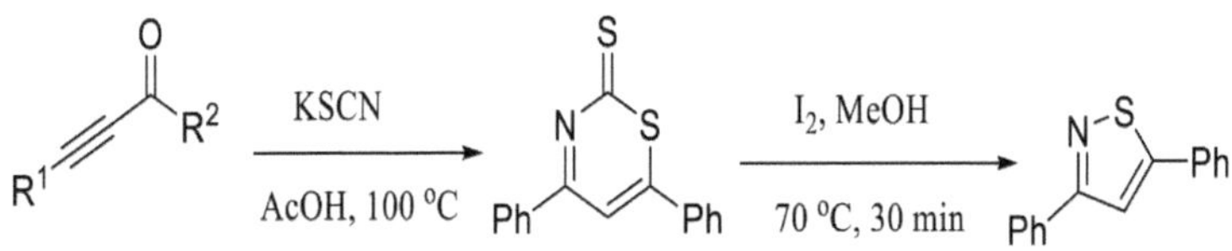

54 55 56

Scheme 3.18 Synthesis of isothiazoles **56**.

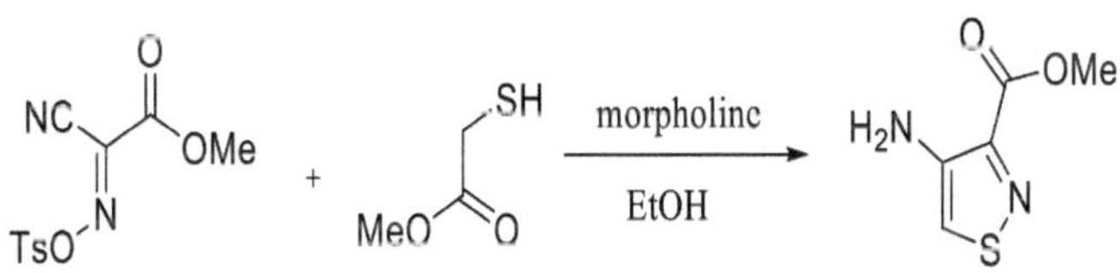

57 58 59

Scheme 3.19 Synthesis of 4-amino-isothiazoles **59** through the cyclization process.

ring rearrangements were seen in transition metal-free environments. Thus, the metal-free route represented a powerful tool in heterocyclic chemistry to obtain important heterocycles. For example, Koutentis *et al.* reported the synthesis of 3-halo-isothiazole-5-carbonitriles **60** by a rearrangement (see Scheme 3.20).[28]

The synthesis of 3-substituted benzoisothiazoles using a Wittig-type oxidative heteroatom approach was described by Sun *et al.* The initial material for the cyclization is obtained by *ortho*-metalation of *tert*-butyl aryl sulfoxides **62** and *tert*-butylsulfinyl imines. NBS and acetic acid resulted in a Wittig-type rearrangement, obtaining the benzoisothiazoles product **63.** (see Scheme 3.21).[29]

The synthesis of isothiazoles using a three-component strategy from enaminoesters, sulfur, and bromodifluoroacetamides/esters was reported

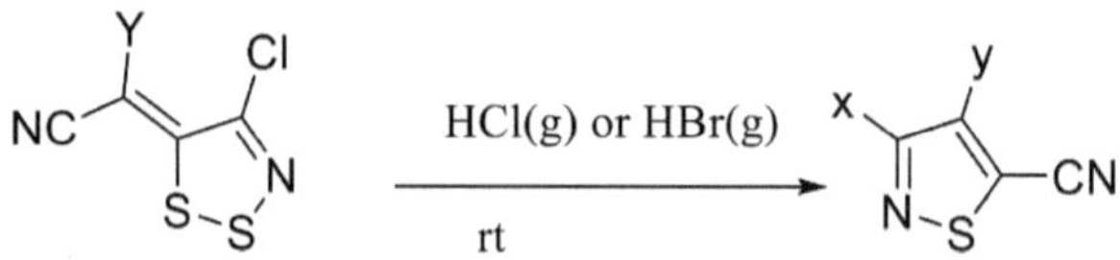

60 61

X = Br, Cl
Y = Br, Cl, CO_2Et, CN

Scheme 3.20 3-Halo-isothiazole-5-carbonitrile synthesis using a ring rearrangement process **60**.

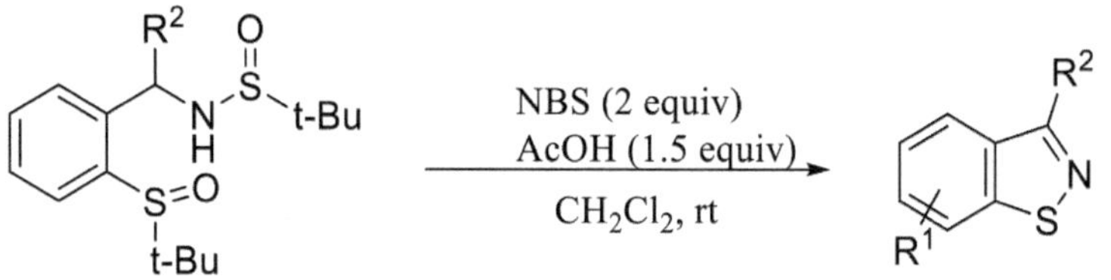

62 63

R^1 = H, F, OCH_3 , **R**^{2} = Ar, HetAr, Alkyl

Scheme 3.21 Synthesis of benzoisothiazoles **63** using an NBS and acetic acid reaction system results in a Wittig-type rearrangement, obtaining the desired product.

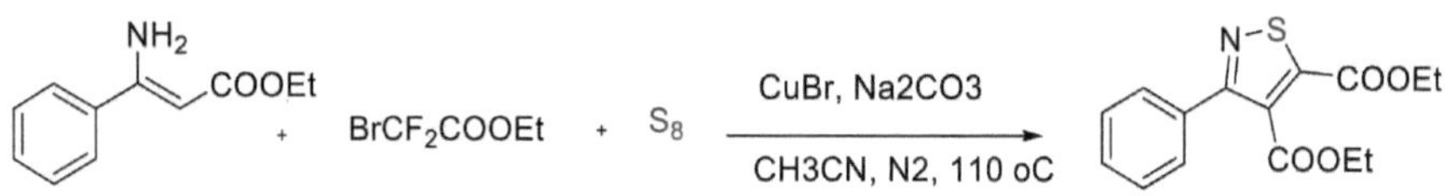

64 65 66

Scheme 3.22 A three-component strategy for the synthesis of isothiazoles **66** using enaminoesters, sulfur, and bromodifluoroacetamides/esters.

by Song *et al.* (see Scheme 3.22).[30] In this method, elemental sulfur was used as a source of sulfur. However, inert conditions were required to yield the corresponding products (**66**).

The synthesis of tricyclic isothiazole oxide derivatives **68** was described by Lamers and Bolm (see Scheme 3.23).[31] In a typical synthesis, **67** and 1 equivalent of *meta*-chloroperoxybenzoic acid (*m*-CPBA) and *N*-chlorosuccinimide react in dichloromethane (DCM) at room temperature in the presence of NaOH.

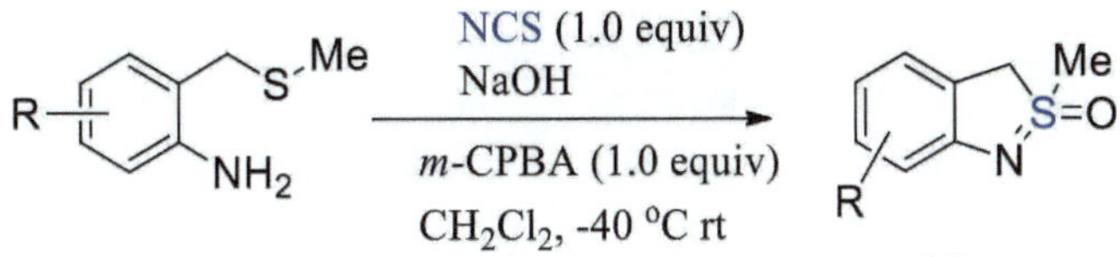

67 68

R = Br, Cl, F, CO_2R, NO_2

Scheme 3.23 Synthesis of isothiazoles **51**, which is scalable and effective and just requires visible light and easily accessible oxime derivatives **50**.

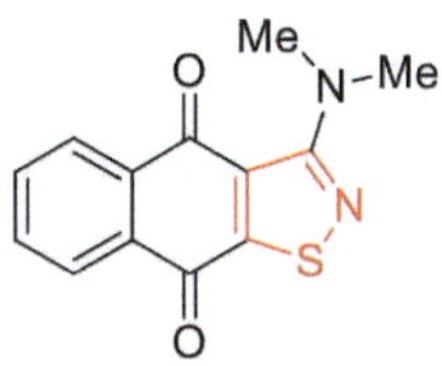

Figure 3.3 HDAC inhibitor (anticancer activity).

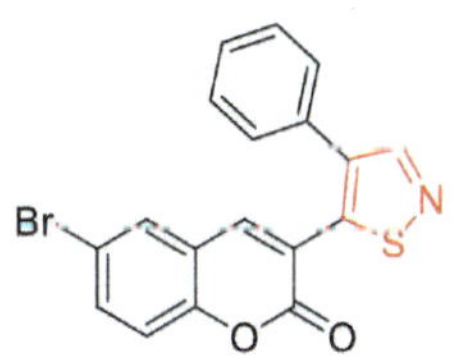

Figure 3.4 Antiproliferative agent.

3.3 Biological Applications

3.3.1 Isothiazole as the Anticancer Agent

Brunton has reported the biological importance of isothiazole-containing drugs in his book. HDAC inhibitors have been highlighted as a novel category of anti-cancer drugs.[32] HDAC inhibitor promotes tumor cell death that's why they used in cancer treatment. Isothiazole derivatives show anticancer activity. (see Figure 3.3).

A sequence of 3-(4-phenylisothiazol-5-yl)-2*H*-chromen-2-one has been discovered by Penta *et al.*, and they illustrated moderate antiproliferative activity of the developed compounds against malignant cell lines.[33] Using etoposide as the reference drug, the developed compounds were tested for *in vitro* cytotoxicity against A549, PC3, SKOV3, and B16F10 cancer cell lines. The IC_{50} values of the tested compounds have been determined and compared with the reference drug over a period of 24 h against B16F10, A549, PC3, and SKOV3 cell lines (see Figure 3.4).

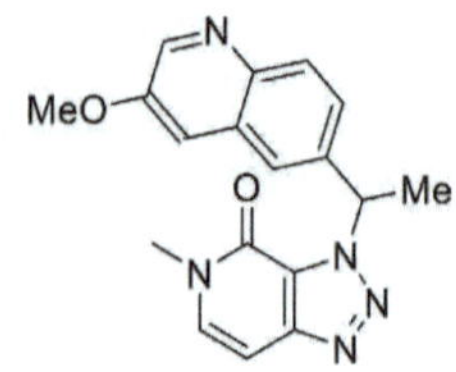

Figure 3.5 Tyrosine kinase c-Met inhibitor.

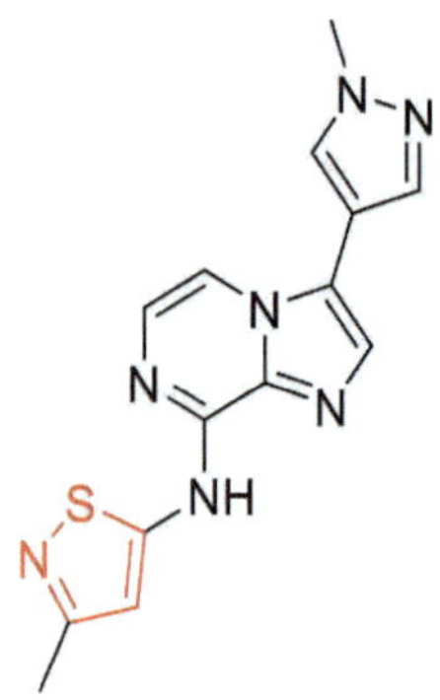

Figure 3.6 Aurora kinase inhibitors.

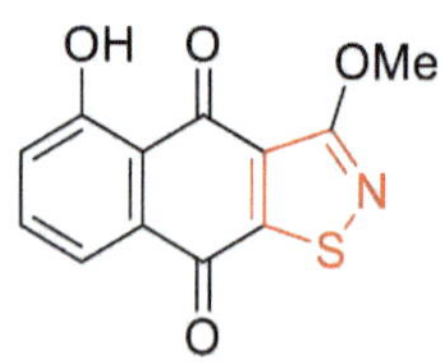

Figure 3.7 Aulosirazole.

The discovery of isothiazole derivatives, which show nanomolar suppression of c-Met kinase activity, has been reported by Bode *et al.*[34] (see Figure 3.5).

Belanger *et al.* demonstrated the synthesis of novel isothiazole derivatives and also a structure–activity relationship (SAR) study of novel derivatives such as Aurora kinase inhibitors[35] (see Figure 3.6).

More than 20 years ago, Moore *et al.* isolated aulosirazole from the blue-green alga *Aulosira fertilissima*, and X-ray crystallography and NMR spectroscopy were used to identify its structure.[36] Aulosirazole exhibits the ability to induce selective tumor cell cytotoxicity (see Figure 3.7).

Moody *et al.* reported the first synthesis of aulosirazole with anticancer activity.[37]

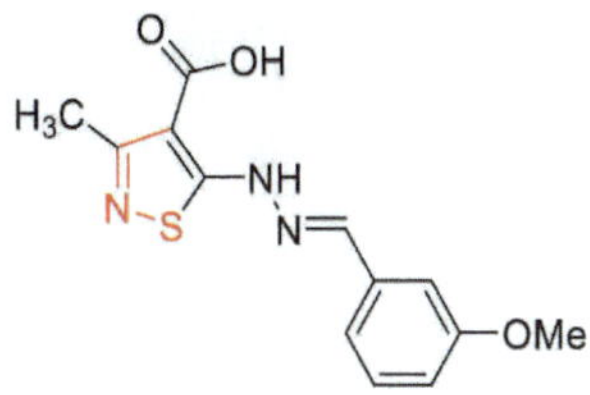

Figure 3.8 Novel anticancer agent.

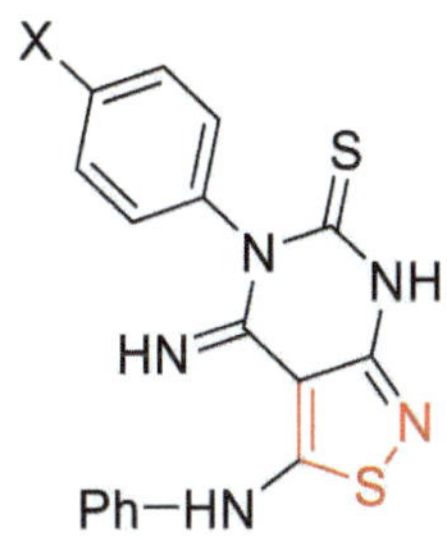

Figure 3.9 Anticancer agent.

Ryng *et al.* synthesized isothiazole derivatives with anticancer activity[38] (see Figure 3.8).

Shiri *et al.* designed and synthesized new derivatives of isothiazolo[3,4-*d*]pyrimidine and studied them against multiple types of cancer cells, including PC3, ACHN, HeLa, HL-60, and MCF7[39] (see Figure 3.9).

Kiselyov *et al.* synthesized a class of pyridin-5-(4*H*)-ones, 3,7-diaryl-6,7-dihydroisothiazolo[4,5-*b*], which exhibit antimitotic activities and significantly restrict the growth of multidrug-resistant (MDR) cell lines[40] (see Figure 3.10).

Nam *et al.* obtained a novel series of isothiazole dioxide derivatives, which exhibited potent T-type calcium channel inhibitor activity[41] (see Figure 3.11).

Kletskov synthesized isothiazole derivatives of comenic acid, which are used in brain tumor chemotherapy[42] (see Figure 3.12).

3.3.2 Isothiazole as the Antidiabetic Agent

Zhang *et al.* designed and produced phenylpropanoic acids based on isothiazoles as GPR120 agonists.[43] In C57BL/6DIO mice, the synthesized chemical compound reduces plasma glucose levels during an oral glucose

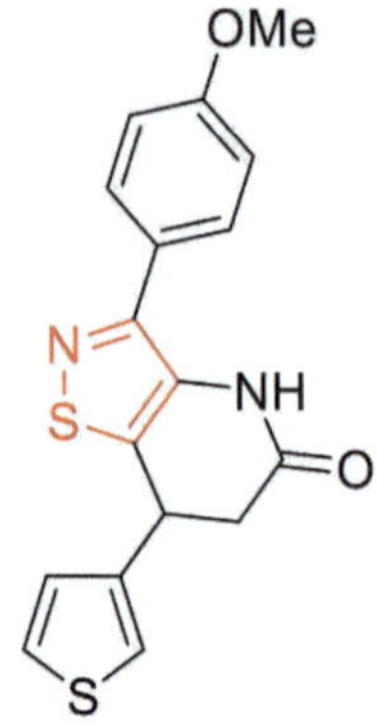

Figure 3.10 Antimitotic agent.

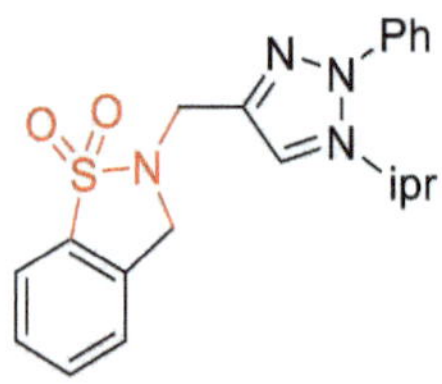

Figure 3.11 T-type calcium channel inhibitors.

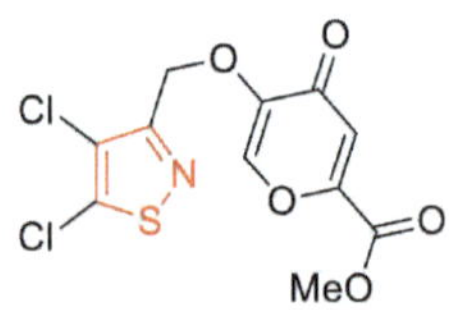

Figure 3.12 Isothiazole moieties.

tolerance test (OGTT). It also exhibits good pharmaceutical properties (see Figure 3.13).

Ulven *et al.* reported structure–activity relationship studies of a non-acidic sulfonamide FFA4 agonist[44] (see Figure 3.14).

3.3.3 Isothiazole as an Antiviral Agent

Isothiazolo[4,3-*b*]pyridines are well known as potent and selective cyclin G-associated kinase (GAK) inhibitors with promising antiviral activity, which were designed and synthesized by Jonghe *et al.*[45] (see Figure 3.15).

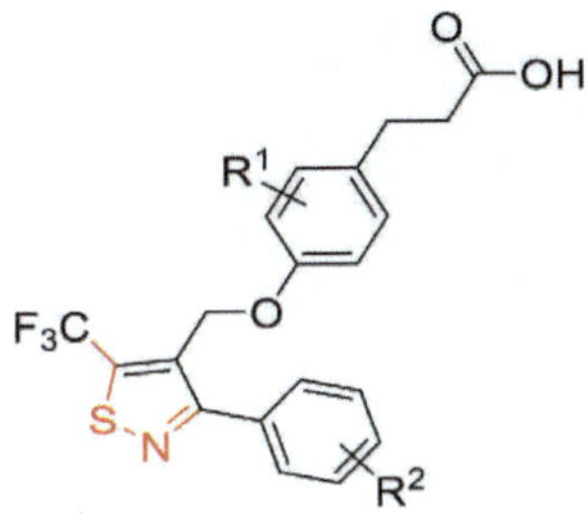

R^1 = 2-F, 4-Cl
R^2= 3,5-di-F

Figure 3.13 Isothiazole as the antidiabetic agent.

F

Figure 3.14 Isothiazole as the antidiabetic agent.

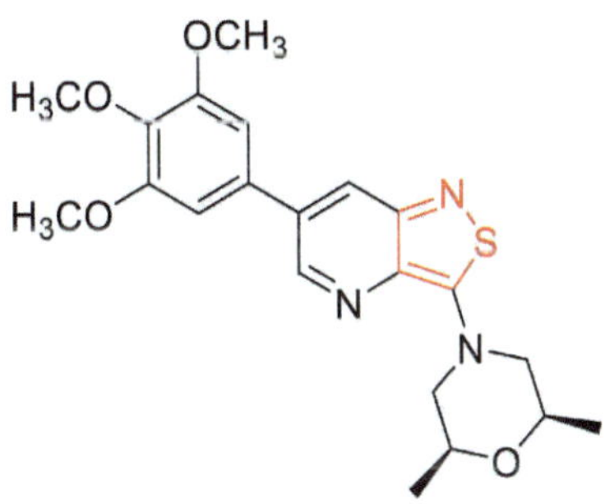

Figure 3.15 Selective GAK inhibitors.

According to Einav *et al.*, isothiazolo[5,4-*b*]pyridines are strong, ATP-competitive, and selective GAK inhibitors which play a crucial role in antiviral action mechanisms[46] (see Figure 3.16).

3.3.4 Isothiazole with Neurological Activity

Clarksville reported in his patent that isothiazole derivatives showed neurogenic activity[47] (see Figure 3.17).

It has been demonstrated by Fisher *et al.* that isothiazole carboxamides have strong mGluR1 antagonist activity and are a powerful allosteric

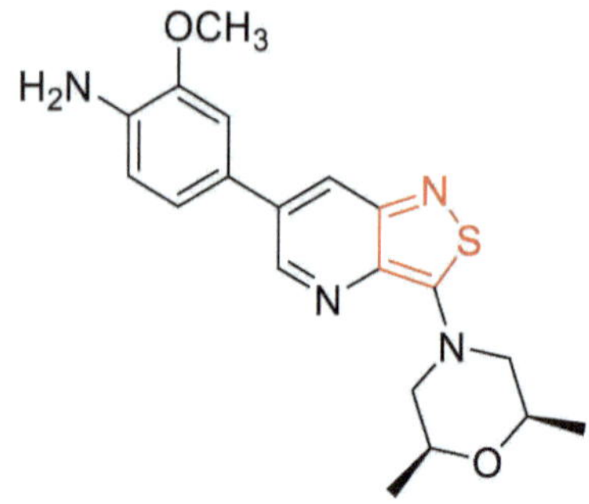

Figure 3.16 Selective GAK inhibitors.

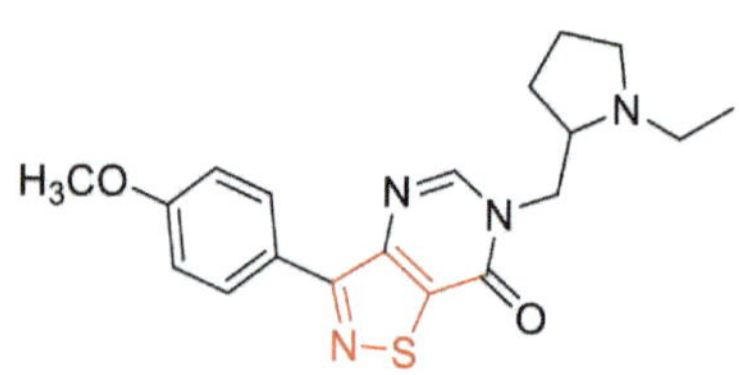

Figure 3.17 Stimulating neurogenic agent.

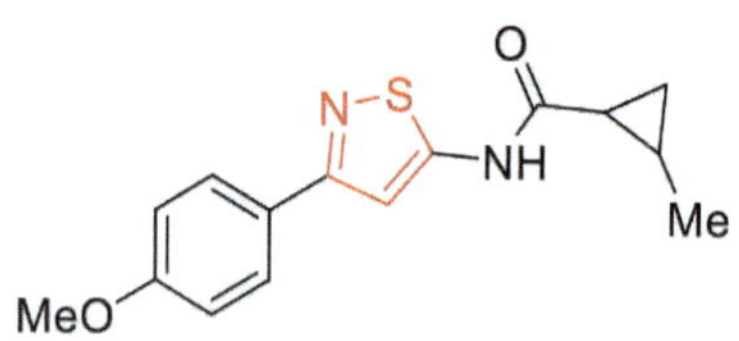

Figure 3.18 Potent mGluR1 antagonists.

agent.[48] Finally, the authors synthesized a novel series of isothiazole amides with potent mGluR1 antagonist activity (see **Figure 3.18**).

Oh *et al.* reported dipyridines (see Figure 3.18) exhibiting comparable neuroprotective effects against MPP+ (1-methyl-4-phenylpyrimidium)-induced neurotoxicity in neuroblastoma SH-SY5Y cells.[49]

The SARM1 enzymatic activity was selectively inhibited by an inventive class of potent and permanent isothiazole inhibitors discovered by Krauss *et al.*, protecting rat and human axons *in vitro*[50] (see Figure 3.19).

The derivatives of isothiazoles show promising effects in the treatment of neurodegenerative diseases of the peripheral and central nervous system.

Zhang and Jin discovered that the benzo[*c*]isothiazole group exhibits utility in organic solar cells and serves as a material for electron transport[51] (see Figure 3.20).

Figure 3.19 Isothiazole derivative.

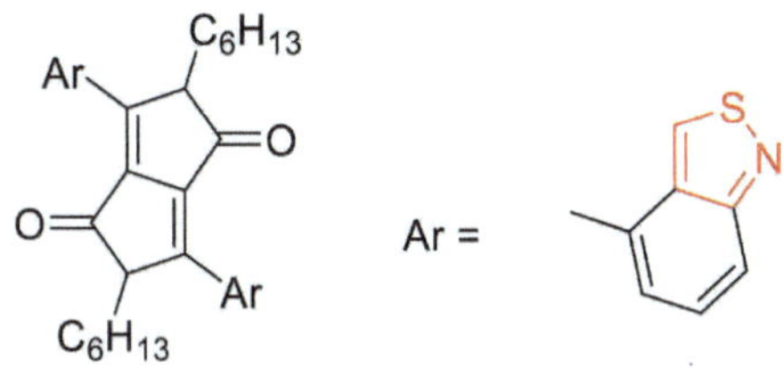

Figure 3.20 Material for electron transport.

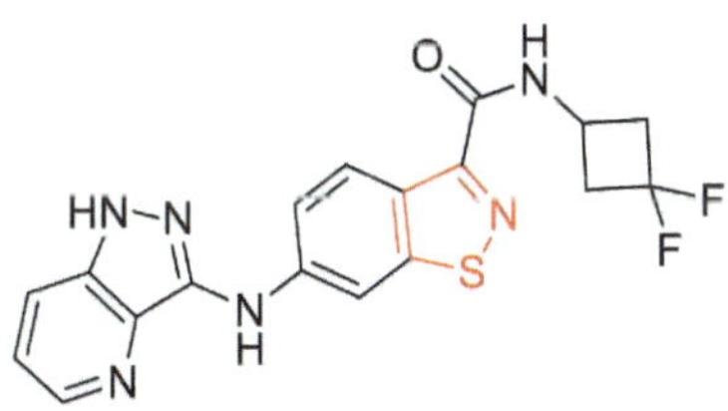

Figure 3.21 A drug used in the treatment of Parkinson's disease.

Hopkins *et al.* reported the finding, biological description, and structure–activity relationship of a novel active isothiazole derivative for treating Parkinson's disease[52] (see Figure 3.21).

3.3.5 Isothiazole as Antimicrobial/Antibacterial/Antifungal Agents

Qiang *et al.* synthesized novel 3-(2-(heterocyclo-2-ylthio)ethoxy)benzo[*d*]isothiazoles.[53] The resulting compounds have good antibacterial properties *in vitro*; the lowest inhibitory doses range from 4 to 32 μg mL^{-1} (see Figure 3.22).

Figure 3.22 Isothiazole as the antimicrobial agent.

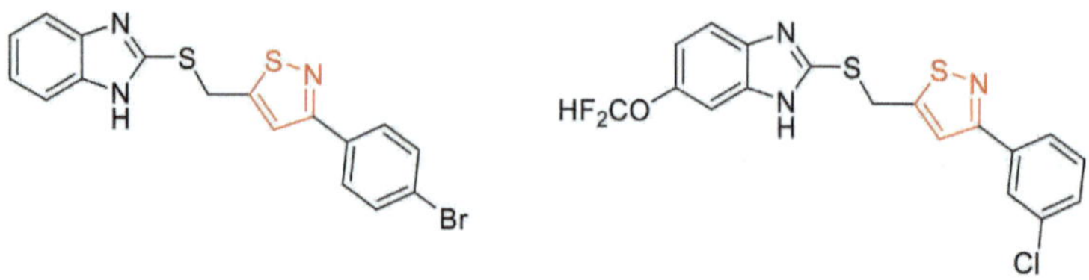

Figure 3.23 Isothiazole as the antimicrobial agent.

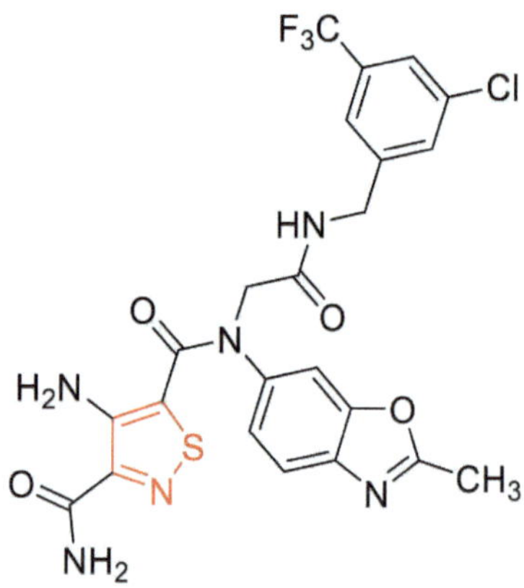

Figure 3.24 Isothiazole-3,5-dicarboxamide derivative.

Lokanatha Rai *et al.* obtained 5-{[(1*H*-benzo[*d*]imidazol-2′-yl)thio]methyl}-3-aryl isothiazole derivatives, and significant antibacterial activity against both Gram-positive and Gram-negative bacteria was demonstrated by the generated drugs[54] (see Figure 3.23).

Sriram *et al.* investigated the pharmaceutically viable enzyme to develop a novel class of medications that are highly resistant against *Pseudomonas aeruginosa* (Pa).[55] They developed novel Pa DNA GyrB inhibitors through additional optimization by following the synthesis of new *Pseudomonas aeruginosa* gyrase B inhibitors. *Pseudomonas aeruginosa* are the bacteria (germ) found in the environment causing infections in blood and pneumonia (see Figure 3.24).

Kodama *et al.* isolated the compounds from the culture of *Streptomyces* sp., which exhibit antibiotic activity[56] (see Figure 3.25).

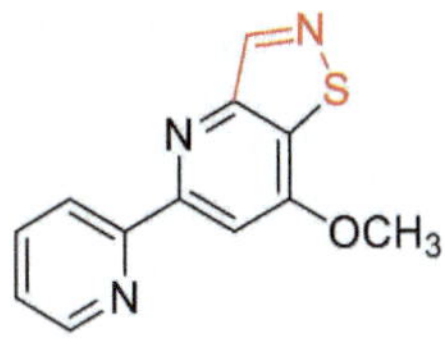

SF2738F

Figure 3.25 Neuroprotective isothiazolopyridine derivative.

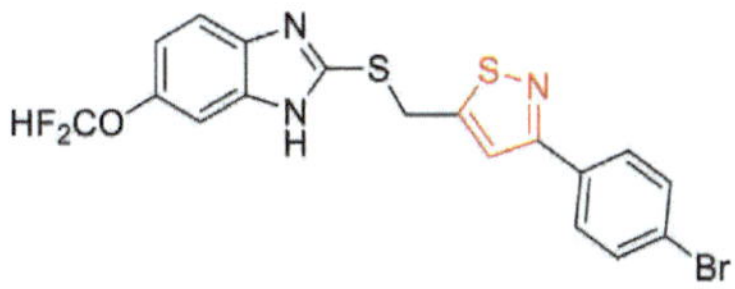

Figure 3.26 Antifungal agent.

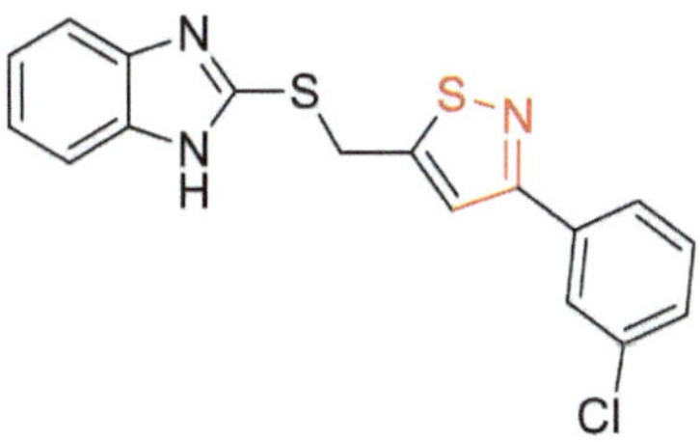

Figure 3.27 Antioxidant.

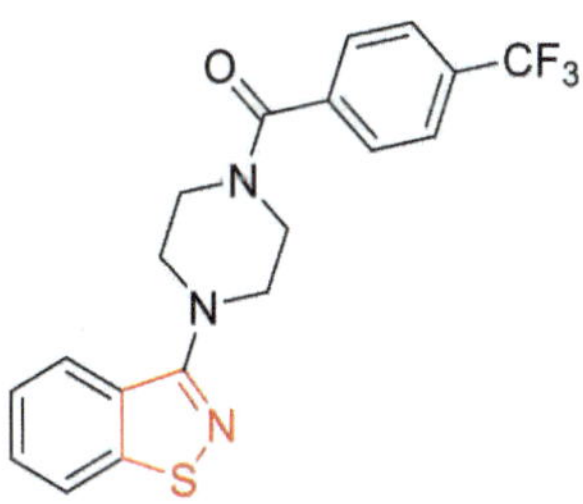

Figure 3.28 Isothiazole as the antifungal agent.

Davis explains the antifungal and antioxidant activities of the following compounds[57] (see Figures 3.26 and 3.27).

Mishra and Chundawat synthesized piperazine derivatives.[58] With minimum inhibitory concentration (MIC) values of 250 μg mL^{-1}, the discovered compounds demonstrated strong inhibitory action against the fungal strain *C. albicans* (see Figure 3.28).

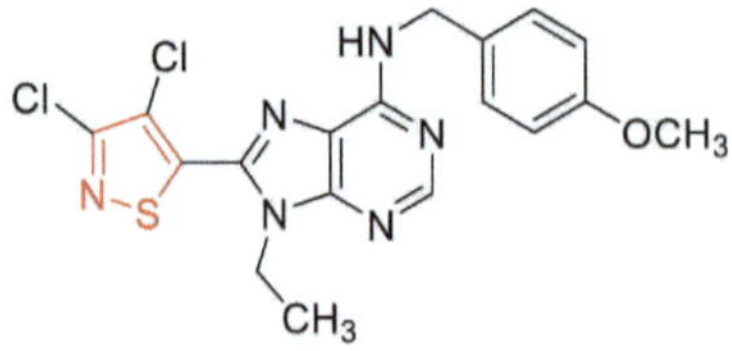

Figure 3.29 Isothiazole as the fungicide.

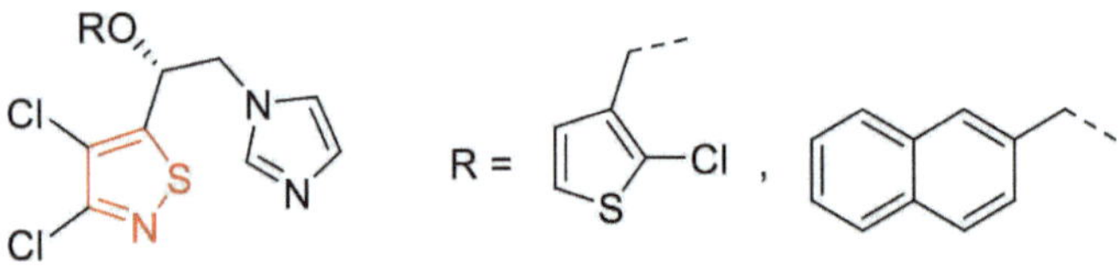

Figure 3.30 Isothiazole as the fungicide.

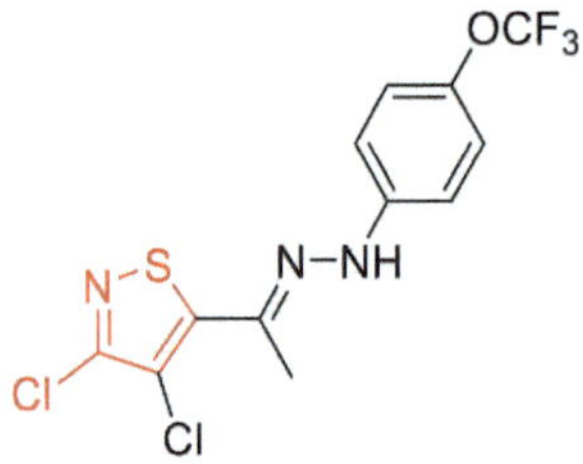

Figure 3.31 Isothiazole as the fungicide.

3.3.6 Isothiazole as Fungicides/Pesticides/Plant Protectors

Fan *et al.* obtained novel isothiazole-purine derivatives.[59] Their study revealed that the synthesized compound shows fungicidal activity (see Figure 3.29).

Wang *et al.* synthesized novel 3,4-chloroisothiazole-based imidazole as fungicides[60] (see Figure 3.30).

Fan *et al.* designed and synthesized ferimzone derivatives, which showed fungicidal activity with EC_{50} values ranging from 1.17 to 3.84 μg mL^{-1} against *Rhizoctonia solani*[61] (see Figure 3.31).

Himmler described a one-pot reaction to synthesize isotianil from isatoic anhydride.[62] In 2020, Zhang *et al.* discovered the synthesis, composition, and spectroscopic characteristics of isotianil, a bactericide (see Figure 3.32).

Bousquet *et al.* recognized brassilexin, a new phytoalexin isolated from the leaves of *Brassica juncea* (Cruciferae), which has antifungal properties[63,64] (see Figure 3.33).

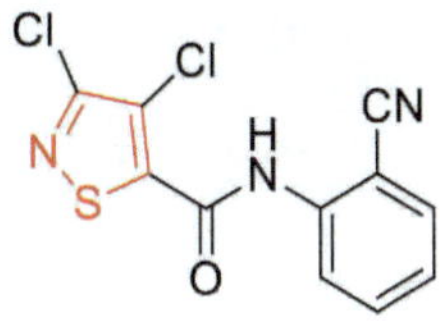

Figure 3.32 Isotianil, a bactericide.

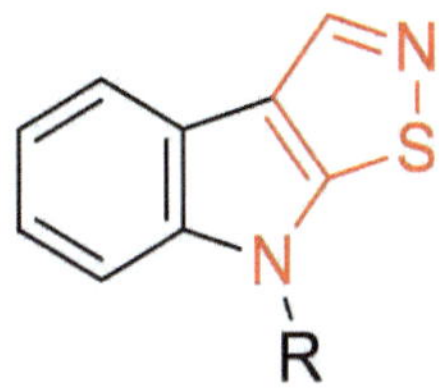

Figure 3.33 Brassilexin (R = H) and sinalexin (R = OMe).

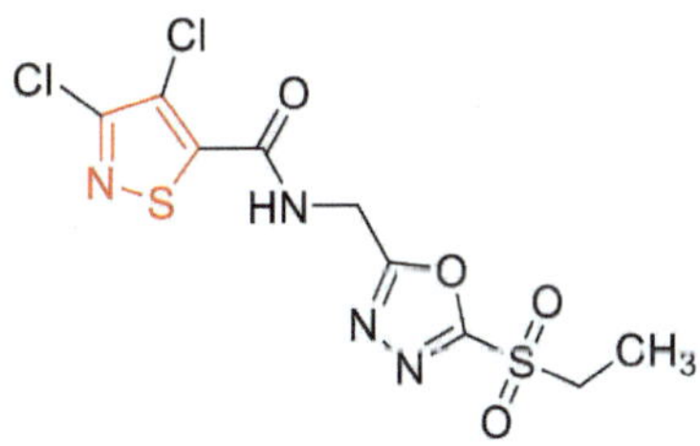

Figure 3.34 Isothiazole as pesticides.

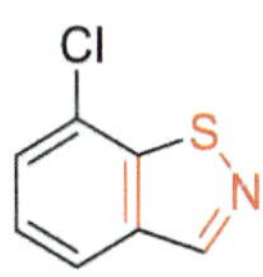

Figure 3.35 Benclothiaz.

Song *et al.* designed and synthesized novel compounds containing the 3,4-dichloroisothiazolylamide moiety.[65] A potent antibacterial action was demonstrated by the 3,4-dichloroisothiazolamide moiety, which may be studied further and used to create commercial pesticides (see Figure 3.34).

Ricci *et al.* synthesized benclothiaz first.[66,67] The nematicidal activity of benzodiazepines was identified in 1991 by Ciba-Geigy (now Syngenta). It was discovered that benclothiaz exhibited the strongest nematicidal activity overall (see Figure 3.35).

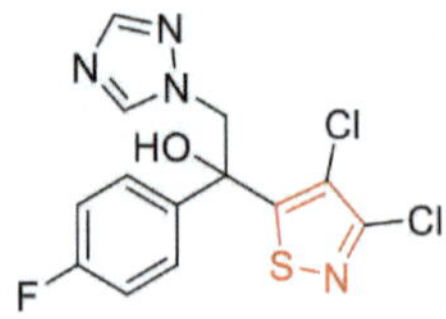

Figure 3.36 Antifungal activity.

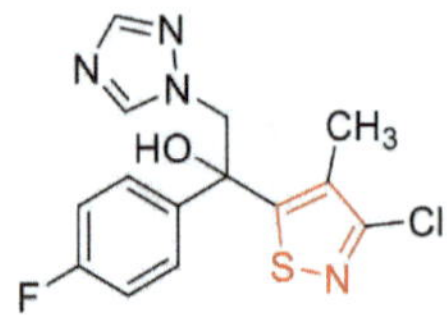

Figure 3.37 Anti-TMV activity.

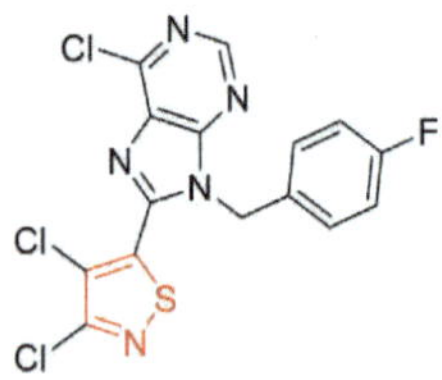

Figure 3.38 Isothiazole as the fungicide.

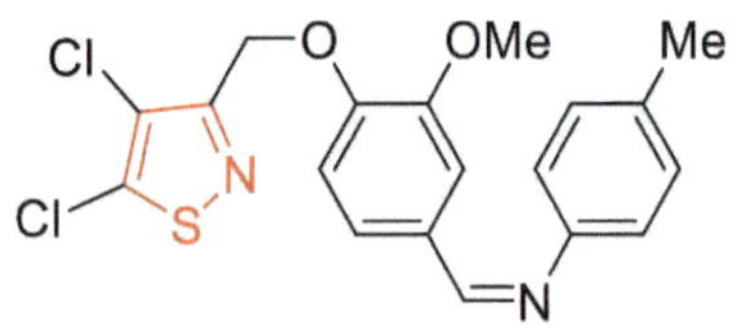

Figure 3.39 Insecticidal isothiazole derivatives.

Zhao *et al.* synthesized a number of unique 1,2,4-triazole substances based on 3,4-dichloroisothiazole that demonstrate fungicidal and *in vivo* anti-tobacco mosaic virus (TMV) properties[68] (see Figures 3.36 and 3.37).

Zhi-Jin *et al.* synthesized the above-mentioned isothiazole compounds, which exhibit fungicidal activity[69] (see Figure 3.38).

Zolotar *et al.* synthesized a vanillin-isothiazoline scaffold for the demonstration of insecticidal activity[70] (see Figure 3.39).

The synthesis of a novel plant protection agent and an assessment of its SAR-inducing action are reported by Fan *et al.*[71] As new plant protection agents in the twenty-first century, isothiazole-5-carboxamide derivatives are

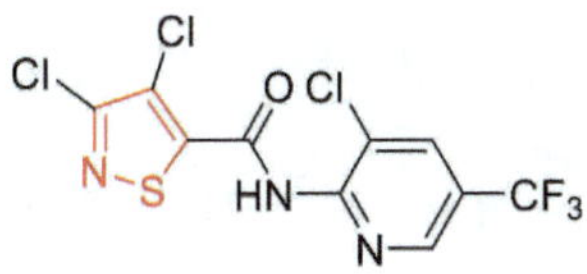

Figure 3.40 Novel plant protection agent.

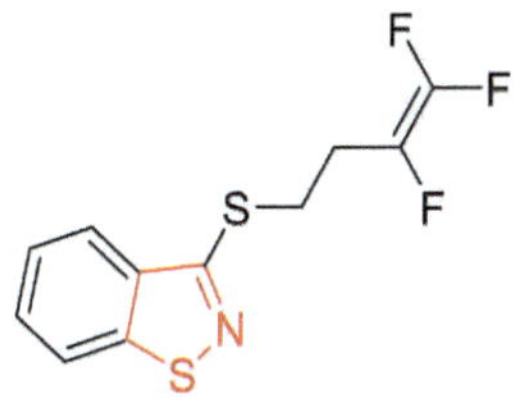

Pesticide

Figure 3.41 Benzoisothiazole derivative acts as pesticides.

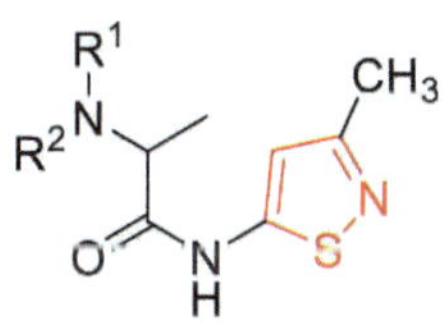

Figure 3.42 Analgesic agent.

thought to be the most active novel plant elicitor candidates (see Figure 3.40).

Zhou and Wang focused on an overview of the fungicidal activities of 3,4-dichloroisothiazole. 3,4-Dichloroisothiazole derivatives have diverse biological activities including plant immune-inducing activity.[72]

Huebl and Pieroh reported that the benzothiazole derivative acts as a pesticide[73] (see Figure 3.41).

Uyeo *et al.* have reported a series of 5-(aminoacylamido)-3-methylisothiazole for their analgesic action[74] (see Figure 3.42).

George *et al.* found that isothiazole derivatives are the newer antipsychotic agents for schizophrenia disease.[75] Lurasidone, a benzothiazole derivative, is the antipsychotic drug that was approved by the US Food and Drug Administration (FDA) in 2010 (see Figure 3.43).

Barilli *et al.* reported the synthesis of benzo[*d*]isothiazole derivatives, which show gastrointestinal prokinetic activity in rodents[76] (see Figure 3.44).

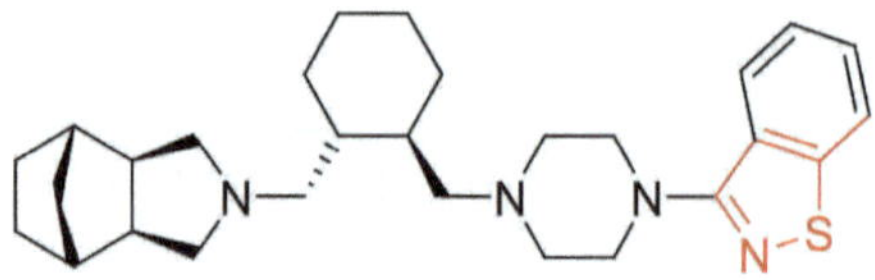

Figure 3.43 Lurasidone (antipsychotic medicine).

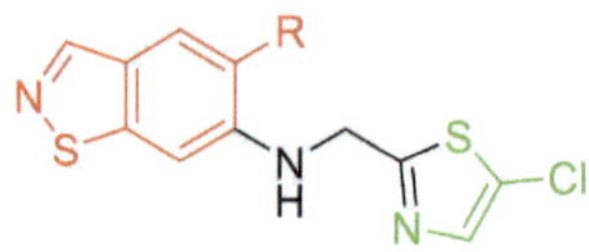

R = H, Cl

Figure 3.44 Gastrointestinal prokinetic activity.

3.4 Conclusion

In conclusion, a broad range of methodologies have been published for the synthesis of isothiazole derivatives. Using unique methodologies, isothiazole derivatives can be synthesized with moderate-to-high yields. By adding suitable substituent groups to the isothiazole ring, a variety of new compounds were designed and synthesized to improve their biological activities. Future research may thus primarily concentrate on isothiazole compounds due to their unique properties and promising applications.

Abbreviations

CHK1 inhibitors	Checkpoint kinase-1 inhibitors
DABCO	1,4-Diazabicyclo[2.2.2]octane
DCE	1,2-Dichloroethane
DCM	Dichloromethane
DMF	Dimethylformamide
DMSO	Dimethyl sulfoxide
DPPH	2,2-Diphenyl-1-picrylhydrazyl
DTP/SiO_2	Silica-supported dodecatungstophosphoric acid
eq./equiv.	Equivalent
FFA4	Free fatty acid receptor
GAK	Cyclin G-associated kinase
GPR agonists	G-protein-coupled receptor agonists
h	Hour
HDAC	Histone deacetylase
K_3PO_4	Tripotassium phosphate

mCPBA	*meta*-Chloroperoxybenzoic acid
MDR	Multidrug resistant
MGluR	Metabotropic glutamate receptors
MIC	Minimum inhibitory concentration
Na_2S	Sodium sulfide
NBS	*N*-Bromosuccinimide
NEt_3	Triethyl amine
NPs	Nanoparticles
PBMC	Peripheral blood mononuclear cell
PTSA	*p*-Toluenesulfonic acid
RT	Room temperature
SAR	Structure–activity relationship
SARM	Selective androgen receptor modulator
SbF6	Fluoroantimonic acid
TMV	Tobacco mosaic virus

References

1. (a) J. A. Builla, J. J. Vaquero and J. Barluenga, in *Modern Heterocyclic Chemistry*, Wiley-VCH, Weinheim, 2011. (b) J. A. Joule and K. Mills, in *Heterocyclic Chemistry*, Wiley-Blackwell, Chichester, 5th edn, 2010. (c) A. R. Katritzky, C. A. Ramsden, E. F. V. Scriven and R. J. K. Taylor, in *Comprehensive Heterocyclic Chemistry III*, Elsevier Ltd. 2008.
2. (a) R. D. Taylor, M. MacCoss and A. D. G. Lawson, *J. Med. Chem.*, 2014, **57**, 5845–5859. (b) C. Cabrele and O. Reiser, *J. Org. Chem.*, 2016, **81**, 10109–10125. (c) T. Y. Zhang, *Heterocycl. Chem.*, 2017, **121**, 1–12.
3. (a) A. S. S. Hamad Elgazwy, *Tetrahedron*, 2003, **59**, 7445–7463. (b) L. M. T. Frija, A. G. L. Pombeiro, M. L. Kopylovich, *Coord. Chem. Rev.*, 2016, **308**, 32–55. (c) O. M. Mulina, A. L. Ilovaisky, A. O. Terent'ev, *Eur. J. Org. Chem.*, 2018, 4648–4672. (d) S. Chakroborty, C. Bhanja, S. Jena, *Heterocycl. Commun.*, 2013, **19**, 79–87. (e) K. A. Scott, J. T. Njardarson, *Top. Curr. Chem.*, 2018, **376**(1),5. (f) E. A. Ilardi, E. Vitaku, J. T. Njardarson, *J. Med. Chem.*, 2014, **57**, 2832–2842. (g) D. H. R. Barton, S. V. Ley, P. D. Magnus, *Chem. Soc., Chem. Commun.*, 1975, **20**, 855–856.
4. C. A. Lipinski, F. Lombardo, B. W. Dominy and P. J. Feeney, *Adv. Drug. Delivery. Rev.*, 2001, **46**, 3–26.
5. (a) R. Slack and K. R. H. Wooldridge, *Adv. Heterocycl. Chem.*, 1965, **4**, 107. (b) M. Davis, *Adv. Heterocycl. Chem.*, 1972, **14**, 43–98.
6. A. F. Abdel-Magid, *ACS Med. Chem. Lett.*, 2015, **6**, 1097–1098.
7. M. J. Fisher, R. T. Backer, V. N. Barth, K. E. Garbison, J. M. Gruber, B. A. Heinz, S. Iyengar, S. P. Hollinshead, A. Kingston, S. L. Kuklish, L. Li, E. S. Nisenbaum, S. C. Peters, L. Phebus, R. M. A. Simmons and E. van der Aar, *Bioorg. Med. Chem. Lett.*, 2012, **22**, 2514–2517.

8. D. B. Belanger, P. J. Curran, A. Hruza, J. Voigt, Z. Meng, A. K. Mandal, M. A. Siddiqui, A. D. Basso and K. Gray, *Bioorg. Med. Chem. Lett.*, 2010, **20**, 5170.
9. S. Soni, S. Koley and M. S. Singh, *Tetrahedron Lett.*, 2017, **58**, 2512–2516.
10. D. Anand, O. P. Patel, R. K. Maurya, R. Kant and P. P. Yadav, *J. Org. Chem.*, 2015, **80**, 12410–12419.
11. C. E. Blunt, C. Torcuk, Y. Liu, W. Lewis, D. Siegel, D. Ross and C. J. Moody, *Angew. Chem. Int. Ed.*, 2015, **54**, 8740–8745.
12. H. Xie, G. Z. Li, F. Zhang, F. H. Xiao and G. J. Deng, *Green Chem.*, 2018, **20**, 827–883.
13. P. Barton, *Tetrahedron Lett.*, 2018, **59**, 815–817.
14. B. Seo, Y. G. Kim and P. H. Lee, *Org. Lett.*, 2016, **18**, 5050–5053.
15. B. B. Liu, H. W. Bai, H. Liu, S. Y. Wang and S. G. Ji, *J. Org. Chem.*, 2018, **83**, 10281–10288.
16. H. Yuan and Z. Sun, *Synlett*, 2019, **30**(16), 1904–1908.
17. S. Q. Wang, B. L. Hu and X. G. Zhang, *Adv. Synth. Catal.*, 2019, **361**(6), 1459–1462.
18. B. B. Liu, W. B. Cao, F. Wang, S. Y. Wang and S. J. Ji, *J. Org. Chem.*, 2018, **83**(18), 11118–11124.
19. J. Li, J. Li, X. Li, Q. Liu, L. Chen, Y. Huang and Y. Lia, *Adv. Synth. Catal.*, 2020, **362**(4), 1–11.
20. G. Huang, J. Li, X. Ji, L. Chen, Q. Liu, X. Chen, Y. Huang and Y. Li, *Chem. Commun.*, 2020, **56**, 5763–5766.
21. V. Dwivedi, M. Rajesh, R. Kumar, R. Kant and M. S. Reddy, *Chem. Commun.*, 2018, **53**, 11060–11063.
22. M. B. Djukić, M. S. Jeremić, I. P. Filipović, O. R. Klisurić, R. M. Jelić, S. Popović, S. Matić, V. Onnis and Z. D. Matović, *ChemistrySelect*, 2020, **5**(37), 11489–11502.
23. C. Nagaraju, S. H. Ashok, S. Shamanth, S. C. Nagarakere, M. P. Sunilkumar, R. K. Subbegowda and K. Mantelingu, *Synth. Commun.*, 2020, **50**(17), 2647–2654.
24. M. J. Cabrera, S. Cembellín, A. H. Salem, M. Berton, L. Marzo, M. C. Maestro and J. Alemán, *Green Chem.*, 2020, **22**, 6792–6797.
25. S. Moon, Y. Nishii and M. Miura, *Org. Lett.*, 2021, **23**(1), 49–53.
26. V. Dwivedi, M. Rajesh, R. Kumar, R. Kant and M. S. Reddy, *Chem. Commun.*, 2017, **53**, 11060–11063.
27. A. R. Rovira, A. Fin and Y. Tor, *Chem. Sci.*, 2017, **8**, 2983–2993.
28. A. S. Kalogirou, I. C. Christoforou, H. A. Ioannidou, M. J. Manos and P. A. Koutentis, *RSC Adv.*, 2014, **4**, 7735–7748.
29. F. Xu, Y. Chen, E. Fan and Z. Sun, *Org. Lett.*, 2016, **18**(11), 2777–2779.
30. X. Ma, X. Yu, H. Huang, Y. Zhou and Q. Song, *Org. Lett.*, 2020, **22**(14), 5284–5288.
31. P. Lamers and C. Bolm, *Org. Lett.*, 2018, **20**(1), 116–118.

32. L. L. Brunton, B. Chabner, L. S. Goodman and B. C. Knollman, in *Goodman & Gilman'S The Pharmacological Basis Of Therapeutics*, McGraw Hill, New York, 12th edn, 2011.
33. S. R. Ambati, S. Gudala, A. Sharma, S. Penta, V. L. Reddy, Y. Bomma, V. R. Janapala and S. Pola, *J. Heterocycl. Chem.*, 2017, **54**(4), 2333–2341.
34. C. M. Bode, A. A. Boezio, B. K. Albrecht, S. F. Bellon, L. Berry, M. A. Broome, D. Choquette, I. Dussault, R. T. Lewis and M. H. J. Lin, *Bioorg. Med. Chem. Lett.*, 2012, **22**(12), 4089–4093.
35. D. B. Belanger, P. J. Curran, A. Hruza, J. Voigt, Z. Meng, Z. Mandal, M. A. X. Siddiqui, A. D. Basso and K. Gray, *Bioorg. Med. Chem. Lett.*, 2010, **20**(17), 5170–5514.
36. K. Stratmann,, J. Belli,, C. M. Jensen,, R. E. Moore,and G. M. L. Patterson, *J. Org. Chem.*, 1994, **59**(21), 6279–6281.
37. C. E. Blunt, C. Torcuk, Y. Liu, W. Lewis, D. Siegel, D. Ross and J. C. Moody, *Angew. Chem. Int. Ed.*, 2015, **54**(30), 8740–8745.
38. I. Jęśkowiak, M. Mączyński, J. Trynda, J. Wietrzyk and S. Ryng, *Bioorg. Chem.*, 2019, **91**, 103082.
39. H. Khoshniazi, H. Eshghi, M. T. Najjaran, Z. T. Najaran, K. Saadat and A. Shiri, *J. Sulfur Chem.*, 2021, **42**(2), 193–201.
40. V. V. Semenov, B. V. Lichitsky, A. N. Komogortsev, A. A. Dudinov, M. M. Krayushkin, L. D. Konyushkin, O. P. Atamanenko, I. B. Karmanova, Y. A. Strelenko, B. Shor, M. N. Semenova and A. S. Kiselyov, *Eur. J. Med. Chem.*, 2017, **125**, 573–585.
41. J. R. Hong, Y. J. Choi, G. Keum and G. Nam, *Bioorg. Med. Chem.*, 2017, **25**(17), 4677–4685.
42. A. V. Kletskov, V. I. Potkin, I. A. Kolesnik, S. K. Petkevich, A. V. Kvachonak, M. O. Dosina, D. O. Loiko, M. V. Larchenko, S. G. Pashkevich and V. A. Kulchitsky, *Nat. Prod. Commun.*, 2018, **13**(11), 1507–1510.
43. X. Zhang, C. Cai, Z. Sui, M. Macielag, Y. Wang,, W. Yan, A. Suckow, H. Hua, A. Bell, P. Haug, W. Clapper, C. Jenkinson, J. Gunnet, J. Leonardand W. V. Murray, *ACS Med. Chem. Lett.*, 2017, **8**(9), 947–952.
44. C. M. G. Azevedo, K. R. Watterson, E. T. Wargent, S. V. F. Hansen, B. D. Hudson, M. A. Kępczyńska, J. Dunlop, B. Shimpukade, E. Christiansen, G. Milligan, C. J. Stocker and T. Ulven, *J. Med. Chem.*, 2016, **59**(19), 8868–8878.
45. B. M. Gualda, S. Y. Pu, M. Froeyen, P. Herdewijn, S. Einav and S. D. Jonghe, *Bioorg. Med. Chem. Lett.*, 2020, **28**(1), 115–188.
46. E. Bekerman, G. Randall and S. Einav, *J. Clin. Invest*, 2017, **127**(4), 1338–1352.
47. J. K. A. Clarksville, *Patent US8609840 B2*, 2008.
48. M. J. Fisher, R. T. Backer, V. N. Barth, K. E. Garbison, J. M. Gruber, B. A. Heinz, S. Iyengar, S. P. Hollinshead, A. Kingston and S. L. Kuklish, *Bioorg. Med. Chem. Lett.*, 2020, **22**(7), 2514–2517.
49. S. H. Kim, H. Ko, S. H. Bang, S. H. Park, D. J. Kim, S. C. Kwon, S. Y. Kim, J. Shin and D. C. Oh, *Bioorg. Med. Chem. Lett.*, 2011, **21**(19), 5715–5718.

50. T. Bosanac, R. O. Hughes, T. Engber, R. Devraj, A. Brearley, K. Danker, K. Young, J. Kopatz, M. Hermann, A. Berthemy, S. Boyce, J. Bentley and R. Krauss, *Brain*, 2021, **144**(10), 3226–3238.
51. X. Zhang and R. Jin, *Front. Chem. (Lausanne, Switz.)*, 2019, **7**, 122.
52. S. R. Bollinger, D. W. Engers, J. D. Panarese, M. West, J. L. Engers, M. T. Loch, A. L. Rodriguez, A. L. Blobaum, C. K. Jones, A. T. Gray, P. J. Conn, C. W. Lindsley, C. M. Niswender and C. R. Hopkins, *J. Med. Chem.*, 2019, **62**(1), 342–358.
53. F. Wen, X. L. Jie, S. H. Wang, S. Z. Feng, W. X. Hui and L. Qiang, *Chinese J. Struct. Chem.*, 2017, **6**(36), 911–917.
54. U. K. Narayana, C. S. Vicas, V. S. Murthy, S. Y. Kotian, K. Byrappa and K. M. Lokanatha Rai, *J. Chem., Biol. Phys. Sci.*, 2016, **6**(3), 861–868.
55. S. Jogula, V. S. Krishna, N. Medaa, V. Balraju and D. Sriram, *Bioorg. Chem.*, 2020, **100**, 103905.
56. S. Gomi, S. Amano, E. Sato, S. Miyadoh and Y. Kodama, *J. Antibiot.*, 1994, **47**(12), 1385–1394.
57. (a) M. Davis, *Adv. Heterocycl. Chem.*, 1972, **14**, 43–98. (b) M. Sutter and W. Kunz, *Eur. Pat. Appl. EP*, 1991, **454**, 621, Ciba-Geigy AG.
58. V. Mishra and T. S. Chundawat, *Turk. J. Chem.*, 2019, **43**(2), 713–729.
59. W. Weibo, Z. Li, W. Gao, X. Liu, Y. Lv, Z. Hao, L. Tang, K. Li, B. Zhao and Z. Fan, *J. Agric. Food Chem.*, 2021, **69**(32), 9461–9471.
60. L. Chen, B. Zhao, Z. Fan, X. Liu, Q. Wu, H. Li and H. Wang, *J. Agric. Food Chem.*, 2018, **66**(28), 7319–7327.
61. Y. Zhang, J. Li, X. Liu, W. Gao, S. Song, Y. Rong, L. Tan, T. V. Glukhareva, V. A. Bakulev and Z. Fan, *J. Agric. Food Chem.*, 2023, **71**(8), 3705–3718.
62. (a) T. Himmler, *Int. Patent Appl.*, WO2004/002969, 2004. (b) J. Zhanga, J. Jia, B. Wanga, A. Q. Jiaa and Q. F. Zhang, *Russ. J. Org. Chem.*, 2020, **56**(10), 1801–1805.
63. M. Devys, M. Barbier, I. Loiselet, T. Rouxel, A. Sarniguet, A. Kollmann and J. F. Bousquet, *Tetrahedron Lett.*, 1988, **29**(49), 6447–6448.
64. M. Soledade, C. Pedras and K. C. Smith, *Phytochemistry*, 1997, **46**(5), 833–837.
65. J. Xiang, D. Liu, J. Chen, D. Hu and B. Song, *Pestic. Biochem. Physiol.*, 2020, **170**, 104695.
66. A. Ricci, A. Martani, O. Graziani and M. L. Oliva, *Ann. Chim.*, 1860, **53**, 1963.
67. (a) M. Davis, *Adv. Heterocycl. Chem.*, 1972, **14**, 43–98. (b) M. Sutter and W. Kunz, *Eur. Pat. Appl. EP*, 1991, **454**, 621, Ciba-Geigy AG.
68. L. Chen, Q. Wu, Z. Fan, H. Li, J. Li, W. Hu, X. Liu, N. P. Belskaya, T. Glukhareva and B. Zhao, *Chin. J. Chem.*, 2018, **36**(8), 731–736.
69. W. Wei-Bo, L. Xiao-Yu, L. Zhi-Xinyi, G. Wei, L. You, L. Kun, T. V. Glukhareva, T. Liang-Fu and F. Zhi-Jin, *Chin. J. Struct. Chem.*, 2022, **41**, 2202091–2202097.
70. A. V. Kletskov, V. I. Potkin, E. A. Dikusar and R. M. Zolotar, *Nat. Prod. Commun.*, 2017, **12**(1), 105–106.

71. X. Qi, K. Ji, L. Chen, Y. Zhang, N. Zhang, W. Gao, Y. Li, X. Liu and Z. Fan, *Int. J. Mol. Sci.*, 2022, **23**(10), 5348.
72. M. Zhou and W. Wang, *Front. Plant Sci.*, 2018, **9**, 1613.
73. D. Huebl and E. Pieroh, *Patent DE 3837578 A1*, 1990.
74. S. Uyeo, H. Fujimura, A. Asai and Y. Zasshi, *Yakugaku Zasshi*, 1963, **83**, 195.
75. M. George, R. Amrutheshwar, R. P. Rajkumar, S. Kattimani and S. A. Dkhar, *Eur. J. Clin. Pharmacol.*, 2013, **69**, 1497–1509.
76. A. Barilli, L. Aldegheri, F. Bianchi, L. Brault, D. Brodbeck, L. Castelletti, A. Feriani, I. Lingard, R. Myers, S. Nola, L. Piccoli, D. Pompilio, L. F. Raveglia, C. Salvagno, S. Tassini, C. Virginio and M. Sabat, *J. Med. Chem*, 2021, **64**(9), 5931–5955.

CHAPTER 4

Synthesis, Properties, and Biological Applications of Benzothiazoles

G. K. PRASHANTH* (0000-0001-6691-4030)[a], MANOJ GADEWAR[b], SRILATHA RAO[c], MITHUN KUMAR GHOSH[d], K. V. YATISH[e] AND M. MAHADEVA SWAMY[f]

[a]Research and Development Center, Department of Chemistry, Sir M. Visvesvaraya Institute of Technology, Bengaluru, 562157, India; [b]Department of Pharmacology, School of Medical and Allied Sciences, K.R. Mangalam University, Gurgaon, 122103, India; [c]Department of Chemistry, Nitte Meenakshi Institute of Technology, Bengaluru, 560064, India; [d]Department of Chemistry, Medi-Caps University, Indore, India; [e]Department of Chemistry, Navkis College of Engineering, Hassan, 573217, India; [f]Department of PG Chemistry, JSS College of Arts, Commerce and Science, Mysuru, 570025, India
*E-mail: prashanth_chem@sirmvit.edu; prashaanthgk@gmail.com

4.1 Introduction

Heterocyclic compounds are cyclic compounds with structures containing one or more atoms of other elements along with carbon. These non-carbon atoms within the rings are commonly known as heteroatoms. Nitrogen, sulfur, and oxygen are the most prevalent heteroatoms found in heterocyclic compounds. However, in recent years, there has been

S-Heterocycles: Retrospect, Prospects, and Biological Applications
Edited by Shrikaant Kulkarni, Hemantkumar Akolkar, Vijay M. Khedkar and A. K. Haghi

Published by the Royal Society of Chemistry, www.rsc.org

significant research interest in heterocyclic compounds that contain less common heteroatoms, such as phosphorus, tin, boron, silicon, bromine, and others.[1–5] Although there are many heterocyclic compounds with ring configurations comprising three to six carbon atoms, it is noteworthy that those with five or six carbon atoms are much more significant and have been extensively investigated. The heterocyclic compounds are prevalent and hold particular significance owing to the diverse range of physiological activities attributed to this class of substances. Many significant compounds feature heterocyclic rings in their structures. These include a majority of the members of the vitamin B complex, antibiotics, alkaloids, amino acids, chlorophyll and various plant pigments, drugs, genetic materials such as DNA, dyes, enzymes, and many more.

Benzothiazole is a heterocyclic compound characterized as a weak base, and it is notable for its diverse range of biological activities, making it an area of ongoing scientific interest. This compound is extensively present in the realms of bioorganic and medicinal chemistry, where it finds applications in the field of drug discovery. Benzothiazole has an advantageous bicyclic ring structure with a wide array of applications. It exhibits diverse biological activities, including antimicrobial, anticancer, anthelmintic, and antidiabetic properties.[6–15] Furthermore, benzothiazoles have been used in industrial applications as antioxidants and vulcanization accelerators. Notably, during the 1950s, there was intensive research on various 2-amino benzothiazoles for their potential use as central muscle relaxants. Since then, there has been a decrease in active interest in this chemical family among medicinal chemists. The interest among biologists in this chemical series was reignited when the pharmacological profile of riluzole was uncovered. Riluzole, also known as RP-25279, PK-26124, Rilutek, or 6-trifluoromethoxy-2-benzothiazolamine, was observed to have an impact on glutamate neurotransmission in various biochemical, electrophysiological, and behavioral experiments.[16] Following the discovery, benzothiazole derivatives have been the subject of extensive studies, revealing a wide range of chemical reactivity and biological activities. Numerous benzothiazoles, particularly compounds such as 2-aryl-benzothiazole, have garnered significant consideration because of their exclusive structure.[16,17] They have been explored for their applications as radioactive amyloid imaging agents and potential anticancer agents.[18,19] Benzothiazoles, as a bicyclic ring system, have diverse applications. 2-(4-Aminophenyl)benzothiazoles are a new group of powerful and extremely selective anticancer drugs among them.[20,21] They exhibit a distinct profile of cytotoxic responses across various cell lines, making them of particular interest in the field of cancer research.[22,23] Recent reports have revealed that benzothiazole, its derivatives, and bioisosteres have demonstrated antimicrobial properties against a variety of microorganisms, including yeast, such as *Candida albicans*, and both Gram-negative and Gram-positive bacteria, such as *Staphylococcus epidermidis*, *Enterobacter*, *E. coli*, and *Pseudomonas aeruginosa*.[24–27] Fused rings are combined to form

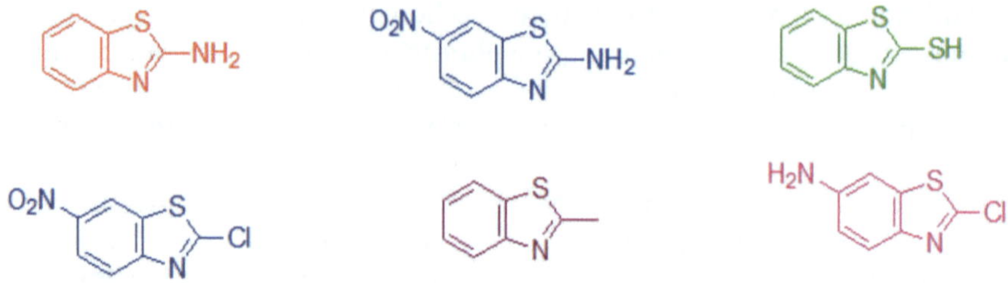

Figure 4.1 Derivatives of benzothiazoles.

benzothiazoles, which are heterocycles with a thiazole component. The sulfur and nitrogen atoms found in the basic structure of thiazole serve as the building blocks for numerous pharmacologically and physiologically active substances (Figure 4.1).

This study aims to investigate the properties, biological activities, and synthesis of novel benzothiazoles. Through this study, we hope to offer a greater comprehension of their complex significance and their ability to address contemporary problems in drug development and materials research.

4.2 Synthesis of Benzothiazoles and Their Derivatives

Over time, benzothiazole synthesis has experienced substantial advancements, leading to a variety of techniques for producing these important chemicals. Benzothiazoles are characterized by their structural fusion of a benzene ring with a thiazole ring, and their synthesis typically revolves around building the thiazole ring onto a suitable precursor. Nevertheless, newer, more efficient, and diverse synthetic routes have surfaced, expanding the range of choices available to researchers for their synthesis.

4.2.1 Synthesis from Isothiocyanates

Tweit documented the synthesis of 2-aminobenzothiazole (Figure 4.2) from isothiocyanates with benzene acting as a catalyst.[28]

4.2.2 Synthesis from Triethyl Orthoformate

Štetinová described the synthesis of 2-oxo-2-pyrimidobenzothiazole-3-carbonitrile through reacting *N*-(2-benzothiazolyl)cyanoacetamide with triethyl

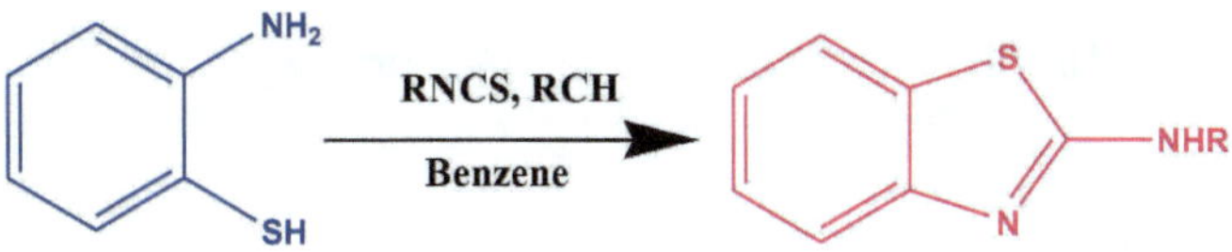

Figure 4.2 Synthesis of 2-aminobenzothiazole from isothiocyanates.

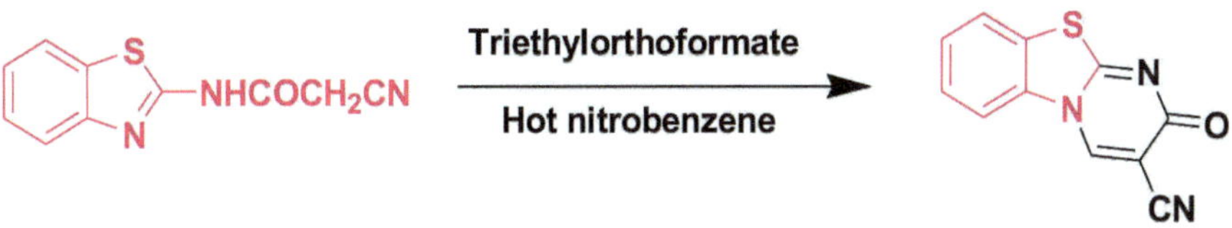

Figure 4.3 Synthesis of 2-oxo-2-pyrimidobenzothiazole-3-carbonitrile from triethyl orthoformate.

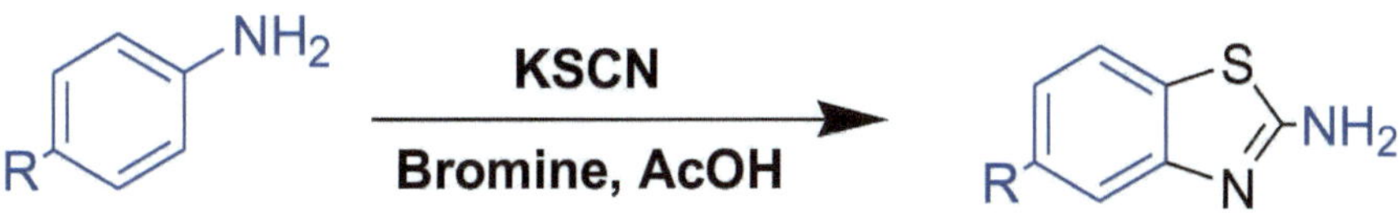

Figure 4.4 Synthesis of 6-substituted-2-aminobenzothiazoles.

orthoformate in hot nitrobenzene, resulting in the formation of the desired compound (Figure 4.3).[29]

4.2.3 Synthesis from Aniline Derivatives

Dogruer *et al.* successfully synthesized a range of 6-substituted-2-aminobenzothiazoles, as shown in Figure 4.4, from *p*-substituted anilines with KSCN and Br.[30]

4.2.4 Synthesis from Hydrazine

Latrofa *et al.* proposed an approach for the synthesis of 1,3,4-triazolobenzothiazole. This synthesis involved a mixture of silica gel in xylene, trimethyl orthoformate, and hydrazine benzothiazole, which resulted in the formation of the desired compound (Figure 4.5).[31]

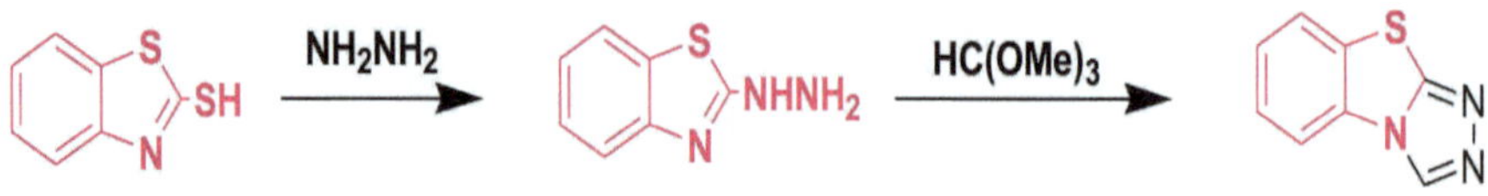

Figure 4.5 Synthesis 1,3,4-triazolobenzothiazole from hydrazine.

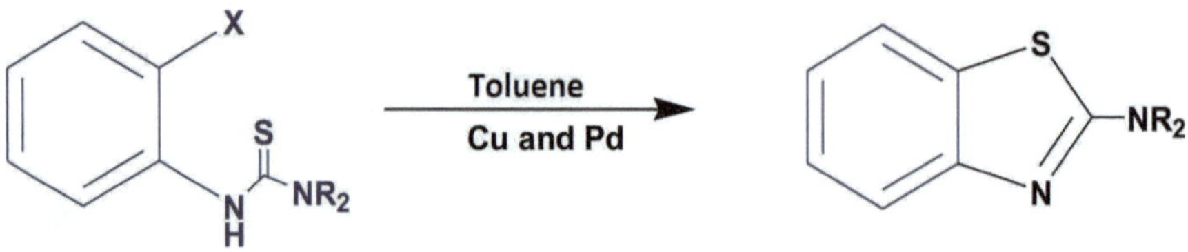

Figure 4.6 2-Aminobenzothiazole formation *via* intramolecular formation of C–S.

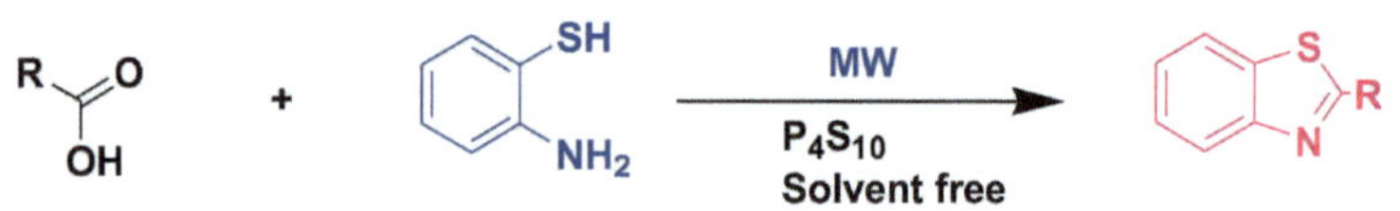

Figure 4.7 Solvent-free synthesis.

4.2.5 Intramolecular Formation of C–S Using Pd and Cu Catalysts

Evindar and Batey successfully documented the synthesis of 2-aminobenzothiazoles, as depicted in Figure 4.6. This synthesis involved an intramolecular carbon–sulfur bond formation process achieved through cross-coupling between thiourea and arylhalide in the presence of Cu and Pd catalysts.[32]

4.2.6 Solvent-free Synthesis

2-Substituted benzothiazoles, as illustrated in Figure 4.7, were synthesized by condensing 2-aminothiophenol with various saturated olefinic fatty acids. This synthesis process was conducted under microwave irradiation in a solvent-free environment, using P_4S_{10} as a catalyst. Notably, this reaction yielded high product yields and was completed in a remarkably short time frame, typically within 3–4 minutes.[33]

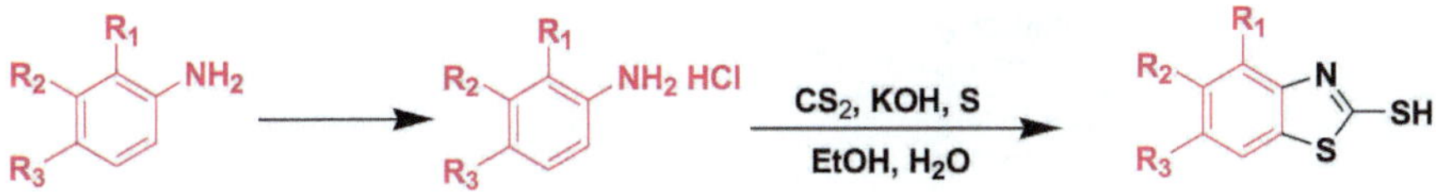

Figure 4.8 Cyclization.

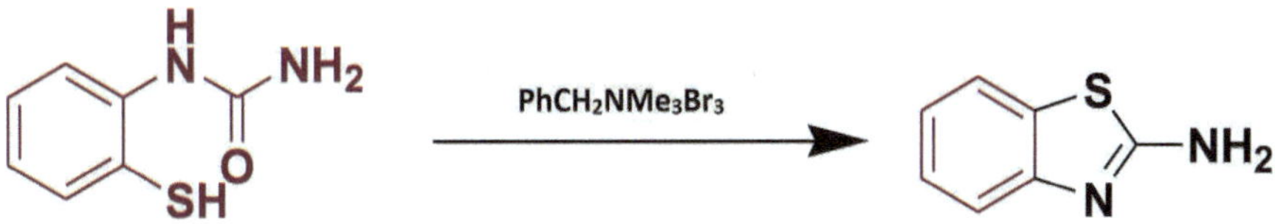

Figure 4.9 One-pot synthesis of 2-aminobenzothiazoles.

4.2.7 Cyclization

The synthesis of substituted 2-mercaptobenzothiazoles, as depicted in Figure 4.8, can be applied to various substituents at the 4th, 5th, and 6th positions of benzothiazoles. The final compound was formed by two steps.

In the first step, anilines are initially transformed into their hydrochloride salts. In the second step, the aniline hydrochloride salts are cyclized to form substituted 2-mercaptobenzothiazoles. This cyclization is achieved by reacting the aniline hydrochloride salt with carbon disulfide in the presence of sulfur under alkaline conditions.[34]

4.2.8 One-pot Synthesis of 2-Aminobenzothiazoles

Jordan *et al.* directly synthesized 2-aminobenzothiazoles, as depicted in Figure 4.9, from substituted arylthioureas. They achieved this synthesis with benzyltrimethylammonium tribromide as a catalyst, which serves as an electrophilic source of Br. This facilitated the transformation of isothiocyanates, tetrabutylammonium thiocyanate, and anilines into 2-aminobenzothiazoles under mild conditions in various solvents, resulting in a good yield. The ease of addition and handling of $PhCH_2NMe_3Br_3$ as a catalyst over Br reduces the possibility of undesired brominated by-product formation. This is one of the main advantages.[35]

4.2.9 Synthesis by Suzuki–Miyaura

An advanced microwave-promoted Suzuki–Miyaura reaction between phenylboronic acid and 2-chlorobenzothiazole was conducted using $Pb(PPh_3)_4$ as a catalyst, as shown in Figure 4.10. Notably, this reaction

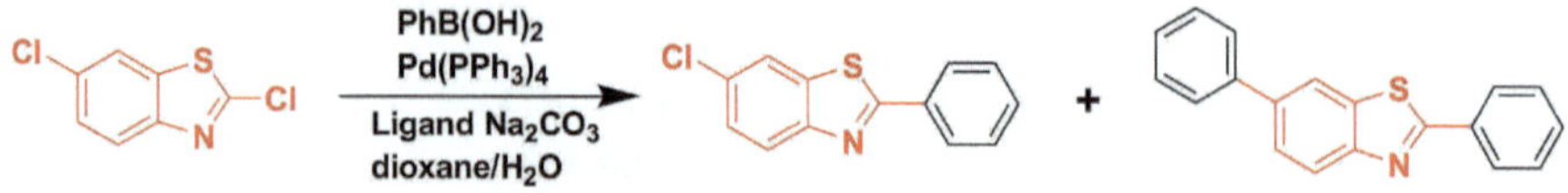

Figure 4.10 Suzuki–Miyaura reaction.

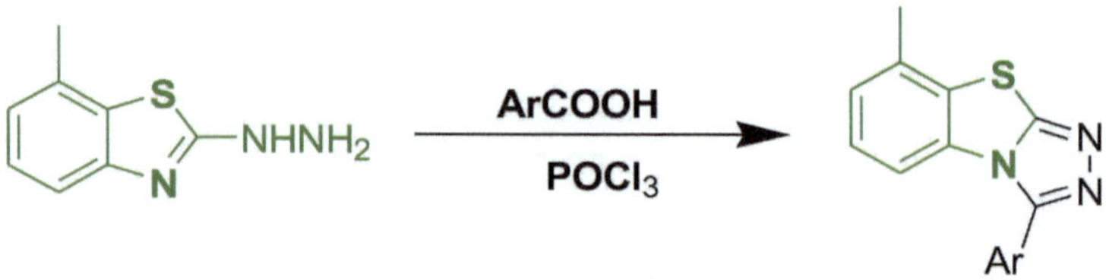

Figure 4.11 Synthesis of fused benzothiazoles.

yielded the adduct with excellent regioselectivity, highlighting the effectiveness of this synthetic approach.[36]

4.2.10 Fused Benzothiazole Synthesis

5-Methyl-3-substituted-1,2,4-triazolo[3,4*b*]benzothiazoles were synthesized and reported by Dong *et al.* (Figure 4.11). To perform this synthesis, 2-hydrazino-4-methylbenzothiazole was treated with a variety of aromatic carboxylic acids in the presence of $POCl_3$ and reflux conditions were maintained for 60 minutes. The reaction mixture was then added to crushed ice and NaOH solution was used to turn the mixture alkaline. After filtering the resultant solid, ethanol was used to recrystallize it and produce the desired product with a good yield.[37]

4.3 Physicochemical Properties of Benzothiazoles

Benzothiazoles, a class of heterocyclic compounds, are valued for a variety of applications in materials science and other fields due to their distinctive and varied physicochemical characteristics. The use of benzothiazoles in industries, including drug research, polymers, and organic electronics, requires an understanding of their characteristics. Benzothiazoles exhibit moderate-to-good solubility in a range of organic solvents.[38] Their application in organic synthesis, especially in reactions involving other organic compounds, is facilitated by their solubility. The substitution pattern on the aromatic ring may affect the solubility of some benzothiazole derivatives. Generally, benzothiazoles have a high melting point and excellent thermal stability.[39] These characteristics make them useful for applications involving high temperatures, such as the production of polymers, where they can be added to improve thermal stability or used as cross-linking agents. Due to their widespread inherent fluorescence, benzothiazole derivatives are valuable in optoelectronics and fluorescence spectroscopy applications.[40] By altering the chemical structure, their fluorescence properties can be adjusted, and they have been used as fluorophores in the development of sensors and imaging agents. They are versatile compounds with applications in a wide range of fields, from materials science to drug development. Their physicochemical properties include solubility, thermal stability, fluorescence, electron density distribution, dipole moment, molecular weight, UV–visible absorption, and chemical reactivity.[41] To fully utilize the potential of benzothiazoles, researchers are still investigating and modifying these characteristics, creating new opportunities for further studies and practical use.

4.4 Biological Activities of Benzothiazoles

Belonging to the heterocyclic chemical family, benzothiazoles have drawn a lot of attention in the field of medicinal chemistry due to their broad variety of biological effects. Benzothiazoles have been studied in great detail as potential candidates for medication development because of their notable pharmacological effects in a variety of settings. The biological activities of benzothiazoles depends on their potency, mechanisms of action, and potential applications as therapeutic agents. In this section, we review the key elements of the biological activities of benzothiazoles.

4.4.1 Antimicrobial Activity

Benzothiazoles have strong antibacterial activities. It has been demonstrated that several chemicals exhibit antibacterial, antifungal, and antiparasitic properties. Nowadays, one of the main challenges faced by chemists is the development of novel and potent antimicrobial drugs. Numerous illnesses, such as typhoid, influenza, amebiasis, AIDS, syphilis, tuberculosis (TB), pneumonia, and common respiratory infections, are mostly caused by microbes. Researchers have looked into the possibilities of treating these significant health issues with the benzothiazole nucleus, which act as an antibacterial agent on numerous occasions. Because of their effectiveness against a variety of diseases, including drug-resistant strains, which researchers have studied, they offer a promising avenue for the development of novel antimicrobial drugs.

A successful attempt to synthesize 2-substituted phenylsulfonamido-6-substituted benzothiazoles (**1**) was made by Hutchinson *et al.* Following that, they conducted investigations into the antibacterial properties of these compounds and discovered that they exhibited significant effectiveness against various bacteria, such as *S. dysenteriae*, *B. subtilis*, and *S.*

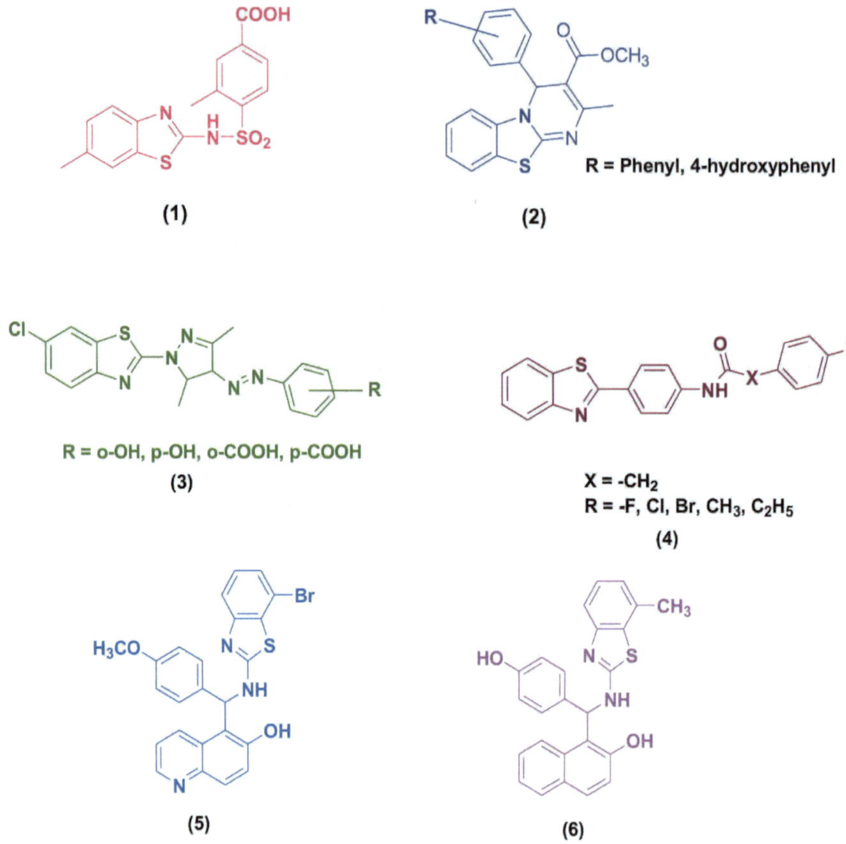

Figure 4.12 Antimicrobial activity.

typhi.[42] 4*H*-Pyrimido[2,1-*b*][1,3]benzothiazole (**2**) derivatives were synthesized by Sahu *et al.*, and screening tests were performed to evaluate their antibacterial effectiveness. These derivatives were tested against both Gram-positive and Gram-negative bacteria, including *B. cereus*, *P. aeruginosa*, *S. typhi*, *P. rettgeri*, *E. coli*, and *S. aureus* (Figure 4.12).[43]

Two new chemical families were proposed by Amir *et al.*: 4-arylazo-1-benzothiazolyl-3,5-dimethylpyrazoles and 4-arylhydrazono-1-benzothiazolyl-3-methylpyrazolin-5-ones (**3**). These compounds formed pharmacophore hybrids by combining pyrazolinone/pyrazole and benzothiazole moieties. Following this, a screening study was carried out to ascertain their antibacterial activity.[44] A novel class of chemicals known as 2[4(4-substituted-benzamido-phenylacetamido)phenyl]benzothiazole (**4**) was reported by Bolelli *et al.* After these compounds were synthesized, their antibacterial and antifungal properties were evaluated. Testing them against a variety of microorganisms, including the yeast *Candida albicans* and drug-resistant isolates of *Escherichia coli*, *Staphylococcus aureus*, *Pseudomonas aeruginosa*,

Bacillus subtilis, and *Klebsiella pneumoniae*, was part of the review process.[44,45] After being synthesized, compound **5** was tested against *B. subtilis*, *S. aureus*, and *E. coli* to determine its antibacterial and antifungal properties. It exhibited significant antibacterial activity and, to a lesser extent, antifungal activity against *Candida albicans* and *Aspergillus niger*. It demonstrated the strongest antibacterial properties, outperforming the other compounds in its ability to combat every type of bacterium that was tested.[46] Compound **6** demonstrated a rather low degree of effectiveness in combating the examined strains of bacteria. This indicates that further research is necessary to develop more potent molecules, and the antibacterial activities of benzothiazole derivatives against bacterial and fungal illnesses are an ongoing field of study.[47]

4.4.2 Anticancer Activity

Benzothiazoles have drawn a lot of attention from researchers due to their potential as anticancer medications. They have been studied for their capacity to inhibit tumor angiogenesis, initiate apoptosis, and halt the growth of cancer cells. The derivatives of benzothiazoles have demonstrated potential in the treatment of several malignancies, such as lung, colon, and breast tumors (Figure 4.13).

Kumbhare *et al.* reported a method for the synthesis of benzothiazolylthiocarbamides (**7**) using catalytic amounts of 4-dimethylaminopyridine. Subsequently, they conducted a chemoselective oxidative cyclization using 1,3-di-*n*-butylimidazolium tribromide, which resulted in the formation of *N*-bisbenzothiazole. These synthesized complexes were then tested against mouse melanoma and human monocytic cell lines. Intriguingly, all the compounds exhibited varying degrees of anticancer activity.[48] A series of benzothiazole derivatives with the 2-thiourea substitution were synthesized, and it was discovered that they exhibited the capability to combat cancer in HeLa and MCF-7 cell lines. Compound **8**, among these derivatives, demonstrated anticancer properties in these cell lines.[42,49] Havrylyuk *et al.* introduced a series of innovative 4-thiazolidinones (**9**) by incorporating the benzothiazole core. They conducted *in vitro* assessments of the anticancer properties of these newly created compounds with the assistance of the National Center Institute. Notably, their research revealed that two of the synthesized compounds exhibited significant activity against various cancer cell lines, including those associated with melanoma, lung cancer, central nervous system cancer, leukemia, colon cancer, breast cancer, prostate cancer, and ovarian cancer.[50] Kini *et al.* heated *o*-aminophenols to high temperatures and performed a reflux reaction by substituting benzoic acids in an environment of polyphosphoric acid. This process resulted in the synthesis of aryl-substituted benzothiazoles (**10**). They subsequently assessed the potential of these compounds as anticancer drugs by testing

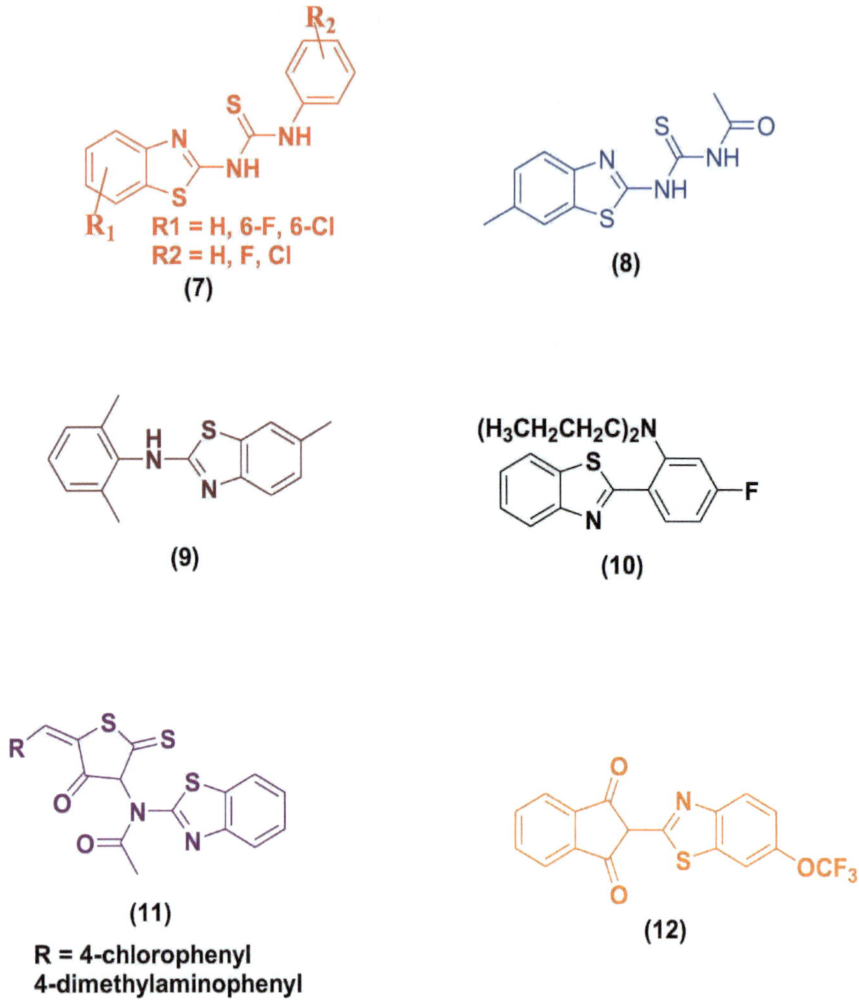

Figure 4.13 Anticancer activity (**7–12**).

them against human cervical cancer cell lines.[51] Noolvi *et al.* conducted an anticancer screening study for the compound 7-chloro-*N*-(2,6-dichlorophenyl)benzothiazol-2-amine (**11**). Their findings demonstrated that this compound exhibited promising results in combating cell lines associated with non-small-cell lung cancer.[52] Jin *et al.* produced γ-aminophosphonates (**12**) that incorporated benzothiazole and a fluorine component using a Mannich-type addition method within an ionic liquid environment, resulting in a good yield and fast reaction time. The novel **12** were subsequently assessed for their potential anticancer properties against A431, PC3, Bcap37, and A375 cell lines *in vitro* using the 3-[4,5-dimethylthiazol-2-yl]-2,5 diphenyl tetrazolium bromide (MTT) assay (Figure 4.14).[53]

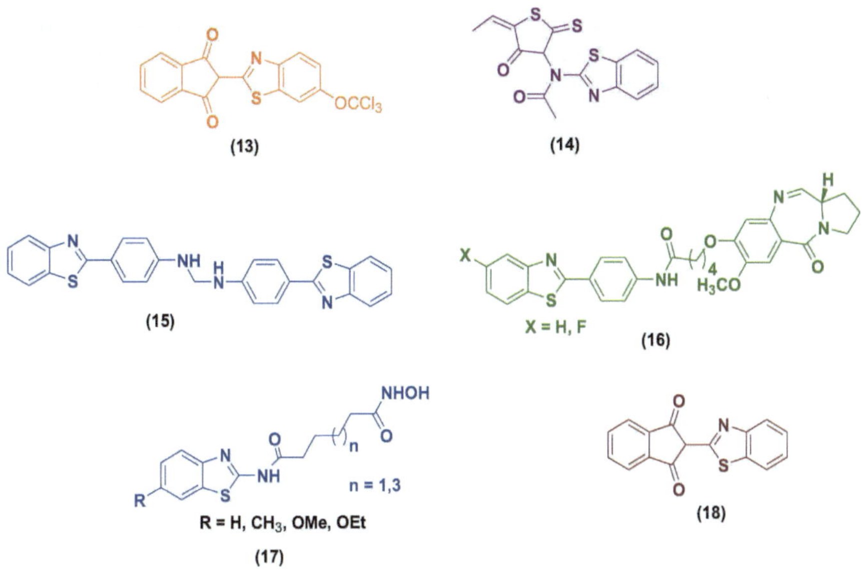

Figure 4.14 Anticancer activity (**13–18**).

Kok *et al.* conducted a study in which they synthesized compounds that included benzothiazole and phthalimide components. These compounds (**13**) were then investigated for their potential anticancer properties in human carcinoma cell lines.[54] Havrylyuk *et al.* conducted an antitumor screening study involving several novel 4-thiazolidinones that featured a benzothiazole component (**14**).[50] Gupta *et al.* reported the derivatives of benzothiazole (**15**) and investigated their anticancer activity. Additionally, an *in silico* pharmacokinetic study suggested that benzothiazole dimers exhibited fewer teratogenic, irritating, and sensitivity properties compared to monomers. According to the quantitative structure–activity relationship (QSAR) study, the ability of benzothiazole derivatives to kill HL-60 cell lines is enhanced by an increase in the number of hydrogen donors.[55] Kamal *et al.* synthesized a series of benzothiazole-linked pyrrolobenzodiazepine conjugates (**16**) with various alkane or alkylamine spacers. Subsequently, they conducted a study to investigate the anticancer activity of these compounds, along with DNA thermal denaturation studies, restriction endonuclease digestion assays, and flow cytometric analysis, particularly in human melanoma cell lines (A375).[56] Oanh *et al.* aimed to recognize new molecules targeting Class-I and Class-II histone deacetylase (HDAC) enzymes. They synthesized two series of benzothiazole-containing analogues of suberoylanilide hydroxamic acid (SAHA) (**17**). The study revealed that several compounds, featuring a 6C-bridge linking the benzothiazole moiety and hydroxamic functional groups, exhibited significant inhibition against HDAC3 and HDAC4 at concentrations as low as μg ml^{-1}. Furthermore, these compounds demonstrated potent cytotoxicity against five cancer cell lines, with an average IC_{50} value as low as 0.81 μg ml^{-1}, making them almost as effective as SAHA.[57] Kok *et al.* described a "one-pot" condensation reaction for synthesizing a phthalic imide derivative (**18**) that includes benzothiazole-containing phthalimide. This compound showed *in vitro* cytotoxic potential against human cancer cell lines. Additionally, the researchers demonstrated that the induced apoptosis on cancer cells by this novel compound involves both caspase-dependent and caspase-independent pathways.[54]

4.4.3 Anti-inflammatory Effects

Benzothiazole compounds can indeed play a role in modifying inflammatory pathways, making them promising candidates for developing anti-inflammatory medications. By targeting specific pathways involved in inflammation, these substances have the potential to be used in the treatment of various inflammatory conditions, including inflammatory bowel disease and rheumatoid arthritis. It is fascinating how understanding the molecular mechanisms of these chemicals opens up possibilities for therapeutic applications (Figure 4.15).

Shafi *et al.* used a click chemistry approach for synthesizing a targeted library of novel bis-heterocycles incorporating 2-mercaptobenzothiazole (**19**). This methodical exploration sheds light on the potential anti-inflammatory properties of these novel bis-heterocycles, providing valuable insights into future therapeutic applications.[58] Among these compounds, compound **20** was found to exhibit a potent anti-inflammatory effect. This suggests that these specifically designed benzothiazoles could hold

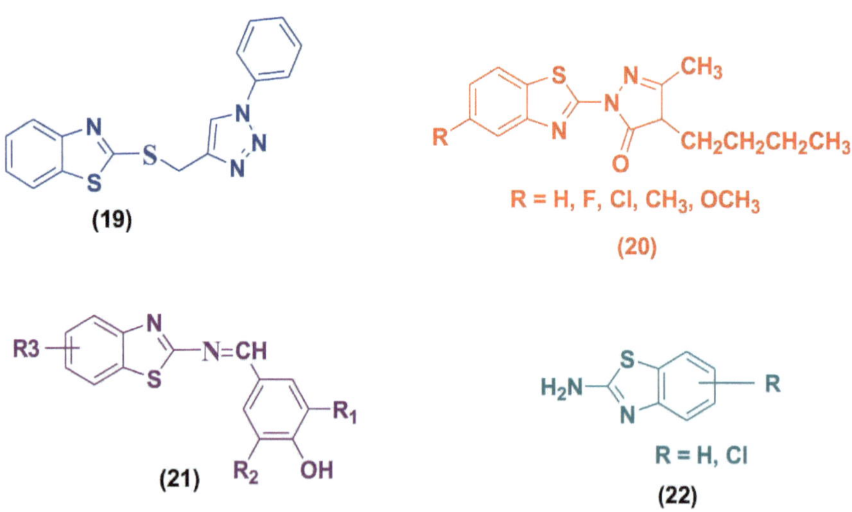

Figure 4.15 Anticancer activity (**19**–**22**).

promise as potential candidates for anti-inflammatory treatments.[59,60] A number of novel thiazolyl/thiazolinyl/benzothiazolyl Schiff bases (**21**) were synthesized by Geronikaki *et al.* These substances were observed to function as lipoxygenase inhibitors, suggesting a possible effect on psoriasis and/or inflammation. The exploration of these novel Schiff bases opens up avenues for developing compounds that could be valuable in addressing conditions related to inflammation and psoriasis.[61] The synthesis of anti-inflammatory drugs based on benzothiazoles led to the testing of a novel 2-amino-benzothiazole derivative, specifically compound **22**, for its anti-inflammatory properties. The study revealed a significant increase in anti-inflammatory activity when 2-aminobenzothiazole was replaced at positions 5 or 4, incorporating electron-withdrawing groups such as Cl, OCH_3, and NO_2. This suggests that specific modifications to the benzothiazole structure can enhance its effectiveness in reducing inflammation, providing valuable insights into drug development in this area.[62]

4.4.4 Antiviral Activity

Benzothiazoles indeed showcase effectiveness against various viral infections, including hepatitis B and C viruses, herpes simplex virus, and HIV. Their ability to hinder the spread of these viruses positions them as promising candidates for developing antiviral medications. By understanding and harnessing the properties of benzothiazoles, researchers can potentially create new and improved models for combating viral infections, providing hope for advancements in antiviral drug development (Figure 4.16).

The discovery of novel benzothiazole sulfonamides, specifically compounds **23** and **24**, revealed their ability to prevent HIV-1 peptidase with an impressive IC_{50} value within the range of 2–3 nM. This innovative

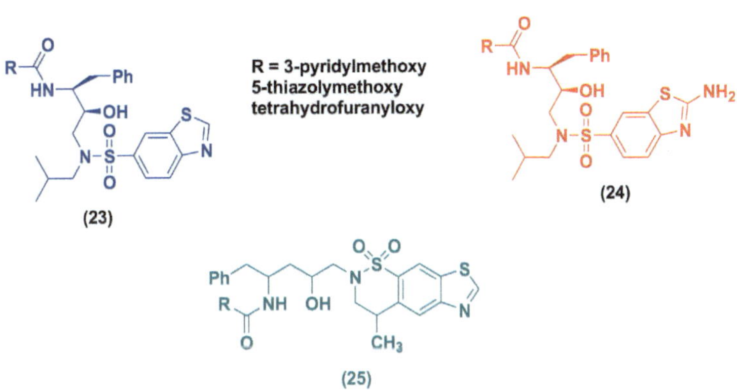

Figure 4.16 Antiviral activity (**23**–**25**).

approach holds promise for developing potent antiviral agents targeting HIV-1 protease.[63] Nagarajan *et al.* demonstrated that replacing the *t*-butylurea moiety with benzothiazolesulfonamide (**25**) resulted in peptidase inhibition through enhanced antiviral activity and potency. This improvement was attributed to the incorporation of various isosteres, counting hydroxyethyl urea at the protease cleavage site. Nearly all of these complexes exhibited decent half-life and bioavailability in mice. Additionally, their studies led to the development of an effective synthetic method for benzothiazole-6-sulfonic acid, achieved through a two-step process initiated from sulfanilamide. This research not only expands our understanding of antiviral agents but also contributes to the efficient synthesis of related heterocyclic compounds.[64]

4.4.5 Antidiabetic Activity

It is fascinating how benzothiazole derivatives are not limited to their roles in antiviral and anti-inflammatory applications. They also exhibit potential in the realm of antidiabetic properties. Their ability to influence glucose metabolism and insulin sensitivity opens up new possibilities for the development of treatments for diabetes and related metabolic disorders. The versatility of benzothiazole compounds in addressing various health conditions underscores their significance in medical research and drug development (Figure 4.17).

The increasing need for effective therapies in achieving optimal glycemic control in diabetes management has led to the synthesis and evaluation of *N*-(6-substituted-1,3-benzothiazol-2-yl)benzenesulfonamide (**26**) derivatives. Moreno-Díaz *et al.* conducted an assessment of the *in vivo* antidiabetic activity of these derivatives in a non-insulin-dependent diabetes mellitus rat model. Additionally, the compounds were evaluated for their impact on 11-HSD1 and PTP-1B enzymes. This study investigated the effects of these

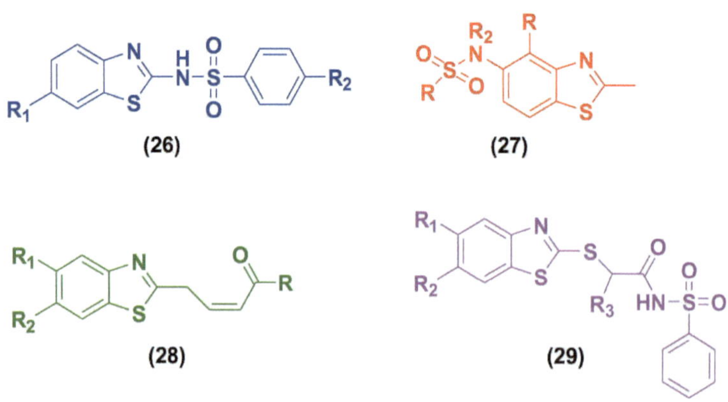

Figure 4.17 Antidiabetic activity (**26–29**).

benzothiazole derivatives on glycemic control and pertinent enzymes to aid in the development of novel diabetes therapeutics.[65] Several new benzothiazole derivatives (**27**) were discovered and synthesized, as reported by Su *et al.* These substances were examined for their ability to inhibit 11-HSD1 of human hepatic microsomes, which is a crucial enzyme. The radioimmunoassay (RIA) method was utilized to measure the inhibitory actions. This study adds to our knowledge of the possible interactions between these benzothiazole derivatives and 11-HSD1, which may help design therapeutic approaches for disorders involving hepatic metabolism.[66] A unique sequence of relieved (*E*)-3-(benzo[*d*]thiazol-2-ylamino)phenylprop-2-en-1-ones (**28**) was discovered by Patil *et al.* Particular attention was paid to the antidiabetic potential of these compounds. The study concentrated on the selective inhibitors of 11β-hydroxysteroid dehydrogenase type 1 (11β-HSD1), which exhibit great promise as therapeutic agents for metabolic disorders, such as type 2 diabetes and obesity. This research contributes to the exploration of potential therapeutic options for managing metabolic disorders through the development of specific inhibitors targeting 11β-HSD1.[67] Navarrete-Vazquez *et al.* conducted research where they synthesized ethyl-2-(6-substituted-benzo[*d*]thiazol-2-ylamino)-2-oxoacetate (**29**) *via* one step. The ability of these substances to inhibit protein tyrosine phosphatase1B (PTP1B) *in vitro* was assessed. Additionally, the *in vivo* hypoglycemic action of the compounds was evaluated, and the results showed a significant reduction in plasma glucose concentration in both the oral glucose tolerance test and the acute normoglycemic model. This effect was comparable to the hypoglycemic drug glibenclamide, suggesting the potential therapeutic relevance of these derivatives in managing glucose levels.[68] A set of phenylsulfonamides (**29**) were synthesized by Ammazzalorso *et al.*, and *in vitro* evaluation was conducted against the agonistic activity of GW7647. A dose-dependent antagonistic profile was exhibited by the best drugs, which all showed an inhibitory effect on peroxisome proliferator-activated receptor α (PPARα) activation. This study illuminates the potential of these phenylsulfonamides as substances that may alter PPARα activity, indicating their use in the context of PPARα activation-related therapeutic interventions.[69]

4.4.6 Antioxidant Activity

A thorough investigation of the possible uses of the compounds is demonstrated by the synthesis and assessment of a range of substances for their antioxidant and anticancer properties, including benzofuranhydrazones, arylbenzimidazoles, benzimidazolidrazones, indole-hydrazones, and benzothiazole analogues. This method gives researchers a comprehensive grasp of the bioactivity of compounds by enabling them to evaluate the compounds across various structural variants and chemical classes. The study of antioxidant and antitumor properties is important because it clarifies the potential of chemicals to reduce oxidative stress and fight cancer cells. This comprehensive assessment approach offers significant contributions to the discovery of molecules with anticancer and antioxidant therapeutic potential. Seeing such thorough research that could result in the identification of novel substances for use in future medical procedures is thrilling.[70] With its potent antioxidant capabilities and advantageous photoprotective or anticancer activity profiles, compound **30** seems to be very special. Its effectiveness against these fungal diseases was demonstrated by its IC_{50} values of 2 g ml^{-1} against the dermatophytes under study. Furthermore, compound **30** demonstrated excellent *in vitro* antioxidant activity assessments utilizing the fluorescence recovery after photobleaching (FRAP) and 2,2-diphenyl-1-picrylhydrazyl (DPPH) tests. Interestingly, it had an IC_{50} value of 9.7 μM against melanoma cells in humans, suggesting that it may have anticancer properties. These findings underscore the versatility and promising therapeutic characteristics of the compound, marking it as a noteworthy candidate for further exploration and development.[70] Tzanova *et al.* showcased an efficient synthesis of three novel benzophenones containing a 1,3-thiazol moiety (**31**). These compounds were subjected to *in vitro* evaluation of their antioxidant activity and tested in three different cell lines: H9c2 cardiomyoblastic cells, cancerous-MCF7, and noncancerous-hTERTHME1 mammary cells. These findings highlight the potential of these benzophenones as antioxidants with possible applications in various cellular contexts.[27,71] Cressier *et al.* contributed to the medicinal field by presenting the novel compounds obtained from thiadiazoles and benzothiazoles (**32**). Interestingly, they tested the radioprotective efficacy of these chemicals in mice. According to the study, several of these chemicals demonstrated potential as effective radioprotectors. This study broadens our knowledge of the radioprotective qualities of thiadiazole and benzothiazole derivatives, creating opportunities for more investigation and development of substances with possible uses in radioprotection (Figure 4.18).[72]

4.4.7 Antitubercular Activity

It is encouraging to learn that compound **33** has strong *in vitro* antitubercular activity against *Mycobacterium tuberculosis*. It implies that this substance has the potential to be investigated and developed further in relation to the treatment of tuberculosis. To find successful therapies for infectious diseases, it is essential to evaluate antimicrobial properties. The remarkable efficiency of compound **33** indicates that it may be a useful weapon to fight against tuberculosis.[73] Pereira *et al.*'s study on the biological tests, Discrete Fourier Transform (DFT) modeling, synthesis, and characterization of two novel metal complexes of 2-(2-thienyl)benzothiazole (BTT, **34**) appears to have a significant influence. The mention of the good efficacy of these metal complexes against *Mycobacterium tuberculosis* raises questions

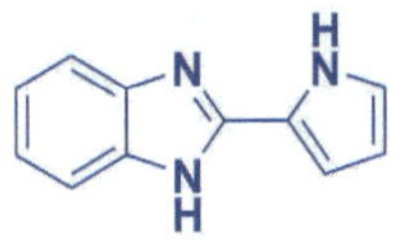

(30)

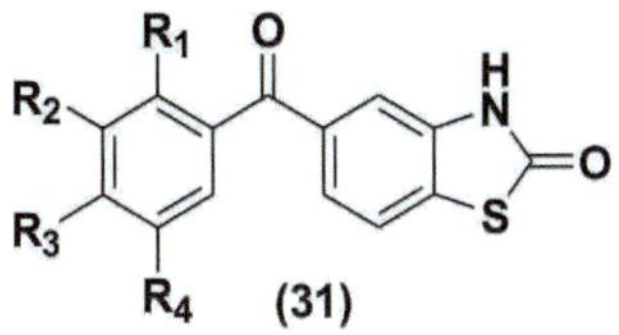

(31)

R1=R2=R4=H; R3= OH
R1=R4=H; R2=R3=OH
R1=R4=OH; R2=R3=H

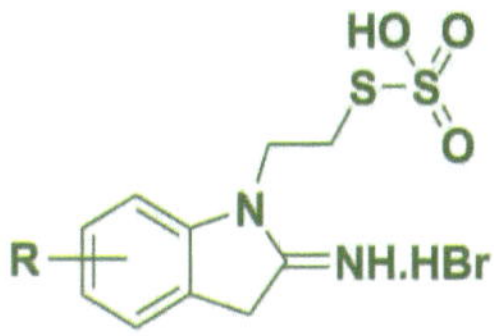

R = H, CH_3, OCH_3

(32)

Figure 4.18 Antioxidant activity (**30**–**32**).

about their potential for use in antituberculosis applications. For these complexes to be used in therapeutic or antimicrobial applications, it is essential to comprehend their composition, characteristics, and biological activities. These metal complexes are thoroughly examined through a mix of theoretical simulations, biological assays, and experimental characterization. This opens up new avenues for investigation and possible applications in the field of tuberculosis treatment. The way that multidisciplinary techniques are advancing the chemical and biological sciences is exciting.[74] Benzothiazoles (**35**) showed the most promising activity in antitubercular screens out of all the compounds examined. This discovery implies that compound **35** would be a good option for more research and development in the search for strong tuberculosis therapies. The discovery of compounds with notable antitubercular activity is crucial in addressing the challenges posed by tuberculosis infections. Telvekar *et al.* synthesized 4-(aryloxy)benzaldehyde (**36**). The utilization of a molecular hybridization approach in their synthesis adds a creative dimension to the design of these compounds.[75] This demonstrates the potential of these derivatives as candidates for further exploration and development in the context of tuberculosis treatment. The molecular hybridization strategy appears to be a successful approach for designing compounds with significant antitubercular activity.[76] The recent patent update revealing the antimycobacterial activity of derivatives of adamantanyl benzothiazoles is noteworthy. Among these derivatives, compound **37** stands out with a remarkable minimum inhibitory concentration (MIC90) of 0.03 g ml^{-1} against H37-Rv. The discovery that compound **37** is isoniazid-equivalent in terms of antimycobacterial activity is particularly significant. These findings underscore the potential of adamantanyl benzothiazoles, especially compound **37**,

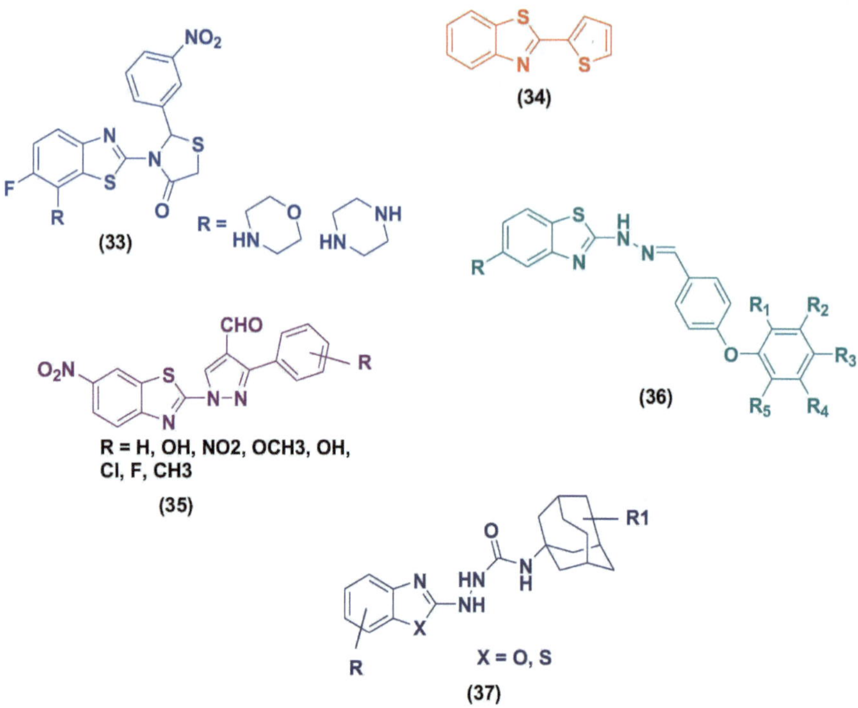

Figure 4.19 Antioxidant activity (**33–37**).

as candidates for further development in the realm of antimycobacterial drugs. Achieving isoniazid-equivalent efficacy is a substantial accomplishment, as isoniazid is a widely used and important drug in the treatment of tuberculosis. The identification of compounds with comparable or superior activity opens avenues for new therapeutic options and addresses challenges associated with antimicrobial resistance (Figure 4.19).[77,78]

4.5 Emerging Trends and Future Prospects

It is exciting to witness the ongoing evolution in the field of benzothiazoles. The emphasis on developing more sustainable synthesis methods aligns with the growing awareness of environmentally friendly practices in research and industry. Exploring bio-based approaches adds another layer of innovation, tapping into the potential of biological sources for benzothiazole synthesis. The continuous discovery of novel drug candidates in this class of compounds showcases their versatility and potential for addressing various health challenges. The expanding application in materials science, with a focus on designing innovative materials, not only broadens the scope of benzothiazoles but also contributes to advancements in optoelectronics, sensors, and polymers. It is clear that benzothiazoles are poised to play a crucial role in shaping the future of both pharmaceuticals and materials science.

4.6 Conclusion

The literature on benzothiazoles indeed showcases the dynamic exploration of their synthetic methods and the diverse biological functions they exhibit. Researchers are adopting innovative approaches, such as utilizing biocatalysts, to improve yields and optimize reaction conditions. This not only contributes to the efficiency of the synthetic processes but also aligns with the growing emphasis on sustainable and bio-friendly methodologies. The wide range of biological activities of the benzothiazole core is particularly fascinating. From anticancer and antibacterial properties to anti-inflammatory, antidepressant, and antidiabetic effects, it reflects the versatility of benzothiazoles in addressing various health challenges. This diversity in biological functions positions benzothiazoles as promising candidates for drug development, offering a multifaceted approach to combatting different diseases. The continuous exploration of benzothiazoles in both synthesis and biological applications highlights their ongoing significance in scientific research.

The promising performance of certain benzothiazole derivatives, surpassing or matching the activity of existing standard medications, bodes well for their potential as therapeutic agents. This suggests that benzothiazole-derived products could indeed become the foundation for upcoming treatments in the market. The versatility and efficacy of benzothiazoles across various biological activities, as seen in their diverse applications, further support the notion that they hold substantial promise for the development of novel and effective therapeutic interventions. It is an exciting prospect to anticipate the emergence of benzothiazole-based treatments that could make a significant impact on healthcare and medical advancements.

References

1. K. Nowicki, P. Pacholak and S. Luliński, *Molecules*, 2021, **26**, 5464.
2. D. D. Roberts and M. G. McLaughlin, *Adv. Synth. Catal.*, 2022, **364**, 2307–2332.
3. L. Guo, J. Tan, S. Kaya, S. Leng, Q. Li and F. Zhang, *J. Colloid Interface Sci.*, 2020, **570**, 116–124.
4. S. L. Suib, J. Přech, E. Szaniawska and J. Čejka, *Chem. Rev.*, 2023, **123**, 877–917.
5. A. Hinz, J. Bresien, F. Breher and A. Schulz, *Chem. Rev.*, 2023, **123**, 10468–10526.
6. A. Abdullah Al Awadh, *Saudi Pharm. J.*, 2023, **31**, 101698.

7. M. Oguz, S. Erdemir and S. Malkondu, *Anal. Chim. Acta.*, 2022, **1227**, 340320.
8. S. Das, H. K. Indurthi, P. Asati, P. Saha and D. K. Sharma, *Dyes Pigm.*, 2022, **199**, 110074.
9. S. A. L. Shaikh, A. C. Mendhe, D. N. Nadimetla, M. R. Biradar, P. Vijayanand, A. L. Puyad, B. R. Sankpal, S. V. Bhosale and S. V. Bhosale, *J. Photochem. Photobiol., A*, 2024, **447**, 115220.
10. K. Shaji, R. Selva Kumar, J. Jose, R. Bhaskar, V. Vetriarasu, S. G. Bhat and S. K. Ashok Kumar, *J. Photochem. Photobiol., A*, 2023, **434**, 114220.
11. Y. Huang, Y. Li, Y. Li, K. Zhong and L. Tang, *New J. Chem.*, 2023, **47**, 6916–6923.
12. S. Muthusamy, K. Rajalakshmi, Y. Seo, M. Xie, P. Kannan, R. S. Lodi, J.-W. Song and Y. Xu, *Dyes Pigm.*, 2022, **207**, 110709.
13. S. Enbanathan, S. Munusamy, D. Jothi, S. Manojkumar, S. Manickam and S. K. Iyer, *RSC Adv.*, 2022, **12**, 27839–27845.
14. D. Jothi, S. Munusamy, S. Manoj Kumar, S. Enbanathan and S. Kulathu Iyer, *RSC Adv.*, 2022, **12**, 8570–8577.
15. Z. Li, J. Wang, X. Peng, Y. Chen and M. Geng, *Inorg. Chim. Acta.*, 2023, **548**, 121378.
16. R. Ali and N. Siddiqui, *J. Chem.*, 2013, **2013**, 1–12.
17. J. E. Mendieta-Wejebe, M. C. Rosales-Hernández, I. I. Padilla-Martínez, E. V. García-Báez and A. Cruz, *Int. J. Mol. Sci.*, 2023, **24**, 9488.
18. Y. Hsu, W. Shi, Y. Tseng, T. Chiu, M. Cheng and K. Chen, *J. Heterocycl. Chem.*, 2023, **60**, 2036–2045.
19. H. Elkady, K. El-Adl, H. Sakr, A. S. Abdelraheem, S. I. Eissa and M. A. El-Zahabi, *Arch. Pharm.*, 2023, **356**, 2300097.
20. J. Zhou, R. Zhao, H. Zhou, S. Yang, F. Tao, Y. Xie, H. Wang and J. Yun, *Front. Pharmacol.*, 2023, **14**, 1196158.
21. B. Sever and H. Ciftci, *Mol. Cell Biochem.*, 2023, **478**, 1099–1108.
22. K. Gopichand, V. Mahipal, N. Nageswara Rao, A. Majeed Ganai and P. Venkateswar Rao, *Results Chem.*, 2023, **5**, 100868.
23. L. Gallego-Yerga, V. Ceña and R. Peláez, *Pharmaceutics*, 2023, **15**, 1698.
24. V. S. K. Cheerala, A. Akhir, D. Saxena, R. Maitra, S. Chopra and S. C. Neelakantan, *RSC Med. Chem.*, 2023, **14**, 1712–1721.
25. A. Zubenko, V. Kartsev, A. Petrou, A. Geronikaki, M. Ivanov, J. Glamočlija, M. Soković, L. Divaeva, A. Morkovnik and A. Klimenko, *Antibiotics*, 2022, **11**, 1654.
26. K. L. F. Hilton, A. A. Karamalegkos, N. Allen, L. Gwynne, B. Streather, L. J. White, K. B. Baker, S. A. Henry, G. T. Williams, H. J. Shepherd, M. Shepherd, C. K. Hind, M. J. Sutton, T. A. Jenkins, D. P. Mulvihill, J. M. A. Tullet, M. Ezcurra and J. R. Hiscock, *J. Mater. Chem. B*, 2023, **11**, 3958–3968.
27. E. N. Djuidje, R. Barbari, A. Baldisserotto, E. Durini, S. Sciabica, J. Balzarini, S. Liekens, S. Vertuani and S. Manfredini, *Antioxidants*, 2022, **11**, 407.

28. R. C. Tweit, *J. Heterocycl. Chem.*, 1970, **7**, 687–688.
29. J. Štetinová, *Molecules*, 1997, **1**, 251–254.
30. D. S. Dogruer, S. Ünlü, M. F. Şahin and E. Yqilada, *Il Farmaco*, 1998, **53**, 80–84.
31. G. Trapani, M. Franco, A. Latrofa, A. Carotti, G. Gerichi, M. Serra, G. Biggio and G. Liso, *Eur. J. Med. Chem.*, 1996, **31**, 575–587.
32. G. Evindar and R. A. Batey, *Org. Lett.*, 2003, **5**, 133–136.
33. D. Matiadis and M. Sagnou, *Int. J. Mol. Sci.*, 2020, **21**, 5507.
34. Y. Murti and D. Pathak, *J. Pharm. Res.*, 2008, **7**, 153–155.
35. A. D. Jordan, C. Luo and A. B. Reitz, *J. Org. Chem.*, 2003, **68**, 8693–8696.
36. Y. Heo, Y. S. Song, B. T. Kim and J.-N. Heo, *Tetrahedron Lett.*, 2006, **47**, 3091–3094.
37. H.-S. Dong, B. Quan, E.-F. Chai and X.-R. Mao, *J. Mol. Struct.*, 2002, **608**, 41–47.
38. A. Trapani, A. Catalano, A. Carocci, A. Carrieri, A. Mercurio, A. Rosato, D. Mandracchia, G. Tripodo, B. Immacolata, P. Schiavone, C. Franchini, E. Mesto, E. Schingaro and F. Corbo, *Carbohydr. Polym.*, 2020, **207**, 720.
39. A. Samide, P. Rotaru, C. Ionescu, B. Tutunaru, A. Moanţă and V. Barragan-Montero, *J. Therm. Anal. Calorim.*, 2014, **118**, 651–659.
40. F. Yan, J. Sun, Y. Zang, Z. Sun, H. Zhang and X. Wang, *J. Iran. Chem. Soc.*, 2020, **17**, 3179–3203.
41. M.-H. Wang, Z.-C. Tan, X.-H. Sun, H.-T. Zhang, B.-P. Liu, L.-X. Sun and T. Zhang, *J. Chem. Eng. Data*, 2005, **50**, 270–273.
42. I. Hutchinson, M.-S. Chua, H. L. Browne, V. Trapani, T. D. Bradshaw, A. D. Westwell and M. F. G. Stevens, *J. Med. Chem.*, 2001, **44**, 1446–1455.
43. P. K. Sahu, P. K. Sahu, J. Lal, D. Thavaselvam and D. D. Agarwal, *Med. Chem. Res.*, 2012, **21**, 3826–3834.
44. M. Amir, S. A. Javed and M. Zaheen Hassan, *Med. Chem. Res.*, 2012, **21**, 1261–1270.
45. K. Bolelli, I. Yalcin, T. Ertan-Bolelli, S. Özgen, F. Kaynak-Onurdag, I. Yildiz and E. Aki, *Med. Chem. Res.*, 2012, **21**, 3818–3825.
46. M. A. Morsy, E. M. Ali, M. Kandeel, K. N. Venugopala, A. B. Nair, K. Greish and M. El-Daly, *Antibiotics*, 2020, **9**, 221.
47. S. K. Maddili, Z.-Z. Li, V. K. Kannekanti, R. R. Y. Bheemanaboina, B. Tuniki, V. K. R. Tangadanchu and C.-H. Zhou, *Bioorg. Med. Chem. Lett.*, 2018, **28**, 2426–2431.
48. R. M. Kumbhare, T. Dadmal, U. Kosurkar, V. Sridhar and J. V. Rao, *Bioorg. Med. Chem. Lett.*, 2012, **22**, 453–455.
49. B. Şahin, A. S. Yağlıoğlu and M. Ceylan, *Org. Commun.*, 2016, **9**, 65–72.
50. D. Havrylyuk, L. Mosula, B. Zimenkovsky, O. Vasylenko, A. Gzella and R. Lesyk, *Eur. J. Med. Chem.*, 2010, **45**, 5012–5021.
51. S. Kini, S. Swain and A. Gandhi, *Indian J. Pharm. Sci.*, 2007, **69**, 46.
52. M. N. Noolvi, H. M. Patel and M. Kaur, *Eur. J. Med. Chem.*, 2012, **54**, 447–462.

53. L. Jin, B. Song, G. Zhang, R. Xu, S. Zhang, X. Gao, D. Hu and S. Yang, *Bioorg. Med. Chem. Lett.*, 2006, **16**, 1537–1543.
54. S. H. L. Kok, R. Gambari, C. H. Chui, M. C. W. Yuen, E. Lin, R. S. M. Wong, F. Y. Lau, G. Y. M. Cheng, W. S. Lam and S. H. Chan, *Bioorg. Med. Chem.*, 2008, **16**, 3626–3631.
55. S. Dutta Gupta, N. H. Moorthy and U. Sanyal, *Int. J. Pharm. Pharm. Sci.*, 2010, **2**, 57–62.
56. A. Kamal, K. S. Reddy, M. N. A. Khan, R. V. C. R. N. C. Shetti, M. J. Ramaiah, S. N. C. V. L. Pushpavalli, C. Srinivas, M. Pal-Bhadra, M. Chourasia, G. N. Sastry, A. Juvekar, S. Zingde and M. Barkume, *Bioorg. Med. Chem.*, 2010, **18**, 4747–4761.
57. D. T. K. Oanh, H. V. Hai, S. H. Park, H.-J. Kim, B.-W. Han, H.-S. Kim, J.-T. Hong, S.-B. Han, V. T. M. Hue and N.-H. Nam, *Bioorg. Med. Chem. Lett.*, 2011, **21**, 7509–7512.
58. S. Shafi, M. M. Alam, N. Mulakayala, C. Mulakayala, G. Vanaja, A. M. Kalle, R. Pallu and M. S. Alam, *Eur. J. Med. Chem.*, 2012, **49**, 324–333.
59. K. Arun, A. K. Shakyaa and H. H. Siddiquib, *Indian J. Heterocycl. Chem.*, 2016, **25**, 243–249.
60. N. Kumar, A. Chowdhary, O. Gudaparthi, N. Patel, S. Soni and P. Sharma, *Indian J. Chem. - B Org. Med. Chem.*, 2014, **53**, 1269–1274.
61. A. Geronikaki, D. Hadjipavlou-Litina and M. Amourgianou, *Il Farmaco*, 2003, **58**, 489–495.
62. P. Venkatesh and S. N. Pandeya, Synthesis, characterisation and anti-inflammatory activity of some 2-amino benzothiazole derivatives, *Int. J. ChemTech Res.*, 2009, **1**(4), 1354–1358.
63. Nagaraj, K. C. Chaluvaraju, M. S. Niranjan, and S. Kiran, 1, 3, 4-Oxadiazole: a potent drug candidate with various pharmacological activities, *Int. J. Pharm. Pharm. Sci.*, 2011, **3**(3), 9–16.
64. A. S. Nagarajan, S. Kamalraj, J. Muthumary and B. S. Reddy, Synthesis of biologically active benzothiazole substituted thiazolidinone derivatives via cyclization of unsymmetrical imines, *Chemistry*, 2009, DOI: 10. 1002/CHIN.201011128.
65. H. Moreno-Díaz, R. Villalobos-Molina, R. Ortiz-Andrade, D. Díaz-Coutiño, J. L. Medina-Franco, S. P. Webster, M. Binnie, S. Estrada-Soto, M. Ibarra-Barajas, I. León-Rivera and G. Navarrete-Vázquez, *Bioorg. Med. Chem. Lett.*, 2008, **18**, 2871–2877.
66. X. Su, N. Vicker, M. Trusselle, H. Halem, M. D. Culler and B. V. L. Potter, *Mol. Cell. Endocrinol.*, 2009, **301**, 169–173.
67. V. S. Patil, K. P. Nandre, S. Ghosh, V. J. Rao, B. A. Chopade, B. Sridhar and S. V. Bhosale, *Eur. J. Med. Chem.*, 2013, **59**, 304–309.
68. G. Navarrete-Vazquez, M. Ramírez-Martínez, S. Estrada-Soto, C. Nava-Zuazo, P. Paoli, G. Camici, J. Escalante-García, J. L. Medina-Franco, F. López-Vallejo and R. Ortiz-Andrade, *Eur. J. Med. Chem.*, 2012, **53**, 346–355.

69. A. Ammazzalorso, A. Giancristofaro, A. D'Angelo, B. D. Filippis, M. Fantacuzzi, L. Giampietro, C. Maccallini and R. Amoroso, *Bioorg. Med. Chem. Lett.*, 2011, **21**, 4869–4872.
70. E. N. Djuidje, R. Barbari, A. Baldisserotto, E. Durini, S. Sciabica, J. Balzarini, S. Liekens, S. Vertuani and S. Manfredini, *Antioxidants*, 2022, **11**(2), 407.
71. T. Tzanova, M. Gerova, O. Petrov, M. Karaivanova and D. Bagrel, *Eur. J. Med. Chem.*, 2009, **44**, 2724–2730.
72. D. Cressier, C. Prouillac, P. Hernandez, C. Amourette, M. Diserbo, C. Lion and G. Rima, *Bioorg. Med. Chem.*, 2009, **17**, 5275–5284.
73. B. S. Sathe, E. Jaychandran, V. A. Jagtap and G. M. Sreenivasa, *Int. J. Pharma Bio Sci.*, 2010, **1**(4), 495–500.
74. G. A. Pereira, A. C. Massabni, E. E. Castellano, L. A. S. Costa, C. Q. F. Leite, F. R. Pavan and A. Cuin, *Polyhedron*, 2012, **38**(1), 291–296.
75. Z. Wang, X.-H. Shi, J. Wang, T. Zhou, Y.-Z. Xu, T.-T. Huang, Y.-F. Li, Y.-L. Zhao, L. Yang, S.-Y. Yang, L.-T. Yu and Y.-Q. Wei, *Bioorg. Med. Chem. Lett.*, 2011, **21**, 1097–1101.
76. V. N. Telvekar, V. K. Bairwa, K. Satardekar and A. Bellubi, *Bioorg. Med. Chem. Lett.*, 2012, **22**, 649–652.
77. A. Kamal, M. A. H. Syed and S. M. Mohammed, *Expert Opin. Ther. Pat.*, 2015, **25**, 335–349.
78. R. Yadav, D. Meena, K. Singh, R. Tyagi, Y. Yadav and R. Sagar, *RSC Adv.*, 2023, **13**, 21890–21925.

CHAPTER 5

Synthesis, Properties, and Biological Applications of 2,4-Thiazolidinediones

MUBARAK H. SHAIKH*[a], BAPURAO B. SHINGATE[b], SACHIN P. KUNDE[c], HEMANTKUMAR N. AKOLKAR[d], VIJAY M. KHEDKAR[e] AND DATTATRAYA N. PANSARE[f]

[a]Department of Chemistry, Radhabai Kale Mahila Mahavidyalaya, Ahmednagar, Maharashtra, 414001, India; [b]Department of Chemistry, Dr. Babasaheb Ambedkar Marathwada University, Aurangabad, Maharashtra, 431004, India; [c]Department of Chemistry, R. B. Narayanrao Borawake College, Shrirampur, Ahmednagar, Maharashtra, 413709, India; [d]Department of Chemistry, Abasaheb Marathe Arts and New Commerce, Science College, Rajapur, Ratnagiri, Maharashtra, 416702, India; [e]School of Pharmacy, Vishwakarma University, Pune, Maharashtra, 411048, India; [f]Department of Chemistry, Deogiri College, Station Road, Aurangabad, Maharashtra, 431005, India
*E-mail: mubbarakshaikh01@gmail.com

5.1 Introduction

2,4-Thiazolidinedione (TZD) (**1a**) (Figure 5.1) is a widely valued scaffold used for the synthesis of medicinally active substances. A heterocycle that contains sulphur atoms is a flexible pharmacophore exhibiting a wide variety of pharmacological effects. It has been demonstrated that TZD

S-Heterocycles: Retrospect, Prospects, and Biological Applications
Edited by Shrikaant Kulkarni, Hemantkumar Akolkar, Vijay M. Khedkar and A. K. Haghi

Published by the Royal Society of Chemistry, www.rsc.org

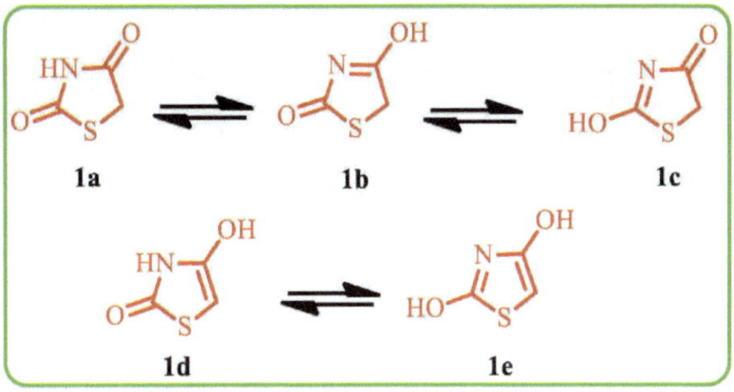

Figure 5.1 Tautomeric structures of TZD.

exhibits biological activities against a wide variety of targets that are of interest to medicinal chemists.[1]

When stored below 30 °C, TZD is a bench-stable, white, crystalline solid with a melting point of approximately 123–125 °C.[2] It is sporadically soluble in organic solvents, such as water, methanol, ethanol, dimethyl sulfoxide, and diethyl ether.[3] Owing to the existence of an α-hydrogen and two carbonyl groups, TZD can occur as a sequence of tautomers (**1a–e**) (Figure 5.1).[2]

5.1.1 Importance of 2,4-Thiazolidinedione Analogues in Medicinal Chemistry

Rosiglitazone **2**, darglitazone **3**, rivoglitazone **4**, pioglitazone **5**, ciglitazone **6**, englitazone **7**, and netoglitazone **8** are some of the insulin-sensitizing medications that belong to the class of TZDs (Figure 5.2).

In addition to the well-established antidiabetic effects,[4] in multiple *in vitro* and *in vivo* studies,[5] TZDs have shown antimicrobial **9**,[6] aldose reductase inhibitory **10**,[7] antimicrobial **11**,[8] anti-inflammatory **12**,[9] oncostatic **13**,[10] anticancer **14**,[11] and tuberculostatic activities **15**[12] supporting their potential contribution to cancer therapy (Figure 5.3).

The members of the nuclear hormone receptor superfamily, which are ligand-activated transcription factors, include peroxisome proliferator-activated receptors (PPARs). Three PPAR isotypes are known to exist: PPARα, PPARβ, and PPARγ. At first, it was believed that PPAR activity was restricted to glucose balance and lipid metabolism. Subsequent research revealed that PPAR activity controls apoptosis, cell differentiation and proliferation, and inflammatory responses.[13] The antagonistic relationship between the pro-inflammatory cytokine activity and PPARα originally revealed the involvement of PPARγ in inflammatory processes. Furthermore, a number of PPARγ agonists suppress macrophage activation.[14] Most of the phar-

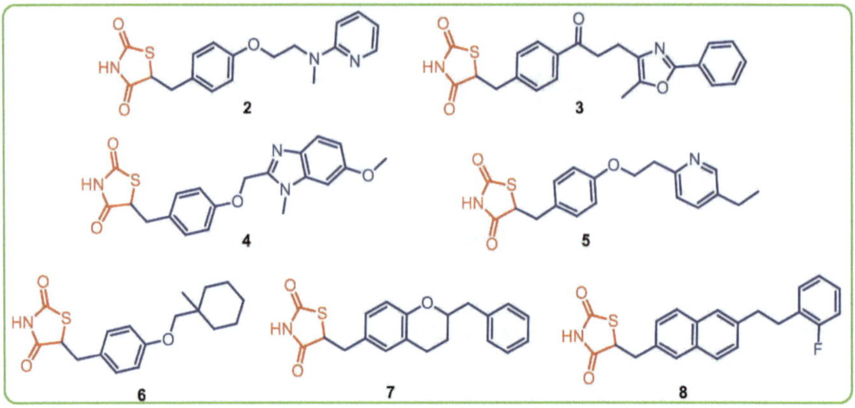

Figure 5.2 Structures of the TZD class of insulin-sensitizing drugs.

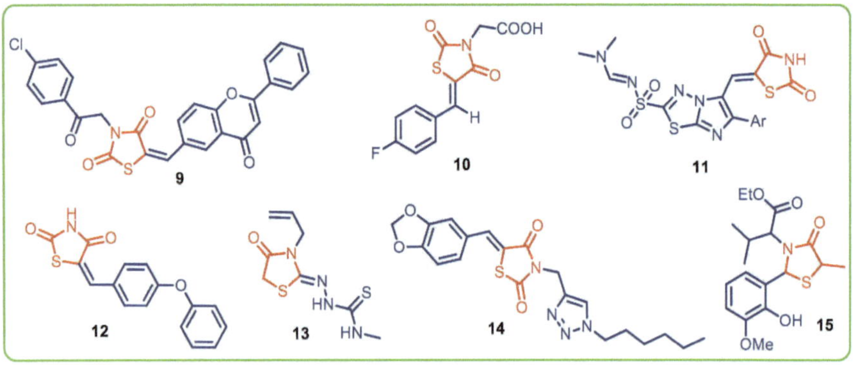

Figure 5.3 Structures of bioactive TZD derivatives.

macological effects, including anti-inflammatory properties, of the two thiazolidines are caused by their shared TZD structure.[15]

5.2 Recent Advances in Synthesis

The study by Kallenberg[16] was the first to report on the synthetic techniques used to produce the TZD core. This process produces appropriate alkyl thiocarbamate (**17**) *in situ* by reacting carbonyl sulphide (**16**) with ammonia using KOH as a base, which then interacts with α-halogenated carboxylic acid (**18**). The produced thiocarbamate was cyclized in an acidic environment to obtain the required TZD (**1a**) with a 90% yield (Scheme 5.1).[16]

The refluxing of α-chloroacetic acid (**18**) and thiourea (**19**) using water as a solvent for an extended duration (~12 h) is a typical method for synthesizing TZD[17] (Scheme 5.2).

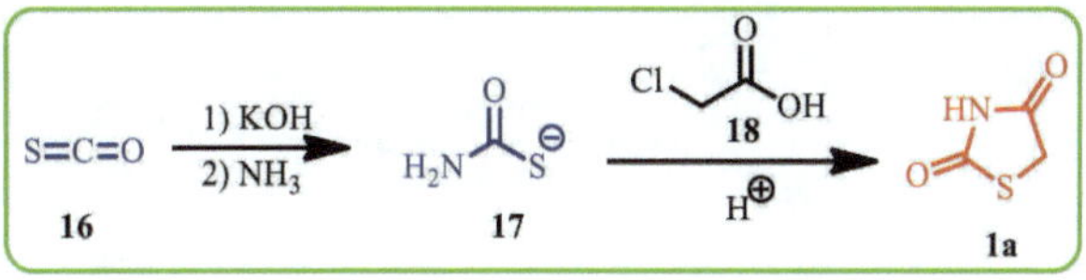

Scheme 5.1 Synthesis of TZD using carbonyl sulphide.

Shaikh *et al.* recently synthesized a small focused library of 1,2,3-triazole-based TZD derivatives with a 90–92% yield utilizing the same technology, which was effectively produced using the click chemistry strategy (Scheme 5.3).[18]

1,1,3,3-Tetramethylguanidinium lactate ([TMG][Lac]) ionic liquid-catalysed synthesis of TZD derivatives **21** with a 91–98% yield using ultrasound was reported by Suresh *et al.*[19] (Scheme 5.4).

Ethyl chloroacetate (**25**) was reacted with thiosemicarbazone (**26**) to produce 2-hydrazino-4-thiazolidinone (**27**) using NaOEt as a base. This reaction is the third widely used synthetic process. The product can then be refluxed with dilute HCl to produce the required TZD (**1a**) (Scheme 5.5).[20]

Compound **25** is also used to synthesize the TZD core utilizing a fourth widely used approach. In this process, **1a** is produced when the reaction between compound **25** and potassium thiocyanate is acidified (Scheme 5.6). Because hazardous hydrogen cyanide (HCN) gas is released as a by-product during this procedure, extreme caution must be exercised.[21]

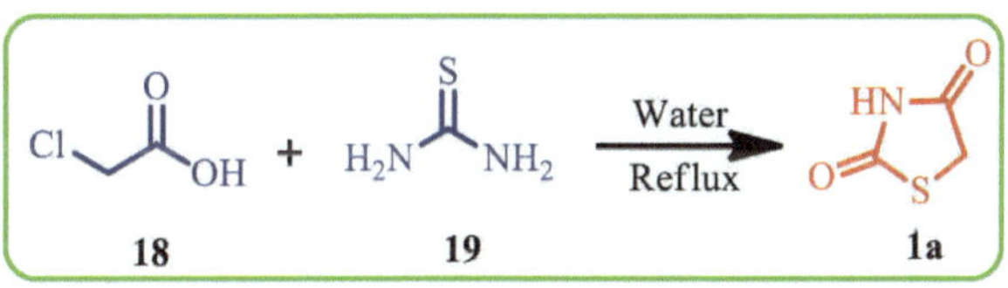

Scheme 5.2 Synthesis of TZD using thiourea.

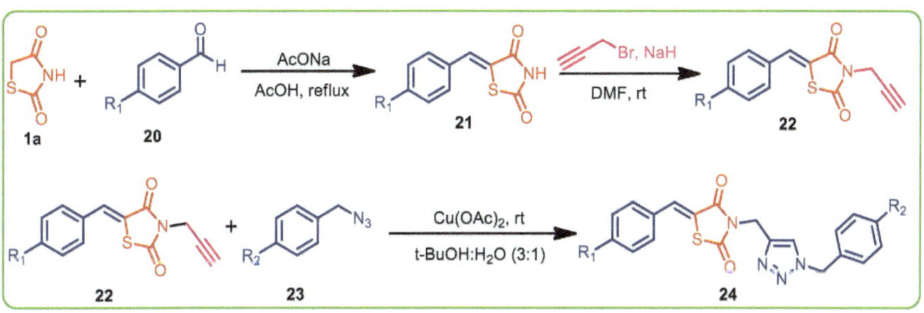

Scheme 5.3 Synthesis of triazole-based TZD derivatives.

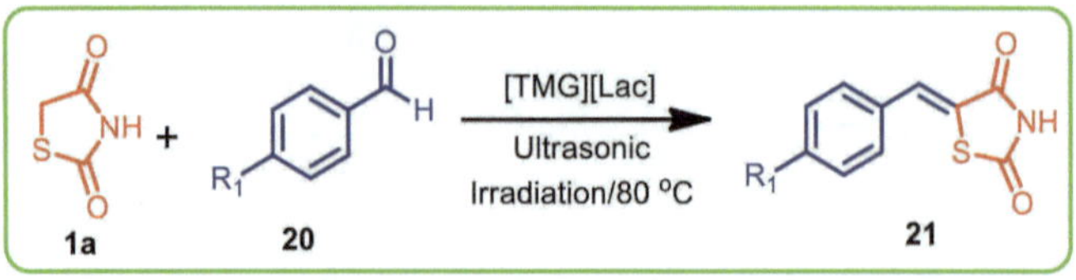

Scheme 5.4 Use of ultrasound for the synthesis of 2,4-thiazolidinedione derivatives.

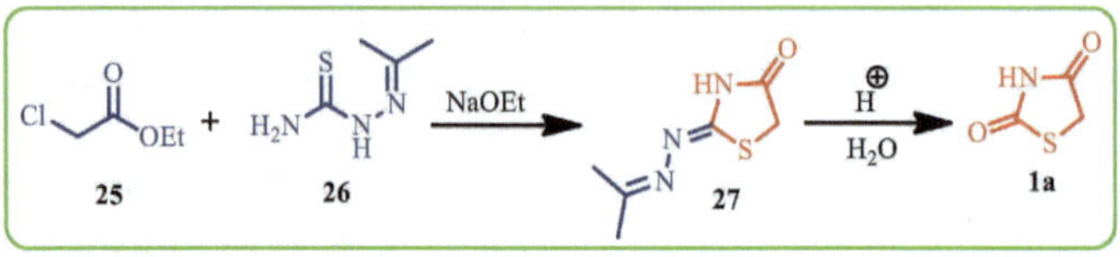

Scheme 5.5 Synthesis of TZD using thiosemicarbazone and ethyl chloroacetate.

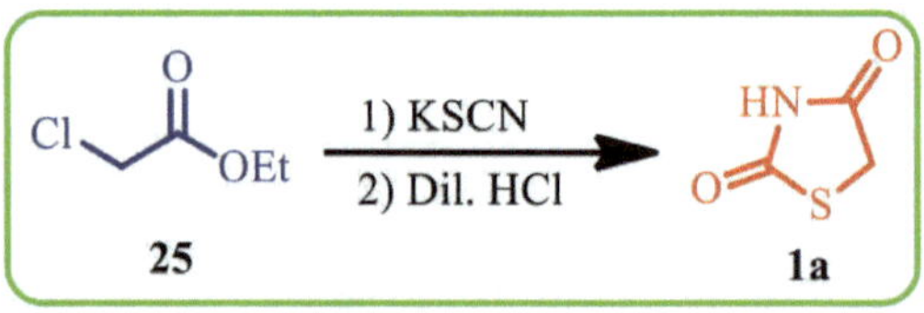

Scheme 5.6 Synthesis of TZD using potassium thiocyanate and ethyl chloroacetate.

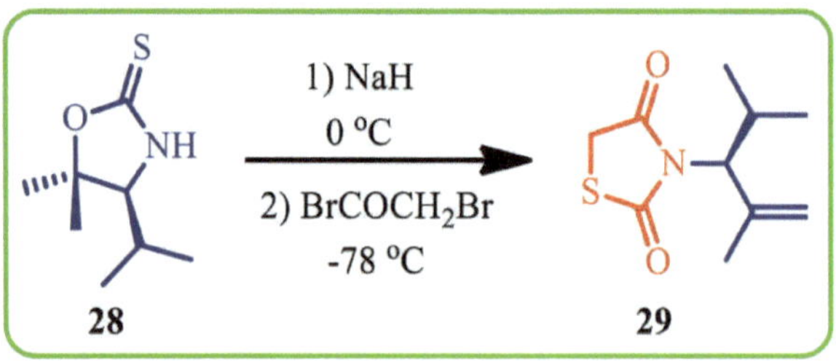

Scheme 5.7 Preparation of TZD.

Through the induction of a shift of the isopropyl group occupying the C4 position to the nitrogen, Mendoza *et al.* were able to transform an oxazolidine-2-thione core moiety into the one found in TZD. Before the dropwise addition of bromoacetyl bromide at −78 °C, **28** was treated with sodium hydride at 0 °C in CH_2Cl_2. The product (**29**), as oil, was produced with a 67% yield (Scheme 5.7).[22]

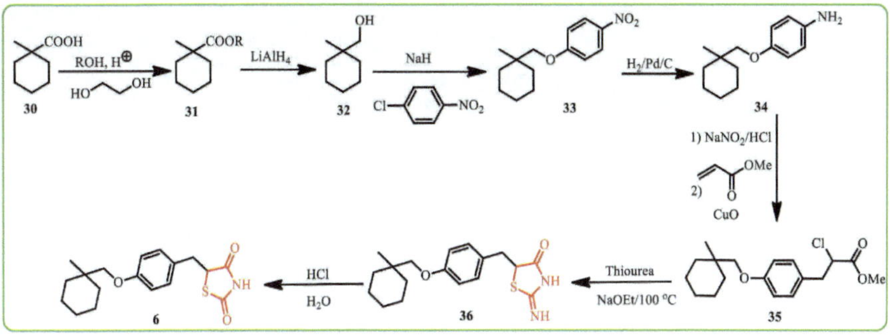

Scheme 5.8 Synthetic route for the ciglitazone drug.

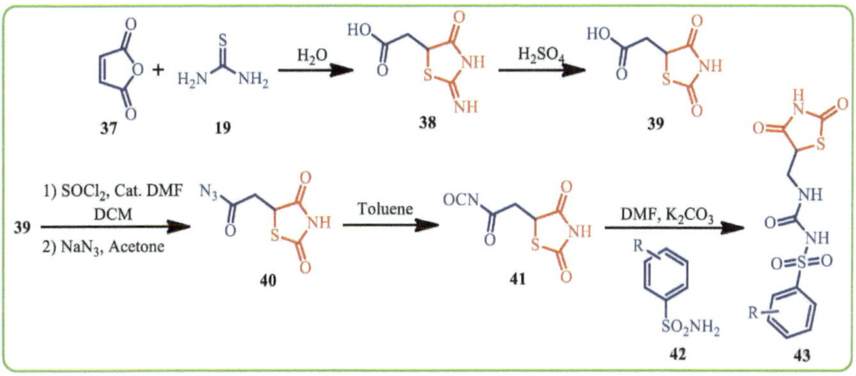

Scheme 5.9 Chain of TZDs produced using maleic anhydride and thiourea.

The Japanese pharmaceutical company Takeda produced ciglitazone (**6**) in 1982[23] by esterifying 1-methylcyclohexanecarboxylic acid (**30**) to synthesize **31**, followed by its reduction with $LiAlH_4$ and condensation with 4-chloronitrobenzene to produce the nitro derivative **33**. Compound **33** is reduced with palladium on carbon (Pd/C), diazotized with $NaNO_2$/aqueous hydrochloric acid producing diazonium chloride, and then coupled with α,β-unsaturated ester to form α-chloro ester **35**. In the presence of NaOEt, the latter interacted with thiourea to produce 2-iminothiazolidin-2-one derivative **36**, which upon acid hydrolysis gave rise to ciglitazone **6** (Scheme 5.8).

Jawale *et al.*[24] produced a novel chain of TZDs by treating thiourea with water and maleic anhydride with a 73–76% yield (Scheme 5.9).

The synthesis of benzimidazole-containing TZDs (**49**) with a 66–87% yield was reported by Darwish *et al.*[25] The benzimidazole ring was produced from *o*-phenylenediamine by condensation with glycolic acid using the Phillips technique, producing compound **45**. Compound **46** was produced by alkylating **45** with various alkyl halides in *N*,*N*-dimethylformamide (DMF) in the presence of sodium hydride. Aldehydes **47** were then

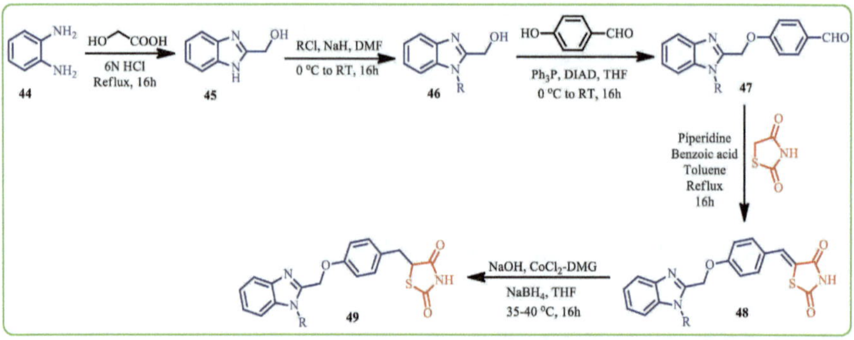

Scheme 5.10 Synthesis of benzimidazole-containing TZDs.

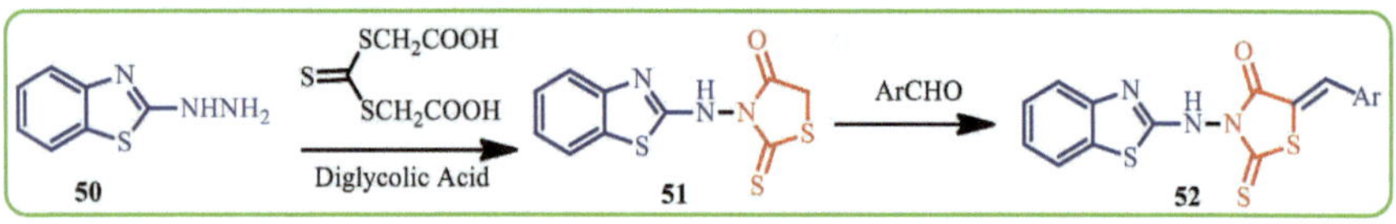

Scheme 5.11 Synthesis of benzothiazole-substituted 4-TZDs.

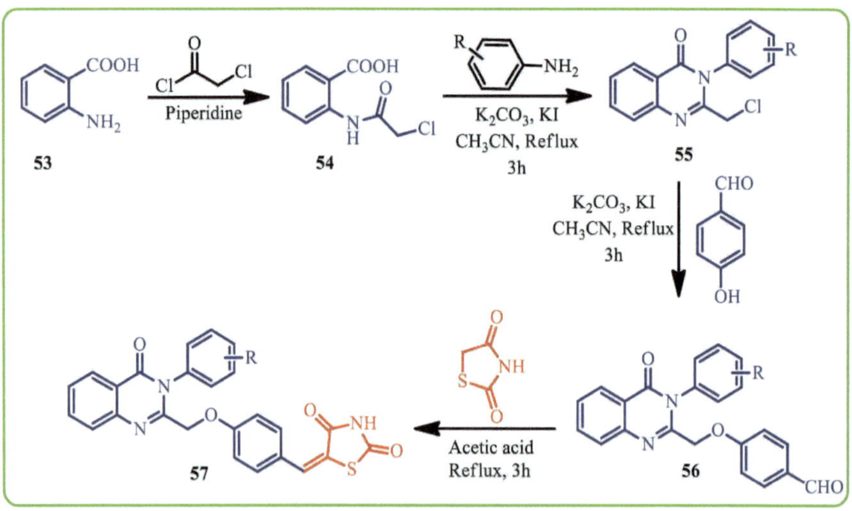

Scheme 5.12 Synthesis of the novel 2-TZD scaffold.

produced by coupling **46** with 4-hydroxybenzaldehyde under Mitsunobu conditions. Ultimately, the target compound **49** was obtained *via* Knoevenagel condensation of **47** with TZD and reduction using $NaBH_4$ in the presence of cobalt with dimethylglyoxime (Co/DMG) (Scheme 5.10).

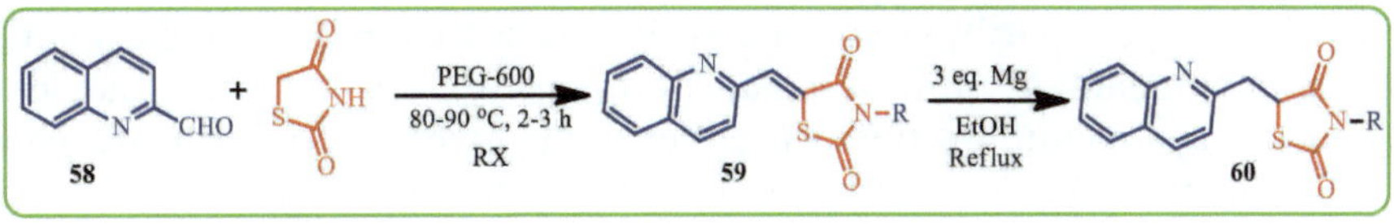

Scheme 5.13 Preparation of quinolinylidinethiazolidine-2,4-diones.

The synthesis of benzothiazole-4-TZD hybrids (**52**) using the Knoevenagel condensation method with a 59–78% yield was described by Havrylyuk *et al.*[26] (Scheme 5.11).

A novel set of 2-TZD scaffolds was developed by Metwally *et al.*[27] with a 38–48% yield (Scheme 5.12). Here, two thiazolidinedione scaffolds were synthesized which differ in the position of the thiazolidinedione ring. When the thiazolidinedione ring is present in the terminal position of the molecule it is termed as an *A*-thiazolidinedione ring. When the thiazolidinedione ring is present in the middle position of the molecule it is termed as a *B*-thiazolidinedione ring.

The polyethylene glycol-600-catalysed synthesis of quinolinylidinethiazolidine-2,4-diones (**60**) with a 75–90% yield as possible anti-hyperglycaemic drugs was reported by Riyaz and Dubey.[28] In this process, polyethylene glycol served as the reaction medium while quinoline-2-carbaldehyde (**58**), TZD, and alkyl halides reacted to yield **59**, which then undergoes reduction with Mg-EtOH to produce **60** (Scheme 5.13).

5.3 Medicinal Chemistry Applications

5.3.1 Antidiabetic Effects

The hallmark of diabetes mellitus (DM) is elevated blood sugar levels over an extended period of time.

There are three types of diabetes:

1. Type-1 diabetes, where the pancreas is unable to generate insulin, was previously known as "juvenile diabetes" or "insulin-dependent diabetes mellitus".
2. Type-2 diabetes, a disease where cells do not react to insulin, was previously known as "non-insulin-dependent diabetes mellitus".
3. The third primary form, gestational diabetes, is caused by elevated blood sugar levels in pregnant women who have never had diabetes before.[29]

Diabetes is mostly caused by either insufficient insulin production, the body's incapacity to utilize it, or a combination of both (insulin is a hormone that controls the metabolism of fats, carbohydrates, and proteins). It is typically a chronic syndrome with varying clinical manifestations, including cardiovascular, hypertensive, renal, and neurological issues. It is a condition in which either the pancreas secretes insufficient insulin or the cells do not react to the insulin that is released, preventing the cells from absorbing blood glucose. Its symptoms include excessive dehydration, fatigue, hunger, and frequent urination. It can be managed by altering eating habits, controlling nutrition, taking medication orally, and, in certain cases, receiving insulin injections.[9] In the development of new drugs, the thiazole moiety is an important heterocyclic unit. A review of the literature reveals that numerous investigations on the synthesis of TZDs have been conducted.

5.3.1.1 PPARγ Agonism

TZDs function as activators, known as PPARs, namely PPARγ, which are extensively utilized in the management of type-2 diabetes. Other TZD compounds, including glitazones, have also shown PPARγ agonism with less adverse effects and/or increased potency. For example, the activation of PPARγ requires the ethylene bridge between the oxygen atoms of the oxime moiety and the aromatic group at the *meta*- or *para*-position of acetophenone oxime or acetylpyridine oxime in TZD derivatives (**61**, IC_{50} = 0.19 μM).[30] The conjugated TZD with substituted chromones through a methylene linkage (**63**) and quercetin through a diether linkage (**62**)[31] showed strong PPARγ agonism with decreased hepatotoxicity. Compound **64**, which is the product of the conjugation of the TZD core with the oxadiazole ring, shows no toxicity to the liver or heart.[32] Benzoxazinone (**65**),[33] benzothiazole (**66**),[34] and chalcones (**67**)[35] are the other conjugates of TZD that have the ability to effectively inhibit PPARγ agonism and reduce blood sugar levels (Figure 5.4).

5.3.1.2 Antidiabetic Agents Other Than PPARγ Agonism

Compounds **68** and **69** (Figure 5.5) and their counterparts with enhanced hypoglycaemic and hypolipidaemic activities were synthesized by Reddy *et al.*[36]

Novel 1,3,4-oxadiazole containing TZD **70** was synthesized by Haider *et al.*[37] (Figure 5.5). Compared to the conventional medications, pioglitazone (gene expression by 1.5-fold) and rosiglitazone (gene expression by 1.00-fold), compound **70** demonstrated significant transactivation of the PPARγ gene (by 2.1-fold) and decreased blood glucose levels.

5.3.2 Anticancer Effects

The anticancer effects stem from the activation of PPARγ, which is contingent upon the type and concentration of tumour cells. The angiogenic activity of TZDs is accomplished by inhibiting endothelial cell motility and

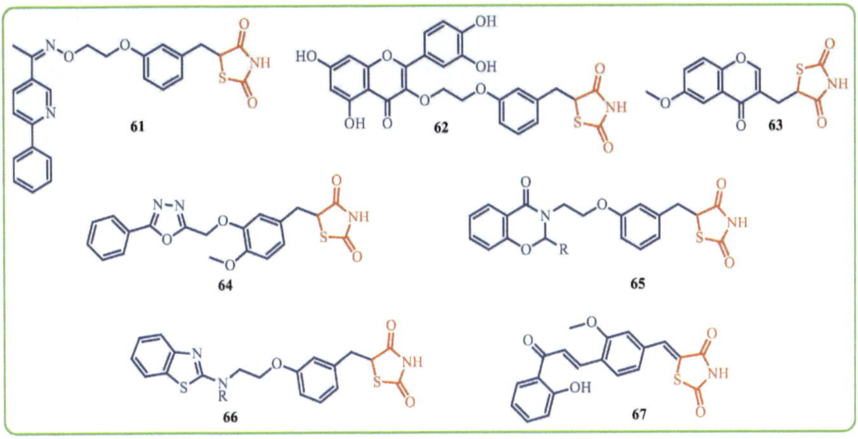

Figure 5.4 TZD derivatives demonstrating PPARγ agonism.

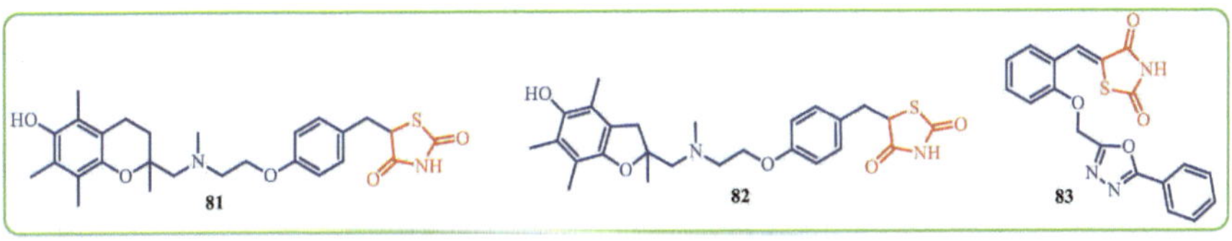

Figure 5.5 TZD derivatives demonstrating antidiabetic activity.

proliferation in addition to reducing the synthesis of vascular endothelial growth factors by tumour cells. As it is believed that PPARγ activation drives the anticancer effect of TZDs, these drugs are currently being evaluated in clinical trials for the treatment of human malignancies expressing high levels of PPARγ.

In Figure 5.6, it is shown that norbornene-2 when combined with 5-ylidene-4-thioxo-2-thiazolidone derivative **71** demonstrated *in vitro* anticancer efficacy against a panel of human tumour cell lines, including SF-268 (cancer of the central nervous system), MCF7 (breast cancer), and NCI-H460 (non-small-cell lung cancer).[38]

A novel class of TZD-lipoid derivatives was synthesized by Chittiboyina *et al.*[39], and their potential anticancer activity against normal and malignant human cell types was assessed. With a half-maximal effective concentration (EC_{50}) of 0.015 μM, compound **72** (Figure 5.6) had the most powerful activity when compared with rosiglitazone and pioglitazone.

The novel hybrid compounds with anticancer activity were tested against the growth of human T-lymphoblastoid (CEM), human cervical carcinoma (HeLa), murine leukaemia (L1210 cell line), and murine mammary carcinoma (FM3A) (Figure 5.6). These molecules, comprising 5-benzylidene thiazolidine-2,4-dione **73**, were designed and synthesized by Romagnoli *et al.*[40]

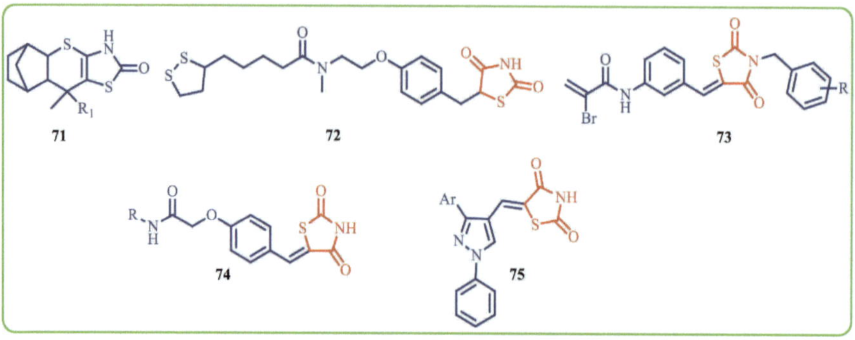

Figure 5.6 TZD derivatives demonstrating anticancer activity.

Patil *et al.*[41] synthesized new hybrid 5-benzylidene-2,4-thiazolidinedione conjugates **74** (Figure 5.6). The synthesized derivatives were tested using the sulforhodamine B (SRB) protein assay for their anticancer activity against human cell lines, such as K-562 (leukaemia), MCF-7 (breast cancer), HepG-2 (hepatoma), PC-3 (prostate cancer), GURAV (oral cancer), and kappa B (KB) (nasopharyngeal cancer).

A new series of 3-(substituted aryl)-1-phenyl-1*H*-pyrazolyl-2,4-thiazolidinedione conjugates **75** (Figure 5.6) were synthesized by Kumar *et al.*[42] and evaluated using conventional Doxil for their cytotoxic activities against lung and breast cancer cell lines.

5.3.2.1 Dual Inhibitors of PI3K-α and MEK

The relevance of the rapidly accelerated fibrosarcoma (RAF)/mitogen-activated protein kinase (MEK)/extracellular signal-regulated kinase (ERK) and phosphoinositide 3-kinase (PI3K)/Akt pathways in the aetiology of human malignancies led to the investigation of various members of these signalling pathways for drug discovery. According to the findings by Engelman *et al.*,[43] it is highly successful in simultaneously attenuating MEK and PI3K, particularly in lung tumours in mice with the Kirsten rat sarcoma virus (KRAS) G12D and H1047R strains. The dual activity (PI3K-α and MEK) of TZD derivatives was reported by Li *et al.*[44] The most effective one was reported to be compound **76** (Figure 5.7). Afterwards, Liu *et al.* also revealed the inhibitory capacity of TZD derivatives on PI3K-α and MEK.[44] While the hydroxyl group in **79** (Figure 5.7) shows equipotent anticancer activity when present at the same position as in **80** (Figure 5.7) in U937 cancer cell lines, modifications to compound **76**, such as replacing the amine moiety of ethylamine with carboxyl **77** (Figure 5.7) and tertiary amino **78** (Figure 5.7) groups, reduced biological activity, expressing the significance of primary amine. These findings revealed that hydrogen-bonding interactions, rather than ionic interactions, were present with amino acid residues in the active site. In the *in vitro* assessment of PI3K-α and MEK1 enzyme inhibition, as

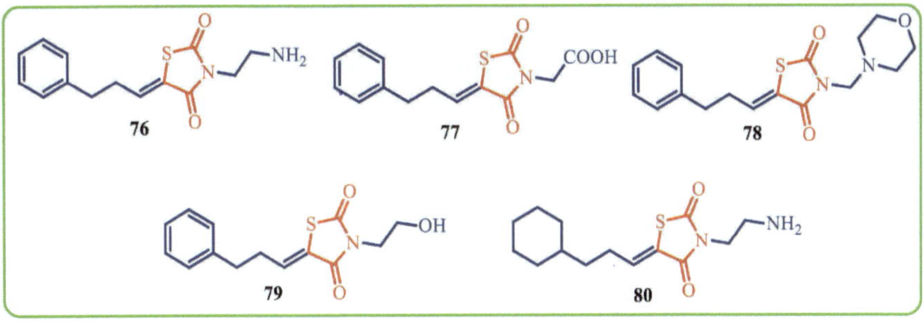

Figure 5.7 TZD derivatives demonstrating dual inhibition for PI3K-α and MEK.

well as in U937 cancer cell lines, compound **80** showed the greatest overall efficacy.

5.3.2.2 RAF Kinase Inhibitor

The protein kinase inhibitor of the cell controls a variety of signalling cascades and has been demonstrated to impede G protein-coupled receptor kinase when it is phosphorylated by protein kinase C.[45] Numerous heterocyclic compounds that block the RAF kinase enzyme, which is necessary for tumour cell survival and proliferation, have been discovered. These inhibitors are preventing the expression of the RAF protein and postponing the Ras–Raf interaction. There are several RAF kinase inhibitors available on the market right now, and they have demonstrated excellent anticancer activity and a very high safety profile.[46] These inhibitory proteins belong to the phosphatidylethanolamine-binding protein family. The RAF kinase inhibitor protein is implicated in physiological processes, such as neural development, cardiac function, and spermatogenesis, and pathological processes, such as diabetic nephropathy and Alzheimer's disease.[47]

In 2018, *N*1-(2-aryl-1,3-thiazolidin-4-one)-*N*3-aryl urea **81** (Figure 5.8) was synthesized by Qi *et al.*[48] and tested for its anticancer efficacy against a panel of kinases, including c-kit, rearranged during transfection (RET), epidermal growth factor receptor (EGFR), Src, type I insulin-like growth factor receptor (IGF-IR), and AXL.

5.3.3 Anti-inflammatory Effects

5.3.3.1 COX Enzyme Inhibitors

Prostaglandin-endoperoxide synthase (PTGS) enzyme, or cyclooxygenase (COX) in particular, is a family of isozymes that is involved in the synthesis of prostanoids and prostaglandins from arachidonic acid, including thromboxane. Non-steroidal anti-inflammatory drug (NSAID)-class medications are the competitive inhibitors of COX. COX enzymes exist in two varieties: COX-1 and COX-2. Both enzymes produce prostaglandins that

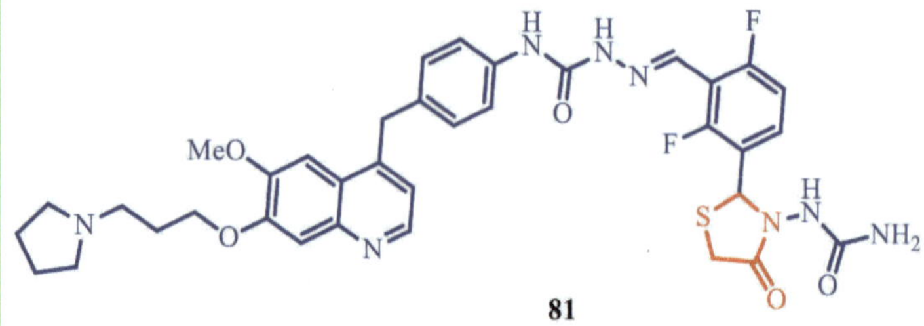

81

Figure 5.8 TZD derivatives demonstrating RAF kinase inhibition activity.

induce pain, inflammation, and fever, but only COX-1 produces prostaglandins that activate platelets and protect the lining of the stomach and intestines.[49] It is believed that COX-2 promotes angiogenesis, tumour tissue invasion, and cancer cell resistance to apoptosis. It is frequently expressed in a variety of cancer types and contributes to genesis of genes and carcinogenesis. Meanwhile, COX-2 contributes to immune evasion and resistance to cancer immunotherapy, which are crucial aspects of innate and adaptive immune responses. The majority of the effects of COX-2 are mediated by the metabolite prostaglandin E2. COX-2 is frequently expressed in a variety of cancer types and contributes to the genesis of genes and carcinogenesis, inducing angiogenesis, tumour tissue invasion, and cancer cell resistance to apoptosis. Prostaglandin E2, a metabolite, mediates the majority of the activities of COX-2. When COX-2 inhibitors are administered to cancer patients, the chance of metastasis is decreased. The prostaglandin cascade and COX-2 are significant players in the inflammatory process that leads to cancer. It has been found that COX-2 inhibitors lessen the likelihood of cancer growth.[50]

After TZD derivatives **82** and **83** (Figure 5.9) were discovered by Ali *et al.* to be powerful analgesic and anti-inflammatory drugs, they hypothesized that COX-2 inhibition is the mechanism behind the biological activity of these molecules.[51] Youssef *et al.*[52] synthesized a number of novel active derivatives of pyrazolyl-2,4-thiazolidinedione **84** (Figure 5.9) and assessed their anti-inflammatory properties *in vitro*.

A group of new compounds of 5-benzylidene thiazolidine-2,4-dione **85** (Figure 5.9) was synthesized and tested for their ability to reduce inflammation *in vitro* by Ha *et al.*[53]

5.3.4 Antioxidant Effects

Atherosclerosis, tumours, and heart diseases are only a few of the illnesses that are mediated by free radicals, which are generated in a variety of biochemical processes and cellular metabolism. In addition to regular biological processes, exposure to various chemicals (polycyclic aromatic

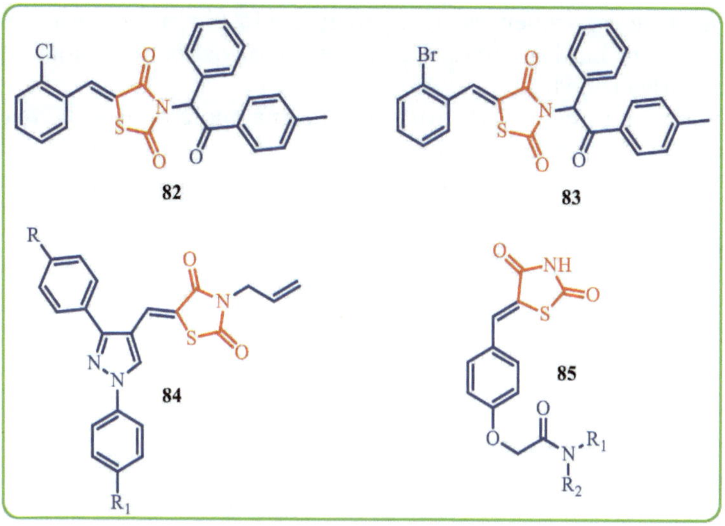

Figure 5.9 TZD derivatives demonstrating COX enzyme inhibitory activity.

hydrocarbons, cadmium, lead, *etc.*), radiation, smoking, and consuming more fattening food can also result in the formation of free radicals. Usually, substances referred to as antioxidants prevent the formation of free radicals. Free radicals are neutralized by antioxidants before they damage the cells. Antioxidants are the natural cleaners and detoxifiers of the body. They transform bodily poisons into innocuous waste products. They absorb harmful cholesterol and shield the body against numerous illnesses, such as cancer and heart attacks. Although synthetic antioxidants, such as butylated hydroxytoluene (BHT) and butylated hydroxyanisole (BHA), are available and are effective antioxidants utilized in many sectors, their use has been restricted due to the possibility of cancer and

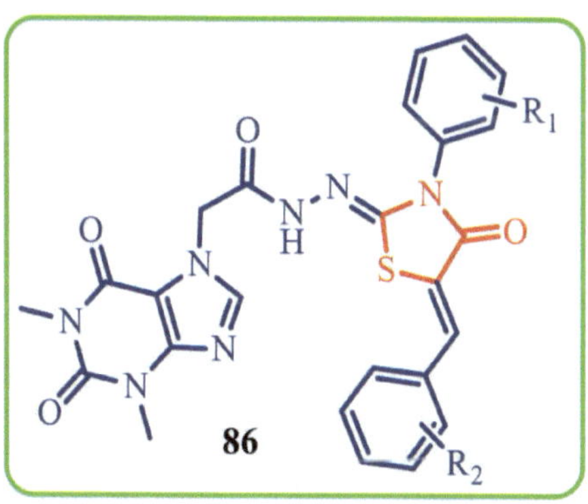

Figure 5.10 TZD derivatives demonstrating antioxidant activity.

other adverse consequences. Thus, their use in pharmaceutical, cosmetic, and food items has been reduced. As a result, oxidation inhibitor chemicals are currently needed.[54]

A chain of unique TZDs with a xanthine moiety **86** was created by Lupascu *et al.*[55] and their antioxidant capacity was assessed using *in vitro* models including the 2,2-diphenyl-1-picrylhydrazyl (DPPH) radical-scavenging assay and the 2,2-azino-bis-(3-ethyl benzothiazoline-6-sulfonic acid) (ABTS) radical-scavenging assay (Figure 5.10).

5.4 TZDs and Their Side Effects

Recently, it has been discovered that PPARγ ligands (TZDs) exhibit anticancer activity in a variety of cancer models by inhibiting tumour angiogenesis and interfering with the cell cycle, proliferation, differentiation, and death. Over the past thirty years, TZDs have drawn a lot of interest due to the possibility of negative effects. These adverse effects were initially noted after glitazones of the first generation were developed as PPARγ inhibitors. Troglitazone-induced liver damage has been linked to the generation of toxic reactive metabolites during hepatic metabolism.[56] Acute liver failure due to the death of liver tissue cells has been connected to this process.[57] Additional mechanisms of induced hepatotoxicity include oxidative stress enhancement, mitochondrial damage,[58] and bile build-up in the liver as a result of bile excretory protein suppression.[58] Specifically, it should be mentioned that troglitazone-related hepatotoxicity is thought to be idiosyncratic rather than dose-dependent.[59] The second most often observed negative effect after the use of the prescribed TZD is weight gain. It is well established that TZDs raise total plasma volume levels *in vivo* and induce oedema.[60] Pre-adipocytes differentiate into tiny fat cells, leading to the redistribution of fat.[61] One of the negative effects of taking TZD that has been discovered more recently is an increased chance of bone fractures. "TZD-induced bone loss" has been identified as the primary cause of this risk, but weight gain may also play a role. Abiogenesis rises as a result of this induced action, while osteoblastogenesis decreases.[62,63] Moreover, insulin directly affects the regulation of osteoblastogenesis, which in turn affects bone formation. The negative effects of elevated risk of bone fractions manifest as TZDs lower insulin levels.[64]

5.5 Conclusion

In the field of medicinal chemistry, the TZD core is a structural motif that is frequently utilized. A wide variety of physiologically active substances used to treat a wide range of illnesses have this structure. It is most widely used in the management of diabetic mellitus. Diabetes mellitus, especially type 2 diabetes, is thought to be one of the main risk factors for cardiovascular disease and death. It has been observed that TZD-containing structures inhibit a wide variety of biological targets, including but not limited to PPARγ, PTP1B, and ALR2. Regretfully, TZDs have historically been linked to detrimental side effects. Significant weight gain, fluid retention, and severe hepatotoxicity are examples of such side effects that are frequently observed. Consequently, the clinical usage of glitazones that had previously been marketed, such as troglitazone and rosiglitazone, was discontinued. Many efforts have been made to functionalize the TZD structure at two primary positions: amidic nitrogen (N3) and activated methylene carbon (C5). The procedures used to fill their roles with substitutes are widely known and have withstood scrutiny for a number of years. Synthetic organic and medicinal chemists can quickly produce a wide range of new derivatives, thanks to their easily replicable approaches.

Abbreviations

ABTS	2,2-Azino-bis-(3-ethylbenzothiazoline-6-sulfonic acid)
Akt	Ak strain transforming
BHA	Butylated hydroxyanisole
BHT	Butylated hydroxytoluene
CH_2Cl_2	Dichloromethane
Co/DMG	Cobalt with dimethylglyoxime
COX	Cyclooxygenasc
dil. HCl	Dilute hydrochloric acid
DM	Diabetes mellitus
DMF	*N,N*-Dimethylformamide
DPPH	2,2-Diphenyl-1-picrylhydrazyl
EC_{50}	Half-maximal effective concentration
EGFR	Epidermal growth factor receptor
ERK	Extracellular signal-regulated kinase
HCN	Hydrogen cyanide
IC_{50}	Half-maximal inhibitory concentration
IGF-IR	Type I insulin-like growth factor receptor
KRAS	Kirsten rat sarcoma virus
$LiAlH_4$	Lithium aluminium hydride
MEK	Mitogen-activated protein Kinase
Mg-EtOH	Magnesium-ethanol
MTT	3-[4,5-Dimethylthiazol-2-yl]-2,5 diphenyl tetrazolium bromide
$NaNO_2$	Sodium nitrite
NaOEt	Sodium ethoxide
NSAID	Non-steroidal anti-inflammatory drugs
Pd/C	Palladium on carbon
PI3K	Phosphoinositide 3-kinase

PPAR	Peroxisome proliferator-activated receptor
PTGS	Prostaglandin-endoperoxide synthase
RAF	Rapidly accelerated fibrosarcoma
RET	Rearranged during transfection
SRB	Sulforhodamine B
Src	Sarcoma
[TMG][Lac]	1,1,3,3-Tetramethylguanidinium Lactate
TZD	2,4-Thiazolidinedione
μM	Micromolar

Acknowledgements

Author MHS is thankful to Principal, Radhabai Kale Mahila Mahavidyalaya Ahmednagar. Author BBS is also grateful to the Head, Department of Chemistry, Dr. Babasaheb Ambedkar Marathwada University Aurangabad for providing necessary assistance.

References

1. A. K. Jain, A. Vaidya, V. Ravichandran, S. K. Kashaw and R. M. Agrawal, *Bioorg. Med. Chem.*, 2012, **20**, 3378.
2. V. Enchev, S. Chorbadjiev and B. Jordanov, *Chem. Heterocycl. Compd.*, 2002, **38**, 1110.
3. 2,4-Thiazolidinedione 2295-31-0, https://www.chemicalbook.com/ChemicalProduct Property_EN_CB6675365.htm.
4. (a) J. M. Lehmann, L. B. Moore, T. A. Smith-Oliver, W. O. Wilkison, T. M. Willson and S. A. Kliewer, *J. Biol. Chem.*, 1995, **270**, 12953. (b) R. Da Ros, R. Assaloni and A. Ceriello, *Diabetic Med.*, 2004, **21**, 1249.
5. (a) M. Li, T. W. Lee, T. S. Mok, T. D. Warner, A. P. Yim and G. G. Chen, *J. Cell. Biochem.*, 2005, **96**, 760. (b) C. W. Shiau, C. C. Yang, S. K. Kulp, K. F. Chen, C. S. Chen, J. W. Huang and C. S. Chen, *Cancer Res.*, 2005, **65**, 1561.
6. (a) M. Tuncbilek and N. Altanlar, *Arch. Pharm. Chem. Life Sci.*, 2006, **339**, 213. (b) O. B. Dundar, O. Ozgen, A. Mentese, N. Altanlar, O. Atli, E. Kendi and R. Ertan, *Bioorg. Med. Chem.*, 2007, **15**, 6012.
7. (a) G. Bruno, C. Costantino, R. Maccari, F. Monforte, R. Ottana and M. G. Vigorita, *Bioorg. Med. Chem.*, 2002, **10**, 1077. (b) S. Suzen and E. Buyukbingol, *Curr. Med. Chem.*, 2003, **10**, 1329.
8. (a) V. V. Mulwad, A. A. Mir and H. T. Parmar, *Indian J. Chem.*, 2009, **48B**, 137. (b) K. R. Alagawadi and S. G. Alegaon, *Arabian J. Chem.*, 2011, **4**, 465.
9. S. Rekha, U. Shantharam and V. Chandy, *Int. Res. J. Pharm.*, 2011, **2**, 81.

10. F. Herrera, J. C. Mayo, V. Martín, R. M. Sainz, I. Antolin and C. Rodriguez, *Cancer Lett.*, 2004, **211**, 47.
11. Y. Chinthala, A. K. Domatti, A. Sarfaraz, S. P. Singh, N. K. Arigari, N. Gupta, S. K. V. N. Satya, J. K. Kumar, F. Khan, A. K. Tiwari and G. Paramjit, *Eur. J. Med. Chem.*, 2013, **70**, 308.
12. A. Verma and S. K. Saraf, *Eur. J. Med. Chem.*, 2008, **43**, 897.
13. B. Sung, S. Park, B. P. Yu and H. Y. Chung, *J. Gerontol., Ser. A*, 2004, **59**, 997.
14. M. Ricote, A. C. Li, T. M. Willson, J. Kelly and C. K. Glass, *Nature*, 1998, **391**, 79.
15. (a) T. M. Willson and W. Wahli, *Curr. Opin. Chem. Biol.*, 1997, **1**, 235. (b) R. A. Komers and A. Vrana, *Physiol. Res.*, 1998, **47**, 215.
16. S. Kallenberg, Stereochemische Untersuchungen der Diketo-thiazolidine (I.), *Ber. dtsch. Chem. Ges. A/B*, 1923, **56**, 316.
17. O. Bozdag, G. Ayhan-Kilcigil, M. Tuncbilek and R. Ertan, *Turk. J. Chem.*, 1999, **23**, 163.
18. M. H. Shaikh, D. D. Subhedar, S. V. Akolkar, A. A. Nagargoje, A. Asrondkar, V. M. Khedkar and B. B. Shingate, *Polycyclic Aromat. Compd.*, 2023, **43**, 3353.
19. K. Suresh and J. S. Sandhu, *Org. Med. Chem. Lett.*, 2013, **3**, 2.
20. J. Taylor, The chloroacetates of S-alkylthiocarbamides, *J. Chem. Soc. Trans.*, 1920, **117**, 4–11.
21. W. Heintz, Beitrage zur Kenntniss der Glycolamidsauren, *Justus Liebigs Ann. Chem.*, 1865, **136**, 213.
22. G. Mendoza, H. Hernández, L. Quintero, M. Sosa-Rivadeneyra, S. Bernès, E. Sansinenea and A. Ortiz, *Tetrahedron Lett.*, 2005, **46**, 7867.
23. T. M. E. Sohda, Y. Imamia, T. Sugiyama and Y. Kawamatsu, *Chem. Pharm. Bull.*, 1982, **30**, 3580.
24. D. V. Jawale, U. R. Pratap, N. Rahuja, A. K. Srivastava and R. A. Mane, *Bioorg. Med. Chem.*, 2012, **22**, 436.
25. K. M. Darwish, I. Salama, S. Mostafa, M. S. Gomaa and M. A. Helal, *Eur. J. Med. Chem.*, 2016, **109**, 157.
26. D. Havrylyuk, L. Mosula, B. Zimenkovsky, O. Vasylenko, A. Gzella and R. Lesyk, *Eur. J. Med. Chem.*, 2010, **45**, 5012.
27. K. Metwally, H. Pratsinis and D. Kletsas, *Eur. J. Med. Chem.*, 2017, **133**, 340.
28. S. N. A. Riyaz and P. K. Dubey, *Indian. J. Chem.*, 2012, **51B**, 1396.
29. Y. Yang, X. Hu, Q. Zhang and R. Zou, *Age Ageing*, 2016, **45**, 761.
30. H. Yanagisawa, M. Takamura, E. Yamada, S. Fujita, T. Fujiwara, M. Yachi, A. Isobe and Y. Hagisawa, *Bioorg. Med. Chem. Lett.*, 2000, **10**, 373.
31. J. Y. Lee, W. H. Park, M. K. Cho, H. J. Yun, B. H. Chung, Y. K. Pak, H. G. Hahn and S. H. Cheon, *Arch. Pharm. Res.*, 2005, **28**, 142.
32. S. Nazreen, M. S. Alam, H. Hamid, M. S. Yar, A. Dhulap, P. Alam, M. A. Pasha, S. Bano, M. M. Alam, S. Haider, C. Kharbanda, Y. Ali and K. K. Pillai, *Bioorg. Med. Chem. Lett.*, 2014, **24**, 3034.

33. G. R. Madhavan, R. Chakrabarti, K. A. Reddy, B. M. Rajesh, V. Balraju, P. B. Rao, R. Rajagopalan and J. Iqbal, *Bioorg. Med. Chem.*, 2006, **14**, 584.
34. R. Jeon, Y. J. Kim, Y. Cheon and J. H. Ryu, *Arch. Pharm. Res.*, 2006, **29**, 394.
35. X. F. Liu, C. J. Zheng, L. P. Sun, X. K. Liu and H. R. Piao, *Eur. J. Med. Chem.*, 2011, **46**, 3469.
36. K. A. Reddy, B. B. Lohray, V. Bhushan, A. S. Reddy, P. H. Kishore, V. V. Rao, V. Saibaba, A. C. Bajji, B. M. Rajesh and K. V. Reddy, *Bioorg. Med. Chem. Lett.*, 1998, **8**, 999.
37. S. Nazreen, M. S. Alam, H. Hamid, M. S. Yar, S. Shafi, A. Dhulap, P. Alam, M. A. Pasha, S. Bano, M. M. Alam, S. Haider, Y. Ali, C. Kharbanda and K. K. Pillai, *Eur. J. Med. Chem.*, 2014, **87**, 175.
38. R. Lesyk, B. Zimenkovsky, D. Atamanyuk, F. Jensen, K. Kieć-Kononowicz and A. Gzella, *Bioorg. Med. Chem.*, 2006, **14**, 5230.
39. A. G. Chittiboyina, M. S. Venkatraman, C. S. Mizuno, P. V. Desai, A. Patny, S. C. Benson and M. A. Avery, *J. Med. Chem.*, 2006, **49**, 4072.
40. R. Romagnoli, P. G. Baraldi, M. K. Salvador, M. E. Camacho, J. Balzarini, J. Bermejo and F. Estevez, *Eur. J. Med. Chem.*, 2013, **63**, 544.
41. V. Patil, K. Tilekar, S. M. Munj, R. Mohan and C. S. Ramaa, *Eur. J. Med. Chem.*, 2010, **45**, 4539.
42. K. S. Kumar, B. M. Reddy and V. H. Babu, *Int. J. Pharm. Pharm. Sci.*, 2014, **6**, 831.
43. J. A. Engelman, L. Chen, X. Tan, K. Crosby, A. R. Guimaraes, R. Upadhyay, M. Maira, K. McNamara, S. A. Perera, Y. Song, L. R. Chirieac, R. Kaur, A. Lightbown, J. Simendinger, T. Li, R. F. Padera, C. García-Echeverría, R. Weissleder, U. Mahmood, L. C. Cantley and K. K. Wong, *Nat. Med.*, 2008, **14**, 1351.
44. Q. Li, J. Wu, H. Zheng, K. Liu, T. L. Guo, Y. Liu, S. T. Eblen, S. Grant and S. Zhang, *Bioorg. Med. Chem. Lett.*, 2010, **20**, 4526.
45. K. Lorenz, M. J. Lohse and U. Quitterer, *Nature*, 2003, **426**, 574.
46. D. Strumberg and S. Seeber, *Onkologie*, 2005, **28**, 101.
47. E. T. Keller, Z. Fu and M. Brennan, *Biochem. Pharmacol.*, 2004, **68**, 1049.
48. B. Qi, Y. Yang, H. He, X. Yue, Y. Zhou, X. Zhou, Y. Chen, M. Liu, A. Zhang and F. Wei, *Eur. J. Med. Chem.*, 2018, **146**, 368.
49. A. Zarghi and S. Arfaei, *Iran. J. Pharm. Res.*, 2011, **10**, 655.
50. N. H. Goradel, M. Najafi, E. Salehi, B. Farhood and K. Mortezaee, *J. Cell. Physiol.*, 2019, **234**, 5683.
51. A. M. Ali, G. E. Saber, N. M. Mahfouz, M. A. El-Gendy, A. A. Radwan and M. A. Hamid, *Arch. Pharm. Res.*, 2007, **30**, 1186.
52. A. M. Youssef, M. S. White, E. B. Villanueva, I. M. El-Ashmawy and A. Klegeris, *Bioorg. Med. Chem.*, 2010, **18**, 2019.
53. Y. M. Ha, Y. J. Park, J. A. Kim, D. Park, J. Y. Park, H. J. Lee, J. Y. Lee, H. R. Moon and H. Y. Chung, *Eur. J. Med. Chem.*, 2012, **49**, 245.
54. L. Feng, K. Lv, M. Liu, S. Wang, J. Zhao, X. You, S. Li, J. Cao and H. Guo, *Eur. J. Med. Chem.*, 2012, **55**, 125.

55. F. G. Lupascu, O. M. Dragostin, L. Foia, D. Lupascu and L. Profire, *Molecules*, 2013, **18**, 9684.
56. K. Kassahun, P. G. Pearson, W. Tang, I. McIntosh, K. Leung, C. Elmore, D. Dean, R. Wang, G. Doss and T. A. Baillie, *Chem. Res. Toxicol.*, 2001, **14**, 62.
57. Y. Yamamoto, M. Nakajima, H. Yamazaki and T. Yokoi, *Life Sci.*, 2001, **70**, 471.
58. M. T. Smith, *Chem. Res. Toxicol.*, 2003, **16**, 679.
59. D. M. Bissell, G. J. Gores, D. L. Laskin and J. H. Hoofnagle, *Hepatology*, 2001, **33**, 1009.
60. M. M. R. Young, L. Squassante, J. Wemer, S. P. van Marle, P. Dogterom and J. H. G. Jonkman, *Eur. J. Clin. Pharmacol.*, 1999, **55**, 101.
61. M. Adams, C. T. Montague, J. B. Prins, J. C. Holder, S. A. Smith, L. Sanders, J. E. Digby, C. P. Sewter, M. A. Lazar, V. K. K. Chatterjee and S. O'Rahilly, *J. Clin. Invest.*, 1997, **100**, 3149.
62. B. Lecka-Czernik, I. Gubrij, E. J. Moerman, O. Kajkenova, D. A. Lipschitz, S. C. Manolagas and R. L. Jilka, *J. Cell. Biochem.*, 1999, **74**, 357.
63. S. O. Rzonca, L. J. Suva, D. Gaddy, D. C. Montague and B. Lecka-Czernik, *Endocrinology*, 2004, **145**, 401.
64. J. Cornish, K. E. Callon and I. R. Reid, *Calcif. Tissue Int.*, 1996, **59**, 492.

Part II: Thiadiazoles

CHAPTER 6

Synthesis, Properties, and Biological Applications of 1,2,4-Thiadiazoles

POPAT MOHITE*[a], NITIN BHOGE[b], SUJATA G. DENGALE[c] AND NIRMALA R. DAREKAR[d]

[a]Department of Pharmaceutical Chemistry, Aldel Education Trust St. John Institute of Pharmacy and Research, Palghar, Maharashtra, India; [b]Department of Chemistry, Shri Dnyaneshwar Mahavidyalaya Newasa, Ahmednagar, Maharashtra, India; [c]Department of Chemistry, S.N. Arts, D.J. Malpani Commerce and B.N. Sarda Science College, Sangamner, Ahmednagar, Maharashtra, 422605, India; [d]Department of Chemistry, Radhabai Kale Mahila Mahavidyalaya, Ahmednagar, Maharashtra, 414001, India

*E-mail: mohitepb@gmail.com

6.1 Introduction

The pharmaceutical sector considers the 1,2,4-thiadiazole moiety to be of great importance due to its diverse biological activity and related therapeutic uses. It is a heterocyclic compound with a five-membered ring containing sulfur and nitrogen atoms. There are various isomers of thiadiazole, including 1,2,4-, 1,2,5-, 1,2,3-, and 1,3,4-thiadiazoles[1,2] (Figure 6.1).

Thiadiazole has a significant impact on the synthesis of many biological target molecules related to various therapeutic categories, such as

S-Heterocycles: Retrospect, Prospects, and Biological Applications
Edited by Shrikaant Kulkarni, Hemantkumar Akolkar, Vijay M. Khedkar and A. K. Haghi

Published by the Royal Society of Chemistry, www.rsc.org

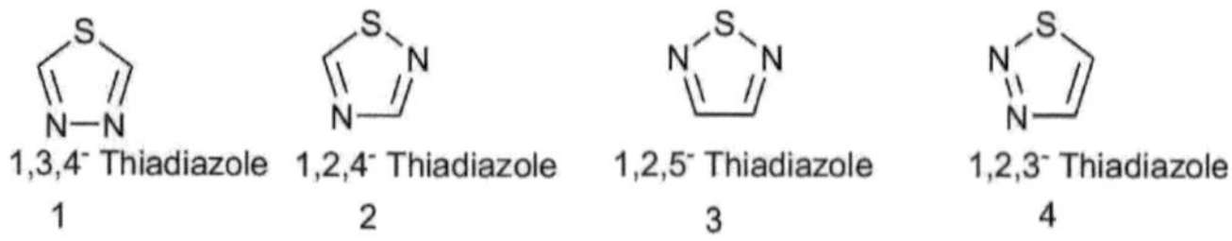

Figure 6.1 Chemical structures of the isomers of thiadiazole.

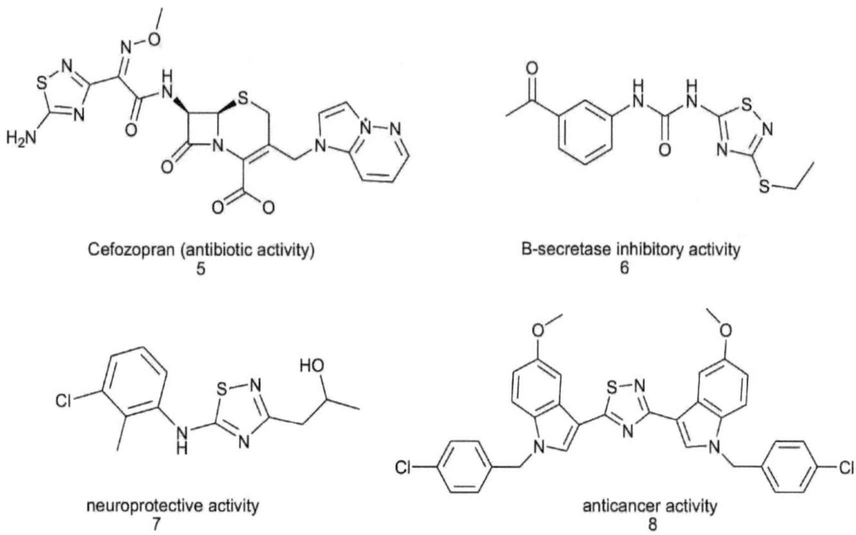

Figure 6.2 Biologically active 1,2,4-thiadiazoles.

fungicides, bactericides, antibiotics, and herbicides. The 1,2,4-thiadiazole derivatives exhibit a wide variety of biological activities among all thiadiazole isomers. To study these biological activities, their synthesis has been explored in the last decade.[3,4] Many researchers have developed different synthetic procedures for 1,2,4-thiadiazole derivatives. Clinically used drugs containing thiadiazole have also been explored[5] (Figure 6.2).

In the present chapter, we studied the biological activities exhibited by substituted 1,2,4-thiadiazoles.

There have been a significant number of scientific investigations involving these 1,2,4-thiadiazoles in recent years. We have chosen to describe the wide range of pharmacological activities documented in the literature because of their significance in both heterocyclic and medicinal chemistry.

1,2,4-Thiadiazole and its derivatives represent a captivating and vibrant area of study within the realm of organic chemistry. This heterocyclic compound, featuring a five-membered ring comprising two carbon, one nitrogen, and two sulfur atoms, serves as a versatile scaffold in the synthesis of diverse chemical entities. The unique arrangement of these atoms in the thiadiazole ring imparts distinctive chemical properties, making it an intriguing subject of exploration for researchers in various scientific domains.

The exploration of heterocyclic compounds has long been a cornerstone in the realm of organic chemistry, with diverse applications in medicinal, agricultural, and materials sciences. Among them, 1,2,4-thiadiazole emerges as a captivating and versatile heterocycle, garnering attention for its unique structural attributes and pronounced biological activities. This chapter delves into the recent advancements in the preparation and biological importance of 1,2,4-thiadiazole, shedding light on the innovative methodologies used for its preparation and the expansive landscape of its biological significance. The structural scaffold of 1,2,4-thiadiazole, characterized by a five-membered ring containing S and N atoms, imparts distinctive physicochemical properties that render it amenable to various synthetic strategies. Recent breakthroughs in synthetic methodologies have propelled the efficient synthesis of diverse 1,2,4-thiadiazole derivatives, expanding the chemical toolbox for researchers in their quest for novel compounds with tailored properties. Beyond the synthetic allure, this chapter explores their myriad biological applications. These applications encompass a broad spectrum of properties, including anticancer, antimicrobial, anti-inflammatory, and antioxidant activities. The unique electronic and steric features of 1,2,4-thiadiazole contribute to its pharmacological versatility, positioning it as a promising scaffold in drug development.[6,7] As we navigate through the recent literature, this chapter aims to elaborate a comprehensive overview of the importance of 1,2,4-thiadiazole in research. By amalgamating synthetic methodologies with a survey of diverse biological activities, it aspires to offer researchers and enthusiasts a holistic understanding of the current landscape and prospects in the dynamic field of 1,2,4-thiadiazole chemistry.

6.2 Recent Advances in the Synthesis of 1,2,4-Thiadiazoles

Zhao *et al.* (2014) developed an environmentally friendly process to produce 3,5-disubstituted 1,2,4-thiadiazole with iodine as a catalyst and O_2 as an oxidant, leading to the oxidative dimerization of thioamides in the presence of water. This procedure was used to produce thioureas and aryl- and alkyl-substituted thioamides (Figure 6.3).[8]

Cheng *et al.* (2012) described an effective oxidative dimerization procedure for converting aryl thioamides **7a–k** into a series of 3,5-disubstituted 1,2,4-thiadiazoles **8a–k**. This process was carried out in dichloroethane (DCE) at RT with 2,3-dichloro-5,6-dicyanobenzoquinone (DDQ) as a mediator, resulting in excellent yields of the products. Interestingly, aryl thioamides with either electron-donating or electron-withdrawing groups on the aromatic ring showed strong reactivity, indicating that the substituent had no discernible electronic effect on the aromatic ring. This process provides a flexible way to produce broadly applicable 3,5-disubstituted 1,2,4-thiadiazole (Figure 6.4).[9]

Mariappan *et al.* (2016) reported the synthesis of 3-substituted-5-arylamino-1,2,4-thiadiazole derivative **10** using phenyliodine(III) bis(trifluoroacetate) (PIFA) *via* the oxidative intramolecular formation of an S–N bond in imidoyl thioureas. This cyclization process features a metal-free approach, with a very short reaction time, excellent yield (70–87%), and easily accessible starting material (Figure 6.5).[10]

Kumar *et al.* (2011) reported a revised procedure for the synthesis of 3,5-bis(indolyl)-1,2,4-thiadiazole derivatives (**14 a–k**). They have described the conversion of equivalent indole-3-carboxaldehydes **11 a–k** into indole-3-carbonitriles **12 a–k** with a good yield by reacting them with hydroxylamine hydrochloride in the presence of formic acid and sodium formate. Indole-3-thiocarboxamides **13 a–k** are then produced by reacting indole-3-carbonitriles with NaSH and $MgCl_2$ in the presence of dimethyl formamide (DMF). When indole-3-thiocarboxamides were further treated with iodobenzene diacetate (IBD) in anhydrous dichloromethane, moderate-to-good yields (50–70%) of bis(indolyl)-1,2,4-thiadiazoles **14 a–k** were obtained (Figure 6.6).[11]

In 2017, Chai *et al.* described a novel procedure for the one-pot synthesis of 3-alkyl-5-aryl-1,2,4-thidiazole using iodine, which produced a moderate-to-good yield of the corresponding derivatives. This procedure involves the

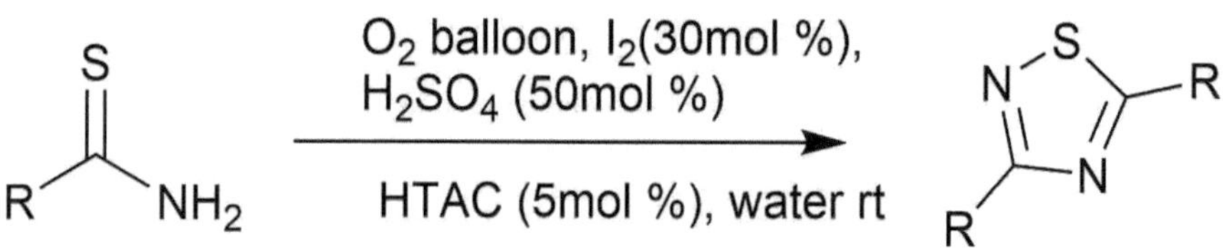

Figure 6.3 Iodine-catalyzed synthesis of 1,2,4-thiadiazole derivatives using amide.

11 12

Figure 6.4 DDQ-catalyzed synthesis of 1,2,4-thiadiazole derivatives using amide.

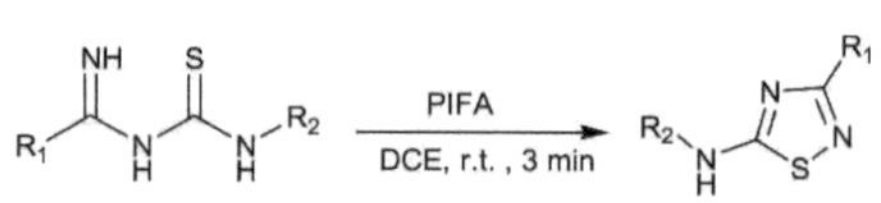

13 14

Figure 6.5 Synthesis of 1,2,4-thiadiazole derivatives *via* PIFA catalysis.

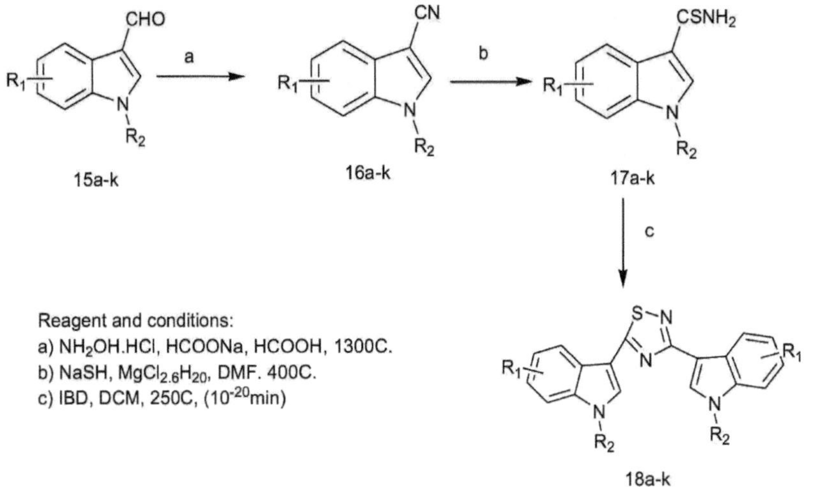

Figure 6.6 Synthesis of 3,5-bis(indolyl)-1,2,4-thiadiazole derivatives from indole-3-carboxaldehydes.

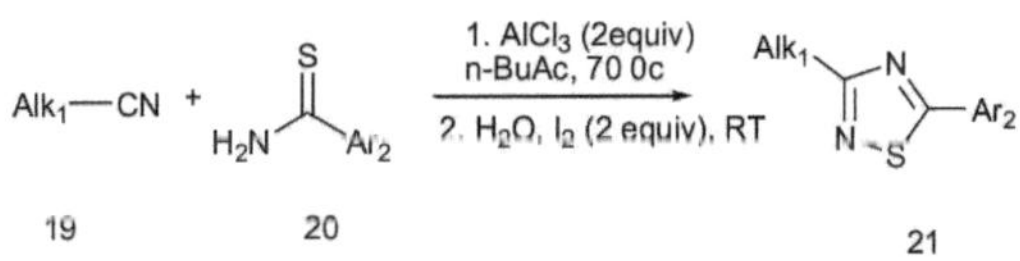

Figure 6.7 Iodine-catalyzed synthesis of 1,2,4-thiadiazole derivatives from the corresponding amide.

intermolecular addition of thioamides to nitriles in the presence of a Lewis acid ($AlCl_3$) and then the intramolecular oxidative coupling of N–H and S–H bonds with I_2 as the mediator. This process is an effective way to synthesize unsymmetrically disubstituted 1,2,4-thiadiazole derivatives (Figure 6.7).[12]

Mayhoub *et al.* (2011) described a highly effective method for the production of 3,5-diaryl-1,2,4-thiadiazole derivatives using methyl bromocyanoacetate. It was found that this procedure worked well for gram-scale reactions with simple workup, high yields, and quick reaction times, as well as for sensitive groups, such as –OH and ortho derivatives (Figure 6.8).[13]

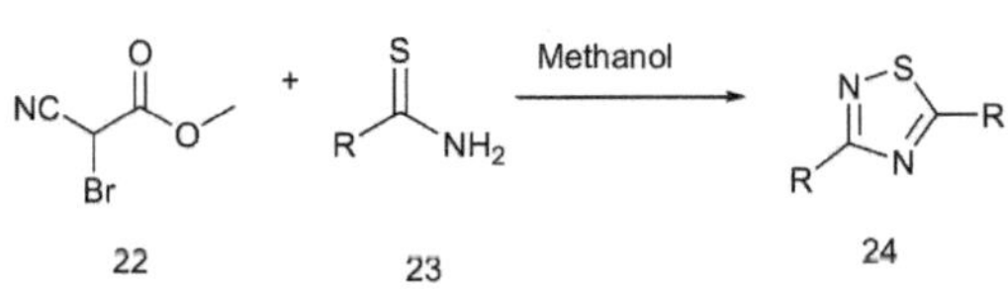

Figure 6.8 Synthesis of 1,2,4-thiadiazole derivatives from the corresponding amide and methyl bromocyanoacetate.

Halimehjani *et al.* (2019) published a highly effective and environmentally friendly oxidative dimerization process catalyzed by I_2 and oxone, yielding 3,5-bis(alkylthio)-1,2,4-thiadiazoles from *S*-alkyl dithiocarbamates. The key features of these protocols are high product yields; transition metal-free procedures; use of mild, easily accessible, affordable, and ecologically friendly oxidants; and use of green solvents, such as water and ethanol (Figure 6.9).[14]

Yadav *et al.* (2013) have shown the successful use of eosin Y as a photoredox catalyst and O_2 as an oxidant in the aerobic oxidative cyclization of primary thioamides to produce 1,2,4-thiadiazoles in the presence of visible light. This methodology offers several benefits over the current 1,2,4-thiadiazole synthetic methods, including high efficiency, sustainability, and the use of environmentally benign reagents, such as air and visible light. In addition, the reaction may withstand a large number of functional groups in a one-pot process (Figure 6.10).[15]

Chauhan *et al.* (2018) have shown the oxidative dimerization of thioamide into 1,2,4-thiadiazoles using *tert*-butyl nitrite in dichloromethane as a practical and successful method. The key features of this approach include its outstanding yields, tolerance for room temperature reactions, good functional groups, a wide choice of substrates, and metal-free conditions (Figure 6.11).[16]

Liu *et al.* (2021) have investigated the possibility of using sulfur, benzamidine hydrochloride, and phenylacetic acid as substrates for producing asymmetrical 3,5-disubstituted 1,2,4-thiadiazoles with moderate-to-good yields and copper as a catalyst. The synthesis of 1,2,4-thidiazoles was achieved *via* cyclization using phenylacetic acid as a carbon source (Figure 6.12).[17]

Pragathi *et al.* (2020) have described a method for producing chalcone-fused 2-quinoline-benzimidazole-1,2,4-thiadiazoles and investigated their

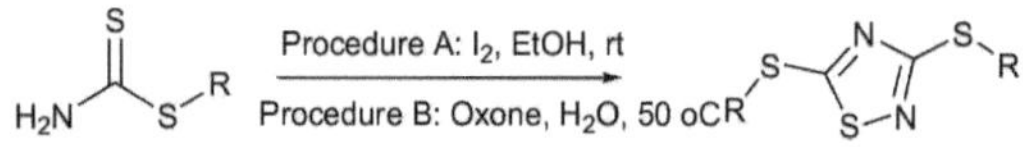

25 26

Figure 6.9 Synthesis of 1,2,4-thiadiazole derivatives from the corresponding *S*-alkyl dithiocarbamates.

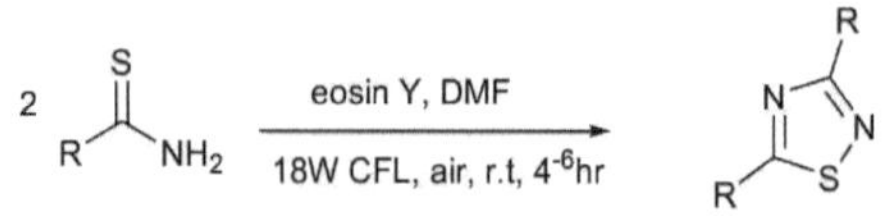

27 28

Figure 6.10 Synthesis of 1,2,4-thiadiazole derivatives from the corresponding amide using the eosin Y catalyst.

anticancer properties. The 2-oxoquinoline-3-carbaldehyde (**34**) derivatives were reacted with 4-methylbenzene-1,2-diamine in the presence of $Na_2S_2O_5$ in dimethyl formamide and refluxed for 6 h to yield compound **35**. Furthermore, it was reacted with 4-methyl formylbenzamidine hydrochloride in the presence of potassium phosphate and sulfur in DMSO at 130 °C for 12 h to yield 4-(5-(2-(1,2-dihydro-2-oxoquinolin-3-yl)-1*H*-benzo[*d*]imidazol-5-yl)-1,2,4-thiadiazol-3-yl)benzaldehyde (**36**). It then underwent an aldol condensation reaction with acetophenone derivatives in a piperidine base at reflux temperature for six hours to yield the final compounds **37** (Figure 6.13).[18]

Chacko *et al.* (2018) reported the synthesis of novel nitrogen-fused imino-1,2,4-thiadiazole derivatives. They synthesized *N*-fused-imino-1,2,4-thiadiazolo isoquinoline derivatives from 3-aminoisoquinolines and isothiocyanates using montmorillonite K-10 *via* the oxidative technique for N–S bond formation. The montmorillonite K-10 catalyst for this broad-ranging technique is non-toxic, low-cost, and environmentally friendly (Figure 6.14).[19]

Yu *et al.* (2018) found that 2-aminopyridine/amidine underwent an oxidative [3+2] annulation with isothiocyanate when Cu was present as a catalyst. It has been suggested that various *N*-fused 1,2,4-thiadiazoles can be synthesized *via* a Cu-catalyzed oxidative annulation reaction of 2-aminopyridine/amidine with isothiocyanate. This method using C–N/N–S bond formations produces a range of 5-amino/imino-substituted 1,2,4-thiadiazole derivatives using a Cu/O_2 catalytic system (Figure 6.15)[20]

Yang *et al.* (2020) examined the development of N–S bonds through electro-oxidative intramolecular dehydrogenation using imidoyl thioureas in the synthesis of 3-substituted-5-amino-1,2,4-thiadiazoles. This electrochemical approach offers a straightforward and effective way to synthesize

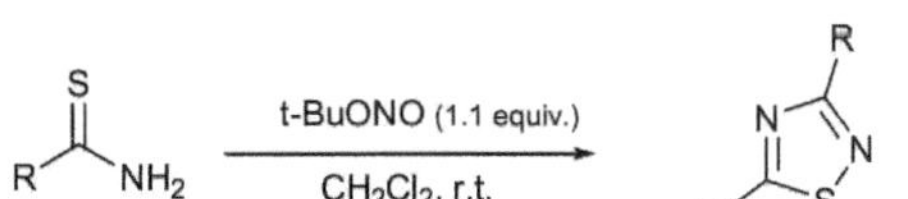

29 30

Figure 6.11 Synthesis of 1,2,4-thiadiazole derivatives from the corresponding amide in the presence of *tert*-butyl nitrite catalyst.

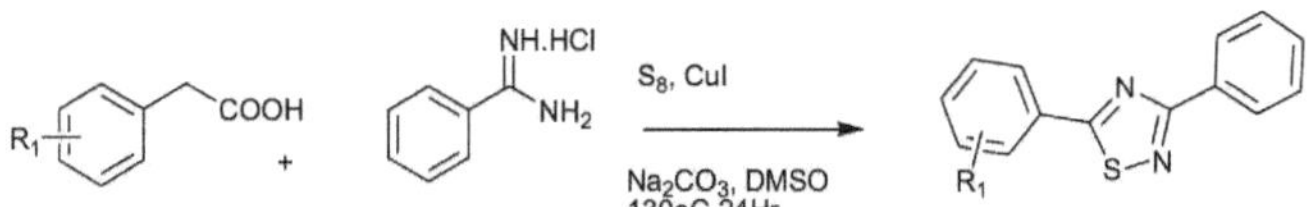

31 32 33

Figure 6.12 Synthesis of 1,2,4-thiadiazole derivatives using phenylacetic acid, benzamidine hydrochloride, and sulfur.

1,2,4-thiadiazole derivatives with very good yields using a wide range of substrates under moderate circumstances without the need for metal catalysts or stoichiometric oxidants (Figure 6.16).[21]

Chai *et al.* (2018) found a way to synthesize 3-aryl-5-amino-1,2,4-thiadiazoles *via* the one-pot reaction using imidates and thioureas. The procedure

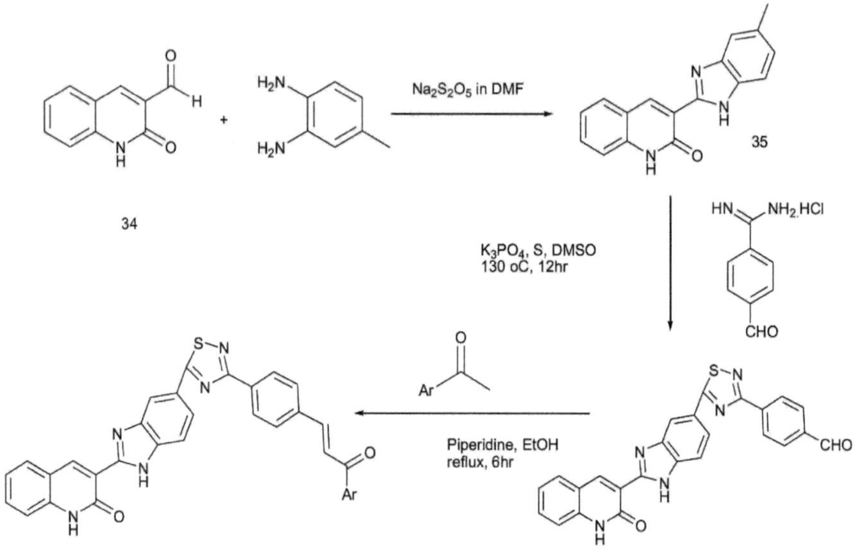

34 35 37 36

Figure 6.13 Synthesis of fused chalcone from 2-oxoquinoline-3-carbaldehyde and 2-quinoline-benzimidazole-1,2,4-thidiazoles.

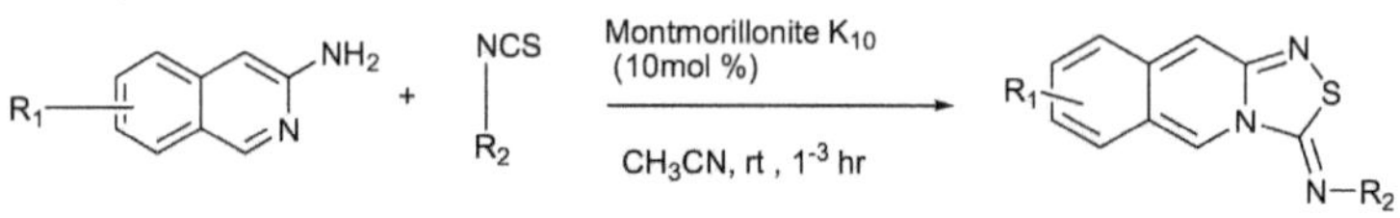

38 39 40

Figure 6.14 Synthesis of novel nitrogen-fused imino-1,2,4-thiadiazole derivative.

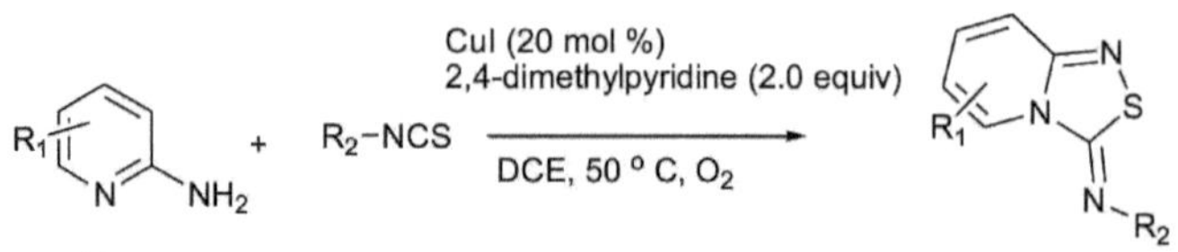

41 42

Figure 6.15 Synthesis of 1,2,4-thiadiazole compounds with 5-amino/imino substituents.

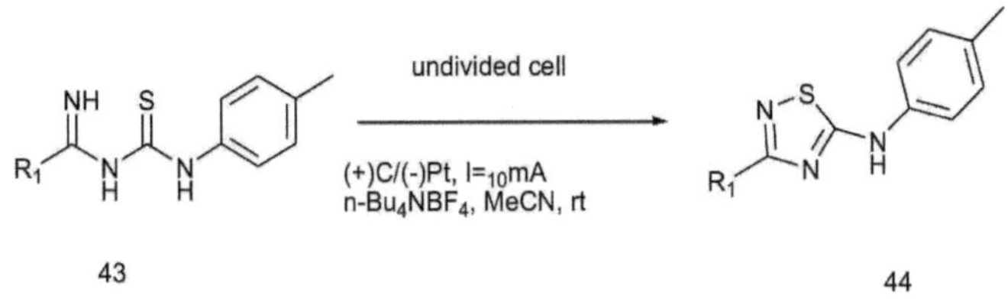

Figure 6.16 Synthesis of 3-substituted-5-amino-1,2,4-thiadiazole using imidoyl thioureas.

uses inexpensive, ecologically benign molecular iodine as the oxidant and easily accessible imidates and thioureas as the starting ingredients. The *N*-carbamothioylamidine intermediate is formed *via* a sequential base-mediated nucleophilic addition–elimination process in the one-pot reaction. The appropriate product is then obtained through I_2-mediated oxidative heterocyclization (Figure 6.17).[22]

An eco-friendly process for isothiocyanate-based 3-substituted-5-amino-1,2,4-thiadiazoles and 4,5-disubstituted/*N*-fused 3-amino-1,2,4-triazoles is provided by Jatangi *et al.* (2018). In this metal-free technique, oxidative C–N and N–S bond formations occur in aqueous medium by I_2. Water is an eco-friendly solvent in this effective and straightforward procedure, which uses non-toxic and economical molecular iodine as the catalyst (Figure 6.18).[23]

Putta *et al.* (2018) developed the first organocatalytic method for synthesizing the rarely studied 1,2,4-thiadiazole scaffolds using the corresponding carboxamides as substrates. In the presence of catalytic vitamin B_3 or thiourea under aerobic conditions, two different conditions

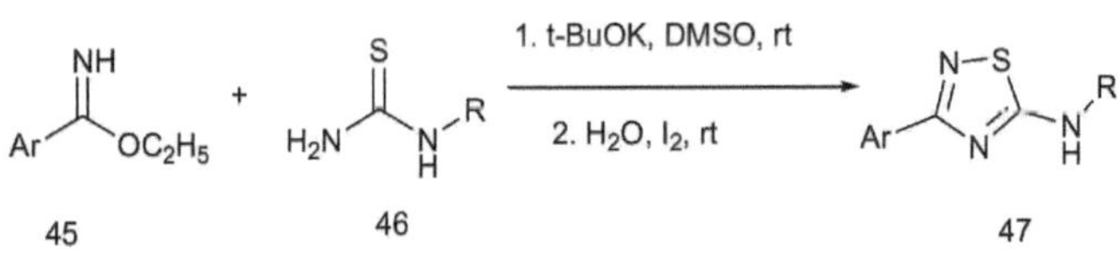

Figure 6.17 Synthesis of 3-aryl-5-amino-1,2,4-thiadiazoles *via* the one-pot reaction with imidates and thioureas.

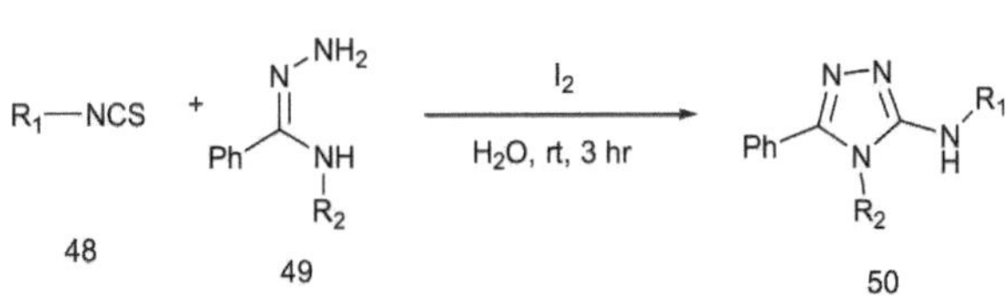

Figure 6.18 Synthesis of 3-substituted-5-amino-1,2,4-thiadiazoles and 4,5-disubstituted/*N*-fused 3-amino-1,2,4-triazoles from isothiocyanates.

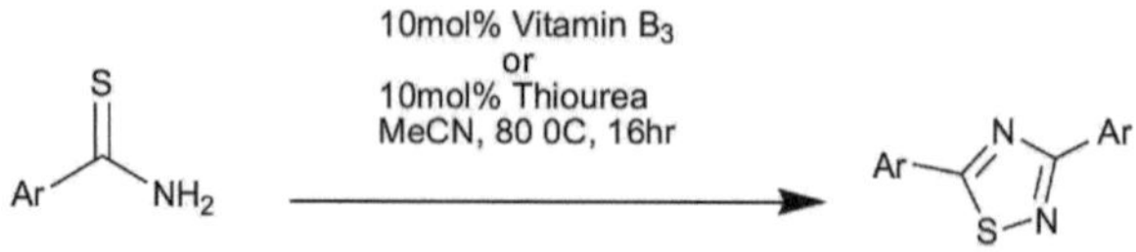

Figure 6.19 Synthesis of 1,2,4-thiadiazoles from amide in the presence of catalytic vitamin B_3 or thiourea under aerobic conditions.

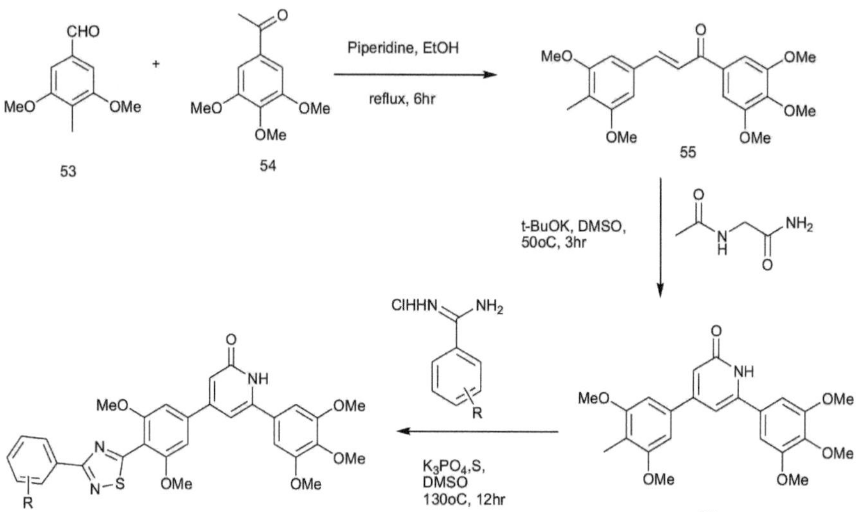

Figure 6.20 Synthesis of fused pyridone derivatives with 1,2,4-thiadiazole using aromatic aldehyde and acetophenone derivative.

were used to realize the transformations. The developed techniques have drawbacks related to earlier approaches that were documented and the expected products are obtained with very high selectivity and yields and no harmful byproducts (Figure 6.19).[24]

Perupogu *et al.* synthesized a novel series of 1,2,4-thiadiazole-fused pyridone derivatives in 2020. To ascertain their structures, they used ^{1}H-NMR, ^{13}C-NMR, and mass spectroscopy. Furthermore, the anticancer efficacy of each compound was tested against a panel of human cancer cells (Figure 6.20).[25]

A new class of heteroarene-fused [1,2,4]-thiadiazoles was reported by Chen *et al.* (2020) *via* the [3+2] oxidative cyclization of 2-aminoheteroarenes and isothiocyanates, which is mediated by iodine. A selective C–N bond is formed at the endo-nitrogen of 2-aminoheteroarenes by this highly regiospecific oxidative [3+2] annulation approach. This is followed by the production of an intramolecular oxidative N–S/N–Se bond. The

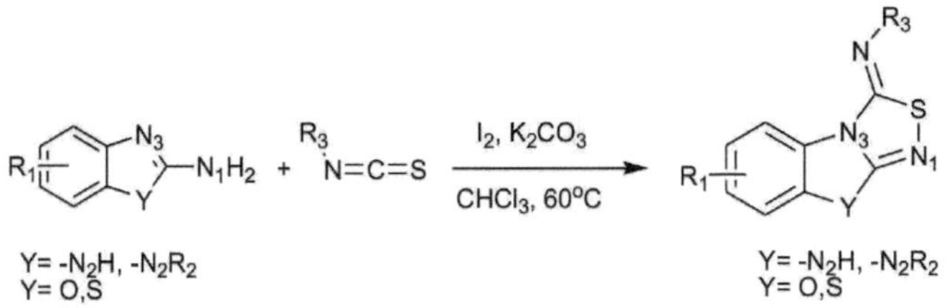

58 59

Figure 6.21 Synthesis of novel heteroarene-fused [1,2,4]-thiadiazoles derived from heteroaminoarenes.

60 TBAP 61

Figure 6.22 Synthesis of 1,2,4-thiadiazoles using tetra(*n*-butyl)ammoniumperoxydisulfate.

cyclization process mediated by iodine involves the formation of regioselective C–N bonds, which are succeeded by oxidative N–S bonds. This synthesis approach can produce novel heteroarene-fused [1,2,4]-thiadiazoles with outstanding yields, as it is flexible enough to work with a wide range of 2-aminoheteroarenes and isothiocyanates (Figure 6.21).[26]

Lee *et al.* disclosed a novel approach in 2020 for the synthesis of 1,2,4-thiadiazoles using tetra(*n*-butyl)ammoniumperoxydisulfate (TBAP) *via* oxidative dimerization of thioamides. With this new oxidative protocol, 1,2,4-thiadiazoles can be practically synthesized from a variety of primary thioamides with excellent yields. It is also very easy to use and highly efficient. The benefits of this approach include TBAP is a stable, easily accessible, and recyclable solid oxidant, use of a broad range of substrates, and excellent yields (Figure 6.22).[27]

6.3 Biological Applications

Several articles reported in the literature address the various biological activities of 1,2,4-thiadiazole derivatives. The most recent and relevant research has demonstrated that these derivatives have a wide range of pharmacological activities that fall under the following categories.

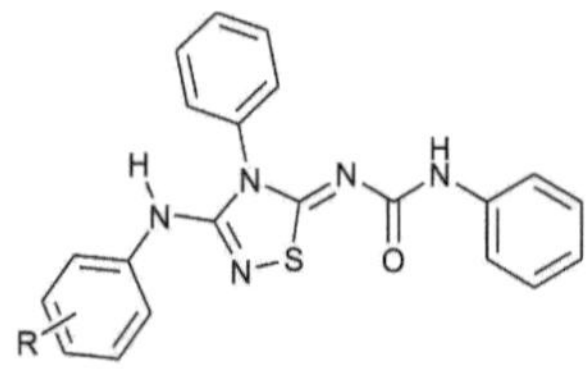

62

Figure 6.23a Structure of 3-(*Z*)-1-phenyl-3-(4-phenyl-3-(phenylamino)-1,2,4-thiadiazol-5(4*H*)-ylidene)urea derivative showing anticonvulsant activity.

6.3.1 Anticonvulsant Activity

The investigation into the anticonvulsant activity of 1,2,4-thiadiazoles holds significance as it explores the importance of these derivatives in the development of novel therapeutic agents for the management of seizures and epilepsy. Understanding their pharmacological properties and mechanisms of action may contribute to the advancement of anticonvulsant drugs, addressing unmet medical needs in neurological disorders.

Siddiqui *et al.* synthesized a series of 1,2,4-thiadiazoles **62** and investigated their anticonvulsant efficacy. The molecule with *para*-chloro substitution showed the highest level of activity in the maximal electroshock (MES) test and partially prevented strychnine seizures, while other compounds in the series were less active (Figure 6.23a).

6.3.2 Anticancer Activity

Exploring the anticancer activity of 1,2,4-thiadiazoles is of paramount importance as it unveils the potential of these compounds as novel agents for cancer therapy. Investigating their mechanisms of action and cytotoxic effects on cancer cells could lead to the evolution of promising anticancer drugs, offering new avenues for the treatment of various malignancies.

Pragathi *et al.* designed, synthesized, and tested all the derivatives of 1,2,4-thiadiazole-1,2,4-triazole having an amide functional group **63** using the MTT assay for anticancer activity against four distinct human cancer cell lines, such as MCF-7, MDA MB-231, A549, and DU-145, for breast cancer, lung cancer, and prostate cancer (Figure 6.23b).[28]

Perlovich *et al.* synthesized novel 1,2,4-thiadiazole derivatives **64** that are powerful neuroprotectors[29] (Figure 6.24).

Oyebamiji *et al.* reported the synthesis of a number of 1,2,4-thiadiazole-1,2,4-triazole derivatives and investigated the anti-epidermal growth factor receptor kinase activities of their analogs, resulting in a reduction in the incidence of human lung cancer[30] (Figure 6.25).

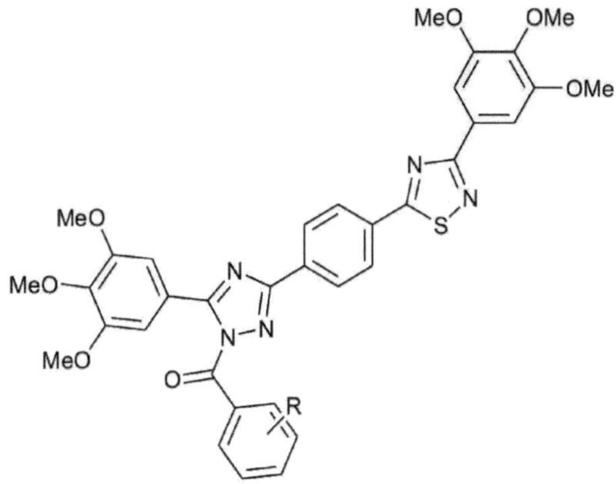

63

Figure 6.23b Structure of (5-(3,4,5-trimethoxyphenyl)-3-(4-(3-(3,4,5-trimethoxyphenyl)-1,2,4-thiadiazol-5-yl)phenyl)-1*H*-1,2,4-triazol-1-yl)(phenyl)methanone exhibiting anticancer activity.

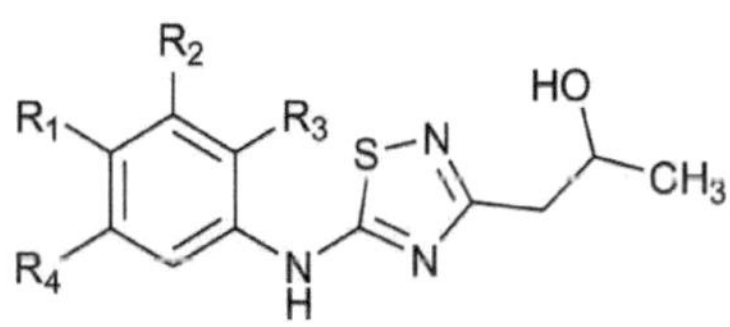

64

Figure 6.24 Structure of 1-(5-(phenylamino)-1,2,4-thiadiazol-3-yl)propan-2-ol showing anticancer activity.

Proshin *et al.* synthesized and screened a series of 3,5-disubstituted-1,2,4-thiadiazole derivatives **69** and **70** for the selectivity of cytotoxic activity toward cancer cells (Figure 6.26).[31]

Sudhakar *et al.* synthesized a series of 1,2,4-thiadiazole derivatives containing 1,2,4-oxadiazole **71**. These compounds showed moderate anticancer efficacy when tested against four different human cancer cell lines, such as MCF-7 (breast cancer), Colo-205 (colon cancer), A549 (lung cancer), and A2780 (ovarian cancer) (Figure 6.27).[32]

Chaithanya *et al.* reported a procedure for the preparation and assessment of 1,2,4-triazole derivatives **72** to study their *in vitro* anticancer activity against a panel of four human cell lines, including MCF-7 (breast cancer), Colo-205 (colon cancer), A549 (lung cancer), and A2780 (ovary cancer). This was performed using the MTT reduction assay (Figure 6.28).[33]

Pragathi *et al.* produced 1,2,4-thiadiazole-1,2,4-triazole derivatives **73** and **74**. Using the MTT reduction assay, all compounds were tested for their

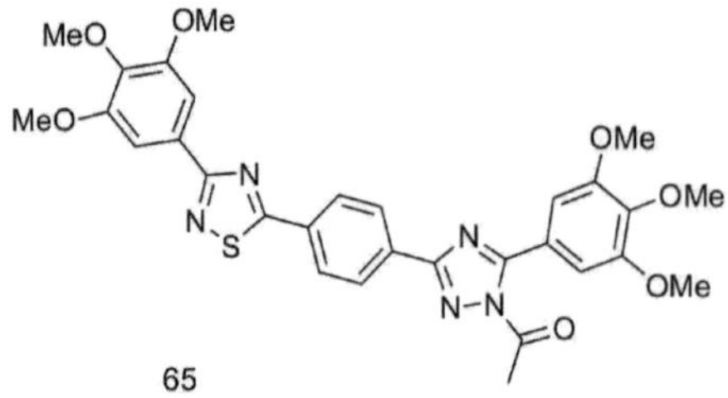

65

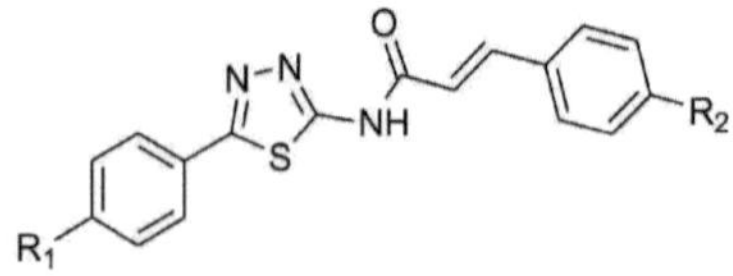

66: R_1, R_2= OCH_3

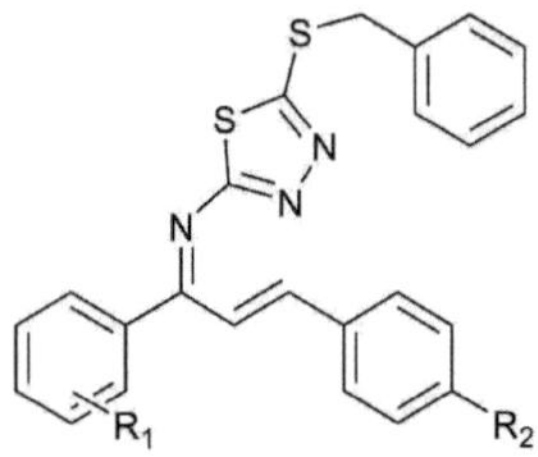

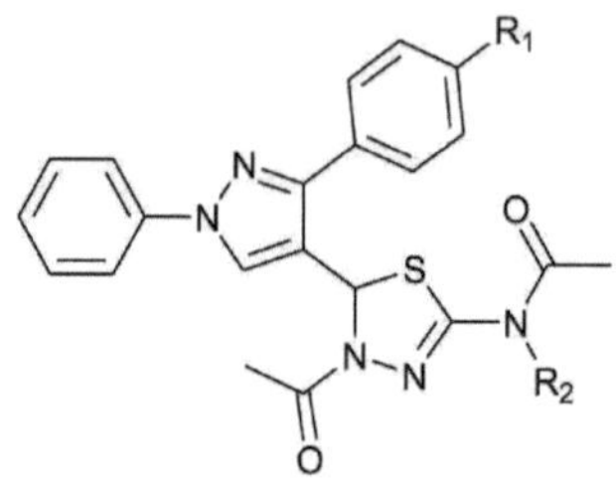

68

67a: R_1, = OCH_3 ,R_2= Cl
67b: R_1= $(CH_3)_2N$, R_2= Cl

Figure 6.25 Structures of 1,2,4-thiadiazole-1,2,4-triazole derivatives showing anticancer activity.

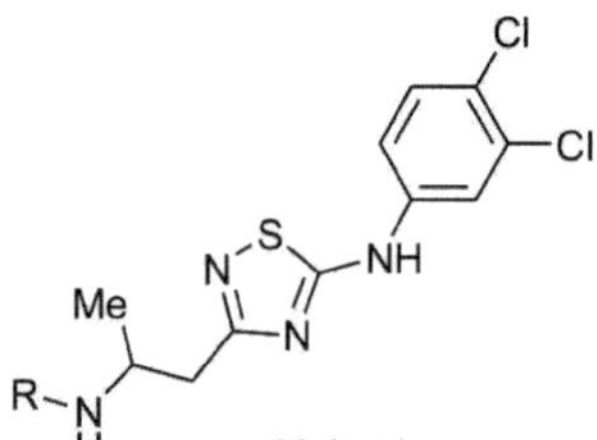

69 (a-g)

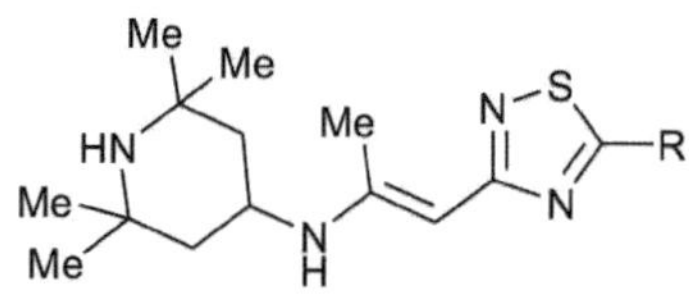

70(a-h)

Figure 6.26 Structures of 3,5-disubstituted 1,2,4-thiadiazole derivatives showing anticancer activity.

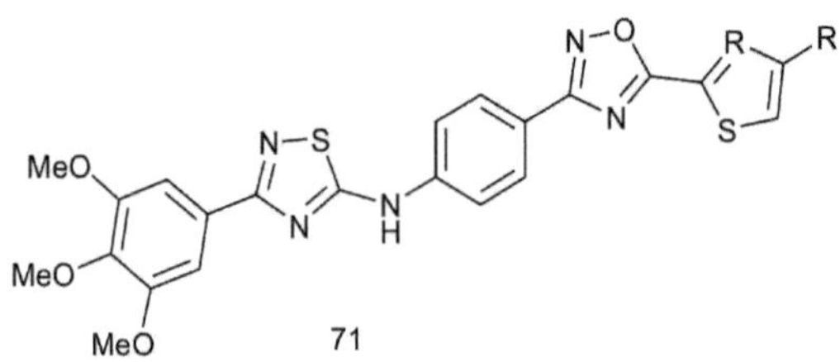

71

Figure 6.27 Structure of 1,2,4-thiadiazole derivatives bearing 1,2,4-oxadiazole with anticancer activity.

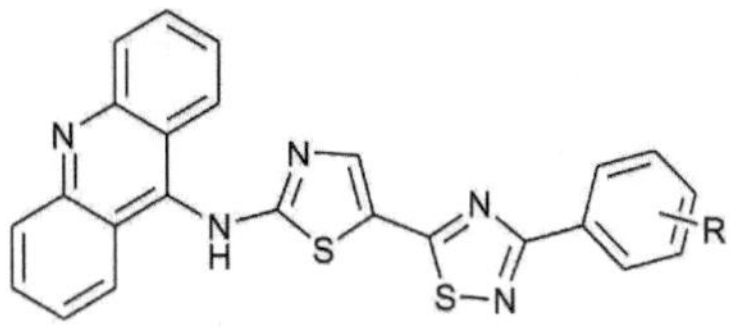

72

Figure 6.28 Structure of 1,2,4-thiadiazole derivatives with anticancer activity.

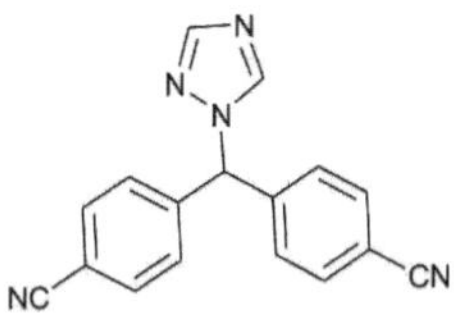

73

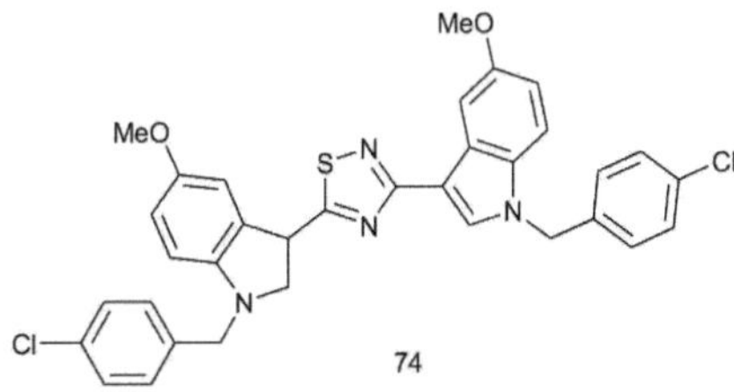

74

Figure 6.29 Structures of 1,2,4-thiadiazole-1,2,4-triazole derivatives showing anticancer activity.

anticancer activities against four distinct cancer cell lines, such as MCF-7 for breast cancer, A549 for prostate cancer, MDA MB-231 for lung cancer, and DU-145 for prostate cancer. Etoposide functions as a standard medication in these tests (Figure 6.29).[34]

A series of novel 1,2,4-thiadiazole-fused imidazo[1,2-*b*]pyridazine derivatives **75** I (R = H, 4-Me, 4-Br, *etc.*) were synthesized by Chidella *et al.* and screened for their anticancer activities against several cancer cell lines, such as breast cancer (MCF-7 and MDA MB-231), lung cancer (A549), and prostate cancer (DU-145) using the MTT assay (Figure 6.30).[35]

Premalatha *et al.* showed that 1,2,4-thiadiazole-fused combretastatin-A4 derivatives had promising anticancer activity against a variety of human cancer cell lines, including MCF-7, A549, MDA MB-231, and HeLa (cervical cancer). Combitastatin-A4 is a clinical drug that is commonly used as a reference (Figure 6.31).[36]

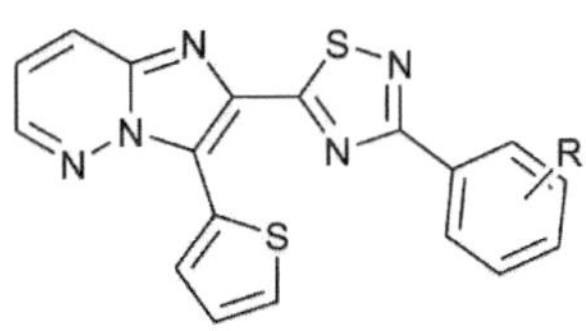

75

Figure 6.30 Structure of 1, 2,4-thiadiazole-fused imidazo[1,2-*b*]pyridazine derivatives showing anticancer activity.

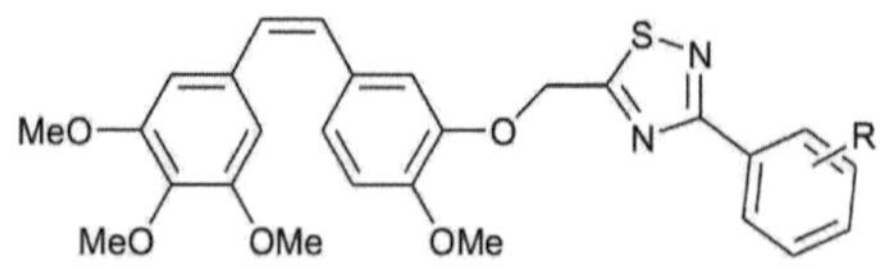

76

Figure 6.31 Structure of 1,2,4-thiadiazole-fused combretastatin-A4 derivatives with anticancer activity.

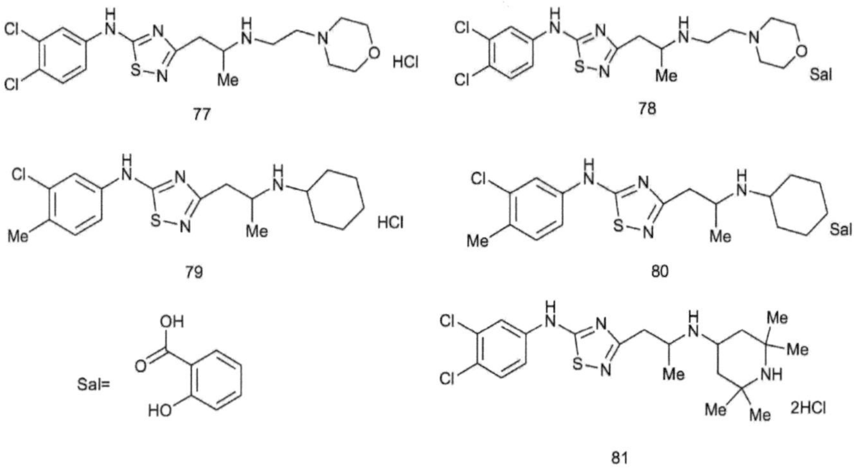

Figure 6.32 Structures of 5-anilino-1,2,4-thiadiazole derivatives showing anticancer activity.

Orlova *et al.* described a process for synthesizing a range of 5-anilino-1,2,4-thiadiazole derivatives **77–81** and screened them for their antitumor activity against leukemia (Figure 6.32).[37]

Pragathi *et al.* synthesized a number of 1,2,4-thiadiazol-1-(benzo[*d*]imidazol-2-yl)quinolin-2-(1*H*)-one chalcone derivatives **82** and evaluated their anticancer activities (Figure 6.33).[38]

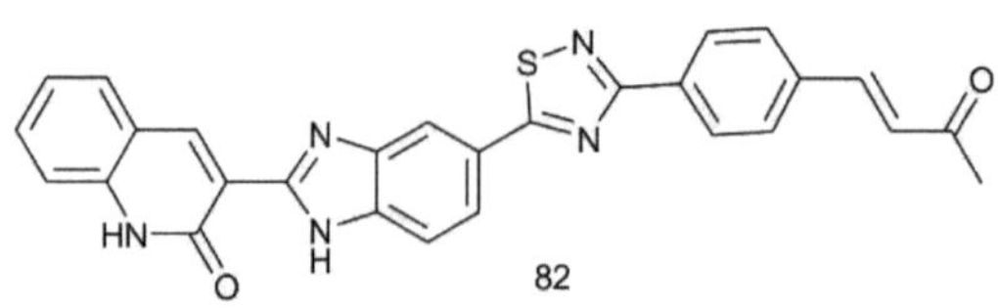

82

Figure 6.33 Structure of chalcone derivatives of 1,2,4-thiadiazol-1-(benzo[*d*]imidazol-2-yl)quinolin-2-(1*H*)-one with anti cancer activity.

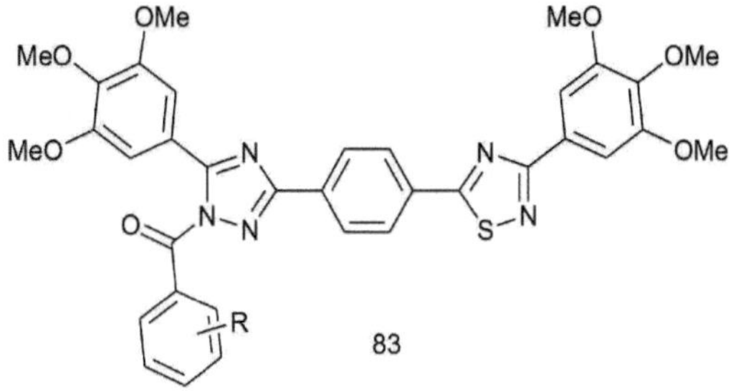

83

Figure 6.34 Structure of 1,2,4-thiadiazole-1,2,4-triazole derivatives showing anticancer activity.

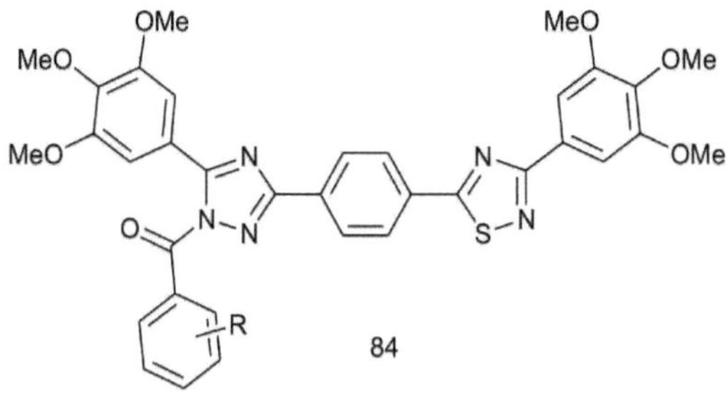

84

Figure 6.35 Structure of 1,2,4-thiadiazole-1,2,4-triazole compounds that possess the amide functionality showing anticancer activity.

Oyebamiji *et al.* synthesized a range of 1,2,4-thiadiazole-1,2,4-triazole **83** analogs to inhibit the downregulation of human lung cancer caused by the epidermal growth factor receptor kinase (Figure 6.34).[39]

Devi *et al.* synthesized 1,2,4-thiadiazole-1,2,4-triazole compounds with the amide functionality and investigated their anticancer activities (Figure 6.35).[40]

Perupogu *et al.* synthesized a series of 1,2,4-thiadiazol-fused benzoxazole-quinazoline derivatives **85** (Ar = 2-thiophenyl, 4-MeCH, 3,5 Cl C H, *etc.*) and screened them for anticancer activity against human cancer cell lines that included MCF-7 (breast cancer), A549 (lung cancer), Colo-205 (colon cancer), and A2780 (ovarian cancer) (Figure 6.36).[41]

Perupogu *et al.* manufactured a number of 1,2,4-thiadiazole-fused pyridone derivatives **86**, where R can be any of the following: H, 4-Me, 4-Cl, *etc.* The MTT assay was used to evaluate the anticancer activity of each compound against a group of human cancer cell lines including A549 (lung cancer), MCF-7 (breast cancer), A2780 (ovarian cancer), and Colo-205 (colon cancer) (Figure 6.37).[42]

Chaithanya *et al.* synthesized a series of 1,2,4-thiadiazole-fused acridine derivatives **87** (R = H, 4-Me, 4-MeO, *etc.*). The MTT reduction assay method was used to evaluate the anticancer activity of each compound against a

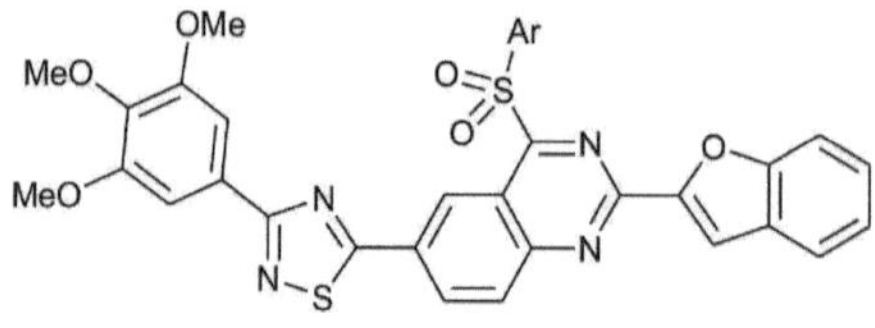

85

Figure 6.36 Structure of 1,2,4-thiadiazole-fused benzoxazole-quinazoline derivatives showing anticancer activity.

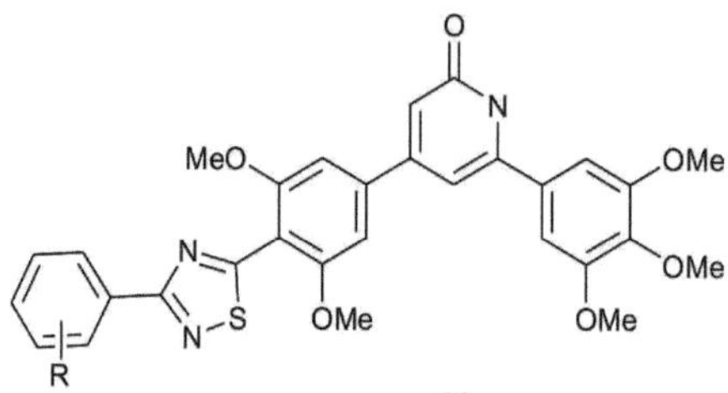

86

Figure 6.37 Structure of 1,2,4-thiadiazole-fused pyridone derivatives showing anticancer activity.

panel of four human cancer cell lines, including MCF-7 (breast cancer), A549 (lung cancer), Colo-205 (colon cancer), and A2780 (ovarian cancer) (Figure 6.38).[43]

Alshabanah *et al.* synthesized coumarin derivatives **88** and studied their anticancer activity (Figure 6.39).[44]

Ashok *et al.* synthesized a series of 1,2,4-oxadiazole-fused 1,2,4-thiadiazole pyrimidines **89** (R = H, 4-Me, 4-Cl, *etc.*). MCF-7 (human breast cancer), A549 (human non-small-cell lung cancer), Colo-205 (human adenocarcinoma colon cancer), and A2780 (human ovarian cancer) were among the human cell lines that were submitted to the MTT assay to determine whether they possess anticancer properties or not (Figure 6.40).[45]

Adepoju *et al.* synthesized 1,2,4-thiadiazole-1,2,4-triazole derivatives **90** and screened them for lung cancer activity (Figure 6.41).[46,47]

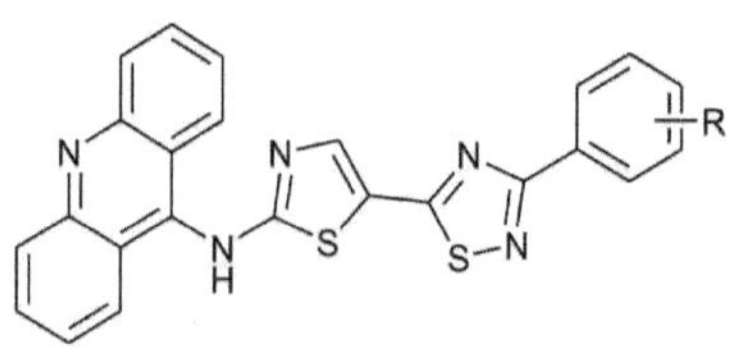

87

Figure 6.38 Structure of 1,2,4-thiadiazole-fused acridine derivatives showing anticancer activity.

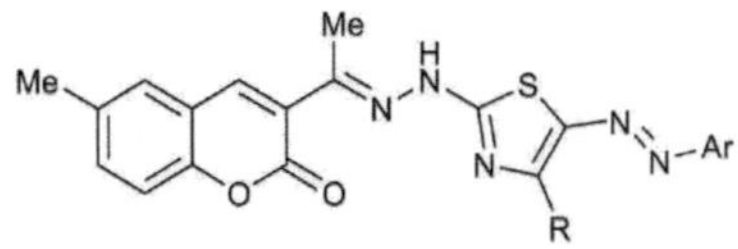

88

Figure 6.39 Structure of coumarin derivatives showing anticancer activity.

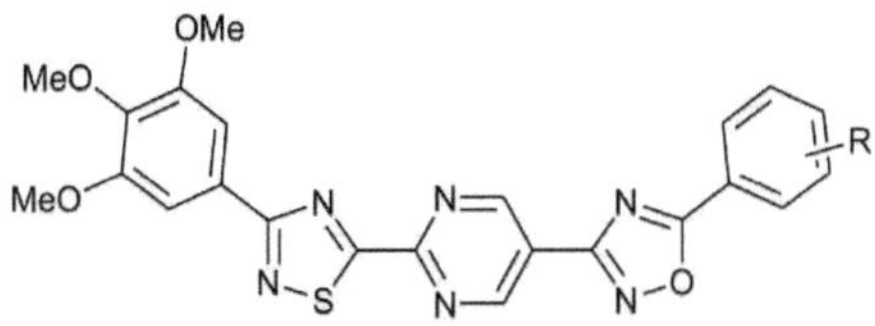

89

Figure 6.40 Structure of 1,2,4-oxadiazole-fused 1,2,4-thiadiazole pyrimidines with anticancer activity.

Ruijter *et al.* synthesized (3-aryl-1,2,4-thiadiazol-5-ylthio)acetamides **91** and **92** and related compounds for the inhibition of mycobacterial infections such as a *Mycobacterium tuberculosis* infection, *e.g.*, in a respiratory system and/or extrapulmonary organs[48] (Figure 6.42).

6.3.3 Central Nervous System Activity

The examination of the central nervous system (CNS) activity of 1,2,4-thiadiazoles is crucial for understanding their potential neurological effects. Investigating their impact on the CNS can provide valuable insights into their pharmacological properties, including neuroprotective or neurotoxic effects, and may contribute to the development of therapeutic interventions for neurological disorders. This research is essential for assessing the safety

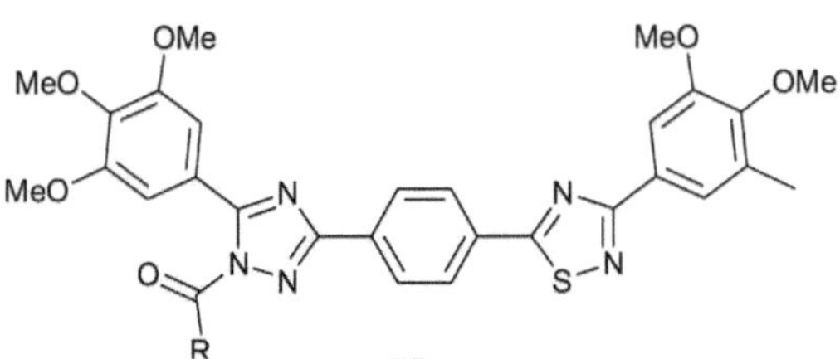

90

Figure 6.41 Structure of 1,2,4-thiadiazole-1,2,4-triazole derivatives showing anticancer activity.

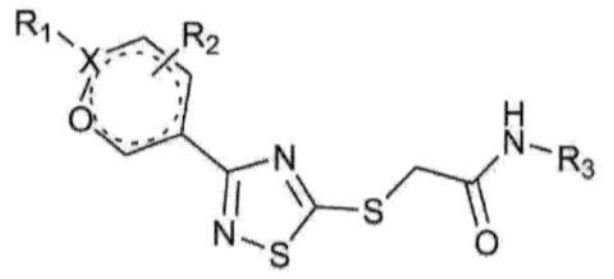

91

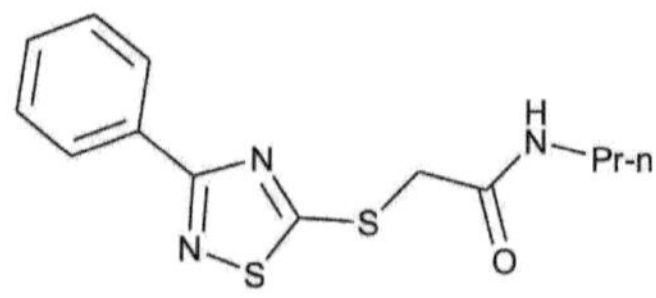

92

Figure 6.42 Structure of (3-aryl-1,2,4-thiadiazol-5-ylthio)acetamide derivatives showing central nervous system activity.

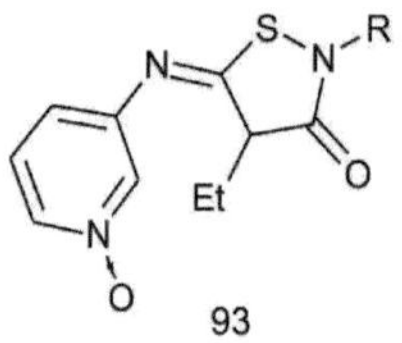

93

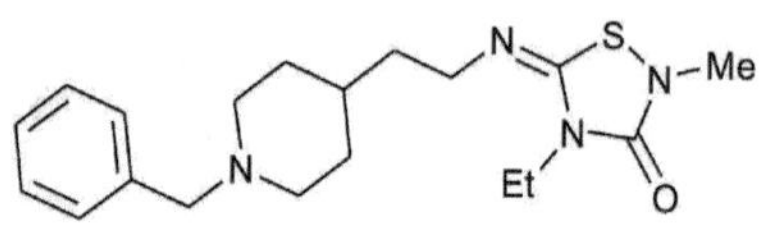

94

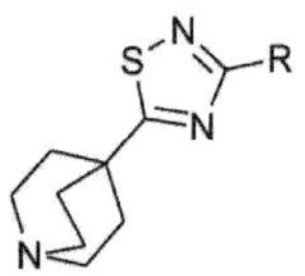

95

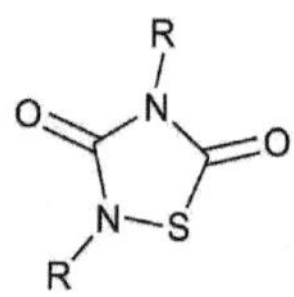

96

Figure 6.43 Structure of 1,2,4-thiadiazole derivatives showing central nervous system activity.

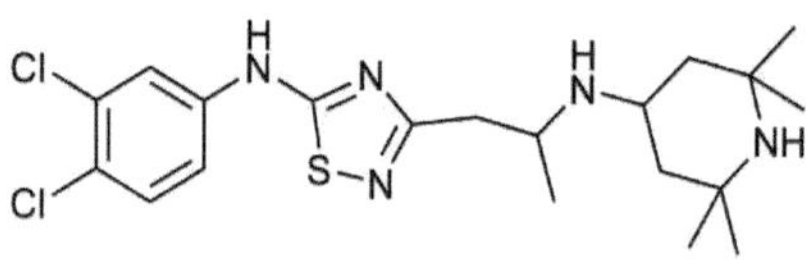

97

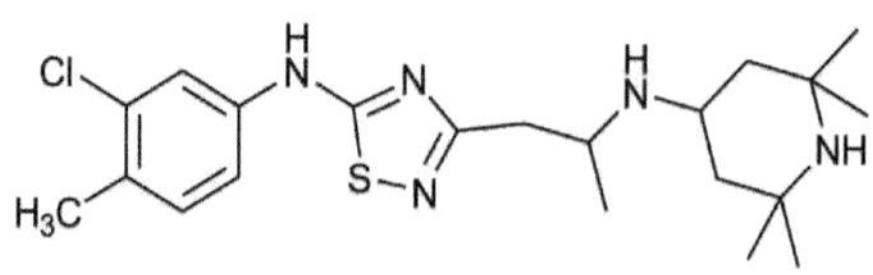

98

Figure 6.44 Structure of 1,2,4-thiadiazole derivatives showing central nervous system activity.

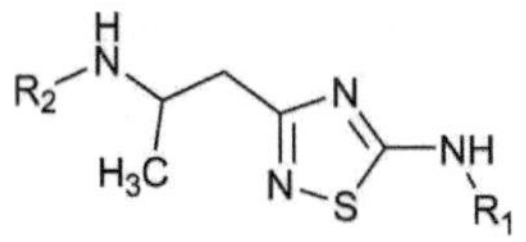

99

Figure 6.45 Structure of 1,2,4-thiadiazole derivatives showing central nervous system activity.

and efficacy of 1,2,4-thiadiazoles as potential candidates for neurological medications.

Castro *et al.* produced a series of 1,2,4-thiadiazole derivatives **93–96**, which were utilized for the purpose of investigating the function of the central nervous system (Figure 6.43).[49]

Volkova *et al.* reported the synthesis of a series of 1,2,4-thiadiazole derivatives **97** and **98** and conducted research on the pharmacologically significant features of putative neuroprotectors based on 1,2,4-thiadiazole (Figure 6.44).[50]

Gupta *et al.* reported the synthesis of numerous 1,2,4-thiadiazole derivatives **99** and discovered them as powerful neuroprotectors. The ability of these substances to block the absorption of Ca^{2+} that is induced by glutamate was studied (Figure 6.45).[51]

6.3.4 Cathepsin B Inhibitor

The exploration of the cathepsin B inhibitor activity of 1,2,4-thiadiazoles is significant as it sheds light on their potential role in modulating protease activity under various pathological conditions, including cancer and neurodegenerative diseases. Understanding their cathepsin B inhibitory effects may unveil new therapeutic possibilities and contribute to the development of targeted therapies for diseases associated with dysregulated proteolysis. This research has implications for drug discovery and the potential treatment of conditions where cathepsin B plays a crucial role.

Leung-Toung *et al.* synthesized a new type of cathepsin B inhibitors with a 1,2,4-thiadiazole heterocycle serving as the thiol-trapping pharmacophore (Figure 6.46).[52]

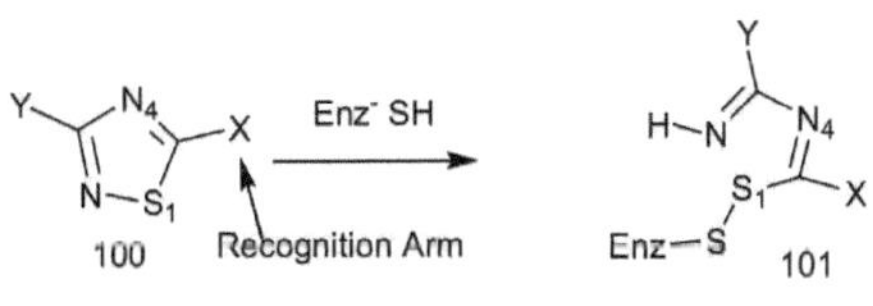

Figure 6.46 Structure of 1,2,4-Thiadiazole heterocycle showing the cathepsin B inhibitor activity.

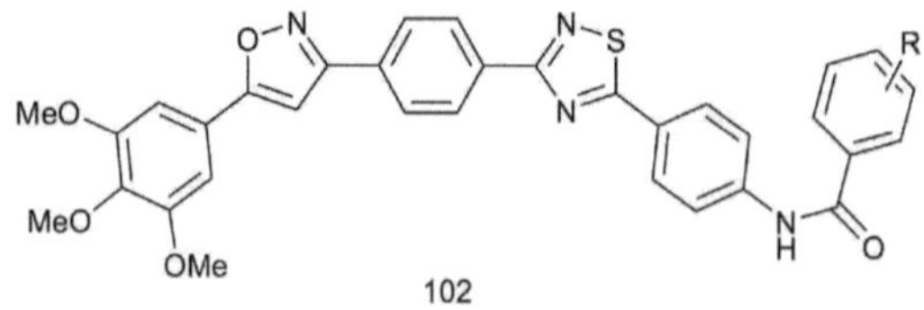

102

Figure 6.47 Structure of amide derivatives containing 1,2-isoxazole and 1,2,4-thiadiazole showing cathepsin B inhibitor activity.

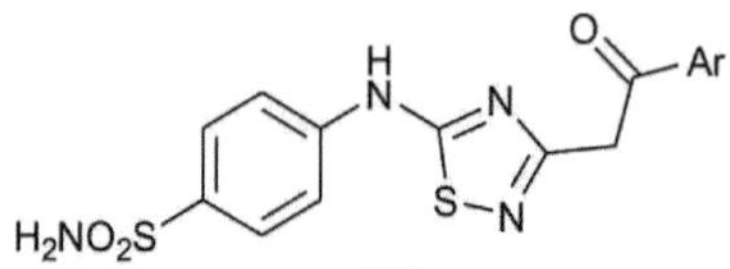

103

Figure 6.48 Structure of benzenesulfonamides bearing pyrazoles and 1,2,4-thiadiazole derivatives showing carbonic anhydrase inhibitor activity.

Shahinshavali *et al.* synthesized a series of amide derivatives containing 1,2-isoxazole and 1,2,4-thiadiazole moieties **102** (R = H, 4-MeO, 4-Br, *etc.*). All products were examined for their anticancer activity against four types of human cancer cell lines, including MCF-7 (breast cancer), Colo-205 (colon cancer), A2780 (ovarian cancer), and A549 (lung cancer). Most of the compounds exhibited good anticancer activity (Figure 6.47).[53]

Kumar *et al.* synthesized novel benzenesulfonamides bearing pyrazoles and 1,2,4-thiadiazoles **103** and screened them for their carbonic anhydrase inhibitor activity (Figure 6.48).[54]

6.3.5 Anticholinesterase Activity and Antioxidant Properties

The investigation into the anticholinesterase activity and antioxidant properties of 1,2,4-thiadiazoles is of great importance, as it explores the potential dual functionality of these compounds. Understanding their ability to inhibit cholinesterase enzymes and act as antioxidants holds promise for therapeutic interventions in conditions involving oxidative stress and cholinergic dysregulation, such as neurodegenerative disorders. This research contributes to the development of multifunctional compounds with potential applications in neuroprotection and antioxidant therapy.

Grigoriev *et al.* synthesized a series of 5-anilino-1,2,4-thiadiazole derivatives **104** with different substituents at position 3 and studied anticholinesterase and antioxidant properties (Figure 6.49).[55]

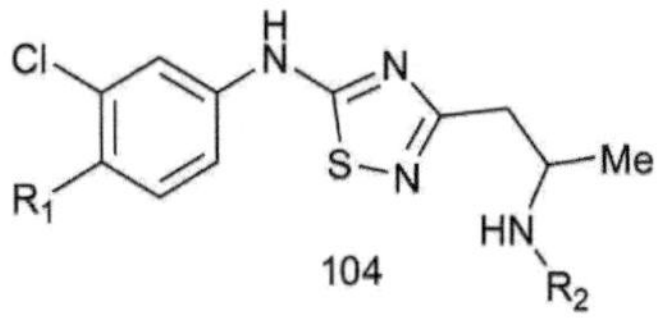

104

Figure 6.49 Structure of 5-anilino-1,2,4-thiadiazole derivatives showing anticholinesterase activity and antioxidant properties.

6.3.6 Alzheimer's Disease

The exploration of 1,2,4-thiadiazoles for Alzheimer's disease is crucial for evaluating their potential therapeutic impact. An investigation into their neuroprotective properties, ability to modulate key pathways involved in Alzheimer's pathology, and potential for mitigating neurodegeneration may contribute to the development of targeted treatments or interventions for this debilitating neurological condition. This research holds promise for advancing our understanding of Alzheimer's disease and exploring novel avenues for therapeutic development.

1,2,4-Thiadiazole derivatives **105** as new potential multifunctional agents for the treatment of Alzheimer's disease were synthesized by Makhaeva *et al.* (Figure 6.50).[56]

Makhaeva *et al.* synthesized a series of conjugates of tacrine with 1,2,4-thiadiazoles **106** as highly effective cholinesterase inhibitors, blockers of *N*-methyl-D-aspartate (NMDA) receptors and antioxidants (Figure 6.51).[57]

Makhaeva *et al.* synthesized a series of new 3-substituted-5-anilino-1,2,4-thiadiazoles **106a–e** containing different substituents at position 3 of the thiadiazole ring, and in particular, their esterase profile and antioxidant properties were studied in neurodegenerative diseases and Alzheimer's disease (Figure 6.52).[58]

Burovikhina *et al.* synthesized 1,2,4-thiadiazole derivatives **107** and studied their neuroprotective potential and high activity in the treatment of Alzheimer's disease (Figure 6.53).[59]

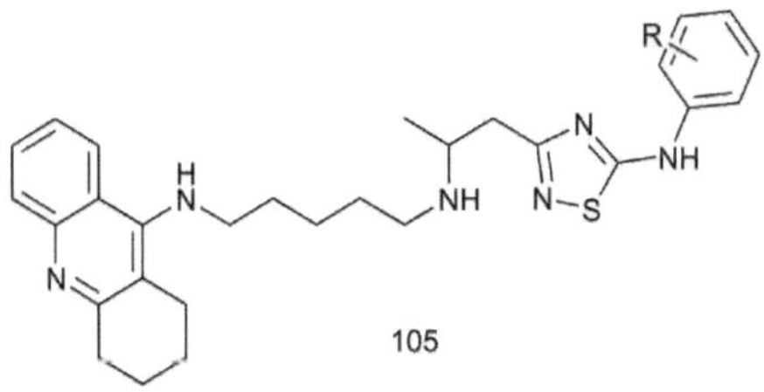

105

Figure 6.50 Structure of 1,2,4-thiadiazole derivatives as agents for the treatment of Alzheimer's disease.

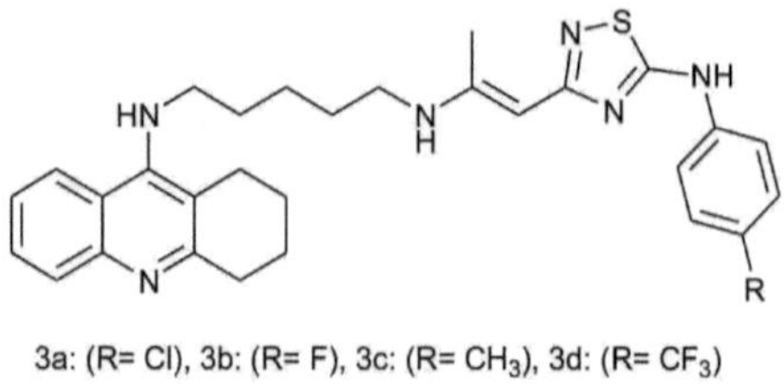

106

Figure 6.51 Structure of conjugates of tacrine with 1,2,4-thiadiazole derivatives showing antioxidant properties.

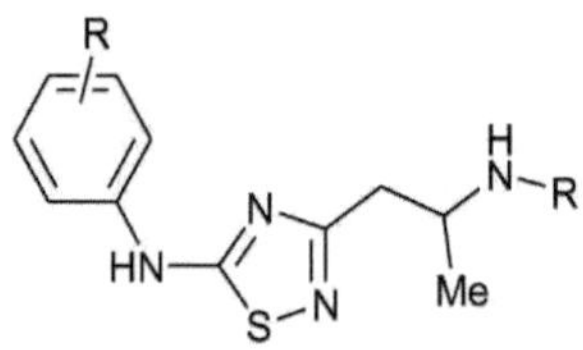

106 a-e

Figure 6.52 Structure of a series of new 3-substituted-5-anilino-1,2,4-thiadiazoles showing antioxidant properties and agents for Alzheimer's disease.

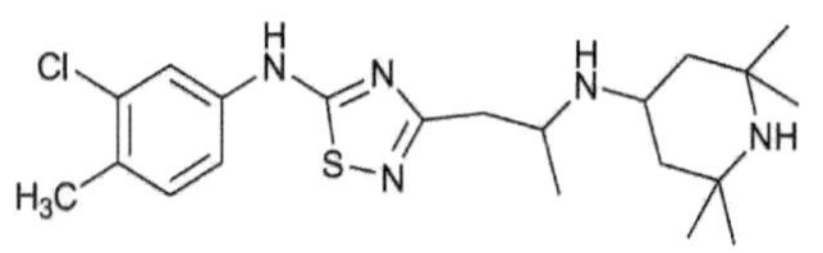

107

Figure 6.53 Structure of 1,2,4-thiadiazole derivatives showing high activity in the treatment of Alzheimer's disease.

6.3.7 Anticonvulsant Activity

Gupta *et al.* synthesized a series of some alkyl ([[(5-imino-4-aryl-4,5-dihydro-1,2,4-thiadiazol-3-yl)carbamoyl]amino)acetate derivatives **108** and **109** (Ar = Ph, 4-chlorophenyl, 4-nitrophenyl, and 4-methylphenyl) having ideal anticonvulsant activity (Figure 6.54).[60]

Gupta *et al.* synthesized a series of newly substituted 1,2,4-thiadiazoles **110** by an appropriate route and screened them for anticonvulsant activity (Figure 6.55).[61]

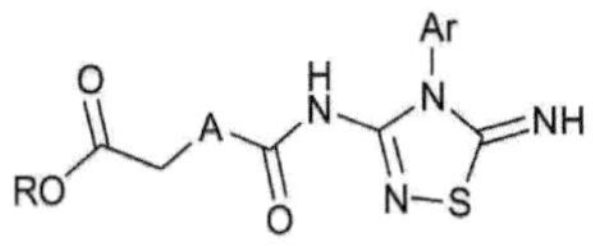

108

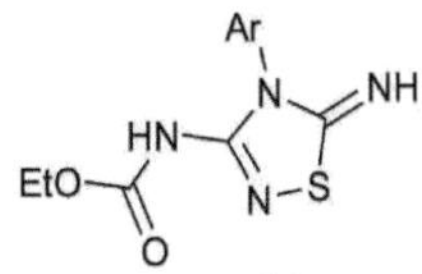

109

Figure 6.54 Structure of alkyl ([(5-imino-4-aryl-4,5-dihydro-1,2,4-thidiazol-3-yl)carbamoyl]amino)acetate derivatives showing anticonvulsant activity.

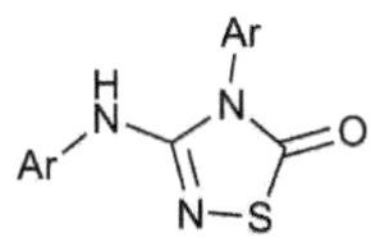

110

Figure 6.55 Structure of substituted 1,2,4-thiadiazole derivatives showing anticonvulsant activity.

6.3.8 Plant Growth Regulator Herbicide

Osminin *et al.* synthesized a series of (*E*) 3-phenyl-2-cyanoprop-2-enethioamides for electrochemical oxidation in an undivided cell in the presence of KBr in an aqueous or aqueous–organic medium, which led to the formation of (2*E*,2′*E*)-2,2′-(1,2,4-thiadiazole-3,5-diyl)bis[3-phenylacrylonitriles] **111** (R = 4 OH, 4-MeO, 2-Cl, or 4-N(Me))[62] (Figure 6.56).

Barber *et al.* synthesized a series of substituted 1,2,4-thiadiazole salts and their use as herbicidally active substances[63] (Figure 6.57).

Krivokolysko *et al.* synthesized a series of substituted 1,2,4-thiadiazoles **113** and studied the herbicide safening effect in the experiments with sunflower seedlings[64] (Figure 6.58).

Hussain *et al.* synthesized a series of substituted 1,3,4-thiadiazole-fused-[1,2,4]-thiadiazole incorporating the 1,4-benzodioxine moiety and studied *in vitro* and *in silico* as potent inhibitors of α-amylase and α-glycosidase[65] (Figure 6.59).

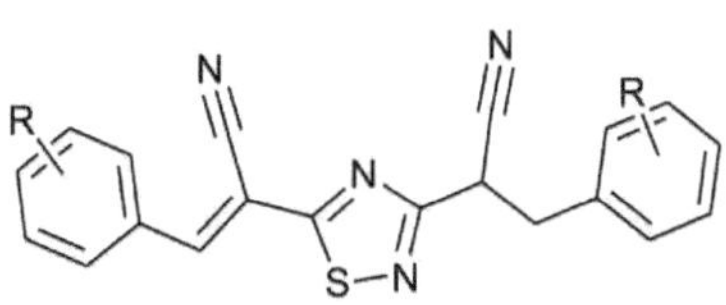

111

Figure 6.56 Structure of (2*E*,2′*E*)-2,2′-(1,2,4-thiadiazole-3,5-diyl)bis[3-phenylacrylonitriles] have plant growth regulator herbicide activity.

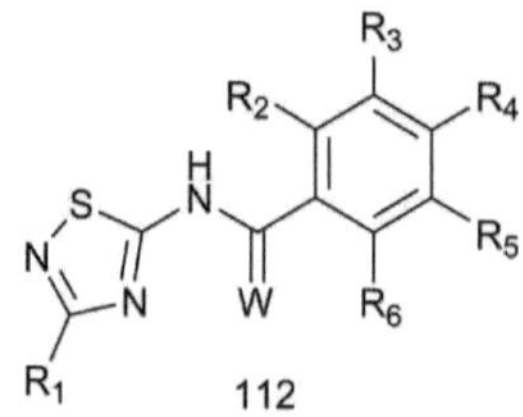

112

Figure 6.57 Structure of 1,2,4-thiadiazole derivatives showing herbicidal activity.

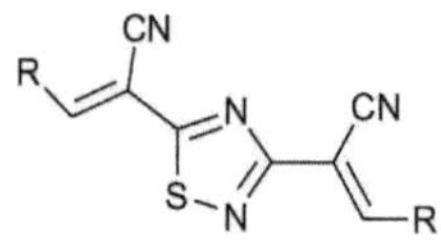

113

Figure 6.58 Structure of 1,2,4-thiadiazoles showing plant growth regulator herbicidal activity.

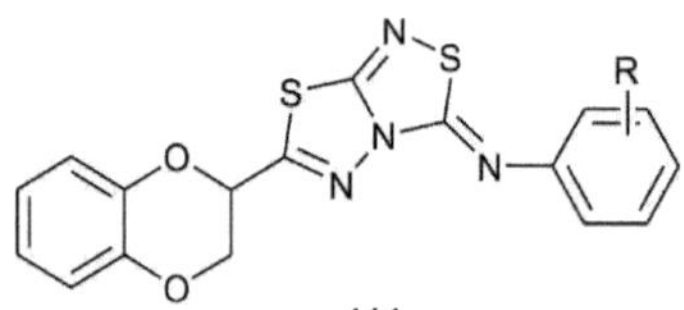

114

Figure 6.59 Structure of substituted 1,3,4-thiadiazole-fused-[1,2,4]-thiadiazole incorporating the 1,4-benzodioxine moiety showing the inhibitory activity of α-amylase and α-glycosidase.

6.4 Conclusion

In conclusion, the recent developments in the synthesis and biological applications of 1,2,4-thiadiazoles underscore their significant role in contemporary chemistry and pharmacology. The versatile synthetic strategies, ranging from traditional methods to green and sustainable approaches, have expanded the accessibility of 1,2,4-thiadiazole derivatives. These advancements not only address the challenges related to synthetic efficiency and environmental impact but also offer diverse structural landscapes for further exploration.

The biological applications of 1,2,4-thiadiazole compounds reveal a broad spectrum of pharmacological activities, including antimicrobial, anti-inflammatory, anticancer, and antioxidant properties. The observed bioactivities substantiate the potential of 1,2,4-thiadiazole as a promising scaffold for drug design and development. The structure–activity relationship studies have provided valuable insights into the molecular features governing the bioactivity of these compounds, paving the way for rational design strategies.

Moreover, the documented pharmacological effects of 1,2,4-thiadiazole derivatives underscore their relevance in addressing various health challenges. The continuous exploration of novel derivatives and formulations holds promise for the development of therapeutic agents with enhanced efficacy, selectivity, and reduced side effects.

As research in this field progresses, further elucidation of the molecular mechanisms underlying the observed biological activities will be crucial. Additionally, efforts toward optimizing synthetic routes, improving scalability, and exploring innovative drug delivery systems will contribute to the translation of 1,2,4-thiadiazole-based compounds from the laboratory to clinical applications. Overall, the recent advancements in the 1,2,4-thiadiazole research signify its enduring importance in medicinal chemistry and its potential to make significant contributions to the pharmaceutical landscape.

Abbreviations

CNS	Central nervous system
DCE	Dichloroethane
DDQ	2,3-Dichloro-5,6-dicyanobenzoquinone
IBD	Iodobenzene diacetate
MIT assay	3-[4,5-Dimethylthiazol-2-yl]-2,5 diphenyl tetrazolium bromide
NMDA receptors	*N*-Methyl-D-aspartate
PIFA	Phenyliodine(III) bis(trifluoroacetate)
TBAP	Tetra (*n*-butyl) ammonium peroxy disulfate

Acknowledgements

The authors are highly thankful to their management for providing the necessary facilities to write this book chapter.

References

1. L. M. T. Frija, A. J. L. Pombeiro and M. N. Kopylovich, *Eur. J. Org. Chem.*, 2017, **2017**, 2670–2682.
2. H. S. Sayiner, M. I. Yilmazer, A. T. Abdelsalam, M. A. Ganim, C. Baloglu, Y. C. Altunoglu, M. Gür, M. Saracoglu, M. S. Attia, S. A. Mahmoud, E. H. Mohamed, R. Boukherroub, N. H. Al-Shaalan, S. Alharthi, F. Kandemirli and M. A. Amin, *RSC Adv.*, 2022, **12**, 29627–29639.
3. J. Matysiak, *Mini-Rev. Med. Chem.*, 2015, **15**, 762–775.

4. T. Anthwal, S. Paliwaland S. Nain, *Chemistry*, 2022, **4**, 1654–1671.
5. A. Tahghighi and F. Babalouei, *Iran. J. Basic Med. Sci.*, **20**, 613–622.
6. M. J. N. Khadri, A. B. Begum, M. K. Sunil and S. A. Khanum, *Results Chem.*, 2020, **2**, 100045.
7. B. A. Bhongade, S. Talath, R. A. Gadad and A. K. Gadad, *J. Saudi Chem. Soc.*, 2016, **20**, S463–S475.
8. J. W. Zhao, J. X. Xu and X. Z. Guo, *Chin. Chem. Lett.*, 2014, **25**, 1499–1502.
9. D. Cheng, R. Luo, W. Zheng and J. Yan, *Synth. Commun.*, 2012, **42**, 2007–2013.
10. A. Mariappan, K. Rajaguru, N. Merukan Chola, S. Muthusubramanian and N. Bhuvanesh, *J. Org. Chem.*, 2016, **81**, 6573–6579.
11. D. Kumar, N. M. Kumar,, K. H. Chang, R. Gupta and K. Shah, *Bioorg. Med. Chem. Lett.*, 2011, **21**, 5897–5900.
12. L. Chai, Y. Xu, T. Ding, X. Fang, W. Zhang, Y. Wang, M. Lu, H. Xu and X. Yang, *Org. Biomol. Chem.*, 2017, **15**, 8410–8417.
13. A. S. Mayhoub, E. Kiselev and M. Cushman, *Tetrahedron Lett.*, 2011, **52**, 4941–4943.
14. A. Z. Halimehjani, A. Sharifiand H. Rahimzadeh, *ChemistrySelect*, 2019, **4**, 2634–2638.
15. V. Srivastava, A. Yadav and L. S. Yadav, *Synlett*, 2013, **24**, 465–470.
16. S. Chauhan, P. Chaudhary, A. K. Singh, P. Verma, V. Srivastava and J. Kandasamy, *Tetrahedron Lett.*, 2018, **59**, 272–276.
17. Y. Liu, Y. Zhang, J. Zhang, L. Hu and S. Han, *Tetrahedron Lett.*, 2021, **65**, 152744.
18. Y. J. Pragathi, D. Veronica, K. Anitha, M. V. B. Rao and R. R. Raju, *Chem. Data Collect.*, 2020, **30**, 100556.
19. P. Chacko and K. Shivashankar, *Synth. Commun.*, 2018, **48**, 1363–1376.
20. W. Yu, Y. Huang, J. Li, X. Tang, W. Wu and H. Jiang, *J. Org. Chem.*, 2018, **83**, 9334–9343.
21. Z. Yang, J. Zhang, L. Hu, L. Li, K. Liu,, T. Yang and C. Zhou, *J. Org. Chem.*, 2020, **85**, 3358–3363.
22. L. Chai, Z. Lai, Q. Xia, J. Yuan, Q. Bian, M. Yu, W. Zhang, Y. Xu and H. Xu, *Eur. J. Org. Chem.*, 2018, **2018**, 4338–4344.
23. N. Jatangi, N. Tumula, R. K. Palakodety and M. Nakka, *J. Org. Chem.*, 2018, **83**, 5715–5723.
24. V. P. R. K. Putta, R. Gujjarappa, N. Vodnala, R. Gupta, P. P. Pujar and C. C. Malakar, *Tetrahedron Lett.*, 2018, **59**, 904–908.
25. N. Perupogu, S. K. Sharif and D. Ramachandran, *Chem. Data Collect.*, 2020, **28**, 100444.
26. J. Y. Chen, M. Selvaraju, Y. T. Lin, S. Dhole, C. Y. Lin and C. M. Sun, *J. Org. Chem.*, 2020, **85**, 5570–5579.
27. K. Lee, R. S. Lee and Y. H. Kim, *Synth. Commun.*, 2020, **50**, 1774–1779.
28. Y. J. Pragathi, R. Sreenivasulu, D. Veronica and R. R. Raju, *Arab. J. Sci. Eng.*, 2021, **46**, 225–232.

29. G. L. Perlovich, A. N. Proshin, T. V. Volkova, L. N. Petrova and S. O. Bachurin, *Mol. Pharmaceutics.*, 2012, **9**, 2156–2167.
30. A. K. Oyebamiji, S. A. Akintelu, O. P. Amao, M. O. Kaka, A. E. Morakinyo, F. A. Amao and B. Semire, *Data Brief*, 2021, **37**, 107234.
31. A. N. Proshin, T. P. Trofimova, O. N. Zefirova, I. V. Zhirkina, D. A. Skvortsov and S. O. Bachurin, *Russ. Chem. Bull.*, 2021, **70**, 510–514.
32. D. G. S. Sudhakar, A. S. Rao and Ch. V. R. Reddy, *Russ. J. Gen. Chem.*, 2019, **89**, 1696–1701.
33. B. Chaithanya, D. PrabhakaraChary and I. V. Kasiviswanath, *Polycyclic Aromat. Compd.*, 2022, **42**, 7754–7777.
34. Y. J. Pragathi, R. Sreenivasulu, D. Veronica and R. R. Raju, *Arab. J. Sci. Eng.*, 2021, **46**, 225–232.
35. K. Chidella, N. Seelam, P. K. R. Cherukumalli, J. Reddy and G. Sridhar, *Chem. Data Collect.*, 2020, **30**, 100554.
36. S. Premalatha and D. Ramachandran, *Chem. Data Collect.*, 2020, **28**, 100481.
37. M. A. Orlova, T. P. Trofimova, A. N. Proshin and A. P. Orlov, *Russ. Chem. Bull.*, 2017, **66**, 932–934.
38. Y. J. Pragathi, D. Veronica, K. Anitha, M. V. B. Rao and R. R. Raju, *Chem. Data Collect.*, 2020, **30**, 100556.
39. A. K. Oyebamiji, S. A. Akintelu, O. P. Amao, M. O. Kaka, A. E. Morakinyo, F. A. Amao and B. Semire, *Data Brief*, 2021, **37**, 107234.
40. E. R. Devi, R. Sreenivasulu, M. V. B. Rao, K. P. Rao, R. V. Nadh and M. Sireesha, *Russ. J. Gen. Chem.*, 2021, **91**, 1105–1111.
41. N. Perupogu, CH. M. Krishna and D. Ramachandran, *Chem. Data Collect.*, 2020, **28**, 100482.
42. N. Perupogu, S. K. Sharif and D. Ramachandran, *Chem. Data Collect.*, 2020, **28**, 100444.
43. B. Chaithanya, D. PrabhakaraChary and I. V. Kasiviswanath, *Polycyclic Aromat. Compd.*, 2022, **42**, 7754–7777.
44. L. A. Alshabanah, L. A. Al-Mutabagani, S. M. Gomha and H. A. Ahmed, *Front Chem.*, DOI:10.3389/fchem.2021.762248.
45. N. Ashok, J. Madhukar and G. Sridhar, *Chem. Data Collect.*, 2021, **32**, 100653.
46. A. J. Adepoju, *Res. J. Pharm., Biol. Chem. Sci.*, 2021, **12**(4), 123.
47. A. K. Oyebamiji, S. A. Akintelu, O. P. Amao, M. O. Kaka, A. E. Morakinyo, F. A. Amao and B. Semire, *Data Brief*, 2021, **37**, 107234.
48. E. Ruijter, M. K. Rong, A. Speer, W. Bitter and S. T. Cole, *World Intellectual Property Organization*, 2021, WO2021040519 A1 2021-03-04.
49. A. Castro, T. Castaño, A. Encinas, W. Porcal and C. Gil, *Bioorg. Med. Chem*, 2006, **14**, 1644–1652.
50. T. Volkova, E. Chibunova, O. Silyukov, A. Proshin and I. Terekhova, *J. Mol. Liq.*, 2017, **247**, 64–69.
51. A. Gupta, P. Mishra, S. N. Pandeya, S. K. Kashaw, V. Kashawand J. P. Stables, *Eur. J. Med. Chem.*, 2009, **44**, 1100–1105.

52. R. Leung-Toung, J. Wodzinska, W. Li, J. Lowrie, R. Kukreja, D. Desilets, K. Karimian and T. F. Tam, *Bioorg. Med. Chem.*, 2003, **11**, 5529–5537.
53. S. Shahinshavali, R. Sreenivasulu, V. R. Guttikonda, D. Kolli and M. V. B. Rao, *Russ. J. Gen. Chem.*, 2019, **89**, 324–329.
54. R. Kumar, A. Kumar, S. Ram, A. Angeli, A. Bonardi, A. Nocentini, P. Gratteri, C. T. Supuran and P. K. Sharma, *Arch. Pharm.*, 2022, **355**(1), 2100241.
55. V. V. Grigoriev, G. F. Makhaeva, A. N. Proshin, N. V. Kovaleva, E. V. Rudakova, N. P. Boltneva, A. V. Gabrel′yan, B. V. Lednevand S. O. Bachurin, *Russ. Chem. Bull.*, 2017, **66**, 1308–1313.
56. G. F. Makhaeva, N. V. Kovaleva, N. P. Boltneva, S. V. Lushchekina, E. V. Rudakova, T. S. Stupina, A. A. Terentiev, I. V. Serkov, A. N. Proshin, E. V. Radchenko, V. A. Palyulin, S. O. Bachurin and R. J. Richardson, *Bioorg. Chem.*, 2020, **94**, 103387.
57. G. F. Makhaeva, V. V. Grigoriev, A. N. Proshin, N. V. Kovaleva, E. V. Rudakova, N. P. Boltneva, I. V. Serkov and S. O. Bachurin, *Dokl. Biochem. Biophys.*, 2017, **477**, 405–409.
58. G. F. Makhaeva, A. N. Proshin, N. P. Boltneva, E. V. Rudakova, N. V. Kovaleva, O. G. Serebryakova, I. V. Serkov and S. O. Bachurin, *Russ. Chem. Bull.*, 2016, **65**, 1586–1591.
59. O. Silyukov, M. Chislov, A. Burovikhina, T. Utkinaand I. Zvereva, *J. Therm. Anal. Calorim.*, 2012, **110**, 187–192.
60. A. Gupta, P. Mishra, S. K. Kashaw, V. Jatav and J. P. Stables, *Eur. J. Med. Chem.*, 2008, **43**, 749–754.
61. A. Gupta, P. Mishra, S. N. Pandeya, S. K. Kashaw, V. Kashaw and J. P. Stables, *Eur. J. Med. Chem.*, 2009, **44**, 1100–1105.
62. V. I. Osminin, A. A. Mironenko, P. G. Dahno, M. A. Nazarenko, A. I. Oflidi, V. V. Dotsenko, V. D. Strelkov, N. A. Aksenov and I. V. Aksenova, *Russ. J. Gen. Chem.*, 2022, **92**, 2235–2245.
63. *Substituted (2-heteroaryloxyphenyl)isoxazolines and salts thereof and their use as herbicidal active substances*, 2023, WO2021028421A1.
64. B. S. Krivokolysko, V. V. Dotsenko, N. A. Pakholka, P. G. Dakhno, V. D. Strelkov, N. A. Aksenov, I. V. Aksenova and S. G. Krivokolysko, *J. Iran. Chem. Soc.*, 2023, **20**, 609–628.
65. R. Hussain, M. Shah, S. Iqbal, W. Rehman, S. Khan, L. Rasheed, H. Naz, H. A. Al-ghulikah, E. B. Elkaeed, R. A. Pashameah, E. Alzahrani and A.-E. Farouk, *Front. Chem.*, 2022, **10**, 1023316.

CHAPTER 7

Synthesis, Properties, and Biological Applications of 1,3,4-Thiadiazoles

KALYANI SONAWANE, RUSHIKESH SAID AND NAVNATH HATVATE*

Institute of Chemical Technology, Mumbai Marathwada Campus, Jalna, Maharashtra, 431203, India
*E-mail: nt.hatvate@marj.ictmumbai.edu.in; navnath711@gmail.com

7.1 Introduction

Heterocyclic compounds elicit considerable interest among medicinal chemists due to their diverse spectrum of biological activities and unique chemical properties. Despite the considerable advancements made in the field of heterocyclic ring systems through extensive research, there is a continued endeavor to discover new heterocyclic compounds that exhibit bioactive properties. In the fields of organic and medicinal chemistry, imidazole, oxadiazole, and thiadiazole are commonly encountered in five-membered heterocycles. These chemicals are renowned for their biological activities.[1] In previous studies, the thiadiazole ring has been used to link several molecules, including anti-parasitic and anti-bacterial medicines. Notably, certain pharmaceuticals derived from this approach remain active in clinical applications. In recent years, scholarly investiga-

S-Heterocycles: Retrospect, Prospects, and Biological Applications
Edited by Shrikaant Kulkarni, Hemantkumar Akolkar, Vijay M. Khedkar and A. K. Haghi

Published by the Royal Society of Chemistry, www.rsc.org

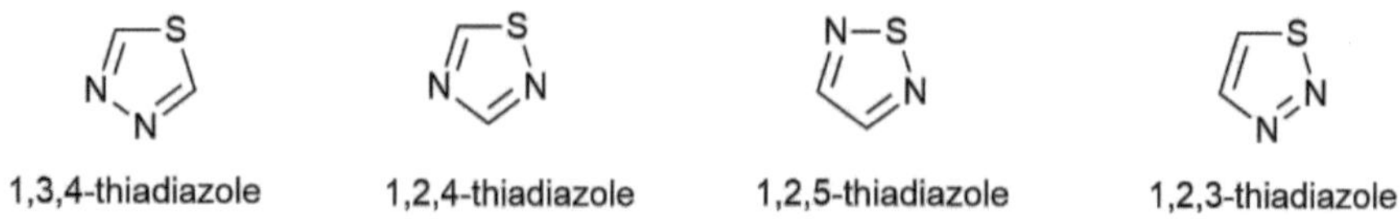

Figure 7.1 Isomers of thiadiazoles.

tions have demonstrated that the thiadiazole ring possesses significant structural significance and exhibits a diverse array of biological activity.[2]

Thiadiazole is classified as a heterocyclic molecule due to its distinctive structure, which consists of a five-membered ring having one sulfur, two nitrogen, and two carbon atoms. The system operates as a donor of two electrons and possesses a domain capable of forming hydrogen bonds.[3]

This chapter deals with the recent advancements in the synthesis and properties of 1,3,4-thiadiazole along with its biological activity.

7.1.1 Isomers of Thiadiazoles

Thiadiazoles exhibit four distinct isomeric structures, such as 1,3,4-, 1,2,4-, 1,2,5-, and 1,2,3-thiadiazoles (Figure 7.1). Among these isomers, 1,3,4-thiadiazoles have received the most extensive research attention. Compounds with thiadiazole rings have been commonly reported to exhibit anti-inflammatory,[4] anti-seizure/anti-epileptic/anti-convulsant,[4] anti-cancer,[4] anti-Alzheimer,[5] anti-viral,[4] anti-tuberculosis,[6] anti-microbial,[4] anti-diabetic,[7] diuretic,[8] anti-obesity,[9] anti-glaucoma,[10] anti-platelet,[11] anti-aging,[9] anti-leishmanicidal,[12] anti-*H. pylori*,[13] and anti-fungal activities.[4]

7.1.2 Chemical Properties and Characteristics of 1,3,4-Thiadiazoles

The 1,3,4-thiadiazole scaffold has experienced notable progress subsequent to the discovery of phenylhydrazine by Emil Fischer and hydrazine by Theodor Curtius in the latter half of the 19th century. The initial discovery of the first 1,3,4-thiadiazole compound was made in 1882 by Fischer. However, it was not until 1890 that Freund and Kuh presented a thorough explication of the structural characteristics of this particular ring system. The research initiated by Busch and his colleagues in 1894 resulted in substantial expansion over the initial decades of the 20th century. The pharmaceutical industry has experienced continuous progress, resulting in the identification of new drugs and enhanced synthetic methodologies within the realm of 1,3,4-thiadiazole chemistry. In the last ten years, various research collectives have conducted investigations on the potential tautomerism of amino-, hydroxy-, and mercapto-1,3,4-thiadiazoles, acquiring substantial novel information.[14]

The sulfur atom in 1,3,4-thiadiazole exhibits an inductive effect and also demonstrates a considerable degree of aromaticity, resulting in its weak basicity. The substance exhibits stability in acidic environments while instability in basic environments.[15] Moreover, it has been shown that 1,3,4-thiadiazoles bearing mercapto substituents, amino substituents, and hydroxyl substituents at positions 2 or 5 demonstrate the presence of tautomeric forms. The nitrogen atoms located at positions 3 and 4 within the ring structure are susceptible to electrophilic attacks. On the contrary, due to the electron-withdrawing characteristics of nitrogen atoms, carbon atoms within the ring exhibit electron deficiency, rendering them largely unreactive toward electrophiles but susceptible to nucleophilic substitution. The reactivity of the ring is generally determined by the presence of different substituents at C_2 and C_5.[16]

Various forms of 1,3,4-thiadiazoles are known to exist, encompassing both aromatic derivatives such as the parent compound 1,3,4-thiadiazole (Figure 7.2a) and 1,3,4-thiadiazole cations (Figure 7.2b) and mesoionic systems (Figure 7.2c). The mesoionic compound has a multifaceted arrangement of diverse heteroatoms, distinct zones of positive and negative charges, a ring consisting of five interconnected atoms with heteroatoms, and conjugated π and p electrons. Mesoionic systems demonstrate a notable combination of high density and substantial polarizability, all the while preserving an overall neutral electron charge. The aforementioned characteristics facilitate the efficient translocation of mesoionic molecules across cellular membranes and their selective interaction with distinct biological targets, exhibiting significant affinities.[4,17]

The scope of this review is to provide a comprehensive exploration of the synthesis, properties, and biological applications of 1,3,4-thiadiazoles. The article delves into the diverse chemical properties and unique biological activities associated with thiadiazole. Additionally, the review covers both conventional and green synthesis methods, offering insights into the advancements in the field. The ultimate goal is to furnish chemists and researchers with a thorough understanding of the current state of knowledge in 1,3,4-thiadiazole chemistry and its potential applications in various biological contexts, encompassing all the biological properties of 1,3,4-thiadiazole and its derivatives.

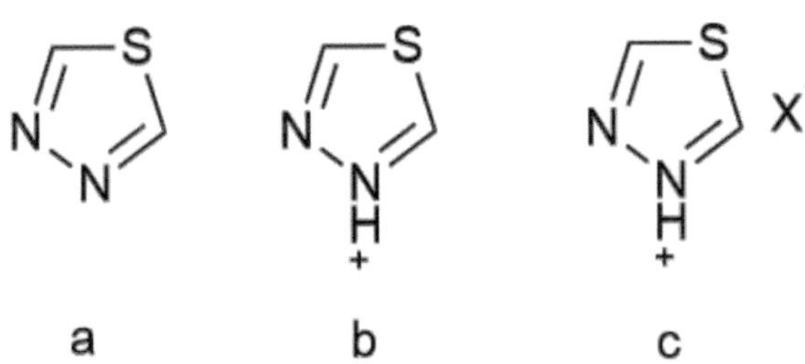

Figure 7.2 Different forms of 1,3,4-thiadiazoles.

7.2 Synthesis of 1,3,4-Thiadiazoles

Advancements in the 1,3,4-thiadiazole chemistry can be attributed to the identification of hydrazine and phenylhydrazines in the latter half of the 19th century. Figure 7.3 illustrates the graphical representation of the numbering scheme applied to the ring system of 1,3,4-thiadiazole. 1,3,4-Thiadiazoles are commonly synthesized using traditional methodologies that entail the cyclization of acylhydrazines, such as *N,N'*-diacylhydrazines and monoacylhydrazines, or by means of the transformation of 1,3,4-oxadiazoles. They can be produced by the utilization of carboxylic acids, aldehydes, hydrazides, thiohydrazides, and nitriles, and oxidations of dihydro-1,3,4-thiadiazole. This investigation provides a concise overview of the contemporary methodologies used for synthesizing the derivatives of 1,3,4-thiadiazole. The prospective advantages of using this classification system to aid chemists in their synthetic endeavors are anticipated.

7.2.1 Conventional Methods for the Synthesis of 1,3,4-Thiadiazoles

In conventional methods, the synthesis of 1,3,4-thiadiazole is characterized by prolonged reaction times, high temperatures, low yields, and the generation of numerous unwanted by-products. Additionally, the use of anhydrous hydrocarbon solvents and sulfurization reagents contributes to the environmental unfriendliness of these methods. These challenges have spurred the exploration of alternative synthetic approaches. One promising strategy involves solvent-free reactions under microwave (MW) irradiation, offering the potential to overcome the drawbacks associated with the traditional methods. This innovative technique enables rapid and uniform heating, leading to shortened reaction times, improved yields, and minimized formation of undesirable by-products while also eliminating the need for harmful solvents and reagents. The adoption of solvent-free microwave-assisted synthesis represents a significant step toward a more sustainable and efficient method for producing 1,3,4-thiadiazoles.[18]

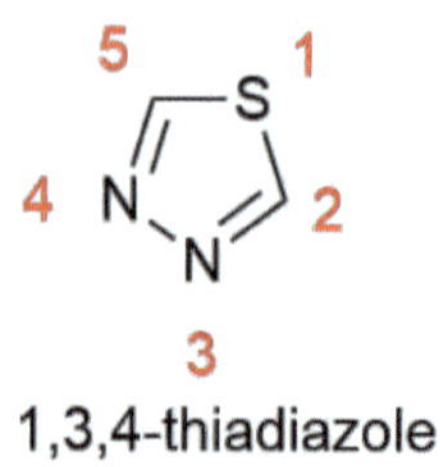

Figure 7.3 Numbering of 1,3,4-thiadiazoles.

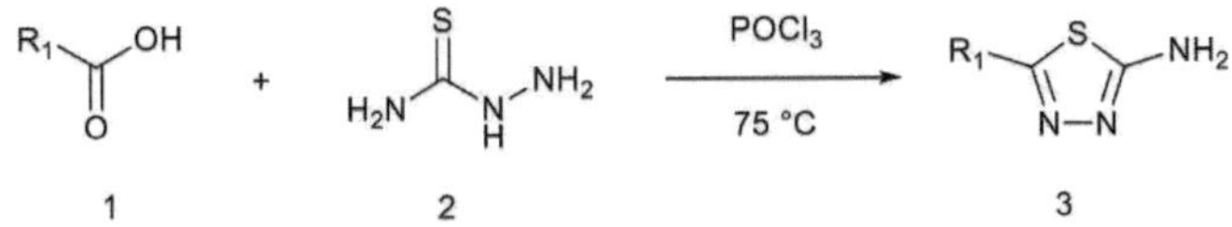

Scheme 7.1 Synthesis of 1,3,4-thiadiazoles from carboxylic acids.

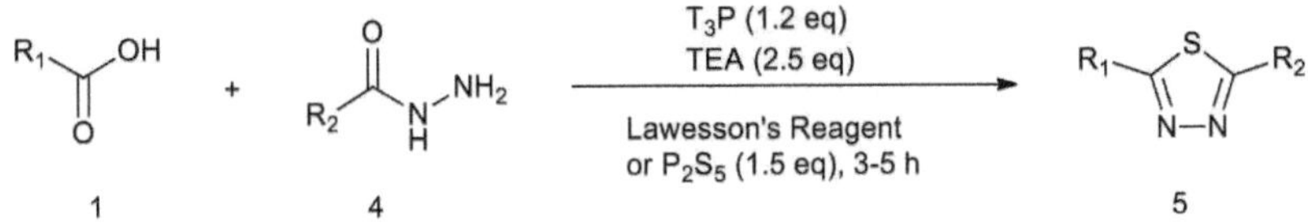

Scheme 7.2 Synthesis of 1,3,4-thiadiazoles from carboxylic acids.

7.2.1.1 *Synthesis of 1,3,4-Thiadiazoles from Carboxylic Acids*

1,3,4-Thiadiazoles can be synthesized by oxidative cyclodehydration of a combination of carboxylic acids **1** and various thiosemicarbazides **2**. This reaction is facilitated by the use of phosphorus oxychloride ($POCl_3$) as the dehydrating agent (Scheme 7.1).[19–21] Both conventional and newer, more environmentally friendly MW-assisted synthesis methods are used,[22] with minor adjustments in the heating time and/or temperature, as well as the amount of dehydrating agents used, in most of the original procedures (whether conventional or MW-assisted).

Over the course of recent decades, various techniques, including one-pot synthesis of 1,3,4-thiadiazoles, have been documented. These approaches offer the advantage of circumventing the laborious nature of multi-step syntheses. Within this set of approaches, certain methods continue to be implemented under challenging circumstances, while others have undergone enhancements. Augustine *et al.* presented a method for synthesizing 1,3,4-thiadiazoles by a one-pot reaction involving carboxylic acids **1** and propylphosphonic anhydride (T_3P) (Scheme 7.2). Notably, T_3P acts as both a coupling agent and a cyclodehydrating reagent. In the majority of cases, it demonstrated notable efficiency and showcased extensive compatibility with various functional groups. Nevertheless, the resultant compounds were found to contain a minor proportion of the by-product 1,3,4-oxadiazole (3–5%). Nonetheless, purification of the product was easily accomplished using chromatography or recrystallization methods.[23]

7.2.1.2 *Synthesis of 1,3,4-Thiadiazoles from Aldehydes*

The production of 1,3,4-thiadiazole derivatives can be accomplished using aldehydes. 2-Amino derivatives of 1,3,4-thiadiazoles were synthesized through a two-step process (Scheme 7.3). First, the condensation of aromatic aldehyde **6** and thiosemicarbazide **2** led to the formation of thiosemicarbazone **7**. Subsequently, thiosemicarbazone **7** underwent

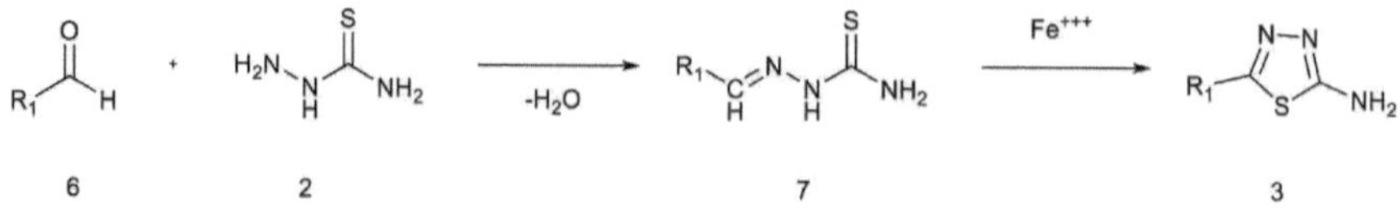

Scheme 7.3 Synthesis of 1,3,4-thiadiazoles from aldehydes.

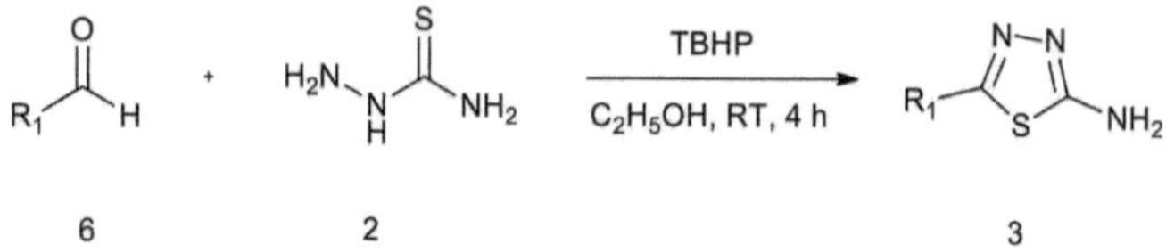

Scheme 7.4 One-pot synthesis of 1,3,4-thiadiazole from aldehydes.

cyclization under oxidative conditions in the presence of $FeCl_3$, producing 2-amino derivatives of 1,3,4-thiadiazole **3**. This method highlights an efficient route to the preparation of valuable heterocyclic compounds with potential applications in various fields.[24]

In another investigation, the one-pot synthesis was successfully conducted using aldehydes **6**, thiosemicarbazide **2**, ethanol, and *tert*-butyl hydroperoxide (TBHP). The reaction mixture was agitated at an ambient temperature for a duration of 4 hours (Scheme 7.4). The first step encompasses the *in-situ* generation of thiosemicarbazones through the reaction of aldehydes with thiosemicarbazide. Following this, ring closure is accomplished through oxidative cyclization, using TBHP as the oxidizing agent, leading to the formation of the final product. This process represents a highly practical, mild, and metal-free one-pot synthesis of amino derivatives of 1,3,4-thiadiazole **3**, which was achieved using readily available starting materials and TBHP. This chemical system has a high degree of tolerance toward diverse functional groups, giving excellent yields.[25]

7.2.1.3 Synthesis of 1,3,4-Thiadiazoles from Hydrazides

The synthesis of 1,3,4-thiadiazoles involves a multi-step process initiated by the reaction of acid hydrazides. The novel derivatives of 1,3,4-thiadiazole were generated through the interaction of substituted isocyanates **8** with acid hydrazides **4** using established methods, as depicted in Scheme 7.5. The conventional synthesis involves two stages: first, substituted isocyanates **8** react with acid hydrazides **4** in dry tetrahydrofuran (THF) utilizing triethylamine (TEA), yielding intermediate derivatives **9**. In the second stage, intermediates **9** undergo a reaction with TEA and *p*-toluenesulfonyl chloride (*p*-TsCl) in *N*-methyl-2-pyrrolidone (NMP) under heating, resulting in the formation of the targeted 1,3,4-thiadiazole derivatives **10**. This

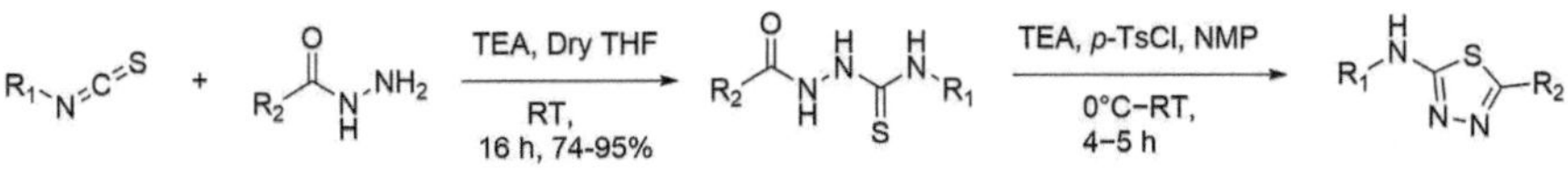

8 4 9 10

Scheme 7.5 Synthesis of 1,3,4-thiadiazoles from acid hydrazides.

synthetic approach demonstrates a methodical pathway to derive novel 1,3,4-thiadiazole compounds.[26]

7.2.1.4 *Synthesis of 1,3,4-Thiadiazoles from Thiohydrazides*

Thiohydrazides exhibit a chemical structure that closely resembles that of thiosemicarbazide and differs solely in the absence of the α-amino group in the former. Therefore, thiohydrazide is commonly used in combination with thiosemicarbazide in the synthesis of 1,3,4-thiadiazoles using similar techniques. Based on the literature, the procedure involves the acylation of thiohydrazides using acid halide, carboxylic acid, or acid anhydride. Alternatively, Schiff base formation can be achieved using aldehydes or hydrazones can be formed using nitriles. 1,3,4-Thiadiazoles are produced by subsequent *in situ* cyclodehydration. The synthesis of **13** was reported by Farrar *et al.*[27] In their study, thiobenzhydrazide **11** was reacted with benzaldehyde to synthesize thiohydrazide derivative **12** (Scheme 7.6a). Additionally, the reaction of thiobenzhydrazide **11** with benzonitrile, using oxidants such as I_2/pyridine and potassium persulfate, led to the generation of *N*-thiobenzoylbenzamidrazone **14** (Scheme 7.6b). The formation of the thiadiazole ring system can be elucidated through the dehydrogenation of thiohydrazide, subsequently leading to the intrastate nucleophilic attack of a thiolate anion on the iminyl or amidinyl carbon atom. This procedure is accompanied by the elimination of either ammonia or hydrogen subsequent to proton transfer. In Scheme 7.6a, it is shown that either an acidic or a neutral environment is likely advantageous, while Scheme 7.6b shows that such conditions have a detrimental impact on the synthesis of thiadiazoles.[27]

7.2.1.5 *Synthesis of 1,3,4-Thiadiazoles from Nitriles*

Nitriles have been found to be present during the synthetic process of thiadiazoles from thiosemicarbazides. Nitriles **15**, when present in a medium containing phenylpropanolamine (PPA), can be easily transformed into iminoesters of PPA. These iminoesters undergo a chemical reaction with thiosemicarbazide, resulting in their conversion into amidazones **16**. Subsequently, a molecule of ammonia is eliminated, resulting in the formation of compounds **17** (Scheme 7.7).[28]

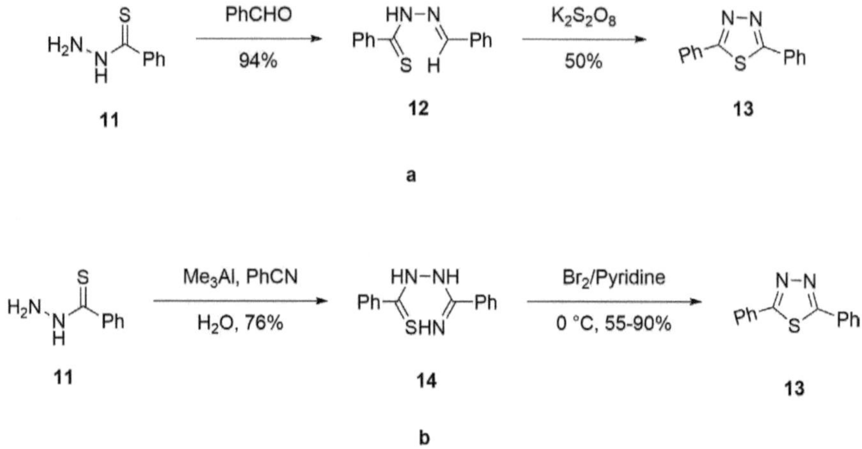

Scheme 7.6 Synthesis of 1,3,4-thiadiazoles from thiohydrazides.

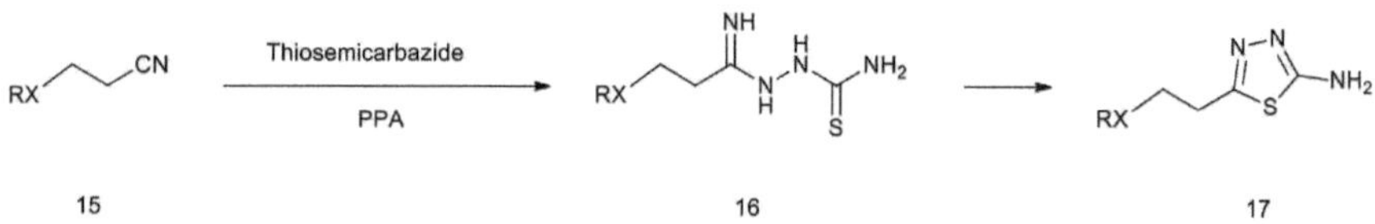

Scheme 7.7 Synthesis of 1,3,4-thiadiazoles from nitriles.

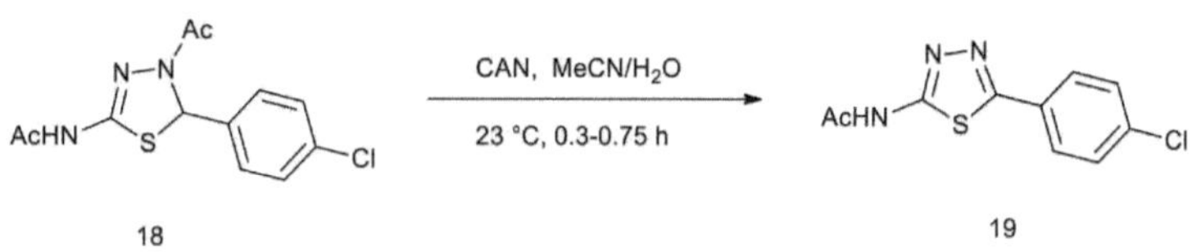

Scheme 7.8 Synthesis of 1,3,4-thiadiazoles by the oxidation of dihydro-1,3,4-thiadiazoles.

7.2.1.6 Synthesis of 1,3,4-Thiadiazoles by the Oxidation of Dihydro-1,3,4-thiadiazoles

The synthesis of 1,3,4-thiadiazoles can be achieved through the oxidation of partially or fully reduced thiadiazoles. The review conducted in Comprehensive Heterocyclic Chemistry III (CHEC-III) examined various approaches for these reactions.[29] Several methods for oxidative diacylation have been reported in the literature, such as the use of cerium(IV) ammonium nitrate (CAN), potassium permanganate, and phenyliodine(III) diacetate (PIDA). Superior yields and pure products can be achieved by utilizing CAN as an oxidizing agent (Scheme 7.8).[30]

7.2.2 Green Methods for the Synthesis of 1,3,4-Thiadiazoles

Numerous techniques for the synthesis of 1,3,4-thiadiazoles have been documented in the literature. The majority of these techniques demonstrate a sequential procedure, commonly involving the cyclization of acid hydrazides, which can be achieved using various reagents, including phosphorus oxychloride, thionyl chloride, or sulfuric acid. These reactions are commonly conducted under rigorous reaction conditions. In recent times, numerous effective methods have been documented pertaining to the formation of 1,3,4-oxadiazoles, with a particular emphasis on their synthesis from easily accessible acid hydrazides and carboxylic acids. However, these methods utilize expensive catalysts and require a prolonged time to complete the reaction.[31]

There are several one-pot synthesis techniques for 1,3,4-thiadiazoles, using mainly thiosemicarbazides and carboxylic acids. These methods include (1) conducting the reaction in ionic liquids,[32] (2) performing the process in the presence of an excess of $POCl_3$,[33] (3) carrying out the reaction using polyphosphoric acid,[34] and (4) utilizing a combination of methanesulfonic acid and phosphorus pentoxide for the reaction.[35] Several known limitations hinder the universal applicability of these one-step synthesis methods. First, the elevated manufacturing expenses of ionic liquids restrict their widespread use.[36] Second, the high toxicity of $POCl_3$ poses a significant challenge. Third, polyphosphoric acid is found to be more suitable for the synthesis of nitriles compared with carboxylic acids. Lastly, methanesulfonic acid can result in the formation of unwanted by-products due to its reactivity. Therefore, the continued relevance lies in advancing cost-effective and secure one-pot techniques for synthesizing 1,3,4-thiadiazole derivatives. Recent research has provided evidence supporting the appropriateness of utilizing polyphosphate ester (PPE) in the production of nitrogen-containing heterocyclic compounds characterized by five-membered rings. This synthesis can be achieved under mild circumstances, namely at temperatures <85 °C.[37] Given the overarching principle underlying the reaction mechanism of PPE, it is reasonable to anticipate the potential applicability of this reagent in facilitating the reaction illustrated in Scheme 7.9.

7.2.2.1 *Microwave-assisted Synthesis of 1,3,4-Thiadiazoles*

The synthesis of the derivatives of 1,3,4-thiadiazoles can be achieved utilizing an environmentally friendly approach by reacting substituted isocyanates **8** with acid hydrazides **4**, using microwave-assisted techniques,

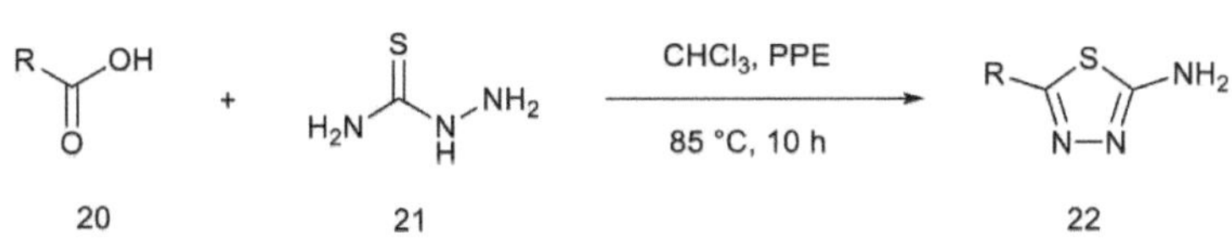

Scheme 7.9 One-pot synthesis of 1,3,4-thiadiazoles from carboxylic acids.

as illustrated in Scheme 7.10. During this process, isocyanates **8** and acid hydrazides **4** undergo a reaction in TEA, dry THF, *p*-TsCl, and NMP. The intended compounds were successfully synthesized with high yields using a one-pot procedure. The kinetics of the reaction was found to be more efficient when subjected to non-classical reaction conditions, notably MW irradiation, as compared to standard classical heating conditions.[26]

Polshettiwar *et al.* synthesized 1,3,4-thiadiazoles from acid hydrazides by a one-step approach utilizing microwave irradiation (Scheme 7.11). A range of aromatic and heterocyclic hydrazides **23** were subjected to a reaction with triethyl orthopropionate, triethyl orthoformate, and triethyl orthobenzoate **24** under the catalytic influence of solid-supported Nafion®NR50 and thionating agent phosphorus pentasulfide in alumina (P_4S_{10}/Al_2O_3). Significantly, this reaction was carried out without using solvents, leading to satisfactory yields of 1,3,4-thiadiazoles.[31]

Feng *et al.* synthesized a series of 1,3,4-thiadiazole derivatives from thiosemicarbazone using a cyclization process in the presence of $FeCl_3$. To enhance efficiency and promote environmental sustainability, the research team opted to use microwave irradiation and ultrasonic techniques for the synthesis of 1,3,4-thiadiazoles. This approach was selected to circumvent the need for organic solvents or acidic conditions, as illustrated in Scheme 7.12. The mixture, including thiosemicarbazone **26**, ferric chloride, and water, was subjected to microwave irradiation at 200 W and ultrasound irradiation at 50 W for a duration of 3 minutes. This treatment resulted in the formation of products with yields ranging from 73% to 90%, as depicted in Scheme 7.12.[38]

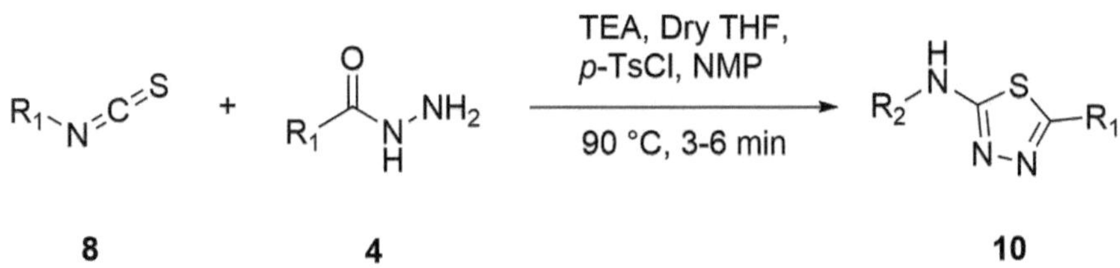

Scheme 7.10 Synthesis of 1,3,4-thiadiazoles from acid hydrazides using the microwave-assisted method.

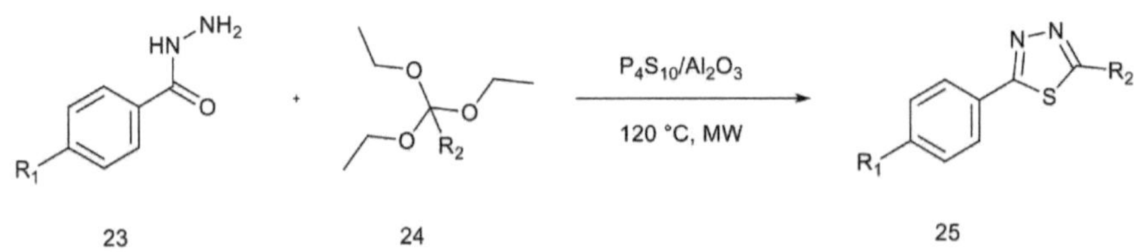

Scheme 7.11 Synthesis of 1,3,4-thiadiazoles from acid hydrazides using the microwave-assisted method without solvents.

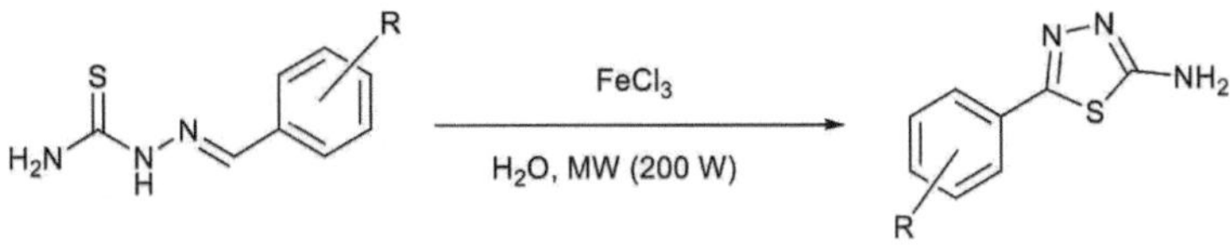

26

27

Scheme 7.12 Synthesis of 1,3,4-thiadiazoles from thiosemicarbazone using microwave and ultrasound irradiation.

7.3 Applications of 1,3,4-Thiadiazoles in Currently Marketed Drugs/Investigational Compounds

The varied pharmacological activities of the 1,3,4-thiadiazole moiety have attracted considerable interest in the field of medicinal chemistry, leading to the investigation of their potential applications in both commercially available drugs and investigational compounds.[39] This heterocyclic scaffold, characterized by a five-membered ring containing three nitrogen and two sulfur atoms, has demonstrated versatile bioactivities, ranging from anti-microbial, anti-inflammatory to anti-cancer properties.[40] Numerous research efforts have led to the development of compounds incorporating the 1,3,4-thiadiazole core, either as standalone drugs or as components in hybrid molecules, showcasing its potential in addressing a spectrum of diseases.

Megazol (**28**), a derivative of the 1,3,4-thiadiazole scaffold, has emerged as a noteworthy anti-protozoal agent with particular efficacy against *Trypanosoma cruzi* and *Trypanosoma brucei* infections. Its promising results in preclinical studies make it a potential candidate for addressing parasitic diseases.[41,42] Azetepa (**29**), also known as CI-973, is a derivative of 1,3,4-thiadiazole, actively investigated for its potential as an anti-cancer agent, particularly in the context of ovarian cancer. Its cytotoxic effects on cancer cells make it a subject of ongoing research in the field of oncology.[43] Methazolamide (**30**), an established carbonic anhydrase inhibitor derived from the 1,3,4-thiadiazole scaffold, finds a wide range of applications in clinical settings for managing conditions, such as glaucoma and certain types of edema. Its ability to modulate carbonic anhydrase activity makes it an essential component in the treatment of these medical conditions.[44]

Acetazolamide (**31**), another derivative of 1,3,4-thiadiazole, stands out as a well-established carbonic anhydrase inhibitor with versatile applications. It is commonly prescribed for glaucoma, altitude sickness, and specific seizure disorders, and hence, the clinical utility of acetazolamide underscores the pharmacological significance of the 1,3,4-thiadiazole core.[45] Glybuzole (**32**), while its particular information is limited, being a 1,3,4-thiadiazole derivative, suggests potential bioactivity. Ongoing research is needed to elucidate its specific applications and mechanisms of action,

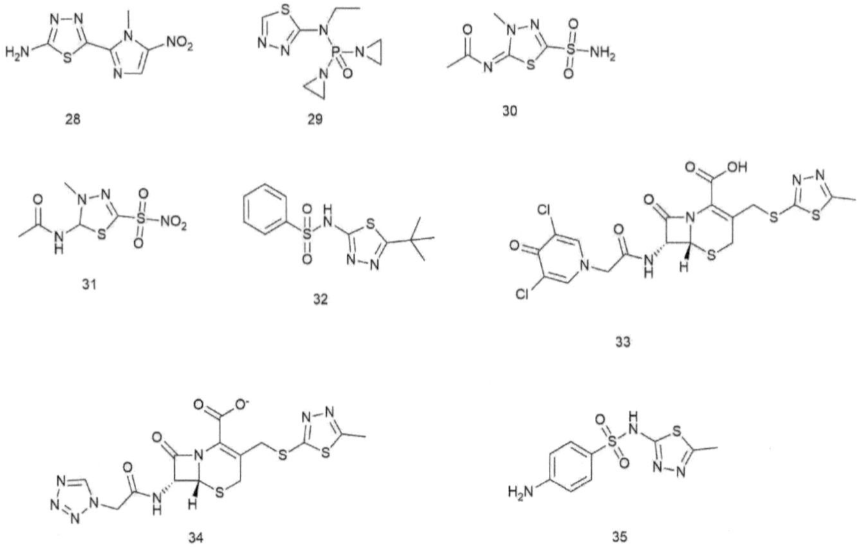

Figure 7.4 Currently marketed drugs containing 1,3,4-thiadiazole.

highlighting its inclusion in the exploration of diverse pharmacological activities associated with the 1,3,4-thiadiazole scaffold.[46]

Cefazedone (**33**) and Cefazolin (**34**), both cephalosporin antibiotics containing the 1,3,4-thiadiazole ring, exemplify the versatility of this scaffold in anti-bacterial applications. These antibiotics are widely prescribed for bacterial infections, especially against Gram-positive bacteria, which underscore the importance of the 1,3,4-thiadiazole core in the development of effective anti-bacterial agents.[47] Sulfamethizole (**35**), a sulfonamide antibiotic featuring the 1,3,4-thiadiazole core, is utilized for the treatment of urinary tract infections. By inhibiting bacterial folate synthesis, sulfamethizole demonstrates the anti-microbial potential inherent in the 1,3,4-thiadiazole scaffold, contributing to its efficacy in combating bacterial infections.[48] The diverse pharmacological applications of 1,3,4-thiadiazole derivatives underscore the significance of this heterocyclic scaffold in drug development, showcasing its adaptability and therapeutic potential across various medical conditions (Figure 7.4).

7.4 Biological Activities of 1,3,4-Thiadiazoles

Recent research has demonstrated that thiadiazole derivatives have a wide range of pharmacological characteristics, such as anti-inflammatory, anti-oxidant, anti-tubercular, anti-microbial, anti-diabetic, and diuretic effects. The development of thiadiazole derivatives has advanced significantly in recent years.[11,49] Numerous scholarly articles have documented the utilization of 1,3,4-thiadiazole derivatives in diverse biological contexts. Recent investigations have shed light on the extensive range of pharmacological activities exhibited by these derivatives, which can be categorized as follows.

7.4.1 Anti-inflammatory Agents

1,3,4-Thiadiazole contributes to the inhibitory activity of anti-inflammatory compounds against cyclooxygenase-2 (COX-2) and potentially other targets, thereby helping to reduce the production of inflammatory mediators involved in various diseases while aiming for selectivity and improved safety profiles. Omar *et al.* designed dual inhibitors of cyclooxygenase-2 and 15-lipoxygenase (15-LOX) by joining 2-amino-1,3,4-thiadiazole and 4-thiazolidine pharmacophores, resulting in more potent compounds that had good gastrointestinal safety and were well tolerated in animal studies. Compound **36** showed the highest cyclooxygenase-2 inhibitory activity with a low IC_{50} (70 nM) value and a high selectivity index of 220.[50]

Shkair *et al.* focused on developing improved analgesic and anti-inflammatory agents by modifying ibuprofen with 2-acetamido-1,3,4-thiadiazole to target COX-2 and reduce side effects. Compound **37** shows the most significant anti-inflammatory efficacy at a rate of 72.5% and a lower risk for causing ulcers (ulcerogenic index of 1.40) compared to ibuprofen. These findings indicate its potential as a safer option.[51] Maddila *et al.* aimed to explore the potential of 1,3,4-thiadiazole derivatives as anti-inflammatory agents to mitigate the negative consequences commonly associated with the prolonged usage of non-steroidal anti-inflammatory drugs (NSAIDs). These compounds were tested in a carrageenan-induced rat paw edema model, featuring hybrid structures combining pyrazole-3-carboxamide and pyrrole-3-carboxamide pharmacophores. Compound **38** emerged as the lead compound, exhibiting remarkable anti-inflammatory activity, with inhibition rates of 77.27% at 3 hours and 81.00% at 5 hours after carrageenan induction, surpassing the performance of the standard NSAID indomethacin.[52]

The study by Mahapatra *et al.* focuses on achieving the synthesis of murrayanine–thiadiazole hybrids by combining murrayanine (obtained from *Murraya koenigii* L.) and thiadiazoles to explore their potential in treating inflammation. The study suggests that murrayanine–thiadiazole hybrid compounds, particularly **39** with specific substituents, exhibit potential anti-inflammatory activities with percentage inhibition of edema of 31.96 ± 2.44 for 1 hour, 43.59 ± 2.23 for 2 hours, and 59.88 ± 2.01 for 3 hours. However, a definitive structure–activity relationship (SAR) remains uncertain due to mixed results.[53] The study reported by Cristina *et al.* addresses the challenges of NSAID therapy, focusing on the need for selective COX-2 inhibitors to minimize gastrointestinal complications.

They introduced a novel approach for the design and synthesis of 2,6-diaryl-imidazo[2,1-*b*][1,3,4]thiadiazole derivatives, combining structural elements from celecoxib and imidazo[2,1-*b*][1,3,4]thiadiazoles with anti-inflammatory properties. Compound **40** was found to be the most potent cyclooxygenase-2 inhibitor and anti-inflammatory drug that, when compared to diclofenac, has a decreased risk of ulcers.[54]

Omar *et al.* investigated the anti-inflammatory properties of 1,3,4-thiadiazole–thiazolidinone hybrids and examined how structural alterations can boost their therapeutic potential. The study highlights the SAR for the dual inhibitors of 15-LOX and COX-2, suggesting that small hydrophobic groups at the 5th position of thiadiazole and specific substitutions at the 5th position of thiazolidinone yield balanced inhibitors. Compound **41**, with a methyl group on thiadiazole and an unsubstituted thiazolidinone, is a potent dual inhibitor with IC_{50} values of 3.11 μM for 15-LOX inhibition and 0.33 μM for COX-2 inhibition.[55] The study performed by Pathak *et al.* explores the formation of thiadiazole hybrids coupled with 1,2,4-triazole compounds as possible agents with anti-inflammatory properties. These hybrid structures combine the pharmacophores of 1,2,4-triazole and 1,3,4-thiadiazole, characterized by their various biological activities. Based on the findings, the most active derivative, **42**, had ideal logP, a better anti-inflammatory activity score, and poor metabolism at several cytochrome P450 sites.[56]

In the study reported by Khan *et al.*, distinct sets of heterocyclic derivatives were strategically formulated and synthesized, drawing inspiration from the fundamental structure of the pharmaceutical compound known as felbinac. The novel compounds contain the biphenyl ring and methylene bridge as present in felbinac, but the carboxyl group is substituted with physiologically functional groups, including 1,3,4-thiadiazole, 1,2,4-triazole, and 1,3,4-oxadiazole. This substitution aims to enhance the efficacy and safety of anti-inflammatory drugs. Notably, compound **43** emerged as the great NSAID with the highest anti-inflammatory (72.87%) and analgesic (65.24%) activities.[57] Ragab *et al.* synthesized novel 2-acetamide-thiadiazole derivatives, incorporating pyran, thiophene, and thienopyrimidine moieties as potential cyclooxygenase-2 inhibitors for anti-inflammatory purposes. Compound **44** exhibited the most fantastic anti-inflammatory activity. Significantly, the aforementioned compounds demonstrated favorable tolerance by the gastric mucosa and did not manifest any detrimental effects on renal and hepatic functionalities in the experimental rat model. Docking studies revealed their similar binding orientation to celecoxib in the COX-2-active site, highlighting their potential as promising analgesic and anti-inflammatory agents.[58]

Gomha *et al.* explored a novel series of fused imidazole, thiazole, and related derivatives by a reaction of 2-bromoacetyl-3-phenyl-1,3,4-thiadiazole with various heterocyclic amines, *o*-phenylenediamine, 2-aminothiophenol, thiourea, thiosemicarbazide, and thiocarbohydrazide. In the study performed on albino rats, compound **45** demonstrated 69.9% inhibition and 213.43% potency compared to the positive control group and standard (indomethacin). These compounds exhibited promising anti-inflammatory and anti-ulcer effects.[59] Gowramma *et al.* designed and synthesized a series of 1,3,4-thiadiazole derivatives and evaluated them utilizing a rat paw edema model to test their anti-inflammatory properties. Compounds **46** exhibited notable anti-inflammatory effects, with 56.32% inhibition, surpassing the drug diclofenac (% inhibition is 65.13%). The study suggests that compounds bearing nitro, methoxy, and hydroxyl phenyl groups in the 2-azetidinone moiety showed significant anti-inflammatory potential.[60] Amir *et al.* synthesized several triazolo-thiadiazole derivatives from 2-naphthoxyacetic acid and evaluated their analgesic and anti-inflammatory effects. Compound **47** exhibited the highest anti-inflammatory activity with 73.24% inhibition, which is more than that of naproxen, and minimized ulcerogenic potential. These compounds were further assessed for hepatotoxicity and histopathological characteristics, showing promise as potential drug candidates for anti-inflammatory purposes.[61]

A group of thiazolidinone compounds based on 5-adamantyl thiadiazole was investigated by Haroun *et al.* in an effort to alleviate the gastrointestinal toxicity linked to conventional NSAIDs. These compounds demonstrated moderate-to-excellent anti-inflammatory activity. Thus, the IC_{50} value of one of the most active compounds **48** as a COX-1 inhibitor was 1.08 μM, which is less than that of ibuprofen (IC_{50} is 12.7 μM) and naproxen (IC_{50} is 40.10 μM). Furthermore, they were found to be reducing the risk of gastric damage and exhibiting no cytotoxicity. These results demonstrate these compounds as promising, innovative, and safer anti-inflammatory drugs with enhanced gastrointestinal characteristics.[62]

These studies presented a diverse range of novel compounds, highlighting their potential as effective anti-inflammatory agents with promising bioactivity profiles. These compounds demonstrated the selective inhibition of cyclooxygenase-1 and cyclooxygenase-2, reduced ulcerogenic effects, and significant analgesic activity. The structural modifications and hybrid designs of these compounds offer a promising pathway for developing safer and more efficient anti-inflammatory agents with favorable bioactivity characteristics, addressing the limitations associated with traditional NSAIDs (Figure 7.5).

7.4.2 Anti-diabetic Agents

In a collective effort by various research groups, diverse molecular architectures, including thiazolidine-4-one, 5-furyl-1,3,4-thiadiazol-2-imine, 1,3,4-thiadiazole-fused-[1,2,4]-thiadiazole, indolin-2-ones, and 1,3,4-thiadiazole derivatives, were synthesized and evaluated for their prospective advantage as α-glucosidase inhibitors. These studies explored novel chemical entities with promising anti-diabetic activities, emphasizing

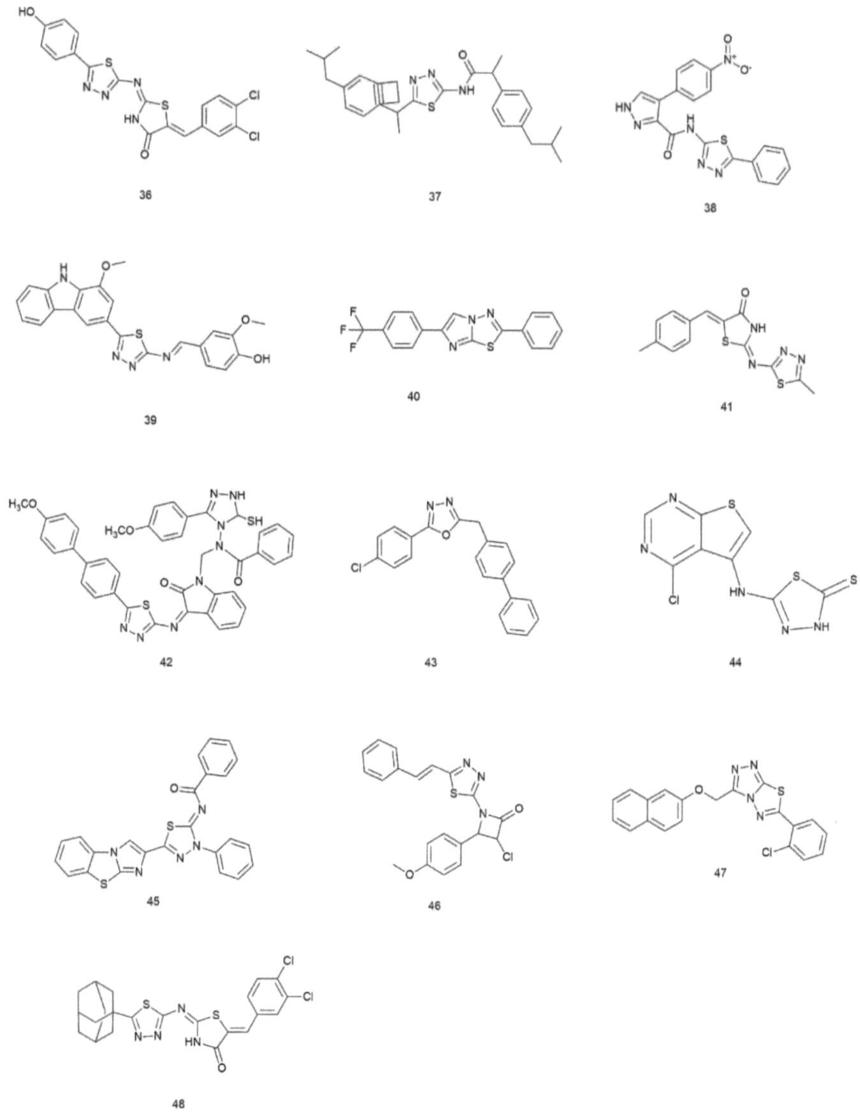

Figure 7.5 1,3,4-Thiadiazole derivatives as anti-inflammatory agents.

molecular design, structure–activity relationships, and *in vitro*/*in vivo* assessments.

Using a one-pot multicomponent procedure, Gummidi *et al.* discovered a novel and effective methodology for the synthesis of a library of thiazolidine-4-one molecular hybrids. The synthesized compounds were assessed for their anti-diabetic activities, revealing that compound **49** demonstrated the outstanding inhibition of α-glucosidase and α-amylase with an IC_{50}

value of 2.59 μM, surpassing the standard inhibitor acarbose by 1.5- and 14-folds, respectively. Structure–activity relationship analysis emphasized the crucial impact of substituents on phenyl rings in determining inhibitory potency.[7] Hameed *et al.* explored a novel approach for mitigating postprandial hyperglycemia in type 2 diabetes mellitus (T2DM) through the synthesis of 5-furyl-1,3,4-thiadiazol-2-imine derivatives. Molecular docking studies with human pancreatic α-amylase and peroxisome proliferator-activated receptor gamma (PPARγ) revealed active binding modes. Compounds **50** demonstrated high docking scores and exhibited significant α-amylase inhibition *in vitro*, correlating with their potential as promising anti-diabetic agents with improved physicochemical properties.[63]

Hussain *et al.* synthesized twenty-five analogs, with each analog being derived from fused 1,3,4-thiadiazole-[1,2,4]-thiadiazole, and evaluated for their anti-diabetic potential. The analogs exhibited diverse inhibition profiles opposing α-glucosidase and α-amylase, with compound **51** (IC_{50} = 0.70 μM) emerging as particularly potent due to the aryl ring undergoing dihydroxy substitutions at the 2nd and 3rd positions. An investigation into the SAR showed the relevance of alterations around the aryl ring.[64] Zhao *et al.* designed and synthesized a novel series of 3-(2-(1,3,4-thiadiazol-2-yl)hydrazono)-1-substituted indolin-2-one derivative. These compounds showed strong inhibitory activity against α-glucosidase, with compound **52** showing the highest inhibitory potential (IC_{50} = 3.12 ± 1.25 μM). Compound **52** was determined to be a non-competitive inhibitor using a kinetic study. Furthermore, molecular docking studies provided the confirmation of its binding interactions with the active site of α-glucosidase. The present study aims to identify a very promising category of powerful α-glucosidase inhibitors, thereby indicating their potential for subsequent exploration in the realm of diabetes treatment.[65]

Vaishnav *et al.* studied a molecular docking approach targeting PPARγ to design a series of 1,3,4-thiadiazole amide and urea derivatives for enhancing insulin sensitivity. Compound **53** shows 87.5% inhibition of blood glucose level, and it seems like it is the most potent molecule for anti-diabetic activity compared with the standard drug pioglitazone (86.3% inhibition). The electron-donating group in compound **54** demonstrated superior efficacy, and the urea series overall exhibited higher effectiveness than the amide series in promoting insulin-guided cellular glucose uptake.[66] Rahim *et al.* synthesized and characterized thiadiazole-based triazole/hydrazone derivatives; the α-glucosidase inhibitory potential was found to be moderate to good, with IC_{50} values varying between 0.40 ± 0.10 μM and 15.40 ± 0.30 μM. Compounds **55** (IC_{50} = 0.40 ± 0.10 μM) have emerged as the most powerful derivative, surpassing the standard drug acarbose. SAR analysis highlighted the significance of the substituent properties and positions on phenyl rings. Additionally, the compound exhibited non-toxicity against the murine fibroblast cell line 3T3.[67]

Taha *et al.* synthesized and evaluated 4-quinolinyl-bearing 1,3,4-thiadiazole-2-amine moieties. The results showed a range of inhibitory effects, as indicated by the IC_{50} values, which varied from 1.30 ± 0.05 to 45.60 ± 0.80 μM for α-amylase inhibition and from 2.70 ± 0.10 to 47.60 ± 0.90 μM for α-glucosidase inhibition. Compounds **56** (α-amylase and α-glucosidase inhibitions of 1.30 ± 0.05 and 2.90 ± 0.10, respectively) emerged as the most potent inhibitors, showing the effects of substituents on the phenyl ring. Notably, fluoro and chloro substitutions contributed to elevated inhibitory potentials.[68]

Li *et al.* developed and assessed new 2,5-disubstituted furan derivatives that included 1,3,4-thiadiazole for their ability to inhibit the enzymes α-glucosidase and β-glucuronidase of *Escherichia coli.* Compound **57** emerged as a potent competitive inhibitor against α-glucosidase (IC_{50} = 0.186 mM), demonstrating key interactions through hydrogen bonding. Derivative **57** exhibited uncompetitive inhibition against *Escherichia coli* β-glucuronidase (EcGUS) (IC50 = 0.082 mM), with molecular docking suggesting the furan ring as a crucial moiety. These results establish the foundation for more *in vivo* research and the generation of bifunctional medications.[69] Gani *et al.* synthesized and subsequently characterized a new series of derivatives. These derivatives were then assessed for their inhibitory effects against α-amylase and α-glucosidase. Compound **65** showed enhanced *in vitro* efficacy compared to the typical acarbose pharmaceutical agent. Additionally, *in vivo* investigations conducted on *Drosophila melanogaster* indicated the potential of compound **58** in reducing glucose levels.[70]

A novel set of oxadiazoles and thiadiazoles were synthesized by Kashtoh *et al.* for their α-glucosidase inhibitory activities. Thiadiazole derivative compound **59** shows the highest activity. Kinetic studies revealed **59** as a competitive inhibitor with a 6.0 ± 0.059 μM K_i value. Crucially, this substance revealed α-glucosidase selectivity while exhibiting negligible activity against phosphodiesterase-I, carbonic anhydrase-II, or cytotoxicity toward the rat fibroblast 3T3 cell line. The study identifies these derivatives as promising candidates for further exploration as α-glucosidase inhibitors, with potential applications in managing post-prandial hyperglycemia and other related complications.[71]

The diverse classes of synthesized compounds demonstrated significant α-glucosidase inhibitory potential, with notable candidates such as compounds standing out for their superior activity. These findings underscore the importance of structural modifications and substitution patterns in influencing inhibitory potency. The selectivity, non-toxicity, and diverse modes of inhibition exhibited by these compounds highlight their potential for further development as effective agents in the management of diabetes and related complications (Figure 7.6).

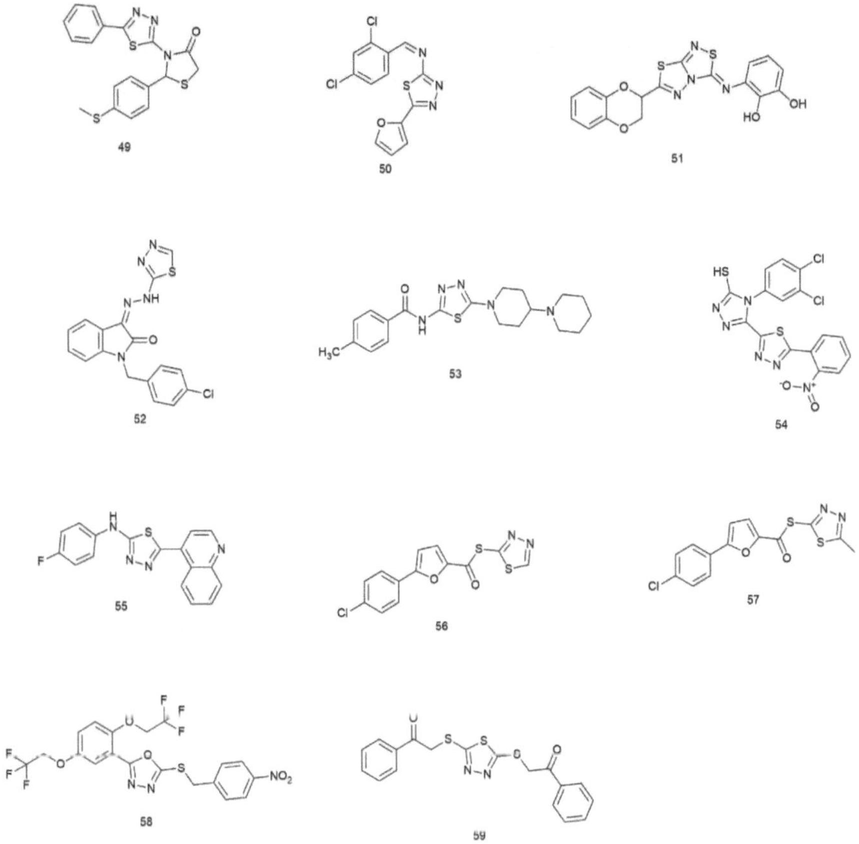

Figure 7.6 1,3,4-Thiadiazole derivatives as anti-diabetic agents.

7.4.3 Anti-seizure/Anti-epileptic/Anti-convulsant Agents

In a comprehensive exploration of 1,3,4-thiadiazole derivatives, researchers have identified promising candidates with notable anti-convulsant properties. From the distinct protection demonstrated by 5-butyl-2-(2,4-dihydroxyphenyl)-1,3,4-thiadiazole to the efficacy of pyrazine-substituted derivatives and the development of safer anti-epileptic drugs through molecular docking, these studies highlight diverse approaches in the quest for effective anti-convulsant agents.

Thirteen 1,3,4-thiadiazole chemical derivatives were screened by Luszczki *et al.* to assess their anti-convulsant potential. The evaluation was centered on their efficacy in reducing the maximal electroshock (MES)-induced seizures in mice. Compound 5-butyl-2-(2,4-dihydroxyphenyl)-1,3,4-thiadiazole (**60**) was the only one to show a clear protective effect; its ED_{50} values ranged from 233 to >500 mg kg^{-1}, and its start times varied. These findings suggest the potential of this compound.[72] According to

Harish *et al.*, the pyrazine-substituted-1,3,4-thiadiazole derivatives were successfully synthesized with good yields by reacting pyrazine-substituted-1,3,4-thiadiazoles with various sulfonyl chlorides. Significant variations in the anti-convulsant effects of several compounds were found when their anti-convulsant activity was assessed using the maximal electroshock seizure method. Compound **61** showed exceptional activity in this regard. Crucially, this substance did not show any neurotoxicity when given at a maximal dose of 100 mg kg^{-1}. The research highlights the potential of 1,3,4-thiadiazole derivatives substituted with pyrazine, particularly those containing groups that remove electrons, as encouraging options for the formation of anti-convulsant medications.[73]

The production of safer and more effective anti-epileptic medications was reported by Sahoo *et al.* This involved the synthesis of derivatives of 1,3,4-thiadiazole, which is guided by molecular docking studies that focused on targeting voltage-gated sodium channels. These compounds exhibited significant anti-convulsant activity in a rat model, surpassing phenytoin in some cases, while neurotoxicity assessments indicated their safety. Microwave-assisted synthesis proved superior, providing better reaction times and higher yields. Compound **62** emerged as a promising lead with noteworthy anti-convulsant properties, showcasing the potential of this approach for future drug development.[74] Toolabi *et al.* focused on synthesizing 2,5-disubstituted-1,3,4-thiadiazoles and evaluating their anti-convulsant and hypnotic activities. Compound **63** emerged as the most potent, showing efficacy comparable to diazepam in both tests. *In vivo* studies and binding assays verified that benzodiazepine receptors are involved in their actions, highlighting the potential of these novel thiadiazole derivatives as candidates for further exploration in the formation of anti-convulsant and hypnotic agents.[75]

Kumudha *et al.* successfully synthesized a series of new compounds, which were further evaluated in terms of their potential anti-convulsant and central nervous system (CNS)-depressive properties. They exhibited notable anti-convulsant efficacy while showing limited central nervous system-depressive properties, therefore presenting themselves as viable contenders for the advancement of potent and specific anti-convulsant medications with diminished adverse reactions. Additionally, their low toxicity suggests potential safety in therapeutic applications. Compound **64** showed minimum toxicity, highlighting its significance as a lead derivative.[76]

The synthesized 1,3,4-thiadiazole derivatives presented in these studies, such as compound **62** with superior anti-convulsant activity, compound **63** showing benzodiazepine receptor involvement, and compound **64** exhibiting excellent efficacy with minimal CNS-depressant effects, underscore the potential for novel and safer anti-epileptic drug development. These findings encourage further exploration and optimization of these

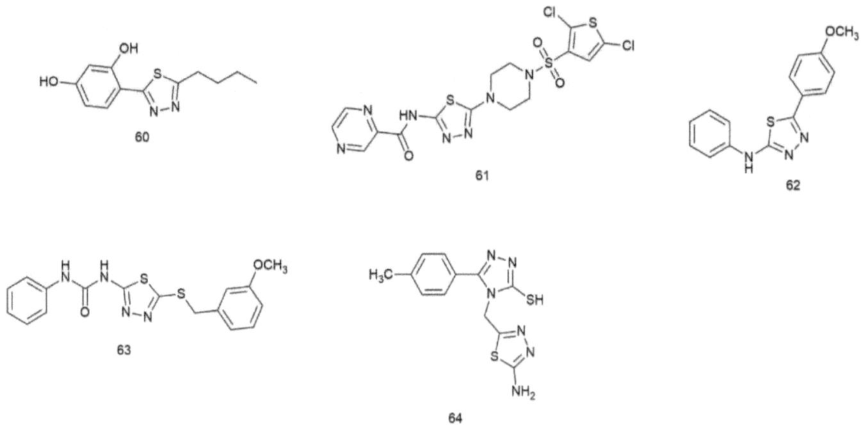

Figure 7.7 1,3,4-Thiadiazole derivatives as anti-seizure agents.

compounds for enhanced therapeutic outcomes in epilepsy management (Figure 7.7).

7.4.4 Anti-cancer Agents

Numerous research teams investigated the biological activity and synthesis of multiple 1,3,4-thiadiazole derivatives, highlighting their potential as attractive candidates for a range of therapeutic uses. The investigations examined various bioactive substances such as Schiff bases, derivatives assessed for their anti-proliferative and anti-cancer properties, and 1,3,4-thiadiazole-adamantane hybrids. To address the difficulties in the treatment of cancer and the production of anti-viral drugs, these compounds were designed and synthesized to discover novel structures with improved pharmacological properties.

In a solvent-free synthesis, 1,3,4-thiadiazole derivatives were successfully synthesized and characterized by Rashdan *et al.* against human breast carcinoma and lung carcinoma, which revealed remarkable anti-cancer potential, with compound **65** standing out for its pronounced anti-lung cancer activity. Compound **65**, selected for its high breast anti-cancer efficacy and safety, demonstrated a sustained release when encapsulated in Na-alginate microspheres, showcasing promise for enhanced chemotherapy with improved drug delivery and hydrophilicity.[77] Through a versatile synthetic approach, Dawood *et al.* synthesized 1,3,4-oxadiazolylimino-1,3,4-thiadiazoles, thiazolidinone, and thiazole derivatives by a reaction of 1,3,4-oxadiazolyl-phenylthiourea. Spectral data confirmed the structures of the synthesized products. Their evaluation against the colon carcinoma cell line (HCT-116) highlighted potent anti-tumor activity, mainly exhibited by the 1,3,4-thiadiazole derivative **66** with an IC_{50} value of 0.73 μg mL^{-1}, positioning it as a promising candidate for further anti-cancer studies.[78]

Janowska *et al.* synthesized a range of 1,3,4-thiadiazole derivatives with varied substituents at the C2 and C5 positions to develop effective anti-cancer medicines. When the biological characteristics of MCF-7 and MDA-MB-231 breast cancer cells were thoroughly evaluated in conjunction with normal fibroblasts, it was discovered that compound **67** significantly demonstrated anti-proliferative activity. This substance showed IC_{50} values of 49.6 and 53.4 μM against MCF-7 and MDA-MB-231 cells, respectively, indicating its potential as a promising lead for additional development in breast cancer therapy.[79] Chowrasia *et al.* conducted a study in which they designed and synthesized a range of fluorinated 3,6-diaryl-[1,2,4]triazolo[3,4-*b*][1,3,4]thiadiazoles. These compounds were subsequently assessed for their potential as anti-cancer agents. Compound **68**, a synthesized chemical, had notable characteristics as a derivative. Its IC_{50} values of 22.1, 19, and 15 μM against the MCF7, SaOS-2, and K562 cell lines, respectively, showed increased anti-proliferative activity. The results of this study demonstrate the promising potential of fluorine-substituted 3,6-diaryl-[1,2,4]triazolo[3,4-*b*][1,3,4]thiadiazoles as competitive candidates for the development of effective anti-cancer drugs.[80] An *N*-(4-(pyrazol-4-yl)thiazol-2-ylimino)-*N*′-phenylthiourea derivative underwent a variety of transformations in this extensive study conducted by Dawood *et al.*, producing a library of 2-(4-(pyrazol-4-yl)thiazol-2-ylimino)-1,3,4-thiadiazole derivatives, thiazolidin-4-one **22**, and 1,3-thiazole **25** derivatives. Significant anti-cancer activity was observed when these compounds were tested against the HepG2, MCF-7, and A549 cancer cell lines. Disubstituted thiourea and 1,3,4-thiadiazole **69** emerged as particularly effective candidates. The research highlights the potential of these compounds in the development of cancer drugs by revealing new tri-heterocyclic systems and offering insights into the SAR.[81]

Gür *et al.* synthesized Schiff bases developed from 5-substituted-1,3,4-thiadiazole-2-amine and assessed their various biological activities. Compound **70** emerged as a multifunctional candidate, demonstrating high DNA-protective ability, potent anti-microbial activity against *S. epidermidis*, and significant cytotoxicity against the MDA-MB-231 and PC-3 cancer cell lines. Compound **71** holds promise for further exploration in developing efficient and targeted therapeutic strategies with minimal cytotoxicity.[82] Wassel *et al.* synthesized 1,3,4-thiadiazolo-adamantane derivatives through molecular hybridization, yielding compounds with possible anti-proliferative effects against MCF-7, HepG-2, and A549 cell lines. Notably, these derivatives exhibited promising inhibitory activity against both EGFR and mutant EGFRL858R-TK, with compound **72** emerging as the most effective against EGFR L858R/T790M. Molecular docking studies supported their binding affinity within the active site of EGFR, showcasing their potential as targeted anti-cancer agents with favorable pharmacokinetic properties.[83] A group of thiazoles comprising the 1,3,4-thiadiazole core and its synthetic derivatives was examined by

Gomha *et al.* The HepG2 liver carcinoma cell line was used to test their anti-cancer properties. Compounds with IC_{50} values ranging from 0.82 to 1.88 μM showed noteworthy anti-cancer activity, surpassing the efficacious medication doxorubicin (IC_{50} = 0.72 μM). Compound **73** (IC_{50} = 0.82 μM) was deemed a viable candidate for more research and development in the treatment of liver cancer.[84]

Altıntop *et al.* conducted a study to evaluate the cytotoxicity of a collection of newly synthesized 1,3,4-thiadiazole derivatives against several human cancer cell lines, including the Bcr-Abl-positive K562 chronic myelogenous leukemia cells, intending to identify powerful anti-cancer drugs. Compound **74** specifically showed a significant inhibitory impact on Abl protein kinase, as evidenced by its IC_{50} value of 7.4 μM. Moreover, it revealed selective activity against K562 cells that express Bcr-Abl. The utilization of molecular modeling simulations has provided valuable insights into the significant involvement of the nitrothiazole moiety in its interactions with essential amino acid residues. These findings hold great promise for the advancement of new kinase inhibitors.[85]

The collective findings underscore the versatility of 1,3,4-thiadiazole derivatives as valuable scaffolds in the field of medicinal chemistry. Compounds such as Schiff bases and 1,3,4-thiadiazolo-adamantane hybrids exhibited multifaceted activities, showcasing DNA-protective ability, anti-microbial potency, and cytotoxicity against cancer cell lines. Additionally, some derivatives demonstrated selective inhibition against Bcr-Abl-positive cells, showing promising efficacy for targeted kinase inhibition. These studies collectively contribute to valuable insights into the design and evaluation of 1,3,4-thiadiazole derivatives, with potential applications in cancer therapy and anti-viral drug discovery (Figure 7.8).

7.4.5 Anti-Alzheimer Agents

A series of investigations, overseen by Alicja Skrzypek and a number of research teams, explored the synthesis and assessment of several 1,3,4-thiadiazole derivatives that have a great deal of promise for treating cognitive decline and cholinesterase inhibition. A variety of compounds are covered by the investigations, such as imidazo[2,1-*b*][1,3,4]thiadiazoles, 2,5-disubstituted-1,3,4-thiadiazoles, and conjugates of 1,3,4-thiadiazole and resorcinol. To find out the efficacy of compounds in inhibiting human carbonic anhydrase (hCA), butyrylcholinesterase (BuChE), and acetylcholinesterase (AChE), a rigorous study of design, production, and evaluation was carried out. This study emphasizes how they might be applied to the treatment of cognitive and neurodegenerative illnesses.

Skrzypek *et al.* performed an efficient synthesis of two novel series of 1,3,4-thiadiazole and resorcinol conjugates, exhibiting modest activity against BuChE and low nanomolar activity against AChE (IC_{50} = 29–76 nM). These compounds exhibited a mixed-type inhibition mechanism, showcased strong affinity in docking simulations for both enzymes, and showed

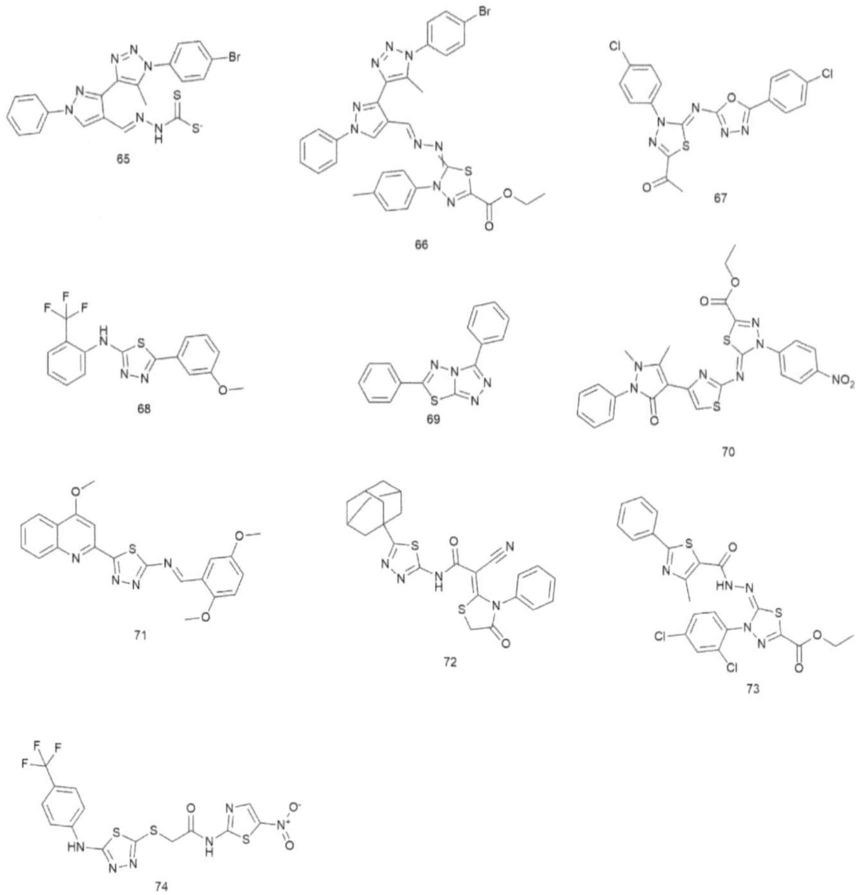

Figure 7.8 1,3,4-Thiadiazole derivatives as anti-cancer agents.

promising suppression of self-induced aggregation in *in vitro* studies. Specifically, chlorophenylaminothiadiazoles (compound **75**) exhibit promising multifunctional properties, making them possible candidates for more research and development as medications to treat Alzheimer's disease. Additionally, they demonstrated antioxidant properties, metal ion chelation, acceptable cytotoxicity, and a favorable absorption, distribution, metabolism, excretion, toxicity (ADMET) profile, indicating their potential as multifunctional cholinesterase inhibitors.[86] In a subsequent investigation by the same authors, novel 1,3,4-thiadiazole derivatives were prepared, and their ability to inhibit acetyl and butyryl cholinesterase was evaluated. The docking simulations highlighted binding interactions in the catalytic substrate site, and these derivatives exhibited favorable drug-likeness properties, suggesting potential for further development as cholinesterase inhibitors with promising pharmacokinetics.[87]

In a comprehensive investigation targeting the treatment of cognitive decline, 2,5-disubstituted 1,3,4-thiadiazole derivatives have been designed, synthesized, and pharmacologically evaluated by Bala *et al.* Molecular docking studies predicted their substantial binding affinity toward rhAChE. Pharmacokinetic research, behavioral evaluations using escape learning methods, step-down passive avoidance, and *in vitro* acetylcholinesterase inhibition assays were conducted. Compound **76** exhibited significant AChE inhibition and demonstrated efficacy in mitigating oxidative stress-induced cognitive decline, showcasing its potential as a lead compound for further exploration in cognitive dysfunction.[88]

Askin *et al.* have designed and synthesized a novel series of imidazo[2,1-*b*][1,3,4]thiadiazole by substitution as 2,6-disubstituted and 2,5,6-trisubstituted compounds and assessed for their inhibitory activities against AChE and hCA I and II isoforms. These substances showed strong inhibitory activity on hCA I, hCA II, and AChE in the low nanomolar range without the presence of the zinc-binding sulfonamide group. Notably, compound **77** exhibited an 18-fold selectivity toward hCA I over acetazolamide. These findings suggest the potential of these novel lead compounds for developing potent hCA and AChE inhibitors.[89] Skrzypek *et al.* developed and assessed a new series of (1,3,4-thiadiazol-2-yl)benzene-1,3-diol derivatives for their anticholinesterase properties. AChE and BuChE were shown to be inhibited by the compounds, with IC_{50} values that vary from >500 to 0.053 μM for AChE and >500 to 0.105 μM for BuChE. Compound **78** is the strongest AChE inhibitor with an IC_{50} value of 0.053 μM and a selectivity ratio of approximately 950 against BuChE. The kinetic studies indicated that compound **78** acts as a mixed-type AChE inhibitor, highlighting its potential as a selective therapeutic agent against AChE.[90]

The studies collectively highlight the promising pharmacological potential of the 1,3,4-thiadiazole derivatives. Derivatives such as chlorophenylaminothiadiazoles show multifunctional properties, making them potential candidates for the treatment of Alzheimer's disease. Additionally, the derivatives of 1,3,4-thiadiazole and 2,5-disubstituted showcased significant AChE inhibition and efficacy against oxidative stress-induced cognitive decline, positioning them as lead compounds for further exploration. The novel imidazo[2,1-*b*][1,3,4]thiadiazoles showed strong inhibitory activity on AChE and hCA, suggesting their potential as lead compounds for developing potent enzyme inhibitors. The mixed-type AChE inhibitor emerged as a promising candidate with selectivity and kinetic attributes, which underscore its potential as a selective therapeutic agent against AChE. These findings collectively provide valuable insights into the evolution of compounds with therapeutic capability in neurodegenerative diseases and cognitive disorders (Figure 7.9).

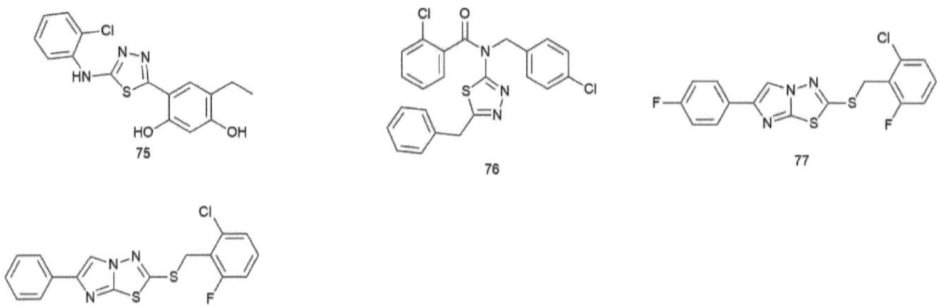

75 76 77 78

Figure 7.9 1,3,4-Thiadiazole derivatives as anti-Alzheimer agents.

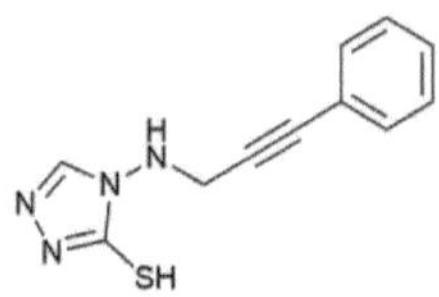

79

Figure 7.10 1,3,4-Thiadiazole derivatives as anti-viral agents.

7.4.6 Anti-viral Agents

The present study conducted by Khramchikhin *et al.* presents a new synthetic methodology for the production of 8-phenyl-[1,2,4]triazolo[3,4-*b*] and 7-benzylidene-[1,2,4]triazolo[3,4-*b*][1,3,4]thiadiazine. In this study, they investigated the interaction system between bi-nucleophilic 3-substituted 4-amino-4*H*-1,2,4-triazole-5-thiols and 3-phenyl-2-propynal to synthesize [1,3,4]thiadiazepine compounds. Among the pool of 20 compounds that were assessed against the influenza H1N1 virus, precisely 50% of these compounds demonstrated a selectivity index equal to or exceeding 10. Significantly, compound **79** exhibited remarkable anti-viral efficacy, showing the highest selectivity index (SI >300). This finding positions it as a prospective candidate warranting further exploration and analysis. The research also offers valuable insights into the preferred mode of interaction with the M2 proton channel, highlighting the potential influence of the ligand structure on the control of target activitythesized and assessed the anti-tube. A total of 91 derivatives of 1,3,4-thiadiazole have demonstrated considerable promise as anti-viral medicines, with notable inhibitory effects against a range of viral strains. The compounds that exhibit structural plasticity and various pharmacological activities provide intriguing possibilities for further study in anti-viral medication development (Figure 7.10).

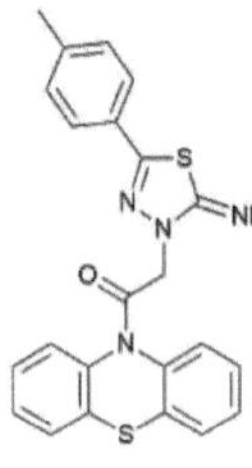

80

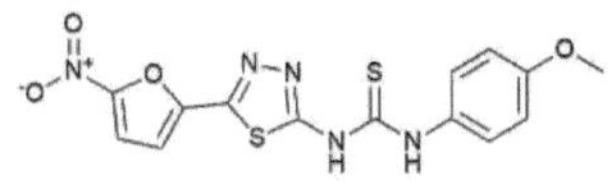

81

Figure 7.11 1,3,4-Thiadiazole derivatives as anti-tuberculosis agents.

7.4.7 Anti-tuberculosis Agents

Ramprasad *et al.* synthesized a unique series of phenothiazine and 1,3,4-thiadiazole derivatives by a single-step chloroamine coupling process. Compound **80** exhibited the highest potency against *Mycobacterium tuberculosis* H37Rv, as determined by evaluation. Its minimum inhibitory concentration (MIC) was measured to be 0.8 μg mL^{-1} (about 1.9 μM). The structure–activity relationship emphasized the importance of alkyl or substituted phenyl groups on the 1,3,4-thiadiazole ring. Surprisingly, the active compounds showed no harmful effects on Vero cells, and through molecular docking experiments involving InhA and CYP121, valuable information was obtained regarding their potential as medicines for combating tuberculosis.[91]

Krishna *et al.* synthesized and assessed the anti-tubercular activity of these new analogs, specifically ((4-methoxyphenyl)carbamoyl)/(5-(5-nitrothiophen-2-yl)-1,3,4-thiadiazol-2-yl)-amide derivatives. Compound **81** exhibited significant inhibitory activity against the enzyme ketol-acid reductoisomerase (KARI) of *Mycobacterium TB*, as evidenced by its K_i values of 2.02, 5.48, and 4.72 μM, respectively. The compounds showed notable efficacy against *Mt H37Rv* in the *in vitro* study, with MIC values as low as 1 μM. Moreover, they revealed promising inhibitory effects in macrophages infected with *Mt* and in nutrient-starved models while demonstrating minimal toxicity toward HEK 293 T-cell lines. The results of this study indicate that these findings have the potential to serve as primary compounds for subsequent refinement and development into medicines for the treatment of tuberculosis.[92] The anti-tubercular potential of 1,3,4-thiadiazole derivatives has been demonstrated, as evidenced by the synthesis of new compounds that exhibit potent inhibitory activity against *Mycobacterium tuberculosis*. Their ability to target specific enzymes and demonstrate efficacy against tuberculosis, as seen in various studies, positions them as valuable candidates for further development in the quest for effective anti-TB drugs (Figure 7.11).

7.4.8 Antimicrobial Agents

Bhinge *et al.* synthesized 1,3,4-thiadiazole derivatives and assessed them for anti-microbial, anti-fungal, and anthelmintic activities. Compound **82**, featuring a *para*-positioned NO_2 group on the phenyl ring, exhibited activity analogous to standard drugs, demonstrating promising efficacy with minimal toxicity. The results of this study highlighted the ability of these synthesized compounds as candidates for developing anti-bacterial and anti-fungal agents, showcasing their significance as pharmacophore models for future drug development in these therapeutic areas.[93] A new series of 1,3,4-thiadiazolyl-sulfanyl-4,5-dihydropyridazin-3(2*H*)-ones with predicted anti-microbial activity was synthesized by El-Badry *et al.* The design involved a strategic combination of thia-Michael addition and cyclization reactions, resulting in functionalized heterocycles. Structural elucidation using various spectroscopic techniques confirmed the successful synthesis of these compounds. Compound **83** shows good anti-bacterial activity. These findings underscore the potential of the designed heterocycles as promising anti-microbial agents, showcasing a methodically tailored approach for the development of new therapeutic compounds.[94]

To evaluate a set of novel 1,3,4-thiadiazole derivatives for their potential anti-bacterial effects against *Xanthomonas axonopodis* pv. *citri* (Xac) and *Xanthomonas oryzae* pv. *oryzae* (Xoo), Wan *et al.* synthesized and evaluated the compounds. Compound **84** exhibited significant inhibitory effects against subsp. *citri* Xac and Xoo. Similarly, compound 2-((2-chloro-1*H*-indol-3-yl)methylene) showed noteworthy inhibitory activity. Compound **85** exhibited remarkable efficacy against Xoo. The results of this study emphasize the potential of these compounds as highly effective anti-bacterial agents, surpassing the efficiency of a commercially available reference compound.[95]

Pund *et al.* conducted additional research in which they successfully synthesized a collection of innovative compounds (designated as AP-1 to AP-10) using (2*S*)-3-(benzyloxy)-2-[(*tert*-butoxycarbonyl)amino]propanoic acid as the initial substrate. A spectrum analysis of the compounds was performed to determine their characteristics, and their anti-microbial activity was assessed using the *in vitro* testing. It is worth mentioning that compound **86**, which includes nitro or methoxy groups, had significant inhibitory activity against Gram-negative bacteria. On the other hand, compounds **87** and **88**, which consist of nitro, methoxy, or hydroxyl groups, remarkably inhibited the growth of Gram-positive bacteria.[96]

1,3,4-Thiadiazole derivatives have demonstrated significant anti-microbial activity, serving as promising agents against a range of microorganisms. Their diverse structural modifications and proven efficacy against bacteria and fungi highlight their potential as valuable candidates for the production of innovative anti-microbial drugs (Figure 7.12).

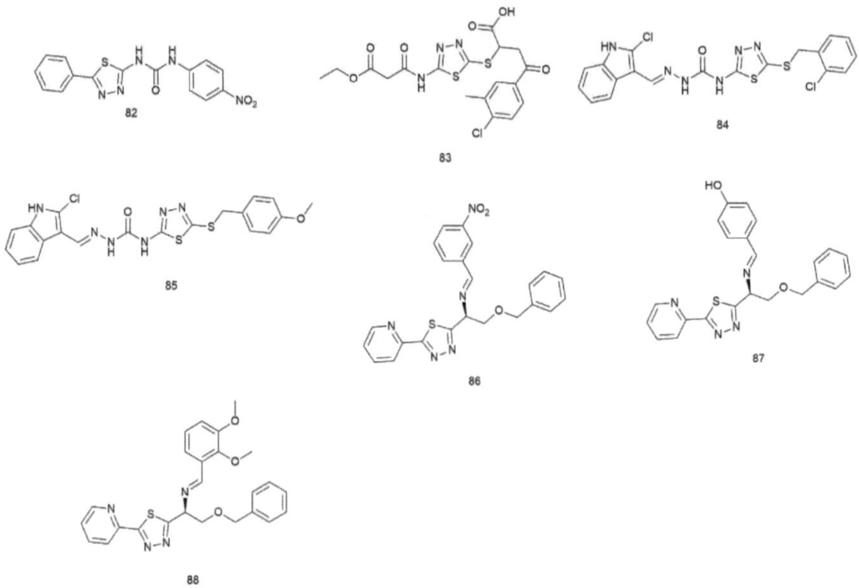

Figure 7.12 1,3,4-Thiadiazole derivatives as anti-microbial agents.

7.4.9 Diuretic Agents

Drapak *et al.* synthesized and evaluated 5-amino-1,3,4-thiadiazole-2-thiol derivatives using a one-pot synthesis method. The synthesized derivatives, particularly 5-benzylthio-1,3,4-thiadiazol-2-amine derivative **89**, exhibited significant diuretic activity in an *in vivo* study using a white rat model. The aforementioned compounds underwent additional scrutiny to assess their kaliuretic, saluretic, and natriuretic characteristics, as well as their capacity for inhibiting carbonic anhydrase. This analysis facilitated the establishment of a structure–diuretic activity relationship. These findings suggest the potential of these derivatives as promising diuretic agents with favorable properties.[97] In the study by Ergena *et al.*, the derivatives of thioate-1,3,4-thiadiazoles were synthesized and characterized, demonstrating promising diuretic activity in *Swiss albino* mice. Notably, 5-methyl-substituted derivatives exhibited significant increases in water and electrolyte excretion compared to negative controls and 5-amino-substituted compounds. The compound with a 4-nitro-substituted benzene ring at the 2-thioate group of 5-methyl-1,3,4-thiadiazole showed the highest diuretic activity, emphasizing the potential of these derivatives as effective diuretic agents. Compound **90** is mentioned as a lead compound.[8]

Studies have demonstrated that the derivatives of 1,3,4-thiadiazole have noteworthy diuretic activity, particularly when assessing their effects on the urine output and electrolyte excretion. The aforementioned compounds, specifically the derivatives of 5-amino-1,3,4-thiadiazole-2-thiol, demonstrate considerable potential as a viable pathway for the advancement of diuretic

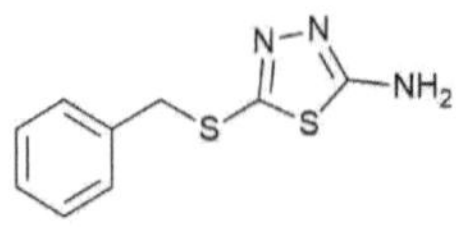

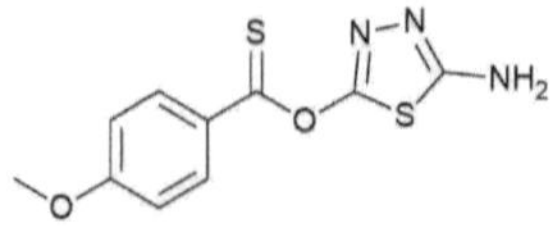

89

90

Figure 7.13 1,3,4-Thiadiazole derivatives as diuretic agents.

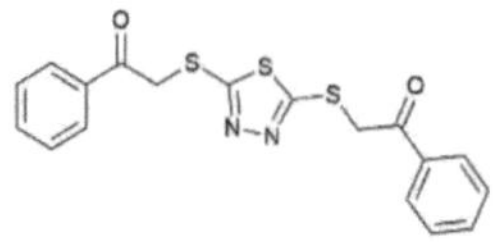

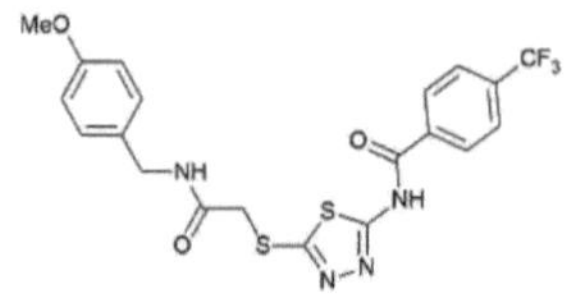

91

92

Figure 7.14 1,3,4-Thiadiazole derivatives as anti-obesity agents.

medicines. These molecules hold promise for addressing medical disorders associated with fluid balance and renal function (Figure 7.13).

7.4.10 Anti-obesity Agents

Kashtoh *et al.* synthesized and evaluated oxadiazoles and thiadiazoles for their α-glucosidase inhibitory activities. Fifteen compounds, including oxadiazoles and thiadiazoles, were identified as potent inhibitors, with thiadiazole-based compound **91** showing competitive inhibition (K_i = 6.0 ± 0.059 μM). These compounds demonstrated selectivity for α-glucosidase, without significant toxicity against carbonic anhydrase-II, phosphodiesterase-I, or the rat fibroblast 3T3 cell line, suggesting their potential as novel α-glucosidase inhibitors for further investigation as anti-obesity agents.[71]

Okada *et al.* synthesized 27 thiadiazole derivatives, which led to the discovery of compound **92**, demonstrating superior GLS1 inhibitory activity with an IC_{50} value of 46.7 μM. Compound **92** emerges as a highly promising and novel GLS1 inhibitor with potential applications in anti-cancer, anti-obesity and anti-aging treatments.[9]

1,3,4-Thiadiazole derivatives have demonstrated potential as anti-obesity agents in various studies, showcasing their ability to modulate key physiological processes related to weight regulation. These compounds hold promise for further exploration and development as novel therapeutic agents for managing obesity and associated metabolic disorders (Figure 7.14).

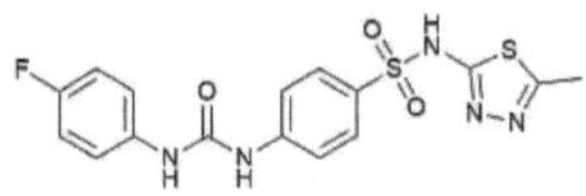

93

Figure 7.15 1,3,4-Thiadiazole derivatives as anti-glaucoma agents.

7.4.11 Anti-glaucoma Agents

In the quest for hCA inhibitors, Mishra *et al.* synthesized and screened a collection of derivatives of *N*-1,3,4-thiadiazol-2-yl-4-(3-substitutedphenylureido)benzenesulfonamide against hCA I, II, and VII isoforms. While the majority of them exhibited weak inhibition, compound **93** demonstrated modest potency against hCA I and VII, with K_i values of 73.7 and 85.8 μM, respectively. These compounds have the potential to act as lead molecules in the development of selective inhibitors that possess secondary sulfonamide functions. Anti-glaucoma therapy may potentially provide assistance (Figure 7.15).[98]

7.4.12 Anti-platelet Agents

Tehrani *et al.* designed and synthesized 2-hydrazinyl-1,3,4-thiadiazole derivatives, characterized by their hydrazone group, as potential anti-platelet agents. Through *in vitro* anti-platelet aggregation assays, compounds **94**, **95**, and **96** exhibited the highest inhibition of COX enzyme, suggesting their potential as potent anti-platelet agents.[99] In a parallel study, Ruel *et al.* synthesized novel 1,3,4-thiadiazole derivatives, with compound **97** showing remarkable *in vivo* anti-platelet activity, indicating its potential as a lead compound for synthesizing new anti-platelet agents (Figure 7.16).[100]

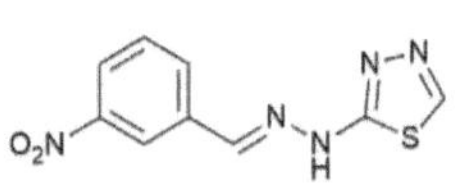

94

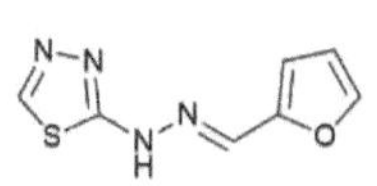

95

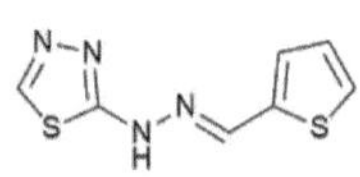

96

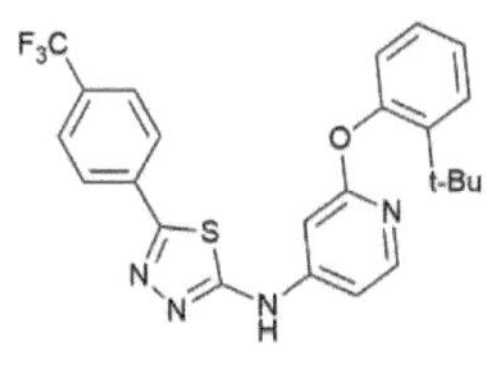

97

Figure 7.16 1,3,4-Thiadiazole derivatives as anti-platelet agents.

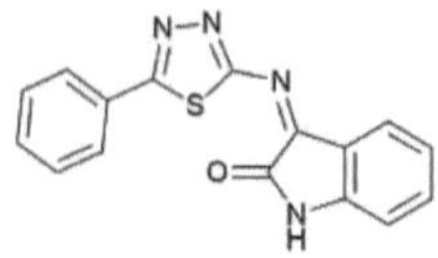

98

Figure 7.17 1,3,4-Thiadiazole derivatives as anti-aging agents.

7.4.13 Anti-aging Agents

SIRT1, a crucial regulator in multiple biosynthetic pathways, plays a pivotal role in various physiological processes by activating peroxisome proliferator-activated receptor-γ and interacting with adenosine-mono-phosphate kinase. Consequently, it has emerged as a significant therapeutic target for a range of health conditions including neurodegenerative diseases, cancer, metabolic disorders, and aging-related interventions. In the quest for potent SIRT1-targeted compounds, Saha *et al.* studied a series of novel 1,3,4-thiadiazole derivatives through virtual screening, homological modeling, and docking. Compound **98** demonstrated superior interaction energy (approximately 6.4 kcal mol^{-1}) compared to resveratrol. Molecular dynamics simulations further confirmed that **98** has excellent conformation fitting stability in the SIRT1-binding site, positioning it as a promising lead compound for future anti-aging drug design (Figure 7.17).[101]

7.4.14 Anti-leishmanicidal Agents

Cinnamic acid derivatives, abundant in nature, exhibit diverse biological activities, including antiseptic, stimulant, carminative, and insecticidal properties. Using hydroxy-cinnamic acids, such as ferulic, caffeic, and chlorogenic acids, exhibiting a variety of biological effects through the phenylpropanoid pathway, scientists have investigated their potential in medication development. Notably, the cinnamoyl moiety is featured in drugs such as ozagrel for ischemic stroke and multifunctional hybrid drugs that conjugate cinnamic acid with a second pharmacophore, showcasing promising results in the treatment of complex diseases, such as Alzheimer's disease, cancer, cardiovascular conditions, infectious diseases, and anti-leishmanicidal agents.[102]

7.4.15 Anti-*H. pylori* Agents

Tabei *et al.* successfully synthesized a series of novel anti-microbial compounds. Specifically, they developed twenty-five new derivatives of 2-(5-(5-nitrofuran-2-yl)-1,3,4-thiadiazol-2-ylimino)thiazolidin-4-one, each of which

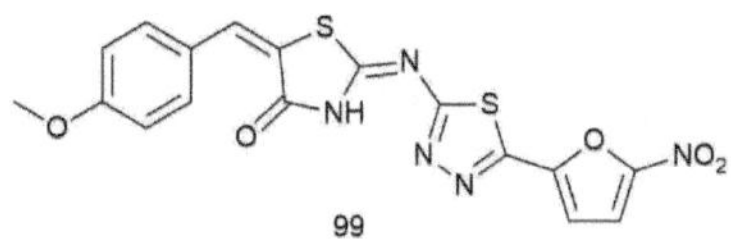

99

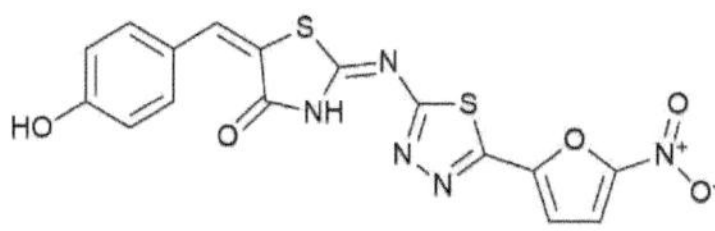

100

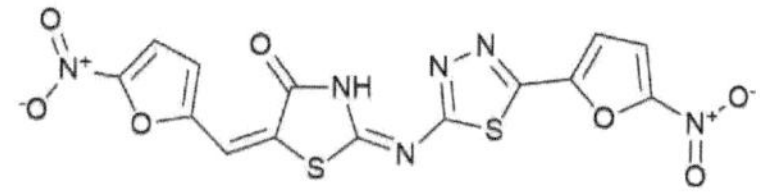

101

Figure 7.18 1,3,4-Thiadiazole derivatives as anti-*H. pylori* agents.

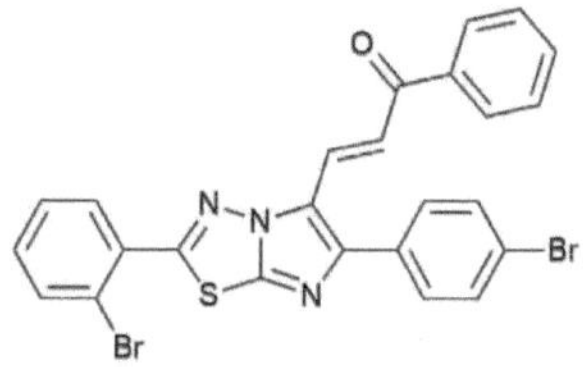

102

Figure 7.19 1,3,4-Thiadiazole derivatives as anti-fungal agents.

possessed an aryl or heteroaryl methylene group at position 5. These compounds demonstrated remarkable potency against Gram-positive bacteria, surpassing ampicillin, while showing no efficacy against Gram-negative strains. Notably, compounds **99**, **100**, and **101** exhibited strong anti-microbial activity against metronidazole-resistant *Helicobacter pylori* strains, suggesting their potential as promising candidates to advance the development of novel pharmaceutical agents targeting antibiotic-resistant microbes. It is imperative to focus on the production of new anti-infectious medications (Figure 7.18).[103]

7.4.16 Anti-fungal Agents

Alwan *et al.* assessed and synthesized 25 newly developed hybrid derivatives of imidazo[2,1-*b*]-1,3,4-thiadiazole with anti-microbial properties. The compounds under investigation are chalcones and Schiff bases that incorporate a 1,3,4-thiadiazole scaffold. Most compounds exhibited anti-fungal activity, with MICs ranging from 1.56 to 100 μg mL^{-1}. Compound **102** demonstrated promising activity against *Cryptococcus neoformans*, with MICs as low as 1.56 μg mL^{-1} (Figure 7.19).[104]

Table 7.1 Some of the patents on 1,3,4-thiadiazoles.[105]

Patent ID	Title	Assignee	Inventor/Author	Grant date M/D/Y
CN-115057831-B	Thiadiazole compound and its application thereof in the preparation of a KKBC 1 protein inhibitor	Shengjing Hospital of China Medical University	Liu Caigang, Yang Yongliang, and Liu Yang	08/04/2023
CN-112279845-B	Aryl- or heteroaryl-substituted thiadiazole compounds and antibacterial application thereof	Institute of Materia Medica of CAMS	Wu Song, Xia Jie, Xue Wenjie, and Zhang Chi	07/28/2023
US-11702414-B2	Thiadiazole IRAK4 inhibitors	Gilead Sciences, Inc.	Stephen Ammann, Elizabeth M. Bacon, Gediminas Brizgys, Elbert Chin, Chienhung Chou, Jeromy J. Cottell, Marilyn Ndukwe, Marina Shatskikh, James G. Taylor, Nathan E. Wright, Zheng-Yu Yang, and Sheila M. Zipfel.	07/18/2023
CN-114195772-B	1,2,4-Oxadiazole derivative containing the 1,3,4-thiadiazole unit and its preparation method and application thereof	Guizhou University	Gan Xiuhai, Liu Dan, Zeng Huanan, Yuan Ting, Wang Zhengxing, Liang Juncheng, Luo Ling, and Geng Wang	06/30/2023
CN-111978312-B	5-Aryl-1,3,4-thiadiazole/1,3,4-oxadiazole-2-amine compound and its preparation method and application thereof	Huanghuai University	Dong Jianghong, Pei Qinglan, Wang Panpan, and Zhang Xiangfei	05/05/2023
CN-114605352-B	Gallic acid derivative containing 1,3,4-thiadiazole amide and its preparation method and application thereof	Jiangsu Ocean University	Liu Weiwei, Jiang Kaijun, Cao joint attack, Shao Zhongbai, Wu Yuyu, Chen Chao, Su Ziqin, Qiu Min, Takada Tian, Liang Xinwen, and Guo Lei	04/25/2023
CN-114671892-B	1,3,4-Thiadiazole derivative containing 7-aminocephalosporanic acid and its preparation method and application thereof	Jiangsu Ocean University	Liu Weiwei, Cao Cao, Qiu Min, Gao Tian, Liang Xinwen, Wu Yuyu, Chen Chao, Su Ziqin, Jiang Kaijun, Shao Zhongbai, and Guo Lei	04/11/2023
CN-111848548-B	5-Substituted-1,3,4-thiadiazole Schiff base derivative and its preparation method and application thereof	Tianjin University of Science and Technology	Wang Meiyi, Peng Yaqi, and Han Junjun	03/28/2023

(continued)

Table 7.1 (*continued*)

Patent ID	Title	Assignee	Inventor/Author	Grant date M/D/Y
CN-113582946-B	3-Aryl-5-thio-1,3,4-thiadiazole-2-thioketone derivative and its preparation method and application thereof	Gannan Normal University	Zheng Luyin, Mei Weijie, Guo Wei, and Wu Yongquan.	01/13/2023
CN-112225733-B	Preparation method of 1,34-thiadiazole pyridine-2-ketone derivative and its application as anti-cancer drugs	Hunan University of Science and Technology	Xie Wenlin, Qi Rui, Tang Ling, and Liu Renzhi	12/09/2022
CN-111440163-B	Pyrazole compound containing methoxy-substituted 1,3,4-thiadiazole unit and its preparation and application thereof	Nantong University	Dai Hong, Ni Yadan, Wang Zhipeng, Qian Cheng, Li Yu, Zhang Yan, Shi Lei, Zhang Haijun, Li Jianhua, Zhou Beibei, Wang Kai, and Xu Hongjuan	09/02/2022
RU-2778090-C1	2-Amino-6-hydroxy-7-(3-oxo dihydroquinoxaline-2-yl)-7*a*-phenyl-pyrrolo[2,1-*b*][1,3,4]thiadiazole-5(7*ah*)-ones with anti-inflammatory activity and their preparation	Federal State Autonomous Educational Institution of Higher Education "Perm State National Research University" (PGNIU)	Irina Vladimirovna, Mashevskaya Jamilya, Nailievna Lukmanova, Danila Yurievich, Apushkin Taisiya, Andreevna Utushkina, Karolina Evgenievna, Yakushina Alexander, and Igorevich Andreev	08/15/2022
CN-112773795-B	Application of the 1,3,4-thiadiazole phenyl furan thiomethyl ester compound in the preparation of the β-glucuronidase inhibitor	Zhejiang University of Technology ZJUT	Wang Hong, Cui Zining, Wei Bin, Zhou Taoshun, and Li Yasheng	07/22/2022
CN-112480031-B	Emodin amide derivatives containing 1,3,4-thiadiazole and its synthesis and application thereof	Jiangsu Ocean University	Liu Weiwei, Cao Lianfeng, Wang Youxian, Liu Shuhao, Shao Zhongbai, Jiang Kaijun, Wu Yuran, Guo Lei, Shi Dahua, and Cao Zhiling	05/20/2022
EP-3724198-B1	2-Oxo-1-pyrrolidinyl imidazothiadiazole derivatives	UCB Biopharma SRL	Laurent Provins and Hugues Chanteux	02/16/2022
JP-6999550-B2	1,3,4-Thiadiazole compounds and their use in the treatment of cancer	Cancer Research Technology Ltd, AstraZeneca AB	Maurice Raymond Verschoyle Finlay, Johannes Wilhelmus Maria Nissink, and Mark David Charles	02/10/2022

(*continued*)

Table 7.1 (*continued*)

Patent ID	Title	Assignee	Inventor/Author	Grant date M/D/Y
US-11230544-B1	Substituted 1,3,4-thiadiazoles as GLS1 inhibitors	Medshine Discovery Inc.	Yonggang Liao, Jun Lin, Chaonan Liu, Changqing Wei, Wenyuan Qian, Chen Zhang, Zhen Gong, Jian Li, and Shuhui Chen	01/25/2022
CN-111747946-B	5-Benzofuran-1,3,4-thiadiazole-2-substituted amide compound and its preparation and application thereof	Hebei University	Li Wan, Li Shengnan, Yan Kecheng, Zhu Huajie, Zeng Taining, and Zhang Xiao	12/31/2021
RU-2761880-C1	Adamantyl-containing derivatives of 1,2,4-triazole and 1,3,4-thiadiazole having monoterpenoid fragments used as the inhibitors of the enzyme tyrosyl-DNA phosphodiesterase 1	Federal State Budgetary Institution of Science Novosibirsk Institute of Organic Chemistry named after N.N. Vorozhtsov Siberian Branch of the Russian Academy of Sciences (NIOH SB RAS), Federal State Budgetary Institution of Science Institute of Chemical Biology, and Fundamental Medicine of the Siberian Branch of the Russian Academy of Sciences (IBFM SB RAS)	Aldar Ayurovich Munkuev, Alexandra Leonidovna Zakharenko, Evgeniy Vladimirovich Suslov, Arina Aleksandrovna Chepanova, Konstantin Petrovich Volcho, Olga Ivanovna Lavrik, and Nariman Faridovich Salakhutdinov.	12/13/2021
RU-2757808-C1	2-Alkylthio-5-(1*H*-1,2,4,-triazole-1-ilmethyl)-1,3,4-thiadiazoles and their production and fungicidal compositions based thereon	Federal State Budgetary Educational Institution of Higher Education "Rus-sian Chemical-Technological University named after D.	Grigory Valerievich Tsaplin, Sergey Vladimirovich Popkov, Anastasia Sergeevna Orlova, and Anna Leonidovna Alekseenko	10/21/2021

(*continued*)

Table 7.1 (*continued*)

Patent ID	Title	Assignee	Inventor/Author	Grant date M/D/Y
		I. Mendeleev" (RHTU named after D. I. Mendeleev)		
CN-109666028-B	1,3,4-Thiadiazole heterocyclic compound having hedgehog pathway antagonist activity	Suzhou Yunxuan Pharmaceutical Co., Ltd.	Zhang Xiaohu, Zheng Jiyue, Ma Haikuo, and Tian Sheng	10/12/2021
KR-102293386-B1	Novel (cyano-dimethyl-methyl)-isoxazoles and -[1,3,4]thiadiazoles	Centrexion Therapeutics Corporation	Doris Ritter, Florian Binder, Henry Dodds, Stefan Georg Müller, Janet Rachel Nicolsson, and Achim Sauer	08/26/2021
US-11033535-B2	Oxadiazoles and thiadiazoles as TGF-β inhibitors	Southern Research Institute, Uab Research Foundation	Mark J. Suto, Vandana V. Gupta, Bini Mathew, and Joanne Murphy-Ullrich	06/15/2021
JP-6873132-B2	1,3,4-Thiadiazole compounds and their use in the treatment of cancer	Cancer Research Technology Ltd, AstraZeneca AB	Maurice Raymond Verschoyle Finlay, Johannes Wilhelmus Maria Nissink, Mark David Charles, and James Matthew Wood	05/19/2021
CN-109369637-B	Novel kojic acid derivative containing 1,3,4-thiadiazole and its application thereof	Hunan University of Science and Technology	Xie Wenlin, Liu Renzhi, and Mei Qihong	04/06/2021
JP-6821680-B2	1,3,4-Thiadiazole compounds and their use in the treatment of cancer	Cancer Research Technology Ltd, AstraZeneca AB	Maurice Raymond Verschoyle Finlay, David Robert Perkins, Johannes Wilhelmus Maria Nissink, Piotr Antoni Laubo, Peter Duncan Smith, and Andrew Bailey	01/27/2021
CN-108570042-B	Indolone derivative containing 1,3,4-thiadiazole and its preparation and application	Hunan University of Science and Technology	Xie Wenlin, Liu Renzhi, and Zhang Huilin	12/15/2020
CN-106883409-B	Poly-2-mercapto-1,3,4-thiadiazole nanoparticle and its synthesis and application thereof	Kunming University of Science and Technology	Huang Shaojun, Ma Chengzhang, Li Chao, Du Ping, and Huang Qiuling	02/18/2020
AU-2015365465-B2	Amido thiadiazole derivatives as NADPH oxidase inhibitors	Genkyotex Suisse Sa	Vivienne Allen, Mark S. Chambers, John M. Ellard, Alastair Hodges, Christopher James Lock, Peter Machin, and Andrew Sharpe	12/12/2019

(*continued*)

Table 7.1 (*continued*)

Patent ID	Title	Assignee	Inventor/Author	Grant date M/D/Y
CN-106674251-B	It can promote triazolone and thiadiazole compounds and their medicinal usage of cardiac muscle mitochondria	ReSource (WuXi) Biological Medicine Technology Co. Ltd, Fudan University	Zhong Tao, Ni Tiehua, Xie Shuying, and Hu Hairong	10/15/2019
CN-106795150-B	1,3,4-Thiadiazole compound and its purposes in treating cancer	Cancer Research Technology Ltd, AstraZeneca AB	Maurice Raymond Verschoyle Finlay, Johannes Wilhelmus Maria Nissink, Mark David Charles, and James Matthew Wood	10/15/2019
US-10308645-B2	Piperidine oxadiazole and thiadiazole orexin receptor antagonists	Merck Sharp & Dohme Corp.	Nigel Liverton, Scott D. Kuduk, Yunfu Luo, Na Meng, and Tingting Yu	06/04/2019
CN-106008557-B	3,6-Diaryl-[1,2,4]triazole and [3,4-*b*][1,3,4]thiadiazole compounds as tumor cell proliferation inhibitors	Shenyang Pharmaceutical University	Zhang Weige, Xu Qile, Wu Yingliang, Sun Maolin, Zuo Daiying, Wang Yueting, Bai Zhaoshi, Bao Kai, and Guan Qi	03/01/2019
US-10160736-B2	1,3,4-Oxadiazole and 1,3,4-thiadiazole derivatives as immunomodulators	Aurigene Discovery Technologies Limited	Pottayil Govindan Nair Sasikumar, Muralidhara Ramachandra, and Seetharamaiah Setty Sudarshan Naremaddepalli	12/25/2018
CN-106866708-B	1,2,4-Triazols[3,4-*b*] -1,3,4-thiadiazole compounds and their application	Tianjin University of Technology	Chen Baoquan, Liu Xiaojia, Tang Rui, Zhang Shuai, Wang Haixin, and Liu Haiying	10/12/2018
CN-104892639-B	3,6-Bis- replaces [1,2,4]triazole[3,4-*b*] [1,3,4]thiadiazole compound and its application thereof	Shanghai Institute of Materia Medica of CAS	Yang Caiguang, Luo Cheng, Gong Shouzhe, Liu Hongchuan, Zhang Ruihan, Zhang Jie, and Jiang Hualiang	10/30/2018
CN-106008623-B	1,2,4-Triazoles containing glucosamine molecules and [3,4-*b*]-1,3,4-thiadiazoles derivatives and its preparation method thereof and purposes	Huaihai Institute of Techology	Liu Weiwei, Cheng Fengchang, Yin Long, Shi Dahua, and Liu Xiujian	07/06/2018
RU-2651572-C2	5-Ethyl-2-amino-1,3,4-thiadiazole derivatives with anesthetic, anti-inflammatory, anti-allergic and analgetic activities	Open Joint Stock Company "All-Russian Scientific Center for the Safety of Biologically Active Substances" (JSC "VNTs BAV")	Sofia Yakovlevna Skachilova, Nikolai Konstantinovich Zheltukhin, Marina Aleksandrovna Demidova, Nikita Sergeevich Popov, Yuri Georgievich Kazaishvili, Alexey Viktorovich Korotonozhkin, Oksana Vladimirovna	04/23/2018

(*continued*)

Table 7.1 (*continued*)

Patent ID	Title	Assignee	Inventor/Author	Grant date M/D/Y
			Proskurina, Marina Vasilievna Olkhovich and German Leonidovich Perlovich	
AU-2016256728-B2	Thiadiazole analogs thereof and methods for treating SMN-deficiency-related-conditions	Novartis AG	Jake Axford, Natalie Dales, and Moo Je Sung	03/29/2018
CN-104277035-B	Acridine-1,3,4-thiadiazole compounds and their preparation and application	Individual	Liu Huagang and Huo Lini	01/04/2017
KR-101667563-B1	1,3,4-Thiadiazole-2-amine derivatives: preparation method thereof and pharmaceutical composition for use in preventing or treating urotensin-receptor activity-related diseases containing the same as an active ingredient	Korea Research Institute of Chemical Technology	Byeong-ho Lee, Gyu-yang Lee, Chae-jo Lim, Ji-hee Seo, Gwang-seok Oh, Jeong-hyeon Lee, and Ju-young Jang	10/28/2016
CN-104098609-B	Imidazo[2,1-*b*]-1,3,4-thiadiazoles containing ferrocene and their preparation method and applications	Shaanxi University of Science and Technology	Liu Yuting, Liang Gangtao, Yin Dawei, Li Xin, Lu Bo, and Zhang Xiaoli	08/17/2016
KR-101608897-B1	Novel 5-aryl-1,3,4-thiadiazole-based hydroxamic acids and their anticancer composition comprising the same as the active ingredient	Chungbuk National University Industry-Academic Cooperation Foundation	Sangbae Han, Youngsoo Kim, Jintae Hong, Hai Nam Nguyen, Thi My Deng Du, Thi Kim Oan Dao, Thi Phuong and Deng Phan	04/05/2016
CN-103232447-B	3-Acetyl-5-acetylimino-2-(*N*-phenothiazinyl)-1,3,4-thiadiazole and its preparation method and application thereof	Shaanxi University of Science and Technology	Yin Dawei, Wang Jie, Zhang Hanli, Liu Beibei, and Liu Yuting	06/03/2015

Table 7.2 Clinical trial updates on 1,3,4-thiadiazoles.[106]

Sr. no.	Study title	Study type	Study phase	Study design	Conditions	Status	ClinicalTrials.gov identifier
1	Combination Drug-Therapy for Patients with Untreated Obstructive Sleep Apnea	Interventional	Phase 2	Randomized	Obstructive sleep apnea (OSA)	Recruiting	NCT04639193
2	Phase I, Open-Label, Study of Tumor Infiltrating Lymphocytes Engineered with Membrane Bound IL15 Plus Acetazolamide in Adult Patients with Metastatic Melanoma	Interventional	Phase 1	NA	Tumor, metastatic melanoma, and melanoma	Recruiting	NCT05470283
3	Comparison of Combined Intravitreal Bevacizumab and Oral Acetazolamide *Versus* Intravitreal Bevacizumab Alone for the	Interventional	Phase 2	Randomized	Macular edema	Active not recruiting	NCT05290948

(*continued*)

Table 7.2 (*continued*)

Sr. no.	Study title	Study type	Study phase	Study design	Conditions	Status	ClinicalTrials.gov identifier
	Treatment of Macular Edema Secondary to Retinal Vein Occlusions						
4	Acetazolamide in Persons with Type 1 Diabetes	Interventional	Phase 1 and phase 2	Non-randomized	Type 1 diabetes	Recruiting	NCT05473364
5	Study of Acetazolamide with Temozolomide in Adults with Newly Diagnosed or Recurrent Malignant Glioma	Interventional	Phase 1	NA	Malignant glioma of brain	Recruiting	NCT03011671
6	Inhibition of Carbonic Anhydrase in Combination with Platinum and Etoposide-based Radiochemotherapy in Patients with Localized Small Cell Lung Cancer	Interventional	Phase 1	NA	Small-cell lung cancer	Recruiting	NCT03467360

(*continued*)

Table 7.2 (*continued*)

Sr. no.	Study title	Study type	Study phase	Study design	Conditions	Status	ClinicalTrials.gov identifier
7	Safety and Efficacy of OBX-115 in Advanced/Metastatic Melanoma Resistant to Immune Checkpoint Inhibitors	Interventional	Phase 1 and phase 2	NA	Tumor skin, metastatic melanoma, and melanoma	Recruiting	NCT06060613
8	Bioavailability, Bioequivalence and Tolerability of IHL-42X Compared to the Reference Drugs	Interventional	Phase 1	Randomized	Obstructive sleep apnea	Recruiting	NCT05857384
9	Vaccine Therapy for the Treatment of Newly Diagnosed Glioblastoma Multiforme	Interventional	Phase 2	Randomized	Glioblastoma multiforme, glioblastoma, malignant glioma, astrocytoma, and grade IV	Active not recruiting	NCT02465268
10	A Feasibility Study Utilizing Immune Recall to Increase Response to Checkpoint Therapy	Interventional	Phase 1	Non-randomized	Melanoma	Recruiting	NCT05077137

(*continued*)

Table 7.2 (*continued*)

Sr. no.	Study title	Study type	Study phase	Study design	Conditions	Status	ClinicalTrials.gov identifier
11	PEP-CMV Vaccine Targeting CMV Antigen to Treat Newly Diagnosed Pediatric HGG and DIPG and Recurrent Medulloblastoma	Interventional	Phase 2	NA	High-grade glioma, diffuse intrinsic pontine glioma, and recurrent medulloblastoma	Not yet recruiting	NCT05096481
12	Filanesib (ARRY-520) in Combination with Pomalidomide and Dexamethasone for Relapsed/Refractory (R/R) Multiple Myeloma (MM) Patients	Interventional	Phase 1 and phase 2	NA	Multiple myeloma	Completed	NCT02384083
13	A Phase 2 Trial of Filanesib in Relapsed/Refractory Multiple Myeloma (AfFIRM)	Interventional	Phase 2	NA	Advanced multiple myeloma	Completed	NCT02092922
14	A Study of Filanesib (ARRY-520) and Carfilzomib	Interventional	Phase 2	Randomized	Advanced multiple myeloma	Completed	NCT01989325

(*continued*)

Table 7.2 (continued)

Sr. no.	Study title	Study type	Study phase	Study design	Conditions	Status	ClinicalTrials.gov identifier
	in Patients with Advanced Multiple Myeloma						
15	Telaglenastat with Radiation Therapy and Temozolomide in Treating Patients with IDH-Mutated Diffuse Astrocytoma or Anaplastic Astrocytoma	Interventional	Phase 1	NA	Astrocytoma, IDH-mutant, grade 2, astrocytoma, IDH-mutant, and grade 3	Active not recruiting	NCT03528642
16	Impact of a Histamine H2 Receptor Antagonist (H2RA) on the Pharmacokinetics (PK) of Telaglenastat in Healthy Subjects	Interventional	Phase 1	Randomized	Drug interaction	Completed	NCT04540965
17	A Study of Telaglenastat (CB-839) in Combination with Palbociclib in Patients with Solid Tumors	Interventional	Phase 1 and phase 2	Non-randomized	Solid tumors, NSCLC, CRC, and KRAS gene mutation	Completed	NCT03965845

(continued)

Table 7.2 (*continued*)

Sr. no.	Study title	Study type	Study phase	Study design	Conditions	Status	ClinicalTrials.gov identifier
18	CB-839 with Everolimus *vs.* Placebo with Everolimus in Participants with Renal Cell Carcinoma (RCC)	Interventional	Phase 2	Randomized	Clear cell renal cell carcinoma	Completed	NCT03163667
19	Glutaminase Inhibition and Chemoradiation in Advanced Cervical Cancer	Interventional	Phase 2	Randomized	Advanced cervical carcinoma, cervical cancer, cervix cancer, and cancer of the cervix	Not yet recruiting	NCT05521997
20	Study of CB-839 in Combination with Paclitaxel in Participants of African Ancestry and Non-African Ancestry with Advanced Triple Negative Breast Cancer (TNBC)	Interventional	Phase 2	Non-randomized	Triple-negative breast cancer (TNBC)	Completed	NCT03057600
21	Study to Evaluate the ECG Effects of Telaglenastat	Interventional	Phase 1	Randomized	Cardiac safety	Completed	NCT04607512

(*continued*)

Table 7.2 (*continued*)

Sr. no.	Study title	Study type	Study phase	Study design	Conditions	Status	ClinicalTrials.gov identifier
	in Healthy Adult Subjects						
22	Testing Whether Cancers with Specific Mutations Respond Better to Glutaminase Inhibitor, Telaglenastat Hydrochloride, Anti-Cancer Treatment, BeGIN Study	Interventional	Phase 2	NA	Advanced malignant solid neoplasm, metastatic malignant solid neoplasm, NF1 mutation-positive malignant peripheral nerve sheath tumor, and unresectable malignant solid neoplasm	Active not recruiting	NCT03872427
23	CANTATA: CB-839 with Cabozantinib *vs.* Cabozantinib with Placebo in Patients with Metastatic Renal Cell Carcinoma	Interventional	Phase 2	Non-randomized	Advanced renal cell carcinoma and metastatic renal cell carcinoma	Completed	NCT03428217
24	Telaglenastat Hydrochloride and Osimertinib in Treating Patients with EGFR-Mutated Stage IV	Interventional	Phase 1 and phase2	NA	Advanced lung non-small-cell carcinoma, metastatic lung non-small-cell carcinoma, and stage	Recruiting	NCT03831932

(*continued*)

Table 7.2 (*continued*)

Sr. no.	Study title	Study type	Study phase	Study design	Conditions	Status	ClinicalTrials.gov identifier
	Non-small Cell Lung Cancer				IV lung cancer AJCC v8		
25	CB-839 HCl in Combination with Carfilzomib and Dexamethasone in Treating Patients with Recurrent or Refractory Multiple Myeloma	Interventional	Phase 1	NA	Recurrent multiple myeloma and refractory multiple myeloma	Active not recruiting	NCT03798678
26	Study of the Glutaminase Inhibitor CB-839 in Hematological Tumors	Interventional	Phase 1	Non-randomized	Non-Hodgkin's lymphoma (NHL), multiple myeloma, Waldenstrom's macroglobulinemia (WM), other B-cell NHL subtypes, including WM, T-cell NHL	Completed	NCT02071888
27	Glutaminase Inhibitor CB-839 and Azacitidine in Treating Patients with Advanced Myelodysplastic Syndrome	Interventional	Phase 1 and phase2	NA	Acute myeloid leukemia with multilineage dysplasia, blasts 20–30% of bone marrow nucleated cells, blasts 20–30% of peripheral	Completed	NCT03047993

(*continued*)

Table 7.2 (*continued*)

Sr. no.	Study title	Study type	Study phase	Study design	Conditions	Status	ClinicalTrials.gov identifier
					blood white cells, chronic myelomonocytic leukemia, high-risk myelodysplastic syndrome, IPSS risk category intermediate-2, and myelodysplastic syndrome		
28	CB-839 + Capecitabine in Solid Tumors and Fluoropyrimidine Resistant PIK3CA Mutant Colorectal Cancer	Interventional	Phase 1 and phase 2	NA	Colorectal cancer, colon cancer, rectal cancer, and solid tumor	Active not recruiting	NCT02861300
29	Novel PET/CT Imaging Biomarkers of CB-839 in Combination with Panitumumab and Irinotecan in Patients with Metastatic and Refractory	Interventional	Phase 1 and phase 2	NA	Colorectal cancer, metastatic colorectal cancer, RAS wild-type colorectal cancer, and refractory colorectal cancer	Active not recruiting	NCT03263429

(*continued*)

Table 7.2 (*continued*)

Sr. no.	Study title	Study type	Study phase	Study design	Conditions	Status	ClinicalTrials.gov identifier
	RAS Wildtype Colorectal Cancer						
30	Study of the Glutaminase Inhibitor CB-839 in Leukemia	Interventional	Phase1	Non-randomized	Acute myeloid leukemia (AML) and acute lymphocytic leukemia (ALL)	Completed	NCT02071927
31	Study of the Glutaminase Inhibitor CB-839 in Solid Tumors	Interventional	Phase 1	Non-randomized	Solid tumors, triple-negative breast cancer, non-small-cell lung cancer, renal cell carcinoma, mesothelioma, fumarate hydratase (FH)-deficient tumors, succinate dehydrogenase (SDH)-deficient gastrointestinal stromal tumors (GIST), succinate dehydrogenase (SDH)-deficient non-gastrointestinal stromal tumors, tumors harboring isocitrate	Completed	NCT02071862

(*continued*)

Table 7.2 (*continued*)

Sr. no.	Study title	Study type	Study phase	Study design	Conditions	Status	ClinicalTrials.gov identifier
					dehydrogenase-1 (IDH1) and IDH2 mutations, and tumors harboring amplifications in the cMyc gene		
32	A Comparative, Pharmacokinetic Study of CB-839 Capsule and Tablet Formulations in Healthy Adults	Interventional	Phase 1	Randomized	Healthy volunteers	Completed	NCT02944435

7.5 Patent Updates on 1,3,4-Thiadiazoles

There are several patents filed on 1,3,4-thiadiazoles. The details of the recent patent updates are presented in Table 7.1.

7.6 Ongoing Clinical Trials

The clinical trial updates for 1,3,4-thiadiazole analogs are provided in Table 7.2. Most of these trials have been conducted to treat several diseases, reflecting the versatility of the compound in therapeutic investigations.

Abbreviations

AChE	Acetylcholinesterase
ADMET	Absorption, distribution, metabolism, excretion, toxicity
BuChE	Butyrylcholinesterase
CAN	Cerium(IV) ammonium nitrate
COX-2	Cyclooxygenase-2
HCA	Human carbonic anhydrase
IC_{50}	Half-maximal inhibitory concentration
KARI	Ketol-acid reductoisomerase
LOX-15	Lipoxygenases15
MES	Maximal electroshock
MIC	Minimum inhibitory concentration
MW	Microwave irradiation
NMP	*N*-Methyl-2-pyrrolidone
NSAID	Non-steroidal anti-inflammatory drugs
PIDA	Phenyliodine(III) diacetate
$PoCl_3$	Phosphorus oxychloride
PPA	Phenylpropanolamine
PPARγ	Peroxisome proliferator-activated receptor gamma
PPE	Polyphosphate ester
p-tsCl	*p*-Toluenesulfonyl chloride
SI	Selectivity index
T2DM	Type 2 diabetes mellitus
T_3P	Propylphosphonic anhydride
TBHP	*Tert*-butyl hydroperoxide
TEA	Triethylamine
THF	Tetrahydrofuran
UI	Ulcerogenic index
Xac	*Xanthomonas axonopodis* pv. *Citri*
Xoo	*Xanthomonas oryzae* pv. *Oryzae*

References

1. M. Szeliga, Thiadiazole derivatives as anticancer agents, *Pharmacol Rep*, October 2020, **72**(5), 1079–1100.
2. D. Kumar, N. Maruthi Kumar, K. H. Chang and K. Shah, Synthesis and anticancer activity of 5-(3-indolyl)-1,3,4-thiadiazoles, *Eur. J. Med. Chem.*, October 2010, **45**(10), 4664–4668.
3. M. Asif, Chemistry, synthesis and progress report on biological activities of thiadiazole compounds - a review, *Mediterr. J. Chem.*, June 2016, **5**(5), 568–588.
4. Y. Li, J. Geng, Y. Liu, S. Yu and G. Zhao, Thiadiazole—a Promising Structure in Medicinal Chemistry, *ChemMedChem*, January 2013, **8**(1), 27–41.
5. M. Hatami, Z. Basri, B. K. Sakhvidi and M. Mortazavi, Thiadiazole – A promising structure in design and development of anti-Alzheimer agents, *Int. Immunopharmacol.*, May 2023, **118**, 110027.
6. B. Hemmateenejad, R. Miri, M. Jafarpour, M. Tabarzad and A. Foroumadi, Multiple Linear Regression and Principal Component Analysis-Based Prediction of the Anti-Tuberculosis Activity of Some 2-aryl-1,3,4-Thiadiazole Derivatives, *QSAR Comb. Sci.*, January 2006, **25**(1), 56–66.
7. L. Gummidi, N. Kerru, O. Ebenezer, P. Awolade, O. Sanni and M. S. Islam, *et al.*, Multicomponent reaction for the synthesis of new 1,3,4-thiadiazole-thiazolidine-4-one molecular hybrids as promising antidiabetic agents through α-glucosidase and α-amylase inhibition, *Bioorg. Chem.*, October 2021, **115**, 105210.
8. A. Ergena, Y. Rajeshwar and G. Solomon, Synthesis and Diuretic Activity of Substituted 1,3,4-Thiadiazoles. Hayat S, editor, *Scientifica*, 8 April 2022, 1–9.
9. T. Okada, K. Yamabe, M. Jo, Y. Sakajiri, T. Shibata and R. Sawada, *et al.*, Design and structural optimization of thiadiazole derivatives with potent GLS1 inhibitory activity, *Bioorg. Med. Chem. Lett.*, September 2023, **93**, 129438.
10. M. Bülbül, R. Kasimoğullari and Ö. I. Küfreviˇoğlu, Amide derivatives with pyrazole carboxylic acids of 5-amino-1,3,4-thiadiazole 2-sulfonamide as new carbonic anhydrase inhibitors: Synthesis and investigation of inhibitory effects, *J. Enzyme Inhib. Med. Chem.*, January 2008, **23**(6), 895–900.
11. T. Anthwal, S. Paliwal and S. Nain, Diverse Biological Activities of 1,3,4-Thiadiazole Scaffold, *Chemistry*, 6 December 2022, **4**(4), 1654–1671.
12. A. Tahghighi and F. Babalouei, Thiadiazoles: the appropriate pharmacological scaffolds with leishmanicidal and antimalarial activities: a review, *Iran. J. Basic Med. Sci.*, June 2017, **20**(6), 613.
13. G. Şerban, 2-amino-1,3,4-thiadiazole derivatives as potential anti-helicobacter pylori agents – at a glance, *Farmacia*, October 2020, **68**(5), 785–791.

14. J. Sandström, Recent Advances in the Chemistry of 1,3,4-Thiadiazoles, *Adv. Heterocycl. Chem.*, 1968, **9**, 165–209.
15. M. Sainsbury, Five-Membered Heterocyclic Compounds with Three Hetero-Atoms in the Ring, in *Rodd's Chemistry of Carbon Compounds*, Elsevier, 1964, pp. 1–209.
16. Y. Hu, C. Y. Li, X. M. Wang, Y. H. Yang and H. L. Zhu, 1,3,4-Thiadiazole: Synthesis, Reactions, and Applications in Medicinal, Agricultural, and Materials Chemistry, *Chem. Rev.*, May 2014, **114**(10), 5572–5610.
17. U. Chiacchio and G. Romeo, Thiadiazoles, in *Modern Heterocyclic Chemistry*, Wiley, 2011, pp. 1253–1400.
18. A. A. Kiryanov, P. Sampson and A. J. Seed, Synthesis of 2-Alkoxy-Substituted Thiophenes, 1,3-Thiazoles, and Related S-Heterocycles via Lawesson's Reagent-Mediated Cyclization under Microwave Irradiation: Applications for Liquid Crystal Synthesis, *J. Org. Chem.*, November 2001, **66**(23), 7925–7929.
19. K. Peng, Y. Li, Y. Bai, T. Jiang, H. Sun and Q. Zhu, *et al.*, Discovery of novel nonpeptide small-molecule NRP1 antagonists: Virtual screening, molecular simulation and structural modification, *Bioorg. Med. Chem.*, January 2020, **28**(1), 115183.
20. S. Ferrari, F. Morandi, D. Motiejunas, E. Nerini, S. Henrich and R. Luciani, *et al.*, Virtual Screening Identification of Nonfolate Compounds, Including a CNS Drug, as Antiparasitic Agents Inhibiting Pteridine Reductase, *J. Med. Chem.*, January 2011, **54**(1), 211–221.
21. R. Rawat and S. M. Verma, Pd 2 (dba) 3 -catalyzed amination of C5-bromo-imidazo[2,1-b][1,3,4] thiadiazole with substituted anilines at conventional heating in Schlenk tube, *J. Sulfur. Chem.*, January 2021, **42**(1), 96–108.
22. X. H. Liu, Y. X. Shi, Y. Ma, C. Y. Zhang, W. L. Dong and L. Pan, *et al.*, Synthesis, antifungal activities and 3D-QSAR study of N-(5-substituted-1,3,4-thiadiazol-2-yl)cyclopropanecarboxamides, *Eur. J. Med. Chem.*, July 2009, **44**(7), 2782–2786.
23. J. K. Augustine, V. Vairaperumal, S. Narasimhan, P. Alagarsamy and A. Radhakrishnan, Propylphosphonic anhydride (T3P®): an efficient reagent for the one-pot synthesis of 1,2,4-oxadiazoles, 1,3,4-oxadiazoles, and 1,3,4-thiadiazoles, *Tetrahedron*, November 2009, **65**(48), 9989–9996.
24. V. Jatav, P. Mishra, S. Kashaw and J. P. Stables, CNS depressant and anticonvulsant activities of some novel 3-[5-substituted 1,3,4-thiadiazole-2-yl]-2-styryl quinazoline-4(3H)-ones, *Eur. J. Med. Chem.*, September 2008, **43**(9), 1945–1954.
25. N. T. Hatvate, S. M. Ghodse and V. N. Telvekar, Metal-free synthesis of 2-aminothiadiazoles via TBHP-Mediated oxidative C-S bond formation, *Synth. Commun.*, February 2018, **48**(3), 285 290.

26. D. Chandra Sekhar, D. V. Venkata Rao, A. Tejeswara Rao, U. Lav Kumar and A. Jha, Design and Synthesis of 1,3,4-Thiadiazole Derivatives as Novel Anticancer and Antitubercular Agents, *Russ. J. Gen. Chem.*, 1 April 2019, **89**(4), 770–779.
27. J. M. Farrar, M. K. Patel, P. Kaszynski and V. G. Young, A New Thiatriazine Isomer: Synthesis, Tautomerism, and Molecular Structure of 3,6-Diphenyl-4H -1,2,4,5-thiatriazine as a Precursor to the 1,2,4,5-Thiatriazinyl Radical,*J. Org. Chem.*, February 2000, **65**(4), 931–940.
28. M. A. Kukaniev, M. D. Osimov, Z. G. Sangov, S. S. Safarov, M. B. Karimov and T. R. Radjabov, Polyfunctional nitriles in the synthesis of derivatives of 1,3,4-thiadiazolo-[3,2-a]pyrimidines, *Chem. Heterocycl. Comp.*, July 2008, **44**(7), 882–885.
29. P. A. Koutentis and C. P. Constantinides, 1,3,4-Thiadiazoles, in *Comprehensive Heterocyclic Chemistry III*, Elsevier, 2008, pp. 567–605.
30. J. Schatz, K. Gogić and T. Benkert, 1,3,4-Thiadiazoles, in *Comprehensive Heterocyclic Chemistry IV*, Elsevier, 2022, pp. 407–447.
31. V. Polshettiwar and R. S. Varma, Greener and rapid access to bio-active heterocycles: one-pot solvent-free synthesis of 1,3,4-oxadiazoles and 1,3,4-thiadiazoles, *Tetrahedron Lett.*, January 2008, **49**(5), 879–883.
32. M. A. Epishina, A. S. Kulikov, N. V. Ignat'ev, M. Schulte and N. N. Makhova, Synthesis of 5-alkyl-2-amino-1,3,4-thiadiazoles and α,ω-bis(2-amino-1,3,4-thiadiazol-5-yl)alkanes in ionic liquids, *Mendeleev Commun.*, November 2011, **21**(6), 331–333.
33. S. I. Farooqi, N. Arshad, P. A. Channar, F. Perveen, A. Saeed and F. A. Larik, *et al.*, Synthesis, theoretical, spectroscopic and electrochemical DNA binding investigations of 1, 3, 4-thiadiazole derivatives of ibuprofen and ciprofloxacin: Cancer cell line studies, *J. Photochem. Photobiol., B*, December 2018, **189**, 104–118.
34. S. M. Golovlyova, Y. A. Moskvichev, E. M. Alov, D. B. Kobylinsky and V. V. Ermolaeva, Synthesis of novel five-membered nitrogen-containing heterocyclic compounds from derivatives of arylsulfonyl- and arylthioacetic and -propionic acids, *Chem. Heterocycl. Compd.*, 2001, **37**(9), 1102–1106.
35. T. Tsuji and K. Takenaka, Convenient Synthesis of 2,7-Disubstituted 5 H -1,3,4-Thiadiazolo[3,2- a]pyrimidin-5-ones and Related Compounds, *Bull. Chem. Soc. Jpn.*, February 1982, **55**(2), 637–638.
36. A. A. Minea, Overview of Ionic Liquids as Candidates for New Heat Transfer Fluids, *Int. J. Thermophys.*, November 2020, **41**(11), 1–15.
37. T. S. Kokovina, S. Y. Gadomsky, A. A. Terentiev and N. A. Sanina, A Novel Approach to the Synthesis of 1,3,4-Thiadiazole-2-amine Derivatives, *Molecules*, August 2021, **26**(17), 5159.
38. H. Feng, X. Ying, Y. Peng, E. V. Van der Eycken, C. Liu and S. Zhao, *et al.*, FeCl3-promoted synthesis of 1,3,4-thiadiazoles under combined microwave and ultrasound irradiation in water, *Monatsh. Chem.*, May 2013, **144**(5), 681–686.

39. M. Bala, P. Piplani, A. Ankalgi, A. Jain and L. Chandel, 1,3,4-Thiadiazole: A Versatile Pharmacophore of Medicinal Significance, *Med. Chem.*, October 2023, **19**(8), 730–756.
40. V. K. Gour, S. Yahya and M. Shahar Yar, Unveiling the chemistry of 1,3,4-oxadiazoles and thiadiazols: A comprehensive review, *Arch. Pharm.*, 2024, **357**(1), 2300328.
41. K. Salomão, E. M. de Souza, S. A. Carvalho, E. F. da Silva, C. A. M. Fraga and H. S. Barbosa, *et al.*, In vitro and in vivo activities of 1,3,4-thiadiazole-2-arylhydrazone derivatives of megazol against Trypanosoma cruzi, *Antimicrob. Agents Chemother.*, May 2010, **54**(5), 2023–2031.
42. G. Chauvière, B. Bouteille, B. Enanga, C. de Albuquerque, S. L. Croft and M. Dumas, *et al.*, Synthesis and Biological Activity of Nitro Heterocycles Analogous to Megazol, a Trypanocidal Lead, *J. Med. Chem.*, 1 January 2003, **46**(3), 427–440.
43. D. Kumar, H. Kumar, V. Kumar, A. Deep, A. Sharma and M. G. Marwaha, *et al.*, Mechanism-based approaches of 1,3,4 thiadiazole scaffolds as potent enzyme inhibitors for cytotoxicity and antiviral activity, *Med. Drug Discov.*, February 2023, **17**, 100150.
44. C. B. Mishra, M. Tiwari and C. T. Supuran, Progress in the development of human carbonic anhydrase inhibitors and their pharmacological applications: Where are we today?, *Med. Res. Rev.*, 21 November 2020, **40**(6), 2485–2565.
45. N. Cimolai, The Neurological Spectrum for Acetazolamide Pharmacotherapy: from Basic Science to Clinical Applications, *SN Compr. Clin. Med.*, 8 December 2021, **3**(12), 2576–2592.
46. R. M. El-Masry, H. H. Kadry, A. T. Taher and S. M. Abou-Seri, Comparative Study of the Synthetic Approaches and Biological Activities of the Bioisosteres of 1,3,4-Oxadiazoles and 1,3,4-Thiadiazoles over the Past Decade, *Molecules*, 22 April 2022, **27**(9), 2709.
47. G. Serban, Future Prospects in the Treatment of Parasitic Diseases: 2-Amino-1,3,4-Thiadiazoles in Leishmaniasis, *Molecules*, 19 April 2019, **24**(8), 1557.
48. S. Rohilla and D. Sharma, Sulfonamides, quinolones, antiseptics, and disinfectants, in *Medicinal Chemistry of Chemotherapeutic Agents*, Elsevier, 2023, pp. 21–63.
49. A. K. Jain, S. Sharma, A. Vaidya, V. Ravichandran and R. K. Agrawal, 1,3,4-Thiadiazole and its Derivatives: A Review on Recent Progress in Biological Activities, *Chem. Biol. Drug Des.*, 24 May 2013, **81**(5), 557–576.
50. Y. M. Omar, H. H. M. Abdu-Allah and S. G. Abdel-Moty, Synthesis, biological evaluation and docking study of 1,3,4-thiadiazole-thiazolidinone hybrids as anti-inflammatory agents with dual inhibition of COX-2 and 15-LOX, *Bioorg. Chem.*, October 2018, **80**, 461–471.

51. M. H. Shkair, A. K. Shakya, N. M. Raghavendra and R. R. Naik, Molecular Modeling, Synthesis and Pharmacological Evaluation of 1,3,4-Thiadiazoles as Anti-inflammatory and Analgesic Agents, *Med. Chem.*, 5 January 2016, **12**(1), 90–100.
52. S. Maddila, S. Gorle, C. Sampath and P. Lavanya, Synthesis and anti-inflammatory activity of some new 1,3,4-thiadiazoles containing pyrazole and pyrrole nucleus, *J. Saudi Chem. Soc.*, September 2016, **20**, S306–S312.
53. D. K. Mahapatra, R. S. Shivhare and A. G. M. Haldar, Novel Schiff's Base Containing Murrayanine- 1,3,4-Thiadiazole Hybrids as Potential Anti-Inflammatory Agents, *Asian J. Res. Chem. Pharm. Sci.*, 25 November 2017, **2**(2), 27–32.
54. A. Cristina, D. Leonte, L. Vlase, L. Bencze, S. Imre and G. Marc, *et al.*, Heterocycles 48. Synthesis, Characterization and Biological Evaluation of Imidazo[2,1-*b*][1,3,4]Thiadiazole Derivatives as Anti-Inflammatory Agents, *Molecules*, 21 September 2018, **23**(10), 2425.
55. Y. M. Omar, S. G. Abdel-Moty and H. H. M. Abdu-Allah, Further insight into the dual COX-2 and 15-LOX anti-inflammatory activity of 1,3,4-thiadiazole-thiazolidinone hybrids: The contribution of the substituents at 5th positions is size dependent, *Bioorg. Chem.*, April 2020, **97**, 103657.
56. P. Pathak, P. K. Shukla, V. Naumovich, M. Grishina, A. Verma and V. Potemkin, 1,2,4-Triazole-conjugated 1,3,4-thiadiazole hybrid scaffolds: A potent ameliorant of carrageenan-induced inflammation by lessening proinflammatory mediators, *Arch. Pharm.*, 2020, **353**(1), e1900233.
57. S. A. Khan, S. M. Imam, A. Ahmad, S. H. Basha and A. Husain, Synthesis, molecular docking with COX 1& II enzyme, ADMET screening and in vivo anti-inflammatory activity of oxadiazole, thiadiazole and triazole analogs of felbinac, *J. Saudi Chem. Soc.*, May 2018, **22**(4), 469–484.
58. F. A. Ragab, H. I. Heiba, M. G. El-Gazzar, S. M. Abou-Seri, W. A. El-Sabbagh and R. M. El-Hazek, Anti-inflammatory, analgesic and COX-2 inhibitory activity of novel thiadiazoles in irradiated rats, *J. Photochem. Photobiol., B*, January 2017, **166**, 285–300.
59. S. M. Gomha, Z. A. Muhammad, H. M. Gaber and M. M. Amin, Synthesis of Some Novel Heterocycles Bearing Thiadiazoles as Potent Anti-inflammatory and Analgesic Agents, *J. Heterocycl. Chem.*, 23 September 2017, **54**(5), 2708–2716.
60. B. Gowramma, K. T. Praveen, R. Kalirajan and B. Babu, Synthesis of Some Novel 2-Azetidinones/4-Thiazolidinones Bearing 1, 3, 4- Thiadiazole Nucleus and Screening for its Anti-Imflammatory Activity, *Lett. Drug Des. Discov.*, 30 June 2016, **13**(7), 676–683.

61. M. Amir, M. WasimHaq and S. Ehtaishamul, Synthesis of some new condensed heterocyclic 6-substituted-1,2,4-triazolo[3,4-b]-1,3,4-thiadiazole derivatives of 2-naphthoxyacetic acid as potent anti-inflammatory agents with reduced ulcerogenicity, *Indian J. Chem.*, 2017, **56B**, 1177–1184.
62. M. Haroun, M. Fesatidou, A. Petrou, C. Tratrat, P. Zagaliotis and A. Gavalas, *et al.*, Identification of Novel Cyclooxygenase-1 Selective Inhibitors of Thiadiazole-Based Scaffold as Potent Anti-Inflammatory Agents with Safety Gastric and Cytotoxic Profile, *Molecules*, 12 April 2023, **28**(8), 3416.
63. S. A. Hameed, J. Varkey and J. Jayasekhar, In Silico Design, Synthesis and In Vitro Antidiabetic Activity of Novel 5-Furyl- 1,3,4-Thiadiazolimines, *Int. J. Pharma. Sci. Res.*, 30 June 2020, **11**(6), 152–159.
64. R. Hussain, M. Shah, S. Iqbal, W. Rehman, S. Khan and L. Rasheed, *et al.*, Molecular iodine-promoted oxidative cyclization for the synthesis of 1,3,4-thiadiazole-fused- [1,2,4]-thiadiazole incorporating 1,4-benzodioxine moiety as potent inhibitors of α-amylase and α-glucosidase: In vitro and in silico study, *Front. Chem.*, 6 October 2022, **10**, 1–17.
65. X. Zhao, X. Zhan, H. Zhang, Y. Wan, H. Yang and Y. Wang, *et al.*, Synthesis and biological evaluation of isatin derivatives containing 1,3,4-thiadiazole as potent a-glucosidase inhibitors, *Bioorg. Med. Chem. Lett.*, December 2021, **54**, 128447.
66. Y. Vaishnav, D. Dewangan, S. Verma, A. Mishra, A. S. Thakur and P. Kashyap, *et al.*, PPAR gamma targeted molecular docking and synthesis of some new amide and urea substituted 1, 3, 4-thiadiazole derivative as antidiabetic compound, *J. Heterocycl. Chem.*, 4 May 2020, **57**(5), 2213–2224.
67. F. Rahim, H. Ullah, R. Hussain, M. Taha, S. Khan and M. Nawaz, *et al.*, Thiadiazole based triazole/hydrazone derivatives: Synthesis, in vitro α-glucosidase inhibitory activity and in silico molecular docking study, *J. Mol. Struct.*, September 2023, **1287**, 135619.
68. M. Taha, A. A. Khan, F. Rahim, S. Hayat, S. Imran and N. Iqbal, *et al.*, Synthesis, in vitro evaluation and molecular docking studies of hybrid 4-quinolinyl bearing 1,3,4-thiadiazole-2-amine as a new inhibitor of α-amylase and α-glucosidase, *J. Mol. Struct.*, June 2023, **1282**, 135173.
69. Y. S. Li, M. He, T. S. Zhou, Q. Wang, L. He and S. J. Wang, *et al.*, 2,5-Disubstituted furan derivatives containing 1,3,4-thiadiazole moiety as potent α-glucosidase and E. coli β-glucuronidase inhibitors, *Eur. J. Med. Chem.*, April 2021, **216**, 113322.
70. R. S. Gani, A. K. Kudva, K. Timanagouda, S. B. H. Mujawar and S. D. Joshi, *et al.*, Synthesis of novel 5-(2,5-bis(2,2,2-trifluoroethoxy)phenyl)-1,3,4-oxadiazole-2-thiol derivatives as potential glucosidase inhibitors, *Bioorg. Chem.*, September 2021, **114**, 105046.

71. H. Kashtoh, S. Hussain, A. Khan, S. M. Saad, J. A. J. Khan and K. M. Khan, *et al.*, Oxadiazoles and thiadiazoles: Novel α-glucosidase inhibitors, *Bioorg. Med. Chem.*, October 2014, **22**(19), 5454–5465.
72. J. J. Luszczki, M. Karpińska, J. Matysiak and A. Niewiadomy, Characterization and preliminary anticonvulsant assessment of some 1,3,4-thiadiazole derivatives, *Pharmacol. Rep.*, June 2015, **67**(3), 588–592.
73. K. P. Harish, K. N. Mohana and L. Mallesha, Synthesis of Pyrazine Substituted 1,3,4-Thiadiazole Derivatives and Their Anticonvulsant Activity, *Org. Chem. Int.*, 4 December 2013, **2013**, 1–8.
74. B. Sahoo, S. Dinda, B. Kumar, J. Panda and P. Brahmkshatriya, Molecular Docking Study, Green Synthesis and Pharmacological Evaluation of 1,3,4-thiadiazole Derivatives as Potential Antiepileptic Agents, *Mini-Rev. Med. Chem.*, 31 December 2013, **13**(14), 2076–2081.
75. M. Toolabi, M. Khoramjouy, A. Aghcheli, A. Ayati, S. Moghimi and L. Firoozpour, *et al.*, Synthesis and radioligand-binding assay of 2,5-disubstituted thiadiazoles and evaluation of their anticonvulsant activities, *Arch. Pharm.*, 27 December 2020, **353**(12), 1–9.
76. D. Kumudha, R. R. Reddy and T. Kalavathi, Synthesis and Evaluation of Some 1, 3, 4-Thiadiazoles Having Substituted 1, 2, 4-Triazole Moiety for Anticonvulsant and Cns Depressant Activity, *World J. Pharm. Pharm. Sci.*, 2019, **3**(9), 728.
77. H. R. M. Rashdan, M. M. Farag, M. S. El-Gendey and M. M. Mounier, Toward Rational Design of Novel Anti-Cancer Drugs Based on Targeting, Solubility, and Bioavailability Exemplified by 1,3,4-Thiadiazole Derivatives Synthesized Under Solvent-Free Conditions, *Molecules*, 27 June 2019, **24**(13), 2371.
78. K. M. Dawood and S. M. Gomha, Synthesis and Anti - cancer Activity of 1,3,4-Thiadiazole and 1,3-Thiazole Derivatives Having 1,3,4-Oxadiazole Moiety, *J. Heterocycl. Chem.*, September 2015, **52**(5), 1400–1405.
79. S. Janowska, D. Khylyuk, A. Bielawska, A. Szymanowska, A. Gornowicz and K. Bielawski, *et al.*, New 1,3,4-Thiadiazole Derivatives with Anticancer Activity, *Molecules*, 10 March 2022, **27**(6), 1814.
80. D. Chowrasia, C. Karthikeyan, L. Choure, S. Sahabjada, M. Gupta and M. Arshad, *et al.*, Synthesis, characterization and anti cancer activity of some fluorinated 3,6-diaryl-[1,2,4]triazolo[3,4-b][1,3,4]thiadiazoles, *Arab. J. Chem.*, May 2017, **10**, S2424–S2428.
81. K. M. Dawood, T. M. A. Eldebss, H. S. A. El-Zahabi, M. H. Yousef and P. Metz, Synthesis of some new pyrazole-based 1,3-thiazoles and 1,3,4-thiadiazoles as anticancer agents, *Eur. J. Med. Chem.*, December 2013, **70**, 740–749.
82. M. Gür, S. Yerlikaya, N. Şener, S. Özkınalı, M. C. Baloglu and H. Gökçe, *et al.*, Antiproliferative-antimicrobial properties and structural analysis of newly synthesized Schiff bases derived from some 1,3,4-thiadiazole compounds, *J. Mol. Struct.*, November 2020, **1219**, 128570.

83. M. M. S. Wassel, Y. A. Ammar, G. A. M. Elhag Ali, A. Belal, A. B. M. Mehany and A. Ragab, Development of adamantane scaffold containing 1,3,4-thiadiazole derivatives: Design, synthesis, anti-proliferative activity and molecular docking study targeting EGFR, *Bioorg. Chem.*, May 2021, **110**, 104794.
84. S. M. Gomha, N. A. Kheder, M. R. Abdelaziz, Y. N. Mabkhot and A. M. Alhajoj, A facile synthesis and anticancer activity of some novel thiazoles carrying 1,3,4-thiadiazole moiety, *Chem. Cent. J.*, 21 December 2017, **11**(1), 25.
85. M. Altıntop, H. Ciftci, M. Radwan, B. Sever, Z. Kaplancıklı and T. Ali, *et al.*, Design, Synthesis, and Biological Evaluation of Novel 1,3,4-Thiadiazole Derivatives as Potential Antitumor Agents against Chronic Myelogenous Leukemia: Striking Effect of Nitrothiazole Moiety, *Molecules*, 27 December 2017, **23**(1), 59.
86. A. Skrzypek, J. Matysiak, M. Karpińska, K. Czarnecka, P. Kręcisz and D. Stary, *et al.*, Biological evaluation and molecular docking of novel 1,3,4-thiadiazole-resorcinol conjugates as multifunctional cholinesterases inhibitors, *Bioorg. Chem.*, February 2021, **107**, 104617.
87. A. Skrzypek, J. Matysiak, A. Niewiadomy, M. Bajda and P. Szymański, Synthesis and biological evaluation of 1,3,4-thiadiazole analogues as novel AChE and BuChE inhibitors, *Eur. J. Med. Chem.*, April 2013, **62**, 311–319.
88. M. Bala, A. Ankalgi and P. Piplani, Design, synthesis and pharmacological evaluation of disubstituted 1,3,4 thiadiazoles derivatives for the treatment of cognitive dysfunction, *J. Mol. Struct.*, November 2023, **1291**, 135951.
89. S. Askin, H. Tahtaci, C. Türkeş, Y. Demir, A. Ece and G. Akalın Çiftçi, *et al.*, Design, synthesis, characterization, in vitro and in silico evaluation of novel imidazo[2,1-b][1,3,4]thiadiazoles as highly potent acetylcholinesterase and non-classical carbonic anhydrase inhibitors, *Bioorg. Chem.*, August 2021, **113**, 105009.
90. A. Skrzypek, J. Matysiak, M. M. Karpińska and A. Niewiadomy, Synthesis and anticholinesterase activities of novel 1,3,4-thiadiazole based compounds, *J. Enzyme Inhib. Med. Chem.*, 1 August 2013, **28**(4), 816–823.
91. J. Ramprasad, N. Nayak and U. Dalimba, Design of new phenothiazine-thiadiazole hybrids via molecular hybridization approach for the development of potent antitubercular agents, *Eur. J. Med. Chem.*, December 2015, **106**, 75–84.
92. V. S. Krishna, S. Zheng, E. M. Rekha, R. Nallangi, D. V. Sai Prasad and S. E. George, *et al.*, Design and development of ((4-methoxyphenyl)carbamoyl) (5-(5-nitrothiophen-2-yl)-1,3,4-thiadiazol-2-yl)amide analogues as Mycobacterium tuberculosis ketol-acid reductoisomerase inhibitors, *Eur. J. Med. Chem.*, May 2020, **193**, 112178.

93. S. D. Bhinge, V. Chature and L. V. Sonawane, Synthesis of Some Novel 1,3,4-Thiadiazole Derivatives and Biological Screening for Anti-Microbial, Antifungal and Anthelmintic Activity, *Pharm. Chem. J.*, 20 September 2015, **49**(6), 367–372.
94. Y. A. M. El-Badry, M. S. Sallam and M. A. A. El-Hashash, Efficient 1,3,4-Thiadiazole-4,5-dihydropyridazin-3(2H)-ones as Antimicrobial Agents, *Chem. Pharm. Bull.*, 2018, **66**(4), 427–433.
95. J. Wan, Y. Gan, W. Hu, J. Meng, K. Tian and X. Li, *et al.*, Design, synthesis and anti-bacterial evaluation of novel 1,3,4-thiadiazole derivatives bearing a semicarbazone moiety, *Phosphorus, Sulfur Silicon Relat. Elem.*, 3 July 2018, **193**(7), 443–450.
96. A. A. Pund, S. S. Saboo, G. M. Sonawane, A. C. Dukale and B. K. Magare, Synthesis of 2,5-disubstituted-1,3,4-thiadiazole derivatives from (2S)-3-(benzyloxy)-2-[(tert -butoxycarbonyl) amino] propanoic acid and evaluation of anti-microbial activity, *Synth. Commun.*, 16 December 2020, **50**(24), 3854–3864.
97. I. V. Drapak, B. S. Zimenkovsky, M. V. Slabyy, S. M. Holota, L. O. Perekhoda and R. V. Yaremkevych, *et al.*, Synthesis and diuretic activity of novel 5-amino-1,3,4-thiadiazole-2-thiol derivatives, *Biopolym. Cell*, 26 February 2021, **37**(1), 33–45.
98. C. B. Mishra, S. Kumari, A. Angeli, S. M. Monti, M. Buonanno and A. Prakash, *et al.*, Design, synthesis and biological evaluation of N -(5-methyl-isoxazol-3-yl/1,3,4-thiadiazol-2-yl)-4-(3-substitutedphenylureido) benzenesulfonamides as human carbonic anhydrase isoenzymes I, II, VII and XII inhibitors, *J. Enzyme Inhib. Med. Chem.*, 2 November 2016, **31**(sup2), 174–179.
99. K. H. M. E. Tehrani, S. Sardari, V. Mashayekhi, M. E. Zadeh, P. Azerang and F. Kobarfard, One pot synthesis and biological activity evaluation of novel Schiff bases derived from 2-hydrazinyl-1,3,4-thiadiazole, *Chem. Pharm. Bull.*, 2013, **61**(2), 160–166.
100. R. Ruel, A. L'Heureux, C. Thibeault, J. P. Daris, A. Martel and L. A. Price, *et al.*, New azole antagonists with high affinity for the P2Y1 receptor, *Bioorg. Med. Chem. Lett.*, June 2013, **23**(12), 3519–3522.
101. V. Raj, S. Saha, A. Rai, M. Singh, D. Kumar and A. K. Sahdev, in *Homology modeling, Molecular Dynamic Simulation and in silico screening of Activator for the Intensification of human sirtuin type 1 (SIRT1) by novel 1, 3, 4-thiadiazole derivatives-A potential antiaging approach*, Proceedings of MOL2NET, International Conference on Multidisciplinary Sciences, MDPI, Basel, Switzerland, 2015, b019, http://sciforum.net/conference/MOL2NET-1/paper/3320.
102. E. Pontiki, A. Peperidou, I. Fotopoulos and D. Hadjipavlou-Litina, Cinnamate Hybrids: A Unique Family of Compounds with Multiple Biological Activities, *Curr. Pharm. Biotechnol.*, 8 January 2019, **19**(13), 1019–1048.

103. A. Tabei, R. Ejtemaei, A. Mahboubi, P. Saniee, A. Foroumadi and A. Dehdari, *et al.*, Synthesis of new 2-(5-(5-nitrofuran-2-yl)-1,3,4-thiadiazol-2-ylimino)thiazolidin-4-one derivatives as anti-MRSA and anti-H. pylori agents, *BMC Chem.*, 27 December 2022, **16**(1), 38.
104. W. S. Alwan, R. Karpoormath, M. B. Palkar, H. M. Patel, R. A. Rane and M. S. Shaikh, *et al.*, Novel imidazo[2,1-b]-1,3,4-thiadiazoles as promising antifungal agents against clinical isolate of Cryptococcus neoformans, *Eur. J. Med. Chem.*, May 2015, **95**, 514–525.
105. *Google Patents*, https://patents.google.com. Accessed 10 Nov 2023.
106. *ClinicalTrials.gov*, https://clinicaltrials.gov. Accessed 11 Nov 2023.

Part III: Thiopyrans, Thiazines and Thiazepines

CHAPTER 8

Synthesis, Properties, and Biological Applications of Thiopyrans

PARNASHABARI SARKAR*, SOURAV SUTRADHAR, DIPANKAR DAS, PRAKASH KURMI AND BISWA NATH GHOSH

National Institute of Technology Silchar, Silchar, Assam, 788010, India
*E-mail: parnashabari_rs@che.nits.ac.in

8.1 Introduction

Thiopyrans, a newly recognized category under heterocyclic compounds, feature a six-membered ring that contains one sulfur atom. They are analogous to pyrans, in which the oxygen atom is replaced by a sulfur atom. The isomers 2*H*-thiopyran (**a**) and 4*H*-thiopyran (**b**) are characterized by differences in the positioning of double bonds and the existence of an sp^3 carbon atom.[1] The hydrogenated forms of these compounds are categorized as 3,4-dihydro-2*H*-thiopyrans (**c**), 3,6-dihydro-2*H*-thiopyrans (**d**), and tetrahydrothiopyrans (**e**), as depicted in Figure 8.1. The existence of thiopyrans has been known for quite some time. However, due to the lack of the formation of a cyclically conjugated 6π electron system compared to benzene or thiophene, they have remained relatively unstable and initially insignificant.[2] These compounds are of theoretical and practical interest due to their electronic and spatial structures (p-electron overlaps, incomplete planarity, and chiral centers at the 2- or 4-positions

S-Heterocycles: Retrospect, Prospects, and Biological Applications
Edited by Shrikaant Kulkarni, Hemantkumar Akolkar, Vijay M. Khedkar and A. K. Haghi

Published by the Royal Society of Chemistry, www.rsc.org

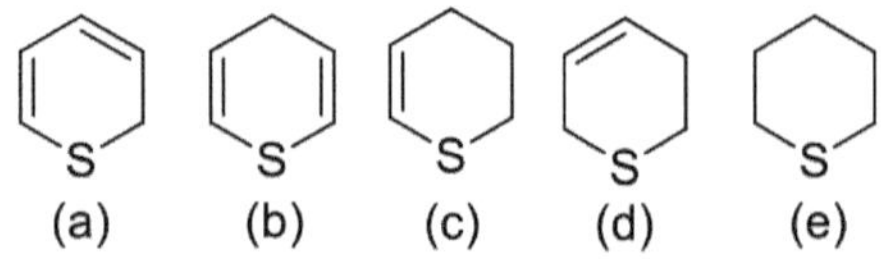

Figure 8.1 Structures of (**a**) 2*H*-thiopyran, (**b**) 4*H*-thiopyran, (**c**) 4*H*-dihydrothiopyran, (**d**) 2*H*-dihydrothiopyran, and (**e**) tetrahydrothiopyran.

with the corresponding substitutions) and various possibilities for their stabilization, for example, through the formation of thiopyrylium cations or thiopyran-2-ones or thiopyran-4-ones.[3] The varied physiological functions associated with this class of compounds contribute to their substantial importance in the realms of organic chemistry and therapeutics.

They are integral components of a variety of natural products and pharmaceutical agents. The diverse physiological roles of thiopyrans make them invaluable in the discovery of therapeutic compounds.

A significant research interest has been prevalent in thiopyran derivatives, which have been explored for their potential therapeutic applications. Precursors, such as thiophenol and thiosalicylic acids, are widely used for synthesizing thiopyran derivatives.[4] On account of multiple substituents present in the thiopyran moiety, this family of sulfur heterocyclic compounds is recognized for a diverse range of therapeutic applications, encompassing antimicrobial, anticancer, antiviral, anti-inflammatory, enzyme inhibitory, and antidiabetic properties. Given their diverse range of activities, investigation on thiopyran derivatives has gained significant interest among research communities as a promising avenue for discovering unique molecules.[1] Ongoing research endeavors continue to delve into the potential applications and properties of thiopyran compounds and their derivatives.

The basic structures of both 2*H*- and 4*H*-thiopyrans, as well as their simple substituted derivatives (see Figure 8.1), are not quite stable under normal conditions, and they are only robust at lower temperatures and under an inert gas atmosphere for a limited time. Under oxidative conditions, thiopyrans function as donors of both protons and hydrides, leading to the formation of thiopyrylium salts. So far, 2-methylene-2*H*- and 4-methylene-4*H*-thiopyrans have only been produced *in situ* and are characterized using spectroscopic methods. Significant improvements in their stability can be achieved by introducing electron acceptor groups, which are involved in mesomeric resonance structures. Methylene thiopyrans substituted with acceptor groups are more stable due to the conjugated arrangement of the double bonds compared to the substituted 2*H*- or 4*H*-thiopyrans. In particular, electron acceptor groups capable of mesomerism in exo-methylene groups demonstrate this effect remarkably (see Figure 8.2).

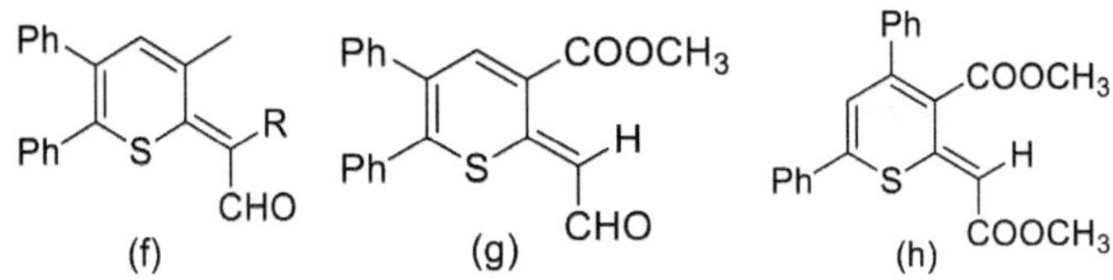

Figure 8.2 Structures of 2-methylene-2*H*-thiopyrans with electron acceptor groups.

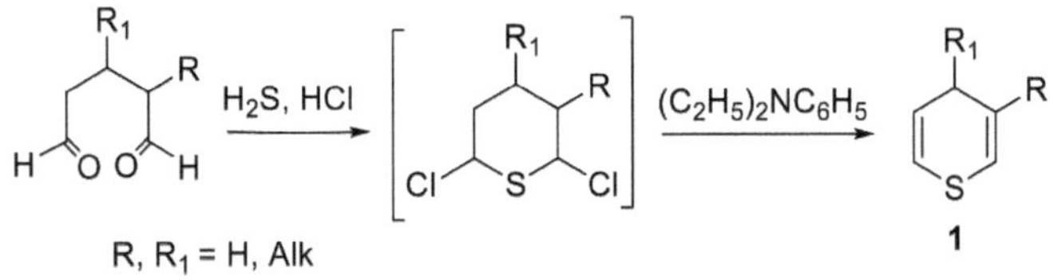

Scheme 8.1

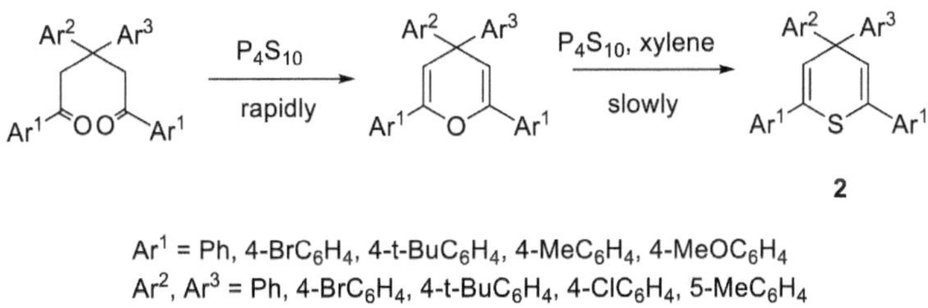

Scheme 8.2

8.2 Developments in the Synthesis of Thiopyran Derivatives

8.2.1 Synthesis of Thiopyran Derivatives Using Glutaraldehyde Derivatives

The synthesis of the simplest 4*H*-thiopyran was achieved by combining H_2S and HCl with glutaraldehyde and heating it in a vacuum with diethylaniline (see Scheme 8.1).[5] This reaction has been applied to the synthesis of substituted thiopyrans and their analogs.[6–12]

Another approach involves the use of P_4S_{10} in boiling xylene as a cyclization agent, leading to the formation of 2,4,4,6-tetraaryl derivatives, and this reaction proceeds *via* the corresponding thiopyrans (see Scheme 8.2).[13,14]

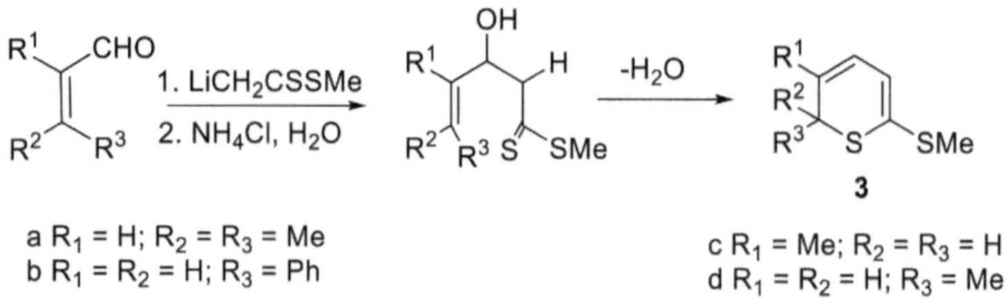

Scheme 8.3

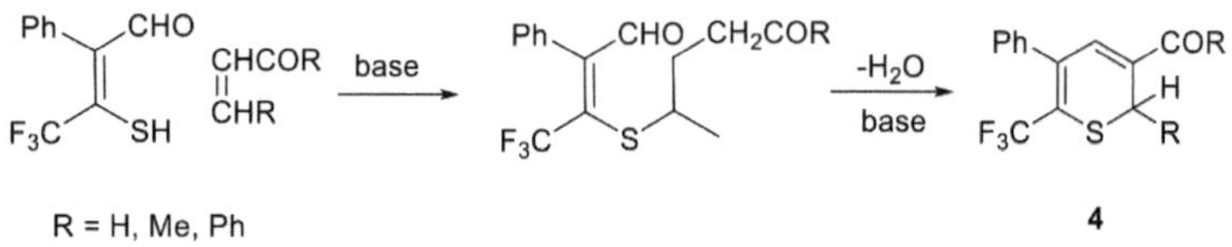

Scheme 8.4

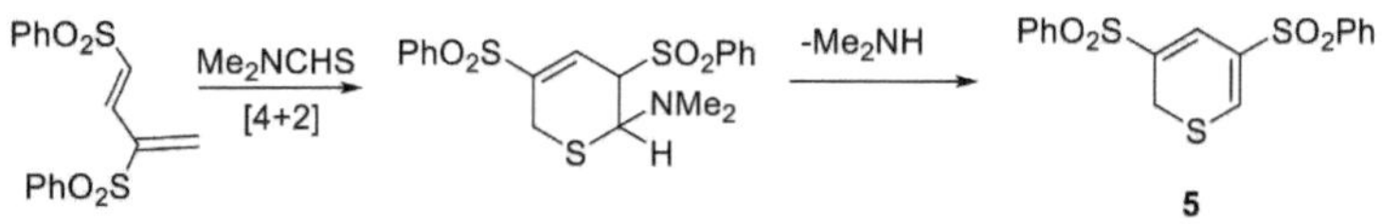

Scheme 8.5

8.2.2 Synthesis of Thiopyrans from Thioenolates

The combination of thioenolates and α,β-unsaturated carbonyl compounds in a cyclocondensation process leads to the production of 2*H*-thiopyrans. The initial sulfur compound used in this process is either methyl-thiol-thione-ethanoate or β-thienolaldehyde (see Scheme 8.3).[15,16]

8.2.3 Synthesis of Thiopyrans from Thioamides

2*H*-Thiopyrans can be synthesized from [4+2] cycloaddition reactions involving a dimethylformamide component, which reacted with the *in situ* generated diene and formed 3,5-disubstituted 2*H*-thiopyran through a dihydro intermediate (see Scheme 8.4).[17] This procedure can be applied to the synthesis of 2-substituted-4,6-bis-benzenesulfonyl-2*H*-thiopyrans (see Scheme 8.5).[18,19]

8.2.4 Hantzsch-like Synthesis of 4*H*-thiopyrans

This method uses cyanothioacetamide as a starting component, and cyclocondensation reactions take place from process A to C under mild

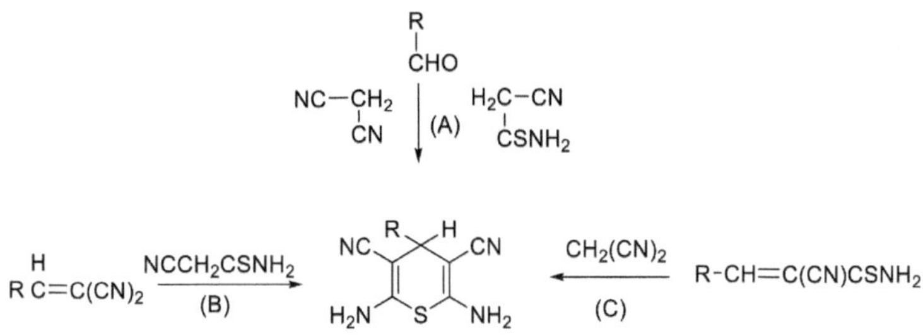

6

Scheme 8.6

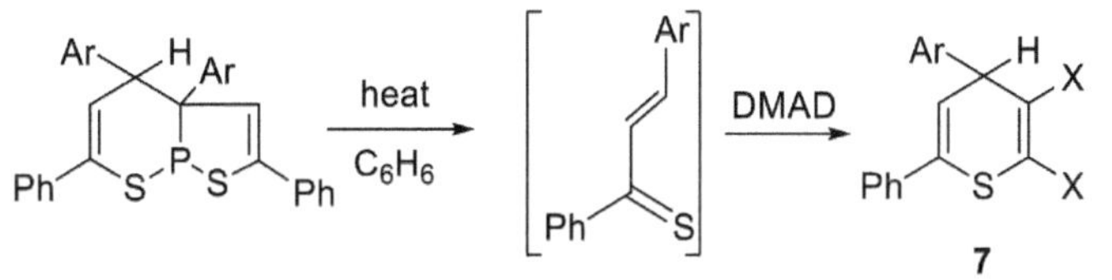

7

Ar = Ph, 4-MeC_6H_4; X = COOMe

Scheme 8.7

conditions, resulting in the formation of products with sufficient yields. The modifications in the process of B and C lead to the formation of 5-(benzothiazole-2-yl) or 5-acetyl derivatives. The overall scheme has been provided in detail in Scheme 8.6.[20–22]

8.2.5 Synthesis of Thiopyrans from Acetylenes

Dimethyl acetylenedicarboxylate (DMAD) undergoes heterocyclizations with α,β-unsaturated thiocarbonyl derivatives, generating a pathway for the synthesis of 4*H*-thiopyrans from enethione precursors. The formation of annulated 4*H*-thiopyrans follows the same steps, as depicted in Schemes 8.7 and 8.8.[23–26]

8.2.6 Synthesis of Thiopyrans from Enamines

Thiopyran derivatives are frequently prepared through the widely used Diels–Alder cycloaddition method involving enaminothiones and electrophilic dienophiles (see Scheme 8.9). Besides the anticipated 2*H*-thiopyrans, 4*H*-isomers are either captured as intermediates or present in the products.[23] Unstable hemiacetal forms as an intermediate in the reaction between enthione 4-$MeOC_6H_4CSCH = CHN(CH_2CH_2)_2$ and enolate MeCH = C(OLi)OEt (Scheme 8.10).[27]

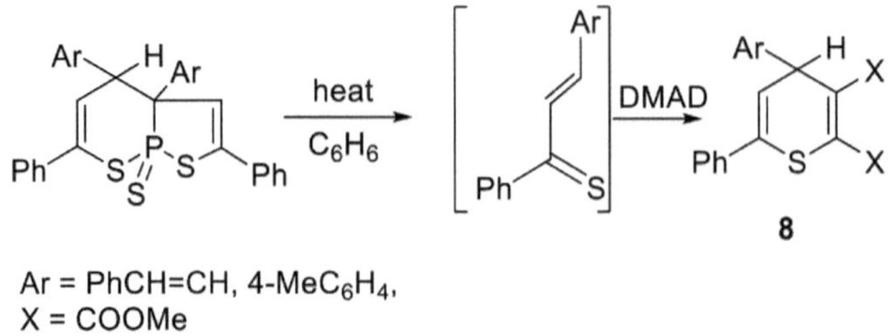

Scheme 8.8

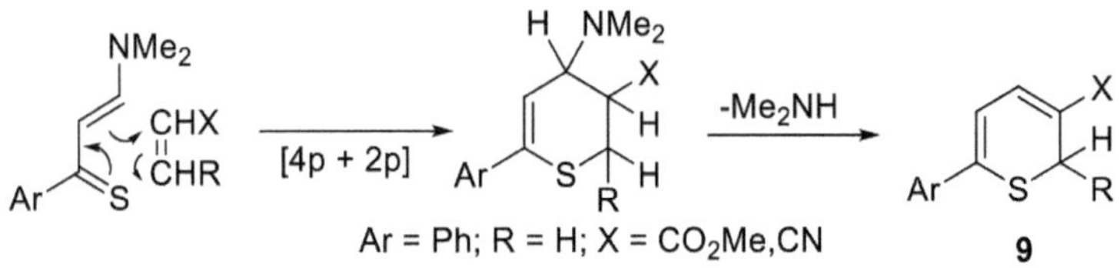

Scheme 8.9

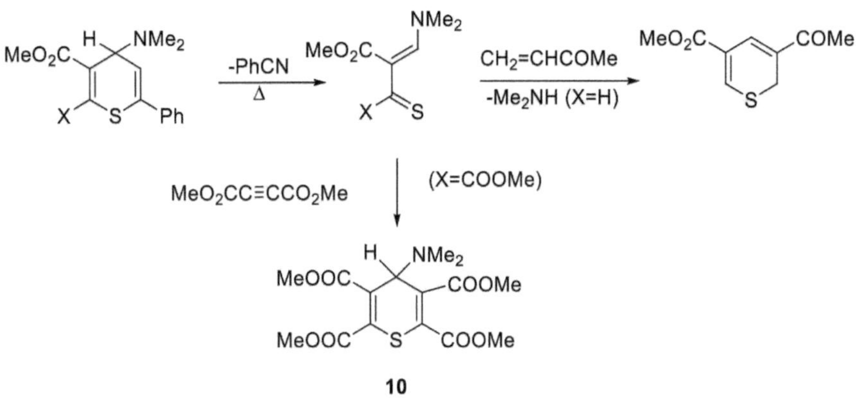

Scheme 8.10

2*H*-Thiopyrans (Ar = Ph; R = H; and X = CO_2Me, CN) can be synthesized through heat- or light-induced cycloaddition–elimination processes, starting with the olefinic components RCH = CHX. However, the presence of intermediates (Ar = Ph, 4-ClC_6H_4, 4-MeC_6H_4; R = Ph, 4$MeOC_6H_4$, 2-furyl; X = NO_2) was observed to impede the elimination process. 2*H*-Thiopyran is obtained through the *in-situ* generation of the starting enaminothione in its preparation.[28]

The application of acetylenic dienophiles typically yields 4*H*-thiopyrans, as observed for **11** and **12** in Scheme 8.11.[29] The primary products obtained in the high-temperature synthesis procedure are the thermodynamically

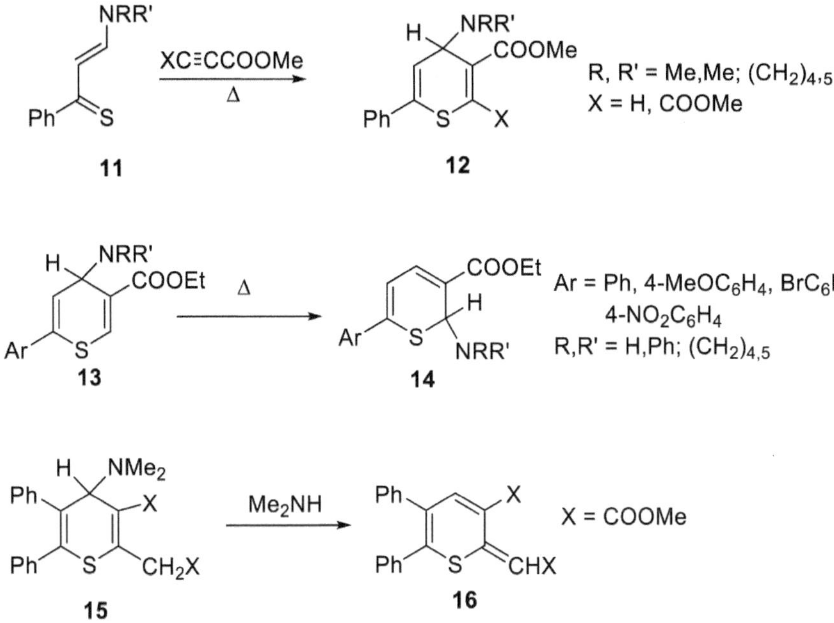

Scheme 8.11

more stable 2-amino-2*H*-isomers **14**. Isotopically labeled ethyl acetylenecarboxylate was used to establish the mechanism of the thermal isomerization from **13** to **14**.[30] 4-Thiopyran **15** formed through the cyclocondensation of PhCSC(Ph)=CHNMe$_2$ (enethione) with MeO$_2$CCH=C = CH–CO$_2$Me (allenic diester) was discovered to be unstable due to its facile conversion to **16**.[31]

8.2.7 Synthesis of Thiopyrans from Thiopyrylium Salts

The reduction of thiopyran salts with LiAlH$_4$ produces a combination of both isomers in different proportions, which is influenced by their structural characteristics (see Scheme 8.12).[32–36] The reaction involving 2,4,6-triphenylthiopyrylium iodide yielded a 1 : 9 ratio of 2*H*- and 4*H*-thiopyrans. The reduction of 2,4,6-triphenylthiopyrylium perchlorate yielded both the corresponding isomeric forms.[37,38]

The introduction of Grignard reagents into thiopyrylium salts leads to the formation of mixtures containing both 4*H*- and 2*H*-isomers, resembling the outcomes obtained with LiAlH$_4$ (see Scheme 8.13). The direction of the nucleophilic attack is contingent on the properties of the reagents. The reaction between thiopyrylium salts and either methyl- or ethyl magnesium bromide results in a combination of 2*H*- and 4*H*-thiopyrans in a 3 : 1 ratio.[39,40]

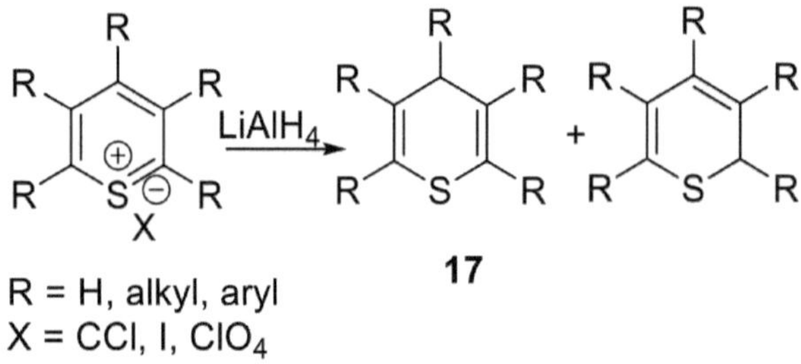

Scheme 8.12

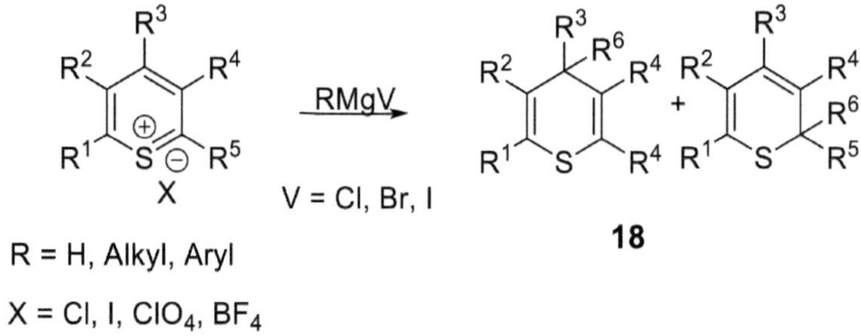

Scheme 8.13

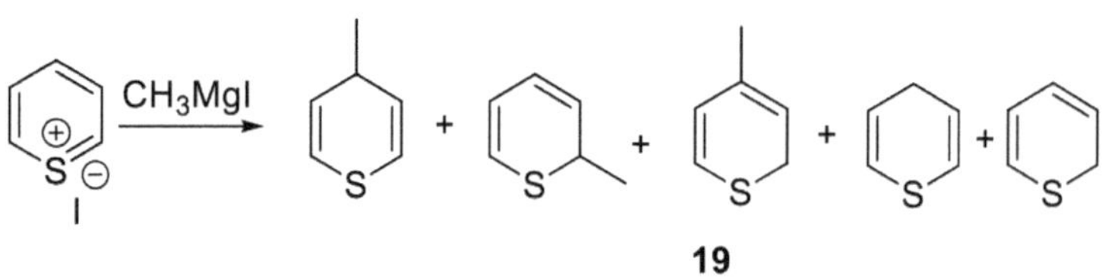

Scheme 8.14

The reaction of benzyl magnesium chloride and perchlorate produced a mixture of isomers, while 2,4,6-triphenylthiopyrylium iodide exclusively yielded 4*H*-thiopyran.[41]

The absence of substituents at the 4-position of thiopyrylium salt leads only to 4*H*-thiopyrans from the Grignard reagent. A reaction of thiopyrylium iodide with methylmagnesium iodide leads to combinations of 4-methyl-substituted-2*H* and -4*H* isomers with 4-methyl-4*H*- and 2-methyl-2*H*-thiopyrans (see Scheme 8.14). Under the reaction conditions, 4-methyl-4*H*-thiopyran undergoes isomerization, resulting in the formation of 4-methyl-2*H*-thiopyran.[42] Additionally, a hydride ion transfer facilitates the reduction of the thiopyrylium cation.

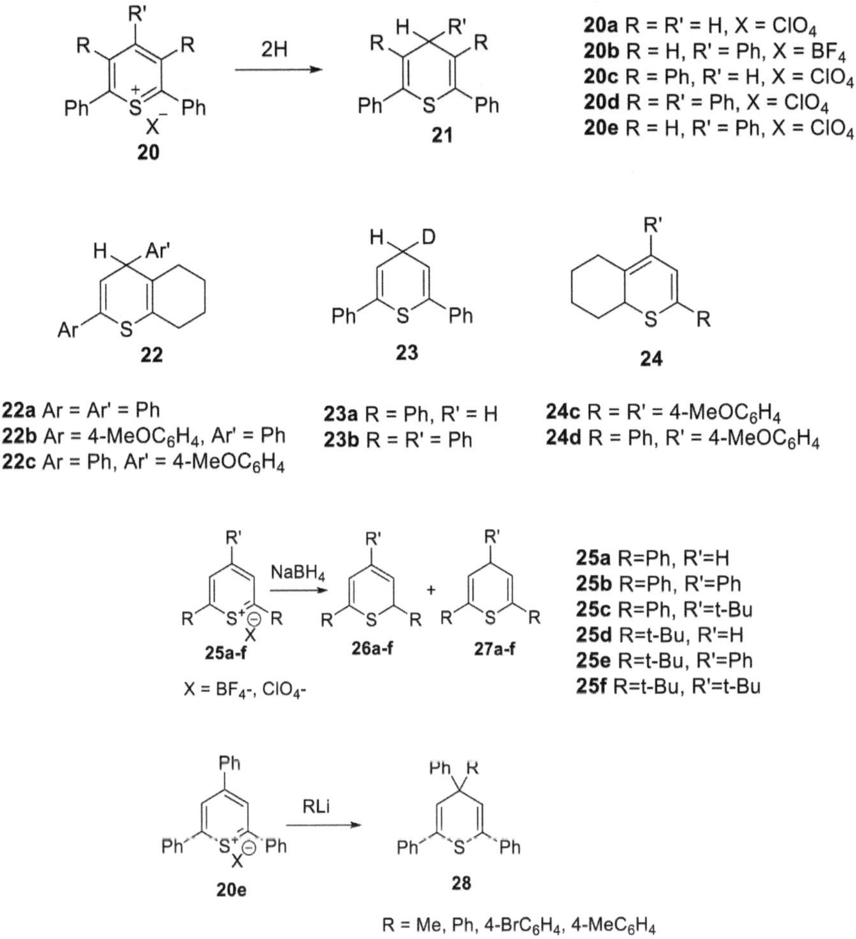

Scheme 8.15

8.2.8 Synthesis of Thiopyrans by Reduction

4*H*-Thiopyran can also be obtained through catalytic hydrogenation of tetrafluoroborate **1b**, giving **2b** (70%) under 10% Pd/C, 30 °C, and H_2/4 MPa.[43] The conversion of **20** into **21** was carried out with hydrides such as $NaBH_4$, $NaBD_4$ in i-PrOH, $LiAlH_4$ in ether, and $LiBH_4$ in tetrahydrofuran (THF) (Scheme 8.15).[44] However, the reduction of the corresponding thiopyrylium trifluoroacetate with $LiAlH_4$ resulted exclusively in 2*H*-thiopyran 5 (R = R′ = 4-MeOC_6H_4).[45,46] The reduction of salts **25d** showed a similar lack of regioselectivity, resulting in the formation of both isomeric thiopyrans **26a–26f** and **27a–27f**.[47]

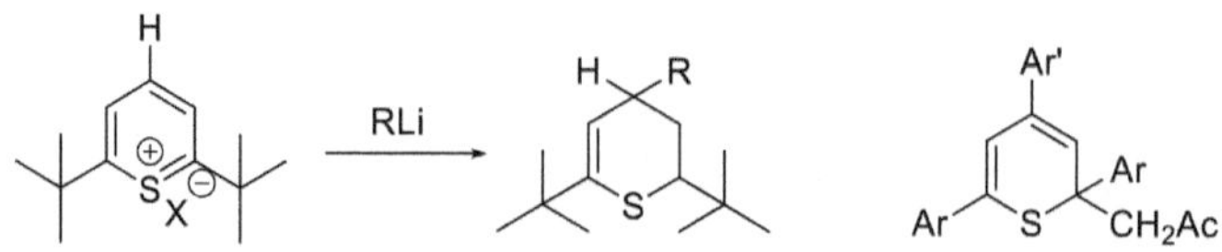

25d

29

30

X= Y = t-Bu, OEt, OMe, Ph
or X = OMe, Y = Ph

a) R = $CH_2OCH(Me)OEt$
b) R = $C(N_2)COOEt$
c) R = $C(N_2)P(O)XY$

Ar, Ar': 4-BrC_6H_4, Ph, 4-ClC_6H_4, Ph-4-MeC_6H_4,
Ph., Ph, 4-ClC_6H_4.Ph, 4-BrC_6H_4., Ph, 4-$MeOC_6H_4$.
Ph, $Me_2NC_6H_4$, Ph, Ph

Scheme 8.16

8.2.9 Synthesis of Thiopyrans by Reactions with C-nucleophiles and Reductive C-substitutions

The conversion of thiopyrylium salts into thiopyrans, involving C-substitutions, can be accomplished through two primary approaches. One-step additions encompass their direct reactions with C-nucleophiles, whereas the two-step procedures entail the initial conversion of salts into nucleophilic intermediates and subsequent attacks with suitable electrophiles (Scheme 8.16).

In the initial method, 2,4,4,6-tetrasubstituted-4*H*-thiopyrans **9** are synthesized from perchlorate **20e** through a reaction with RLi (see Scheme 8.15). Likewise, 4*H*-thiopyrans **29a** and **30b** were found to be produced from salt **25d** using specific reagents such as $LiCH_2OCH(Me)OEt$, which is analogous to previous findings. Meanwhile, phosphorus 4*H*-thiopyran derivatives **29c** were synthesized from a combination of $HC(N_2)P(O)XY$-Et_3N at −78 °C.[48]

The formation of 2-acetonyl-2,4,6-triaryl-2*H*-thiopyrans **30** was carried out through the reaction of the corresponding 2,4,6-triarylthiopyrylium perchlorates with different dialkylammonium salts. Thiopyrylium salts may undergo dimerization to yield thiopyranyl derivatives in the presence of bases under suitable conditions.[49]

In the second method, the sequence starts with (1) lithiating the mixture of **26d** and **27d** with lithium diisopropylamide (LDA) (see Scheme 8.17) and the subsequent reaction with CO_2 to form the respective 4-carboxylic acid and its ethyl ester **31**. (2) Alternatively, the electrochemical reduction of perchlorate **20e** in acetonitrile generates an anionic species, which is subsequently transformed into either a 3 : 1 mixture of isomers **28** (R = *t*-Bu) and **33** or 4*H*-thiopyran **28** (R = $PhCH_2$) through the use of *t*-BuI or $PhCh_2Br$, respectively (see Scheme 8.18).[50]

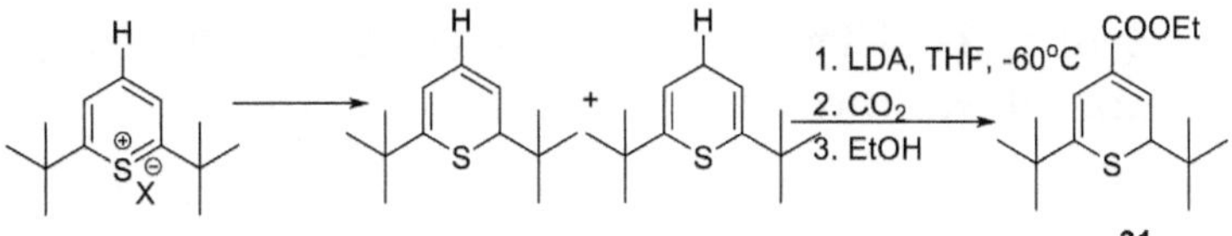

Scheme 8.17

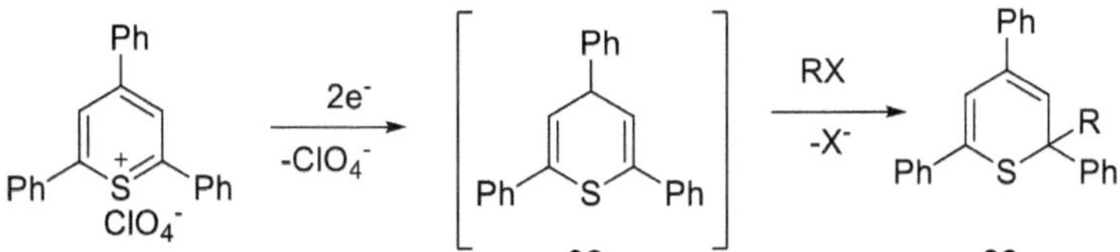

Scheme 8.18

8.2.10 Reactions with Oxygen and Sulfur Nucleophiles

The interactions of sodium methoxide, methanol, and 2,4,6-triarylthiopyrylium perchlorates have been investigated, leading to the transformations as shown in Scheme 8.19. The investigation of these transformations was feasible only within a series of 2,4,6-triaryl-2*H*-thiopyrans **36** (Ar = Ph, 4-BrC_6H_4, 4-$MeOC_6H_4$, 4-ClC_6H_4, 4-MeC_6H_4, and 4-$Me_2NC_6H_4$), which are isolated after heating the corresponding salts with MeONa–MeOH mixtures (see Scheme 8.19).

At −40 °C, the reaction of MeONa with thiopyrylium salts **34** (R′ = H, Me, R = *t*-Bu, and R′ = H, MeO; R = Ph) was found to be kinetically controlled, leading exclusively to the formation of 4*H*-thiopyrans. When R = *t*-Bu, the combination of both 4*H*- and 2*H*-thiopyrans (R′ = Ph, 3-ClC_6H_4, 4-ClC_6H_4, 4-MeC_6H_4, 4-$MeOC_6H_4$, and 4-$Me_2NC_6H_4$) was observed, but only the 2*H*-isomer was formed for R′ = 4-$O_2NC_6H_4$.[51,52]

Establishing the equilibrium composition of thiopyran isomers at 25 to 30 °C was achieved through thermodynamic analyses. Thus, only 2*H*-thiopyran (R = *t*-BuPh) was formed, where R′ was phenyl, or a 3- or 4-substituted phenyl group, and 4*H*-thiopyran when R = Ph and R′ = MeO. 4-Methyl-4*H*-thiopyrans (R = *t*-Bu and Ph) irreversibly transformed into 4-methylene derivatives.[53,54]

8.2.11 Synthesis of Thiopyrans by Reactions with Nitrogen and Phosphorus Nucleophiles

The nucleophilic nature of the reagents, when placed in suitable solvents, plays a pivotal role in determining the reactions of thiopyrylium salts with amines. Under kinetic control at 25 °C, the $BuNH_2$/DMSO reagent was reported to yield mixtures of **38** and **39** (see Scheme 8.20), ultimately resulting in the establishment of equilibrium with the thermodynamically more stable 2*H*-isomers **38** and perchlorates (see Scheme 8.19).[55,56] The adducts of 2*H*-thiopyrans from **40** (R = H, Me; R′ = Ph, 4-BrC_6H_4, 4-ClC_6H_4, 4-Me C_6H_4, and 4-$MeOC_6H_4$) and Me_2NH or morpholine were prepared. Meanwhile, the preparation of thiopyran salts **40** involved the reaction of compound **20** (R = Me, R′ = H, and X = 1/2 $SnCl_6$) with primary amines.[45] The resulting compounds were then tested for both antiviral and antibacterial activities. A reaction of diphenylmethoxyphosphane and 2,6-diphenylthiopyrylium perchlorate (**20a**) with sodium iodide in an acetonitrile medium led to the formation of 4*H*-thiopyran **43** (see Scheme 8.20), which was successfully isolated. The progression from **44** to **45** occurred in a comparable fashion, as depicted in Scheme 8.21.[57]

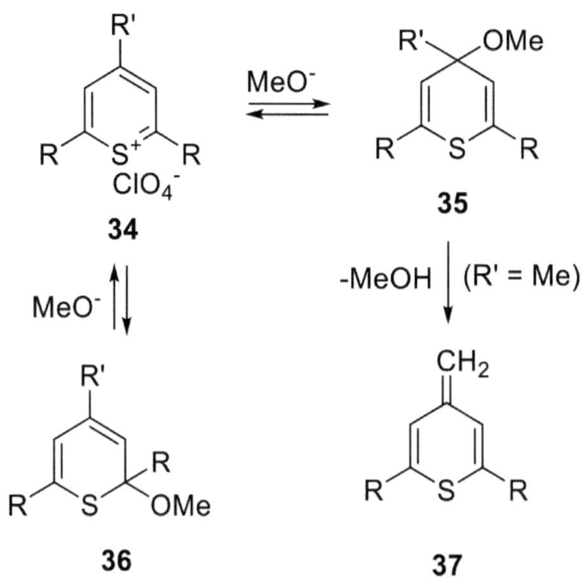

Scheme 8.19

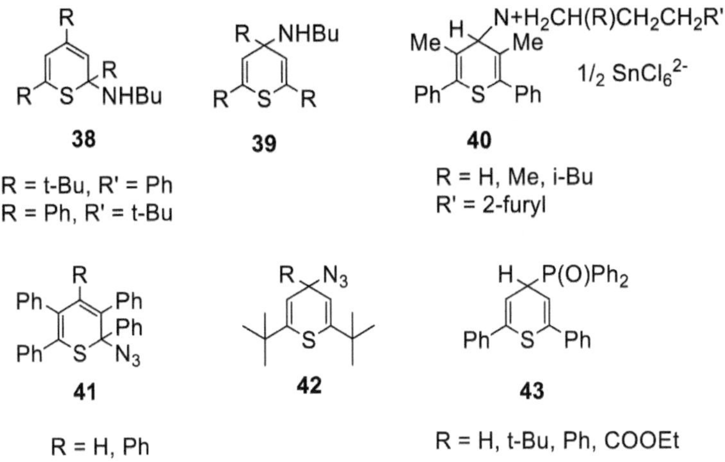

Scheme 8.20

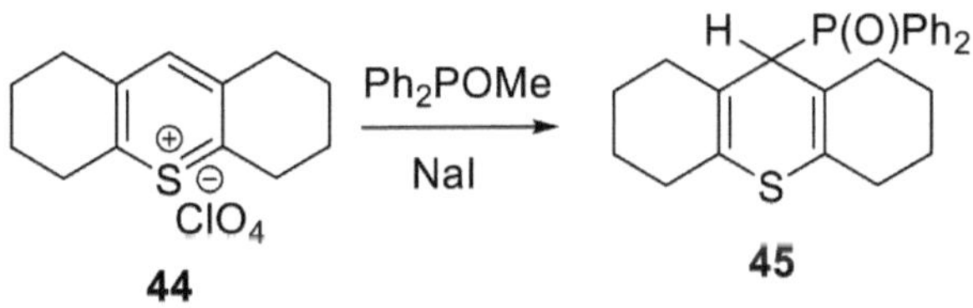

Scheme 8.21

8.2.12 Synthesis of Thiopyrans from Thiopyrones and Similar Cyclic Ketones

The application of diisobutylaluminium hydride (DIBAL-H) in the reduction of 4-thiopyrones, **46a** and **46b**, led to thc formation of 4*H*-thiopyrans **47a** and **47b**, involving the complete removal of oxygen functionalities from the molecules (see Scheme 8.22). The reaction of MeMgI and $(MeO)_2POCH_2Li$ with the carbonyl group in sulfone **48** led to the formation of the corresponding hydroxyl-substituted 4*H*-thiopyrans **49** (see Scheme 8.23). On the other hand, in the presence of the hydrazine group, *S*,*S*-dioxide was converted into the 2*H*-thiopyran derivative (see Scheme 8.24).[58]

The usual scenario involves dihydrothiopyrones as the primary tautomers of hydroxythiopyrans, the synthesis of which involves stabilizing the thiopyran structures through enol ether bonds or other substitution patterns favoring the enol form such as the silylation with an Me_3SiCI–ZnCI complex (TMSCl), triisopropylsilyl trifluoromethanesulfonate (TIPSOTf), *tert*-butyl dimcthylsilyl trifluoromethanesulfonate (TBDMSOTf) in the Et_3N–CH_2Cl_2 complex, TIPSOTf with LDA, or in the phosphorylation with LDA and diethyl chlorophosphite.[59,60] Thus, sequential reactions

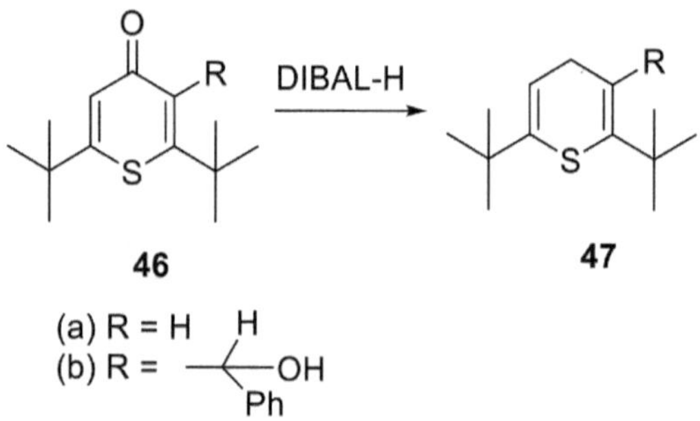

Scheme 8.22

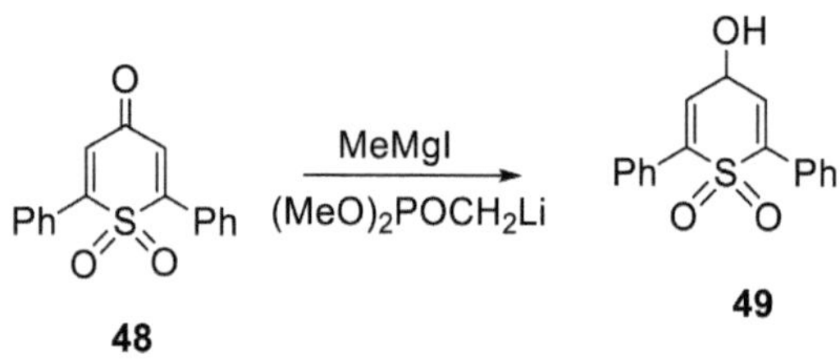

Scheme 8.23

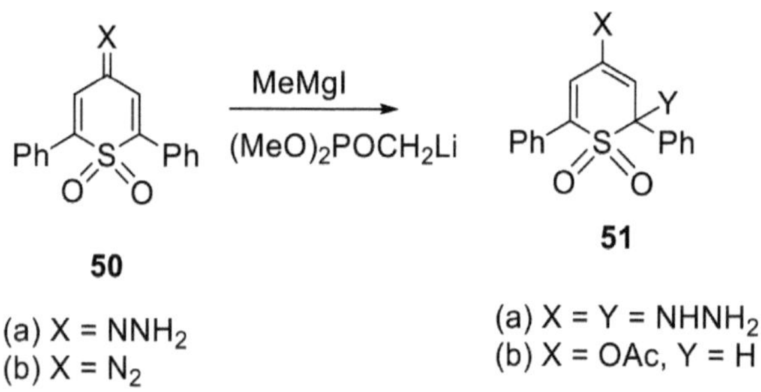

Scheme 8.24

involving 4-substituted-3,6-dihydro-24-thiopyrones, LDA, and TIPSOTf yielded 2*H*-thiopyrans **52a** and **52b** (see Scheme 8.25). Notably, the formation of **52a** led to a problematic mixture with its exocyclic isomer **53**, posing separation challenges. Upon treatment with TIPSOTf or ClPO(OEt)$_2$, 2,3-dihydro-2*H*-thiopyran-3-one engaged in two competing reactions, yielding the anticipated 5-substituted-2*H*-thiopyrans **54a** and **54b**, along with their corresponding 3-substituted isomers **55a** and **55b**, respectively (see Scheme 8.26).[61] Depending on the reaction conditions, the ratios of **54** to **55** varied. The parallel reaction pathways of tetrahydro-4*H*-thiopyran-3,5-dione and TIPSOTf yielded one comprehensible 3,5-disubstituted-2*H*-thiopyran **57**.[61]

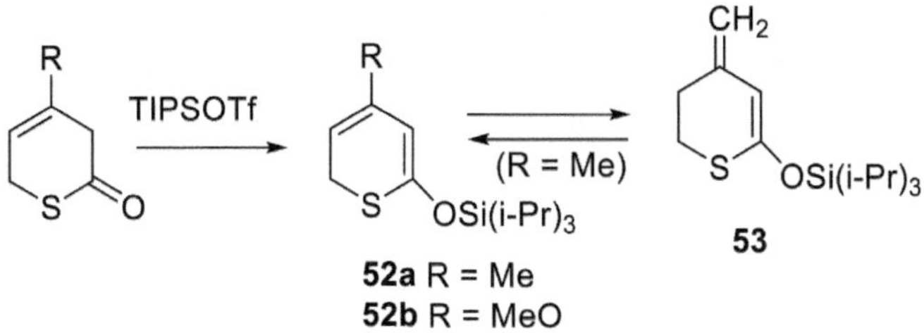

Scheme 8.25

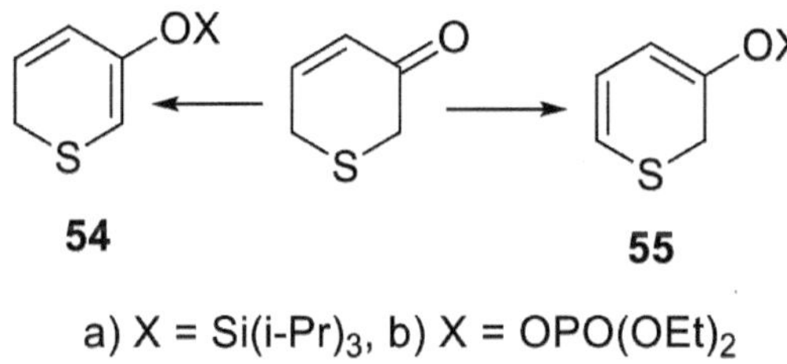

Scheme 8.26

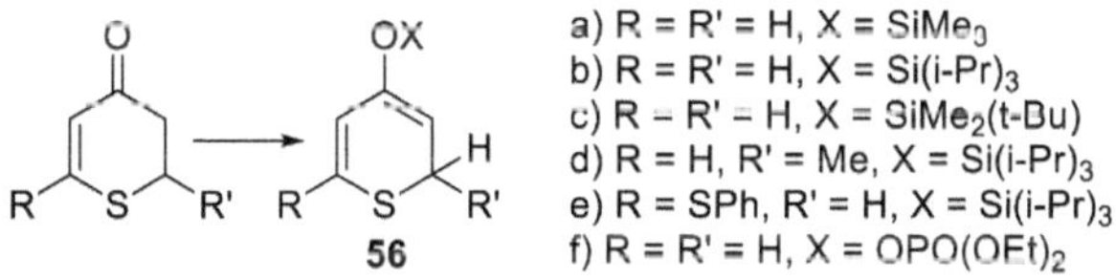

Scheme 8.27

High regioselectivity was observed in the silylation and phosphorylation reactions of different 2,3-dihydro-4*H*-thiopyrones, leading to the formation of 2*H*-thiopyrans **56**(**a–f**) with nearly quantitative yields (see Scheme 8.27).[62]

The reaction of 4*H*-thiopyran-3,5-dione with TIPSOTf led to the formation of 3,5-disubstituted-2*H*-thiopyran **57**. The possibility of obtaining 2*H*-thiopyranoic hydroxyl esters **59** from keto esters **58** is influenced by the heteroatom, and in the presence of sulfur, product **59** is obtained with a favorable yield (see Scheme 8.28).[61]

The Vilsmeier–Haack reaction with 2,3-dihydro-4*H*-thiopyrones offers a more promising route to 2*H*-thiopyran-3-carboxaldehydes, producing chloroaldehydes **60a** and **60b** with yields of 30 and 58%, respectively (see Scheme 8.29).[63]

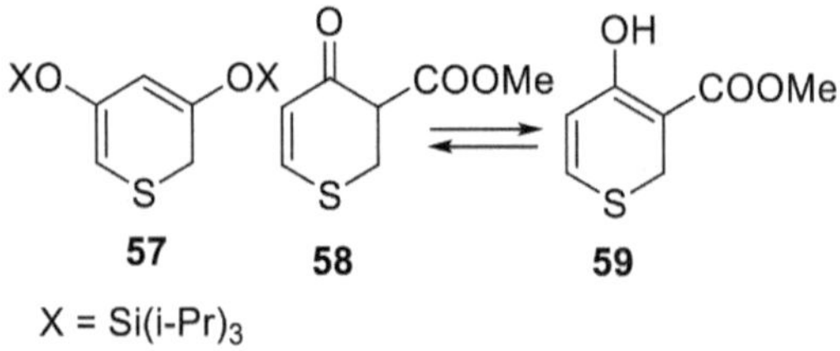

Scheme 8.28

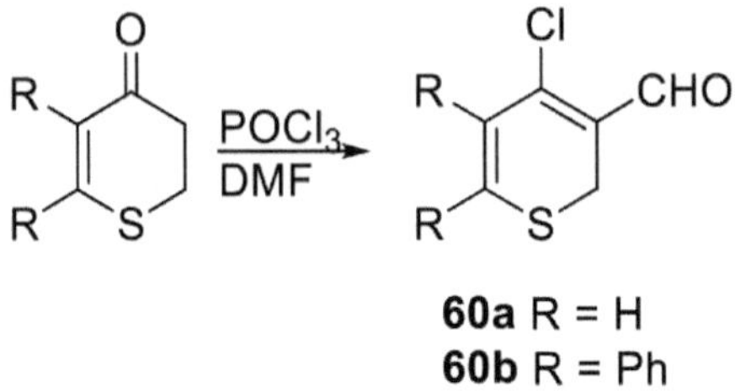

Scheme 8.29

8.2.13 Synthesis of Thiopyrans from Thiophenes

The application of $Rh_2(OAc)_4$ to the thiophene derivative **61** (see Scheme 8.30) revealed the intramolecular generation of 2*H*-thiopyran **62** and its anticipated isomer **63**. The in-depth examinations of the intermolecular Stevenson rearrangement, resulting in the formation of isomers **65** and **66**, have demonstrated that the process advances through thiophene ylides **64**, which are isolable and have three-membered cyclic transition states (see Scheme 8.31). The CO photoextrusion from some 2,3-dihydrothiophen-2-ones gave mixtures containing 2*H*-thiopyrans.[64,65]

8.2.14 Synthesis of Thiopyrans from 1,5-Diketones

The concurrent generation of both isomers is a frequent outcome when 1,5-diketones undergo reactions with H_2S and P_4S_{10}, even though specific 1,5-diketones can readily undergo condensation under these conditions, yielding carbocyclic compounds.[66–70] The extent to which 1,5-diketones react with H_2S under the influence of HCl is notably dependent on the solvent chosen. In methanol, 4*H*-thiopyrans are produced predominantly, whereas mixtures of thiacyclohexanes and thiopyrylium chlorides are produced in the acetic acid medium.

After a short period, the reaction mixture produces thiopyrans, and beyond 48 hours or a longer duration based on the diketone structure, the corresponding thiopyrylium chloride and thiacyclohexane are produced

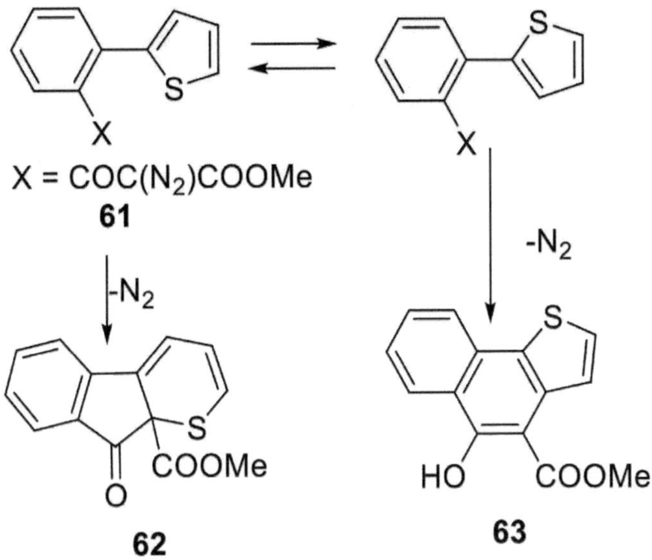

Scheme 8.30

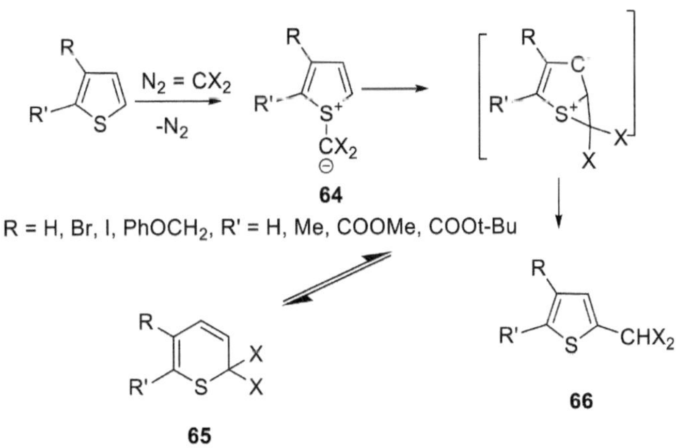

Scheme 8.31

(see Scheme 8.32).[71,72] When subjected to H_2S and HCl in methanol, 1,5-diphenyl-1,5-pentanedione along with its 3-methyl derivative form mixtures containing both thiacyclohexanes and thiopyrylium salts. The compound 1,2,4,5-tetraphenyl-1,5-pentanedione exhibits no reactivity with identical reagents in methanol or acetic acid.[35] Both methanol and acetic acid serve as suitable solvents for a reaction mediated by the presence of substituents at the 2- and 4- or 2-, 3-, or 4-positions in 1,5-diaryl-1,5-pentanediones with H_2S and HCl, culminating in thiopyran formation.

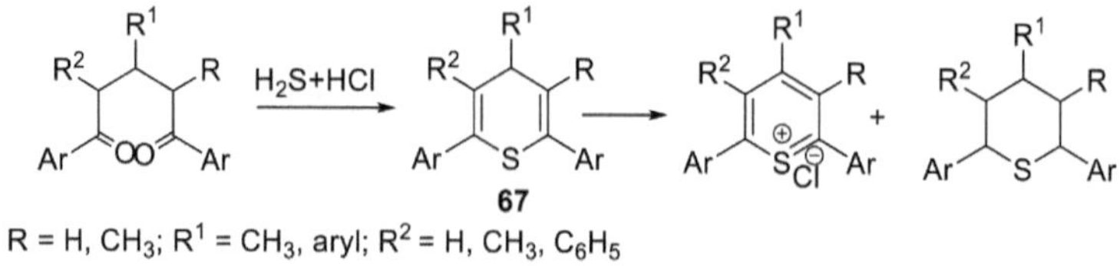

Scheme 8.32

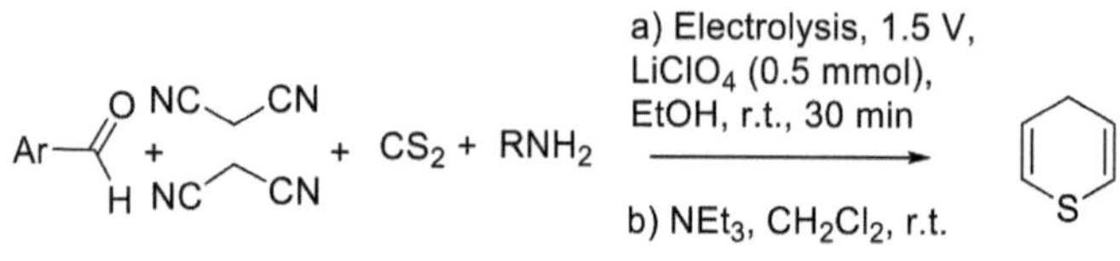

Scheme 8.33

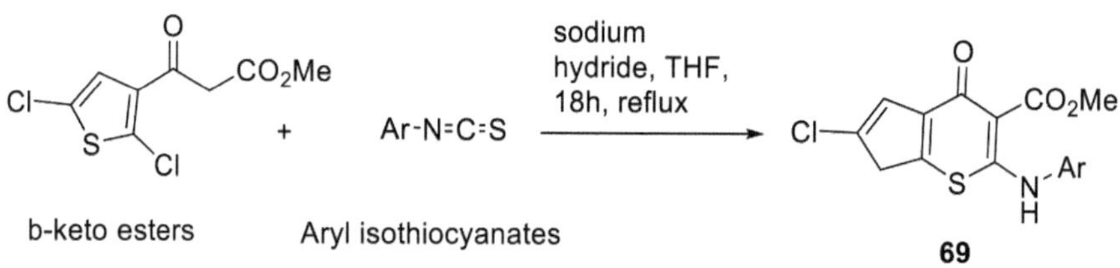

Scheme 8.34

8.2.15 Synthesis of Thiopyrans by Electrosynthesis and with Catalysts

In catalyst-free electrosynthesis, thiopyran derivatives were synthesized by combining aromatic aldehyde and primary amine in the presence of malononitrile and carbon disulfide. The reaction was conducted in ethanol with lithium perchlorate as the electrolyte. Stirring the same reactants with triethylamine (NEt_3) as a catalyst in dichloromethane (DCM) (CH_2Cl_2) at room temperature resulted in the formation of 4*H*-thiopyrans (see Scheme 8.33).[73,74]

8.2.16 Synthesis of Thiopyrans from β-Keto Esters

The synthesis of 4-oxothiopyran derivatives is achieved by reacting aryl-substituted isothiocyanates with β-keto esters under reflux conditions. The reaction occurs in the presence of sodium hydride and dry THF (see Scheme 8.34).[75,76]

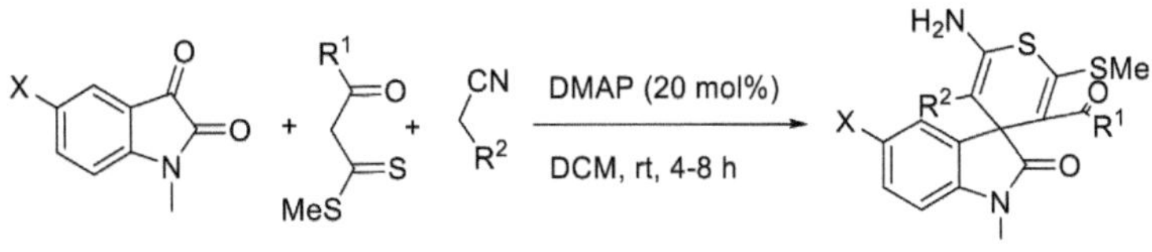

70

Scheme 8.35

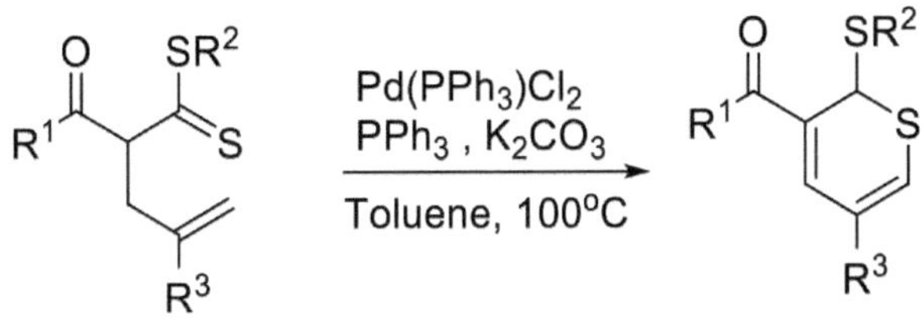

71

Scheme 8.36

8.2.17 Synthesis of Thiopyrans from *N*-methylisatin

For the synthesis of thiopyran derivatives from *N*-methylisatin and β-oxo-dithioester in DCM, 4-dimethylaminopyridine (DMAP) and malononitrile were added and stirred for 4 hours at room temperature. This multi-component reaction resulted in the formation of spiro[indoline-3,4′-thiopyran]-2-one (see Scheme 8.35).[77]

8.2.18 Synthesis of Thiopyrans by Intramolecular Coupling

Using toluene, $Pd(PPh_3)_2Cl_2$, PPh_3, and K_2CO_3 in an intramolecular regioselective CdS coupling, 4*H*-thiopyrans are formed from allyl oxodithioester over a 12-hour reaction period (see Scheme 8.36).[78]

Thiopyrano[2,3-*b*]indole derivatives are formed through a gold-catalyzed reaction, achieving intramolecular hydroarylation and oxidative aromatization of a conjugated system of enyne-tethered indole sulfides (see Scheme 8.37).

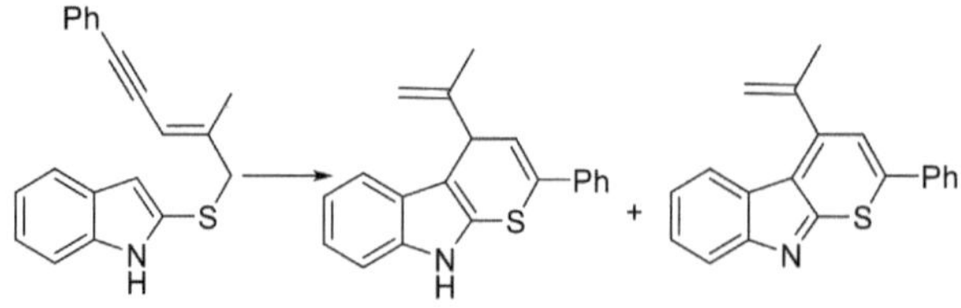

72

Scheme 8.37

8.3 Pharmacological Profile of Thiopyran Derivatives

8.3.1 Antibacterial and Antifungal Activities

In 2019, a series of derivatives of thiopyran and bicyclopyrazolone were reported by Pasha *et al.*, followed by an evaluation for antibacterial and antifungal activities. The antibacterial activities (*in vitro*) of these derivatives were investigated against a series of gram-positive and gram-negative bacterial strains using the Kirby–Bauer disk diffusion method. Reference drugs included nystatin, chloramphenicol, and gentamicin. The results revealed that compound **73** (see Figure 8.3),[79] incorporating an aryl sulfonyl substituent at the N2 site of the pyrazolidinone moiety linked to the thiopyran, showed noteworthy efficacy against *E. coli* and *P. aeruginosa* (negative strain), with growth-inhibitory values of 18.5 and 16.5 (mean), respectively. The most significant activity was recorded for compound **74** against *S. aureus* and *B. subtilis* (positive strains), with values of 14.0 and 15.0, respectively. Compound **74** had a *p*-nitrophenyl substituent at the N2 site of the pyrazolidinone moiety incorporated into the thiopyran. It demonstrated noteworthy antifungal activity against *C. albicans*, with a value of 19.0, surpassing the activity of nystatin, which exhibited a value of 11.3.[79]

In 2017, Pasha *et al.* developed and examined a set of thiopyrans and their oxidized derivatives, exploring their antimicrobial properties. The antifungal potential of the compounds against *C. albicans* was assessed and subsequently compared to antibiotics such as gentamicin, chloramphenicol, and nystatin. Impressively, compounds **75** (IC_{50} = 22.5 mM) and **76** (IC_{50} = 7.3 mM) (see Figure 8.3) emerged as the best within the series, exhibiting heightened antibacterial and antifungal activities against the microorganisms investigated. Structure–activity relationship (SAR) investigations revealed that variations of thiopyran-4-one-*S*,*S*-dioxide with an ester (CO_2Me) group at the third position exhibit improved antibacterial activity.[80]

Hydrogen bonding, initiated by the sulfone group, intensifies the binding affinity of compounds to cell wall elements, which culminates

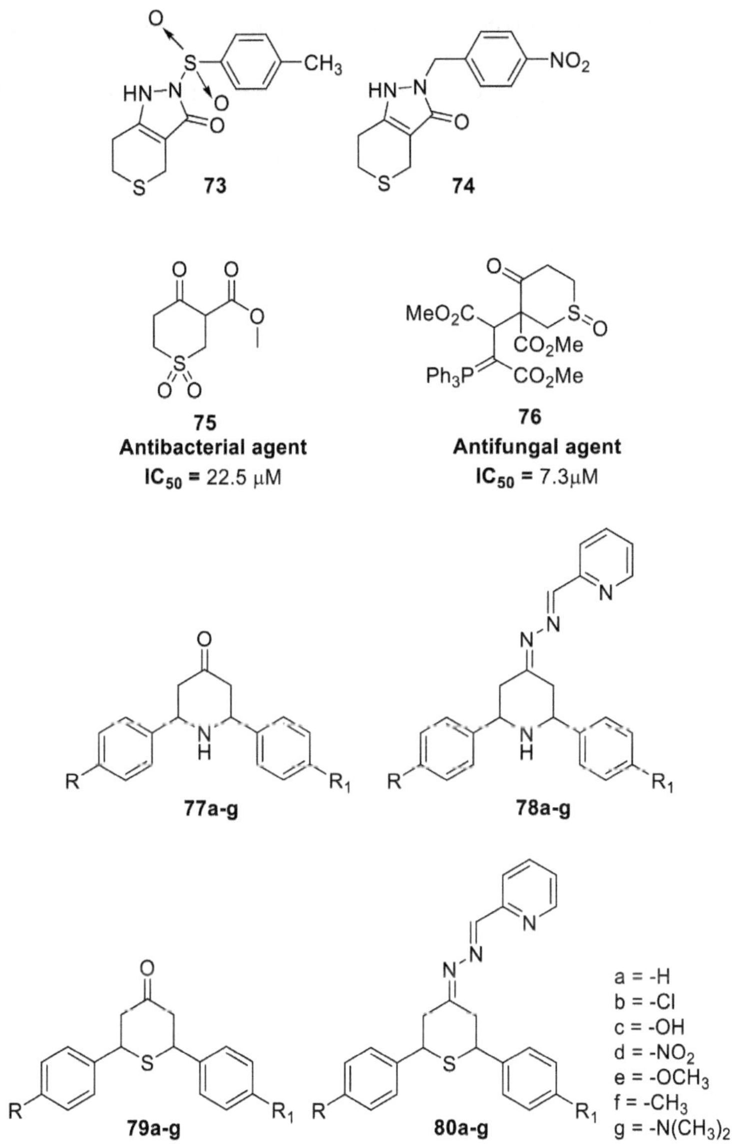

Figure 8.3 Thiopyran compounds as antimicrobial agents.

in bacterial cell death. Additionally, the antifungal activity of sulfoxides, particularly designated as thiopyran *S*-oxides, was evident.

The antibacterial efficacy of compounds **78a–g** and **80a–g** was assessed against diverse human bacterial pathogens. Notably, compound **78d** exhibited substantial antibacterial activity with 4 μg mL^{-1} as the minimum inhibitory concentration (MIC) value against *E. coli*. Similarly, compound **78e** demonstrated heightened antibacterial effectiveness with an MIC of 4 μg mL^{-1} against *K. pneumoniae*, surpassing the performance of ciprofloxacin (MIC: 16 μg mL^{-1}).

Additionally, compounds **77a–g** were investigated against fungal pathogens.[81] High antifungal efficacy was observed for compound **77b** against *C. albicans*, as evidenced by a low MIC of 0.25 μg mL^{-1}, whereas compound **77e** showed relatively similar activity with 0.5 μg mL^{-1}. Compound **77f** exhibited notably higher activity against *M. audouinii* with an MIC of 2 μg mL^{-1} compared to clotrimazole (4 μg mL^{-1}).[81]

8.3.2 Anticancer Activity

In 2017, Ji *et al.* identified compound **81** (see Figure 8.4) as a lead compound and formulated numerous variations as inhibitors targeting p53-MDM2. This process entailed the (1) substitution of the aldehyde with an aromatic amine group; (2) formation of a closed-ring structure by linking the amine side chain with the ester group, giving rise to novel derivatives of spiroindole–thiopyranopyridone; and (3) replacement of the ester and aldehyde groups with a hydroxy group. Subsequently, the compounds designed for the study were synthesized and evaluated for cytotoxic activity (*in vitro*) against A549, MCF-7, and HCT116 cancer cell lines, using compound **81** and Nutlin-3 as reference substances. Notably, among the tested compounds, compound **82** demonstrated heightened activity, exhibiting IC_{50} values from 1.8 to 2.9 mM.

Furthermore, compound **82** demonstrated the ability to induce cell death or apoptosis in A549 cells. In summary, the structure–activity relationship suggested that compounds exhibiting significant antitumor activity typically incorporated bulky hydrophobic groups and a thiopyran nucleus. Conversely, compounds containing hydroxy or hydrophilic groups were associated with a decrease in antitumor activity.[82,83]

2*H*-Thiopyrano[2,3-*d*]thiazole-6-carbaldehyde derivatives synthesized by Lozynskyi *et al.* were tested against SF-268, NCIeH460, and MCF-7 (cancer cells) cell lines. Among these, two compounds demonstrated noteworthy

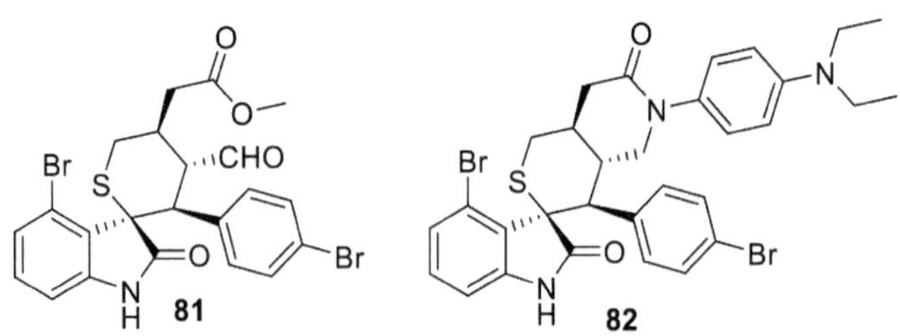

81 p53-MDM2 inhibitor **IC$_{50}$** = 3.55 mM

82 Antitumor agent **IC$_{50}$** = 2.9 mM

Figure 8.4 Thiopyran compounds as anticancer agents.

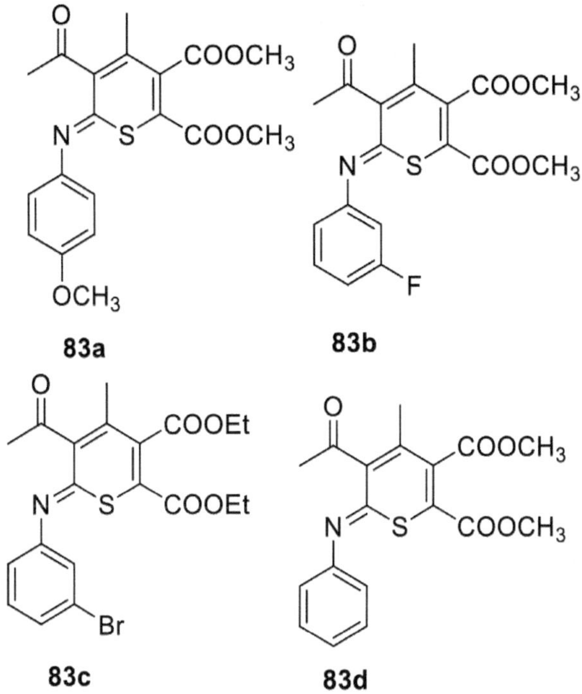

Figure 8.5 Thiopyran compounds as anticancer agents.

antitumor properties, prompting further evaluation against a series of 60 cancer cell lines obtained from various types of cancer, including colon, breast, lung, leukemia, and others. In the context of leukemia treatment, compound 7-phenyl-2-oxo-7-phenyl-3,5,6,7-tetrahydro-2*H*-thiopyrano[2,3-*d*]thiazole-6-carbaldehyde stood out for its remarkable activity, with a mean GI_{50} : TGI ratio of 1.26 : 25.22 mM (GI_{50} is the concentration of a drug that causes a 50% reduction in the growth of cancer cells and TGI is the concentration of drug that yields no net growth over the course of the assay. TGI represents the cytostatic state).[84]

A collection of derivatives of 6*H*-thiopyran-2,3-dicarboxylate, namely **83a–d** (see Figure 8.5), were assessed for their anticancer activity against HCT-15 and MCF-7 (colon and breast) cell lines using the sulforhodamine B (SRB) assay. The outcomes revealed that they demonstrated notable cytotoxicity against both cell lines. Moreover, it was observed that these compounds showed high DNA-binding affinity. In both cancer cell lines, the introduction of compound **83a** decreased the cell percentage in the G_0/G_1 phase (a resting phase of the cell cycle, resistant to chemotherapy), accompanied by an increased number of cells in the G_2/M phase (the G_2 phase checkpoint is a cell cycle checkpoint that prevents cells with damaged DNA from entering mitosis; in this phase cells prepare for mitosis).[85]

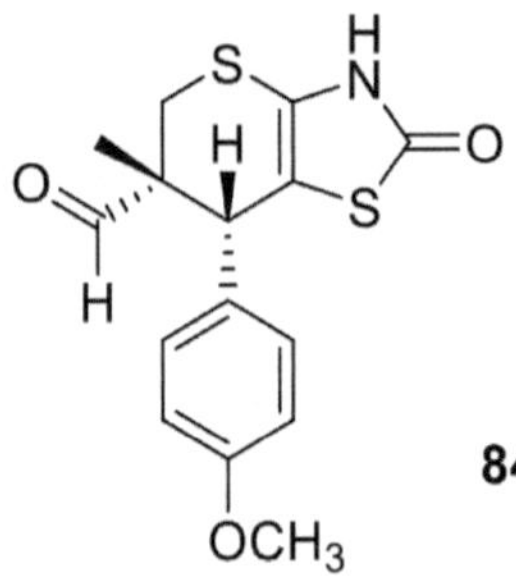

84

Antiviral agent
EC_{50} = 0.07 mM

Figure 8.6 Thiopyrane compounds as antiviral agents.

8.3.3 Antiviral Activity

In 2016, Lozynskyi *et al.* embarked on the synthesis and assessment of newly designed thiopyrano[2,3-*d*]thiazole-6-carbaldehyde derivatives, exploring their prowess as dual-purpose anticancer and antiviral agents. The compounds were assessed for their antiviral activity with the viral capsid antigen enzyme linked immunosorbent assay (VCA ELISA) test against viruses such as Epstein–Barr virus (EBV), herpes simplex virus type 1, and varicella-zoster virus. The VCA ELISA test is a qualitative method that determines the presence of IgM antibodies against the VCA virus in human plasma or serum. Compound **84**, as shown in Figure 8.6, demonstrated outstanding activity against EBV (EC_{50} = 0.07 mM). Additionally, it exhibited moderate activity against the other viruses.[84]

8.3.4 VEGFR-2 Inhibitory Activity

In 2015, Salerno *et al.*[86] reported on a set of benzothiopyrano-[4,3-*d*]pyrimidines designed to function as the inhibitors of vascular endothelial growth factor receptor 2 (VEGFR2). VEGFR2 is a type IV receptor tyrosine kinase and a major target in angiogenesis-related kinases. The inhibitory activity of the compound against the kinase enzyme was evaluated using the kinase insert domain receptor (KDR) inhibition assay, with semaxanib as the reference drug. The findings revealed that compounds with a 2-anilino moiety exhibited substantial inhibition, outperforming semaxanib at a concentration of 7 mM. Additionally, the compounds were evaluated using the human umbilical vein endothelial cell (HUVEC) method, comparing their performance to that of semaxanib. The benzothiopyranopyrimidine compounds featuring the 2-anilino substitution continued to exhibit anticancer activity, showing GI_{50} from 0.35 to 0.74 mM. In both assays, compound **85**, presented in Figure 8.7, exhibited the highest activity level, with an IC_{50} value of 2.7 mM and a GI_{50} value of 0.45 mM.

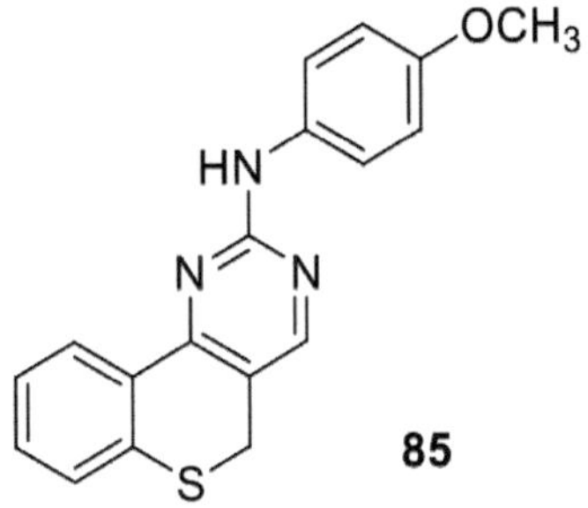

85

VEGFR-2 inhibitor
IC_{50} = 2.7 mM

Figure 8.7 Thiopyran compounds as VEGFR-2 inhibitors.

Additionally, ten human kinases were used to investigate the inhibitory potential of the compound. There was a significant inhibitory activity exhibited by the compound against Aurora kinase and CDC2/CDK1.[86]

The presence of the NH group at the second position on the pyrimidine group, in conjunction with benzothiopyran, facilitated H-bonding with the NH backbone and CO in the hinge area of the enzyme. This specific interaction mechanism contributed significantly to the observed robust VEGFR-2 inhibitory activity.[86]

8.3.5 5-LOX Inhibitory Activity

In 2014, Hu *et al.*[87] synthesized and assessed a series of thiopyrano pyrazole derivatives for their potential as inhibitors of human 5-lipoxygenase (5-LOX). Evaluation of the compounds against human 5-LOX was conducted using the cell-free assay method, with the zilcuton drug used as the standard. Compound **86**, as shown in Figure 8.8, exhibited notable inhibitory activity against 5-LOX, with an IC_{50} value of 5.7 mM. This result closely approached the IC_{50} value of zileuton, recorded at 5.3 mM. Molecular docking also revealed adept binding of the compound to the binding sites in the enzyme. Accordingly, the compound is deemed a promising molecule for the continued advancement of potent 5-LOX inhibitors.[87]

8.3.6 Larvicidal Activity

For larvicidal screening, a variety of tetrahydro-2*H*-thiopyran derivatives, recently synthesized, were evaluated for their impact on urban mosquito larvae, focusing on *Culex quinquefasciatus* sourced from the city drainage system. During the examination, larval mortality was observed at varying

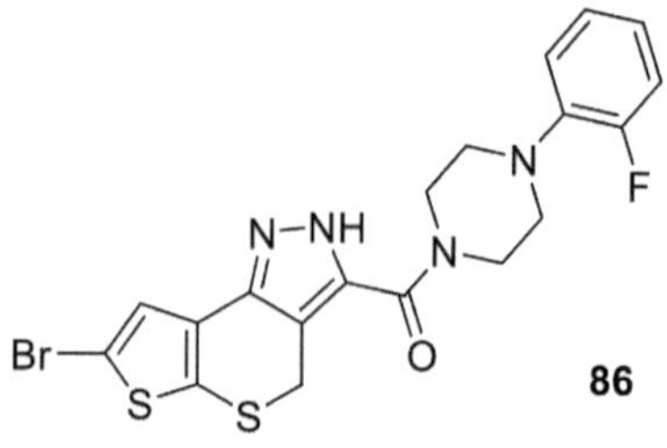

86

5-LOX inhibitor
IC_{50} = 5.7 + 0.9 mM

Figure 8.8 Thiopyran compounds as 5-LOX inhibitors.

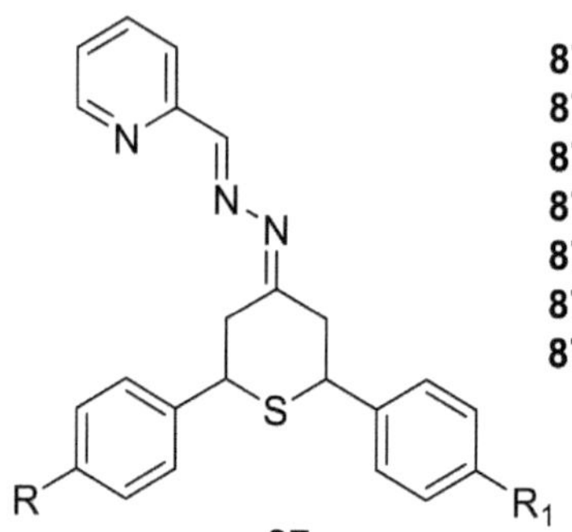

87a: R, R_1 = -H
87b: R, R_1 = -Cl
87c: R, R_1 = -OH
87d: R, R_1 = $-NO_2$
87e: R, R_1 = $-OCH_3$
87f: R, R_1 = $-CH_3$
87g: R, R_1 = $-N(CH_3)_2$

87

Figure 8.9 Thiopyran compounds as larvicides and nematicides.

concentrations of the compounds. A WHO standard protocol (1981) was implemented to gauge the susceptibility of larvae to the compounds. Compound **87e** provided excellent results, achieving 100% larval mortality at a concentration of 20 μg mL^{-1}, surpassing other compounds with an LD_{50} value of 0.8 μg mL^{-1}. Among the compounds assessed, **87c**, **87d**, **87f**, and **87g** (see Figure 8.9) showed average activity with LD_{50} values of 5.7, 1.2, 8.6, and 6.0 μg mL^{-1}, respectively.[81]

8.3.7 Nematicidal Activity

The assessment of the nematicidal activity of the compounds involved the use of juvenile nematodes of *Meloidogyne javanica*.[81] Compounds **87a–g**, as shown in Figure 8.9, were scrutinized for their impact on nematodes, displaying noteworthy activity. The assessment of the toxicity of compounds was carried out at 37 °C during screening. Compound **87e** exhibited exceptional nematicidal activity, reaching absolute mortality at 20 μg mL^{-1} and an LD_{50} value of 3.2 μg mL^{-1}, distinguishing it from other compounds. The activities of compounds **87c, 87d, 87f,** and **87g** fell within a moderate range, showcasing LD_{50} values of 7.8, 5.5, 4.1, and 4.7 μg mL^{-1}, respectively.[81]

8.3.8 Anti-inflammatory Activity

The acute inflammation triggered by formalin is characterized by a potent inflammatory response, leading to significant swelling in the governed set of animals within a few hours of the experiment, with an average volume increase of 47.38%. Treatment with tetrahydrospiro-2*H*-thiopyran derivatives in animals with experimental pathology resulted in diminished exudative reactions, showcasing an anti-inflammatory effect akin to diclofenac sodium, the reference drug. The anti-inflammatory activity in hydrogenated spirothiopyranes depends on the positioning of sulfur within the spirocycle. With sulfur situated at the 4,6′-spiro position relative to the triazinoquinazoline cycle, compound **88** demonstrated anti-inflammatory efficacy similar to diclofenac sodium (see Figure 8.10). Conversely, altering the position of sulfur to the 3,6′-spiro position in the thiopyran cycle (compound **89**) resulted in a notable decrease in activity.[88]

8.3.9 Hypoglycemic Activity

The potent-to-moderate hypoglycemic activity was observed in a series of 4-amino-7,8-dihydro-2-piperazinyl-6*H*-thiopyranol[3,2-*d*]pyrimidines against laboratory mice. Compounds **90** and **91** exhibited appreciable antidiabetic activity. Surprisingly, the addition of a methyl group to compound **92** significantly diminished its activity (see Figure 8.11). The methylated compounds showed weak activity, and the compound with a methoxy group was highly toxic with minimal activity.[89]

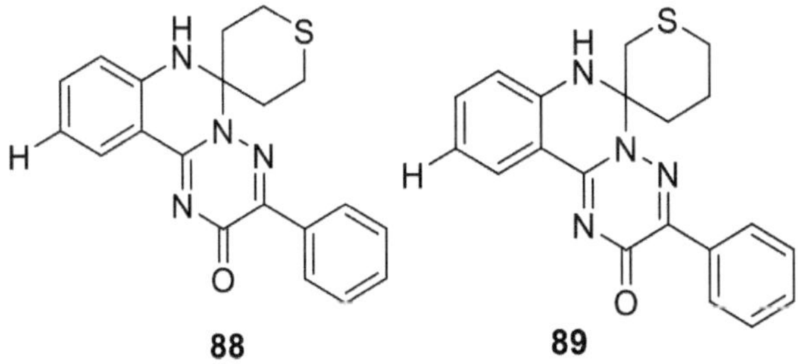

88 **89**

Figure 8.10 Thiopyran compounds as anti-inflammatory agents.

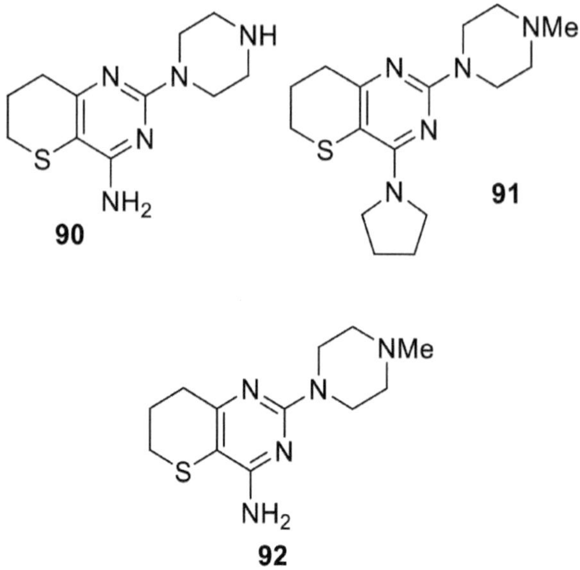

Figure 8.11 Thiopyran compounds showing hypoglycemic activity.

8.4 Conclusion

In recent years, there has been a growing demand for S-heterocycles in the pharmaceutical industry due to their potential therapeutic prowess. The progress in the synthesis and development of S-heterocycles has increased their availability. Thiopyran, a class of one such S-heterocycles, has been under investigation for a long time, and its pharmacological potency has recently been discovered. It has been explored in many areas of pharmacological domains for its antibacterial, antifungal, antiviral, anti-inflammatory, antidiabetic, and even anticancer activities. Thiopyran and its derivatives need to be explored in other areas of biological fields, such as drug-binding studies, and their potency as therapeutic compounds needs to be explored *in vivo*. Thus, despite the advancements made in synthesizing and developing thiopyrans as therapeutic structures, there is a need for research and further development to counteract various challenges and failures, such as the risk of toxicity of S-containing compounds. For example, green and economically viable methods for developing thiopyrans can become assets for producing therapeutic compounds on an industrial scale. The compilation of information on the synthesis and pharmacological advancements of thiopyran and its derivatives might help future research.

Acknowledgements

The authors acknowledge the publishing house for providing an opportunity to write this chapter and NIT Silchar for providing the necessary facilities.

References

1. S. Pathania, R. K. Narang and R. K. Rawal, *Eur. J. Med. Chem.*, 2019, **180**, 486–508.
2. R. Mishra, N. Sachan, N. Kumar, I. Mishra and P. Chand, *J. Heterocyclic. Chem.*, 2019, **55**, 2019–2034.
3. R. Mishra, K. K. Jha, S. Kumar and I. Tomer, *Der. Pharma Chem.*, 2011, **3**, 38–54.
4. R. Mancuso and B. Gabriele, *Molecules*, 2014, **19**, 15687–15719.
5. A. Chaudhary, K. K. Jha and S. Kumar, *J. Adv. Sci. Res.*, 2012, **3**, 3–10.
6. J. Strating and E. Molenaar, *Org. Prep. Proced.*, 1969, **1**, 21–23.
7. J. Degani, R. Fochi and C. Vicenzi, *Tetrahedron Lett.*, 1963, **18**, 1167–1169.
8. J. Degani, R. Fochi and C. Vincenzi, *Gazz. Chim. Ital.*, 1964, **94**, 451.
9. J. Degani and C. Vincenzi, *Boll. Sci. Fac. Chim. Ind. Bologna.*, 1965, **23**, 203.
10. J. Degani and R. Fochi, *Gazz. Chim. Ital.*, 1967, **97**, 397.
11. J. Degani and C. Vincenzi, *Boll. Sci. Fac. Chim. Ind. Bologna.*, 1967, **25**, 51.
12. J. Degani, R. Fochi and C. Vincenzi, *Boll. Sci. Fac. Chim. Ind. Bologna.*, 1965, **23**, 241.
13. M. Schwarz, P. Trška and J. Kuthan, *Collect. Czech. Chem. Commun.*, 1989, **54**, 1854–1869.
14. P. Sebek, R. Hrabal, S. Nešpurek, M. Adamec and J. Kuthan, *J. Chem. Soc., Perkin Trans. 2*, 1992, 1301–1308.
15. A. Guigne and P. Metzner, *Bull. Soc. Chim. Fr.*, 1990, 446.
16. R. Hoffmann, D. Greif, M. Pulst, M. Weissenfels and A. Laurent, *Z. Chem.*, 1990, **30**, 247–248.
17. A. Padwa, B. Harrison, S. S. Murphree and P. E. Yeske, *J. Org. Chem.*, 1989, **54**, 4232–4235.
18. A. Padwa, Y. Gareau, B. Harrison and B. H. Norman, *J. Org. Chem.*, 1991, **56**, 2713–2720.
19. A. Padwa, Y. Gareau, B. Harrison and A. Rodriguez, *J. Org. Chem.*, 1992, **57**, 3540–3545.
20. J. Kuthan, *Adv. Heterocycl. Chem.*, 1983, **34**, 145–303.
21. F. M. Abdel-Galil, S. M. Sherif and M. H. Elnagdi, *Heterocycles*, 1986, **24**, 2023–2048.
22. B. Y. Raid, A. M. Negm, S. E. Abdou and H. A. Daboun, *Heterocycles*, 1987, **26**, 205–238.

23. J. Kuthan, *Adv. Heterocycl. Chem.*, 1983, **34**, 145–303.
24. H. Tanaka and S. Motoki, *Bull. Chem. Soc. Jpn.*, 1986, **59**, 2047–2049.
25. S. Motoki, Y. Terauchi and S. Kametani, *Chem. Lett.*, 1988, **17**, 717–720.
26. S. Motoki, T. Sakai, Y. Matsuo, S. Kametani and T. Saito, *Bull. Chem. Soc. Jpn.*, 1992, **65**, 923–925.
27. J. P. Guemas, M. Lees, A. Reliquet and J. Villieras, *Phosphorus, Sulfur Silicon Relat. Elem.*, 1982, **12**, 325–339.
28. T. Nishio, N. Nakajima and Y. Omote, *J. Heterocycl. Chem.*, 1980, **17**, 405–406.
29. J. B. Rasmussen, R. Shabana and S.-O. Lawesson, *Tetrahedron*, 1981, **37**, 3693–3698.
30. J. B. Rasmussen, R. Shabana and S. O. Lawesson, *Tetrahedron*, 1982, **38**, 1705–1709.
31. M. Pulst, D. Greif, M. Weissenfels, G. J. S. Doad and F. Scheinmann, *J. Chem. Res., Synop.*, 1989, **10**, 300–301.
32. G. Suld and C. Price, *J. Am. Chem. Soc.*, 1961, **83**, 1770–1771.
33. J. Degani and R. Fochi, *Gazz. Chim. Ital.*, 1967, **97**, 397.
34. G. Suld and C. Price, *J. Am. Chem. Soc.*, 1962, **84**, 2090–2094.
35. V. G. Kharchenko and V. I. Kleimenova, *Zh. Organ. Khim.*, 1971, **7**, 613.
36. V. G. Kharchenko, V. I. Kleimenova and A. R. Yakoreva, *Khim. Geterotsikl. Soedin.*, 1970, **7**, 900.
37. C. Price and D. H. Follweiler, *J. Org. Chem.*, 1969, **34**, 3202–3204.
38. C. Price, M. Siskin and C. Miao, *J. Org. Chem.*, 1971, **36**, 794–799.
39. A. Saffieddine, J. Rover and J. Dreux, *Bull. Soc. Chim. Fr.*, 1972, **2**, 703.
40. J. Rover and J. Dreux, *Bull. Soc. Chim. Fr.*, 1972, **2**, 707.
41. O. Chalvet, C. Decoret and J. Dreux, *Bull. Soc. Chim. Fr.*, 1972, **2**, 716.
42. J. Strating and E. Molenaar, *Org. Prep. Procedures.*, 1969, **1**, 21–23.
43. V. G. Kharchenko, O. A. Bozhenova and A. D. Shebaldova, *Zh. Organ. Khim.*, 1982, **18**, 2435.
44. C. H. Chen, J. J. Doney and G. A. Reynolds, *J. Org. Chem.*, 1982, **47**, 680–684.
45. P. L. Desbene, J.-C. Cherton, J.-P. Le Roux and J.-J. Basselier, *Tetrahedron*, 1984, **40**, 3539–3548.
46. S. K. Klimenko, T. I. Tyrina and N. N. Sorokin, *Khim. Geterotsikl. Soedin.*, 1987, **7**, 614.
47. G. Doddi and G. Ercolani, *J. Org. Chem.*, 1991, **56**, 1674–1675.
48. Y. Mori and K. Maeda, *J. Chem. Soc., Perkin Trans. 2*, 1991, 2061–2066.
49. K. Yamamoto, S. Yamazaki, Y. Kohashi and I. Murata, *Tetrahedron Lett.*, 1982, **23**, 3195–3198.
50. K. Yamamoto, A. Matsukawa and I. Murata, *Chem. Lett.*, 1985, **14**, 1119–1122.
51. G. W. Fischer and T. Zimmermann, *Z. Chem.*, 1983, **23**, 333–334.
52. T. Zimmermann and G. W. Fischer, *J. Prakt. Chem*, 1986, **328**, 373–379.
53. M. L. Di Vona, G. Doddi, G. Ercolani and G. Illuminati, *J. Am. Chem. Soc.*, 1986, **108**, 3409–3415.

54. G. Doddi and G. Ercolani, *J. Chem. Soc., Perkin Trans. 2*, 1989, 1393–1396.
55. V. Aceto, G. Doddi and G. Ercolani, *Gazz. Chim. Ital.*, 1989, **119**, 205.
56. I. Murata, *Phosphorus Sulfur*, 1989, **43**, 243–259.
57. V. T. Abaev, I. V. Karsanov, Z. K. Urtaeva, A. F. Blinokhvatov, A. A. Bumber and O. Y. Okhlobystin, *Zh. Obshch. Khim.*, 1990, **60**, 1012.
58. M. R. Detty, *Organometallics*, 1988, **7**, 1122–1126.
59. G. Doddiand G. Ercolani, *J. Org. Chem.*, 1986, **51**, 4385–4390.
60. T. Harayama and M. Otsuyama, *Yakugaku Zasshi*, 1983, **103**, 1096–1099.
61. D. E. Ward, Y. Gai and W. M. Zoghaib, *Can. J. Chem.*, 1991, **69**, 1487–1497.
62. D. E. Ward, W. M. Zoghaib, C. K. Rhee and Y. Gai, *Tetrahedron Lett.*, 1990, **31**, 845–848.
63. N. Vasumathi, D. V. Ramana and S. R. Ramadas, *Synth. Commun.*, 1990, **20**, 2749–2757.
64. H. Storflor, J. Skramstad and S. Nordenson, *J. Chem. Soc., Chem. Commun.*, 1894, 208–209.
65. T. Bowles, R. Jones, A. E. A. Porter, J. A. Rechka, H. S. Rzepa and D. J. Williams, *J. Chem. Soc., Chem. Commun.*, 1985, 1590–1592.
66. E.Knoevenagel, *Justus Liebigs Ann. Chem.*, 1894, **281**, 25–126.
67. E. Knoevenagel, *Justus Liebigs Ann. Chem.*, 1898, **303**, 223–257.
68. P. Rabe and F. Elze, *Justus Liebigs Ann. Chem.*, 1902, **323**, 83–112.
69. P. Rabe, *Justus Liebigs Ann. Chem.*, 1908, **360**, 265–288.
70. M. N. Tilichenko, *Uch. Zap. SGU, Vyp. Khim*, 1959, **71**, 153.
71. V. G. Kharehenko, V. I. Kleimenova, N. M. Kupranets, N. P. Polikarpova and A. R. Yakoreva, *Zh. Organ. Khim*, 1965, **4**, 2054.
72. V. G. Kharchenko, N. M. Kupranets, V. I. Kleimenova, A. A. Rassudova, M. E. Stankevich, N. M. Yartseva and A. R. Yakoreva, *Zh. Organ. Khim*, 1970, **6**, 1119.
73. R. Dubey, V. K. Singh, L. K. Sharma, A. Upadhyay, N. Kumar and R. K. P. Singh, *New J. Chem.*, 2017, **41**, 7836–7839.
74. S. Asadbegi, A. Mobinikhaledi and M. A. Bodaghifard, *Elements*, 2017, **192**, 526–529.
75. L. Z. El-Agha, M. M. El-Abadelah, M. R. Kamal, S. S. Sabri, R. M. Al-As'ad and W. Voelter, *Z. Naturforschung B.*, 2018, **73**, 109–113.
76. S. Kurva, V. Sriramoju, S. Madabhushi and J. B. Nanubolu, *Synth. Commun.*, 2017, **47**, 1702–1707.
77. S. Kurva, V. Sriramoju, S. Madabhushi and J. B. Nanubolu, *Synth. Commun.*, 2017, **47**, 1702–1707.
78. S. Chowdhury, T. Chanda, S. Koley, N. Anand and M. S. Singh, *Org. Lett.*, 2014, **16**, 5536–5539.
79. G. F. Pasha, S. Asghari, M. Tajbakhsh and M. Mohseni, *J. Chin. Chem. Soc.*, 2019, **66**, 660–667.
80. G. F. Pasha, S. Asghari, M. Tajbakhsh and M. Mohseni, *Res. Chem. Intermed.*, 2017, **43**, 7291–7306.

81. I. Akbar, A. Ahamed, I. A. Arif, R. S. Kumar, K. Selva and A. Mailal, *J. Mex. Chem. Soc.*, 2018, **62**, 135–147.
82. C. Ji, S. Wang, S. Chen, S. He, Y. Jiang, Z. Miao, J. Li and C. Sheng, *Bioorg. Med. Chem.*, 2017, **25**, 5268–5277.
83. S. Wang, Y. Jiang, S. Wu, G. Dong, Z. Miao, W. Zhang and C. Sheng, *Org. Lett.*, 2016, **18**, 1028–1031.
84. A. Lozynskyi, S. Golota, B. Zimenkovsky, D. Atamanyuk, A. Gzella and R. Lesyk, *Elements*, 2016, **191**, 1245–1249.
85. F. Tavakolinia, T. Baghipour, Z. Hossaini, D. Zareyee, M. A. Khalilzadeh and M. Rajabi, *Nucleic Acid. Ther.*, 2012, **22**, 265–270.
86. S. Salerno, A. M. Marini, G. Fornaciari, F. Simorini, C. La Motta, S. Taliani, S. Sartini, F. Da Settimo, A. N. García-Argáez and O. Gia, *Eur. J. Med. Chem.*, 2015, **103**, 29–43.
87. J. Hu, W. Zhu, H. Meng, Y. Liu,, X. Wangand C. Hu, *Chem. Biol. Drug Des.*, 2014, **84**, 642–647.
88. O. Kolomoets, O. Voskoboynik, O. Antypenko, G. Berest, I. Nosulenko, V. Palchikov, O. Karpenko and S. Kovalenko, *Acta Chim. Slov.*, 2017, **64**, 902–910.
89. S. Ohno, K. Mizukoshi, O. Komatsu, Y. Kunoh, Y. Nakamura, E. Katoh and M. Nagasaka, *Chem. Pharm. Bull.*, 1986, **34**, 4150–4165.

CHAPTER 9

Recent Developments in the Synthesis and Biological Applications of Thiazine

D. R. THUBE[a], H. N. AKOLKAR[b], N. R. DAREKAR[c], V. M. KHEDKAR[d] AND S. G. KUNDLIKAR*[a]

[a]P.G. and Research Centre, Department of Chemistry, New Arts Commerce and Science College, Parner, District Ahmednagar, Maharashtra, 414302, India; [b]Department of Chemistry, Abasaheb Marathe Arts and New Commerce, Science College, Rajapur, Ratnagiri, Maharashtra, 416702, India; [c]P.G. and Research, Department of Chemistry, Radhabai Kale Mahila Mahavidyalaya, Ahmednagar, Maharashtra, 414001, India; [d]School of Pharmacy, Vishwakarma University, Pune, Maharashtra, 411048, India
*E-mail: saritakundlikar1@gmail.com

9.1 Introduction

Generally, biologically active compounds are derived from heterocyclic structures. Because of the numerous biological characteristics, heterocyclic molecules with thiazine rings have garnered a lot of interest. Compounds with a compact, straightforward thiazine nucleus are used in studies to assess the biological activities of novel products, such as antitumour, antimicrobial, antipsychotic, antifungal, antiviral, and anti-inflammatory properties.

S-Heterocycles: Retrospect, Prospects, and Biological Applications
Edited by Shrikaant Kulkarni, Hemantkumar Akolkar, Vijay M. Khedkar and A. K. Haghi

Published by the Royal Society of Chemistry, www.rsc.org

Thiazines are organic compounds with the molecular formula C_4H_5NS, comprising a six-membered ring containing four carbon, one nitrogen, and one sulfur atoms at different positions, such as 1,2-, 1,3-, and 1,4-thiazines (Figure 9.1).

Numerous chemical synthesis and transformation processes use 1,3-thiazines as reaction intermediates for a variety of chemical reactions.[1,2] A significant number of bioactive chemical molecules with a thiazine moiety are useful in medicine.[3] Thiazine was found in many clinically used drugs, such as BTZ04, chlorpromazine, cefaclor, and PD404182 (Figure 9.2).

Due to the vast range of pharmacological effects associated with thiazine-containing nuclei, such as antitumour,[4] β-secretase inhibitor,[5] antimicrobial,[6] and photosensitizer properties,[7] molecules containing a thiazine moiety have garnered significant attention in the field of medicinal chemistry.

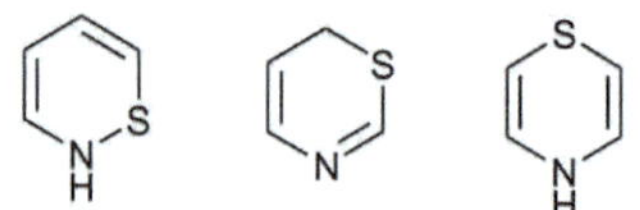

1,2-Thiazine 1,3-Thiazine 1,4-Thiazine

Figure 9.1 Different thiazine derivatives.

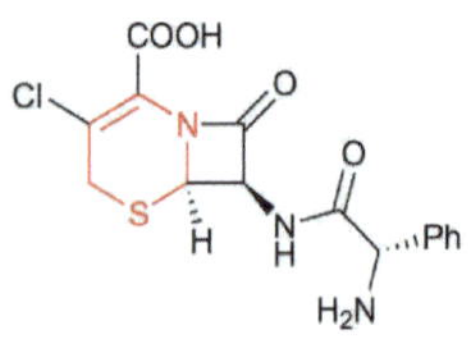

Cefaclor

(Antibiotic)

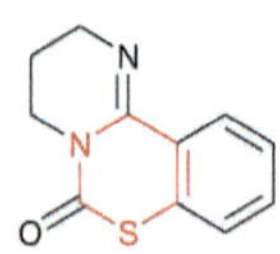

PD404182

(Antiviral)

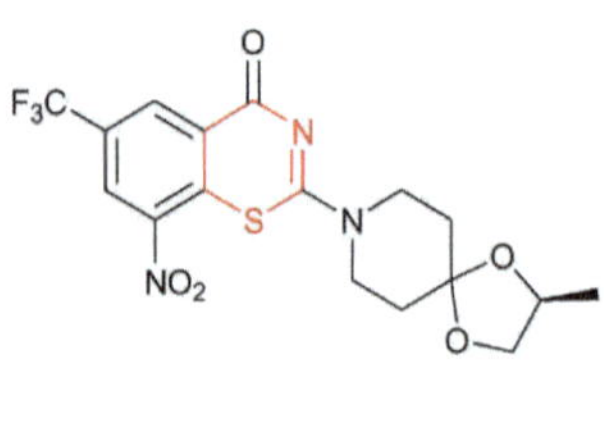

BTZ04

(Antituberculosis)

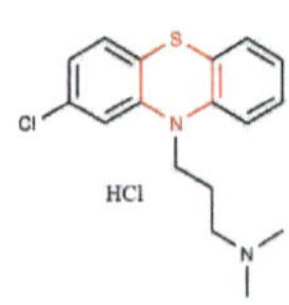

Chlorpromazine

(Antipsychotic Drug)

Figure 9.2 Clinically used drugs containing thiazine structures.

9.2 Recent Advances in the Synthesis of Thiazine

A productive one-pot synthesis of unique benzo[4,5]imidazo[2,1-*b*]pyrimido[4,5-*d*][1,3]thiazine-2,4(3*H*)-diones and benzo[4,5]imidazo[2,1-*b*][1,3]thiazin-4-ones at room temperature was reported by Batooie *et al.*[8] from CH acids, aldehydes, and 2-mercaptobenzimidazole using hexyl triphenyl phosphonium bromide (HTPB) ionic liquid as a solvent and a green promoter. PC3, A431, Panc1, and human dermal fibroblast (HDF) normal cells were utilized as cancer cells in the study to examine the anticancer properties of these compounds (Scheme 9.1).

Ammonium acetate, diethyl 2,2-sulfonyldiacetate, and benzaldehyde at room temperature were reacted in water, as reported by Edayadulla *et al.*,[9] resulting in the formation of 2,6-dicarbethoxy-3,5-diphenyltetrahydro-1,4-thiazine-1,1-dioxide. The produced substances exhibited anticonvulsant properties (Scheme 9.2).

Tozkoparan *et al.*[10] have reacted a series of 1,2,4-triazole-5-thiones with dimethyl acetylene dicarboxylate under reflux conditions in the presence of toluene as a solvent to produce 7*H*-1,2,4-triazolo(3,2-*b*)-1,3-thiazine-7-one 1,3-thiazine derivatives. The synthesized thiazine analogues show intriguing properties, such as antiulcer, analgesic, and anti-inflammatory activities (Scheme 9.3).

Yadav *et al.*[11] have devised a unique three-component technique for the solvent-free synthesis of 1,3-thiazine derivatives from 2-methyl-2-phenyl-1,3-oxathiolan-5-one and 3,6-diaryl-5-mercaptoperhydro-2-thioxo-1,3-thiazin-5-ones in the presence of lithium bromide as a catalyst, which then transformed 2-mercaptoacetic acid and acetophenone into 2-methyl-2-phenyl-1,3-oxathiolan-5-one. The desired product was then

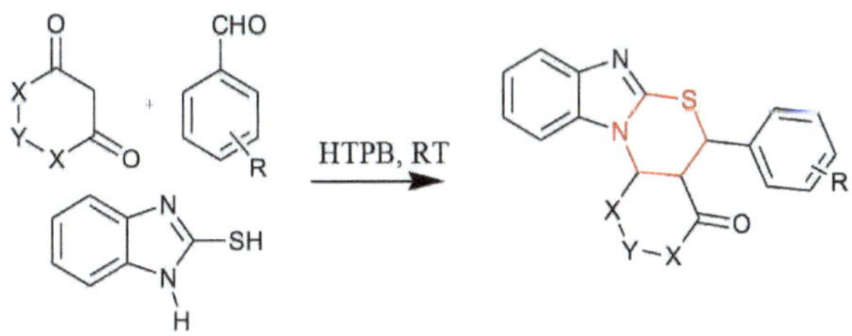

Scheme 9.1 Synthesis of anticancer compounds containing thiazine derivatives.

Scheme 9.2 Synthesis of anticonvulsant compounds containing thiazine derivatives.

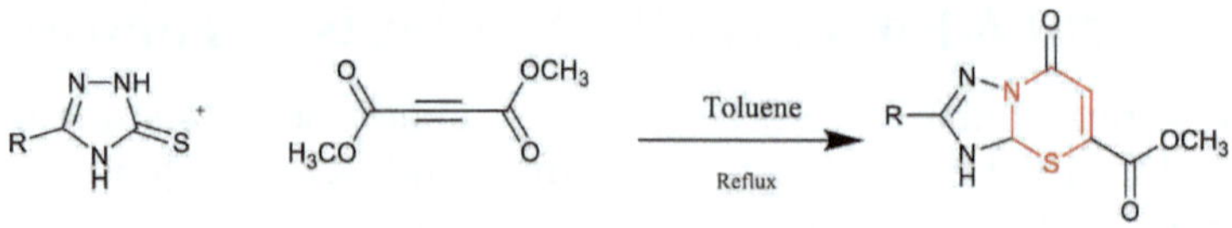

Scheme 9.3 Synthesis of biologically active compounds containing thiazine derivatives.

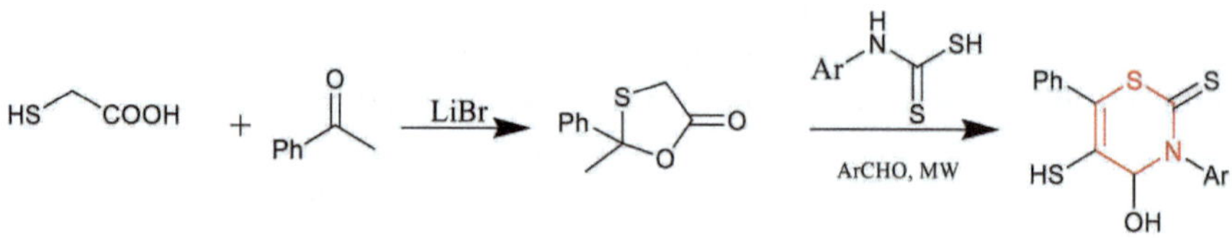

Scheme 9.4 Solvent-free synthesis of 1,3-thiazine derivatives.

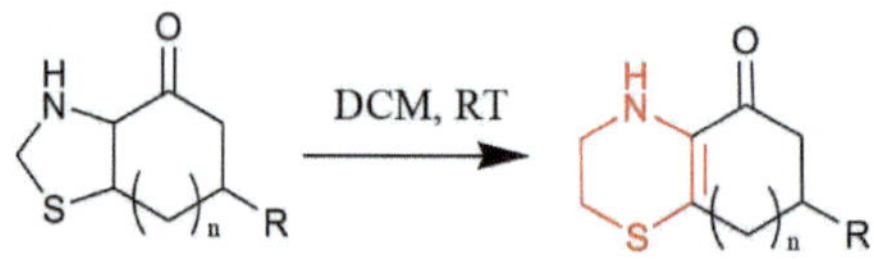

Scheme 9.5 Regioselective formation of dihydro-1,4-thiazine.

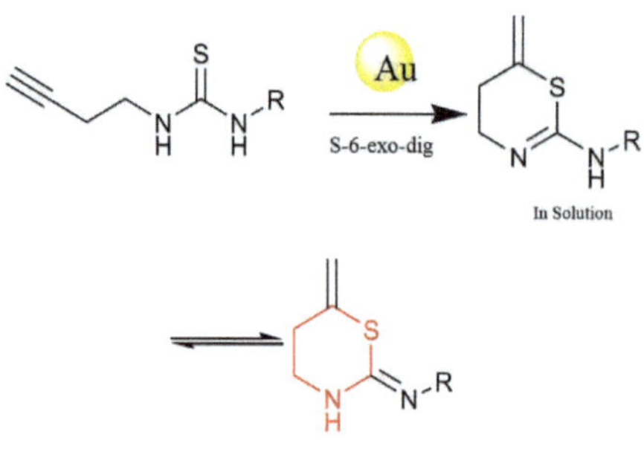

Scheme 9.6 Gold-catalysed 1,3-thiazine formation.

produced by microwave irradiation with benzaldehyde and *N*-aryldithio carbamic acid (Scheme 9.4).

Regioselective formation of dihydro-1,4-thiazine from α-keto spiro-thiazolidine has been reported by Utsukihara *et al.*[12] (Scheme 9.5).

Canudo-Barreras *et al.*[13] reported the gold-catalysed 1,3-thiazine formation and their uncommon tautomer isolation (Scheme 9.6).

Using caesium carbonate as the base in dichloromethane, Wei *et al.*[14] have described the direct one-pot synthesis of functionalized dimethyl 1-aryl-1,9*a*-dihydropyrido[2,1-*c*][1,4]thiazine-1-phosphonates by the treatment of dimethyl (aryl(2-tosylhydrazono)methyl)phosphonates and pyridinium 1,4-zwitterionic thiolates (Scheme 9.7).

Malah *et al.*[15] have reported the synthesis of numerous bis-1,3-thiazine derivatives and their anticancer efficacy *via* condensation reactions of symmetric dialdehydes and 3-substituted amino-2-cyano-3-mercaptoacrylamides in addition to aliphatic ether spacer units. Six distinct human cell lines were used to evaluate the *in vitro* anticancer potential of novel bis-1,3-thiazines: lung A549, colon HCT116, breast MCF-7, prostate PC3, liver HepG2, and normal melanocyte HFB4 (Scheme 9.8).

Al-Difar *et al.*[16] found that 1,3-thiazine derivatives are produced by cyclization of chalcone derivatives and reaction with thiourea in alkaline medium, which occurs *via* Claisen–Schmidt condensation of substituted aromatic acetophenone and aromatic aldehydes (Scheme 9.9).

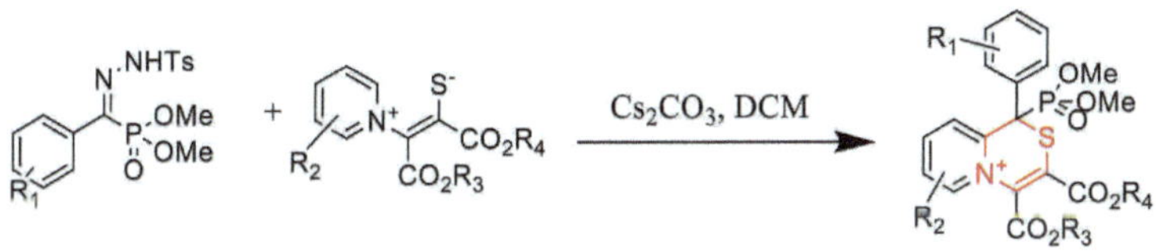

Scheme 9.7 Synthesis of dimethyl 1-aryl-1,9*a*-dihydropyrido[2,1-*c*][1,4]thiazine-1-phosphonates.

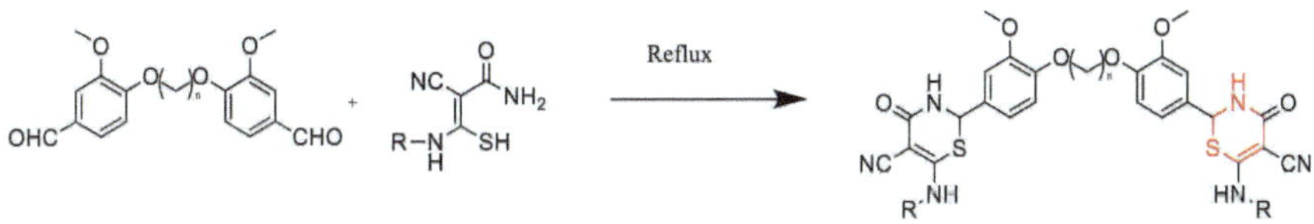

Scheme 9.8 Synthesis of bis-1,3-thiazine derivatives and anticancer efficacy.

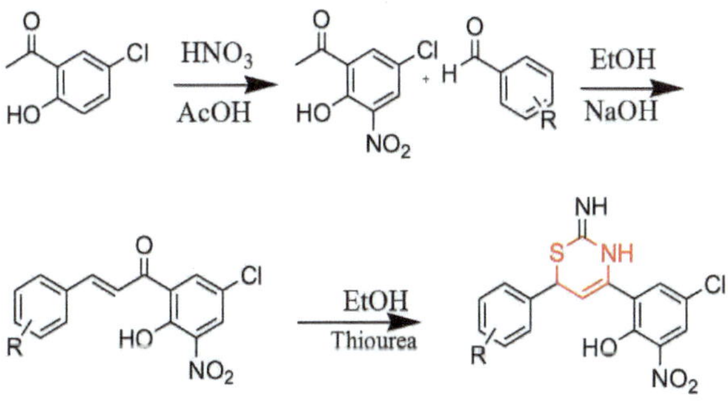

Scheme 9.9 Synthesis of 1,3-thiazine derivatives.

Wang *et al.*[17] produced fluorinated 5,6-dihydro-1,2-thiazine-1-oxides *via* a one-pot domino reaction between *N*-trifluoroacetylated (N-TFA) sulfoximines and 1-trifluoromethyl styrenes. Using the X-ray crystallographic analysis, the structures of the resulting compounds were confirmed (Scheme 9.10).

Legay *et al.*[18] outlined a microwave-assisted three-component process using γ-aminothiols, aromatic aldehyde, and alkene to synthesize 1,3-thiazines. Thioacetamide was used to create *N*-thioacyl imine heterodiene derivatives, and aromatic aldehyde was used to produce different isomers of 1,3-thiazine derivatives by a hetero Diels–Alder reaction with alkenes (Scheme 9.11).

A simple and useful route to the synthetically important six-membered N-, S-, and Se-heterocycles is provided by Muthusamy *et al.*,[19] which is an efficient method for synthesizing 3,4-dihydro-2*H*-1,3-thiazines/3,4-dihydro-2*H*-1,3-selenazines from propargylic alcohols and thiosemicarbazones/thioamides/selenamides (Scheme 9.12).

The synthesis of 5-acyl-4-hydroxy-2*H*-1,3-thiazine-2,6(3*H*)-dione was reported by Ivin *et al.*,[20] which involves the reaction between potassium thiocyanate and acid anhydride with malonic acid (Scheme 9.13).

Nematpour *et al.*[21] devised an ultrasound-assisted method for synthesizing highly functionalized benzo[1,3]thiazine *via* an intramolecular CH

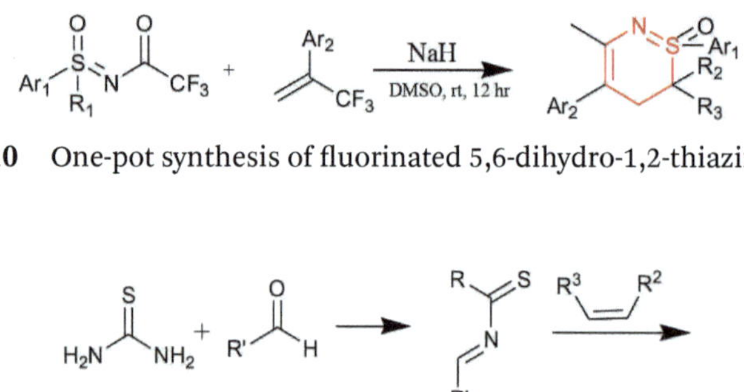

Scheme 9.10 One-pot synthesis of fluorinated 5,6-dihydro-1,2-thiazine-1-oxides.

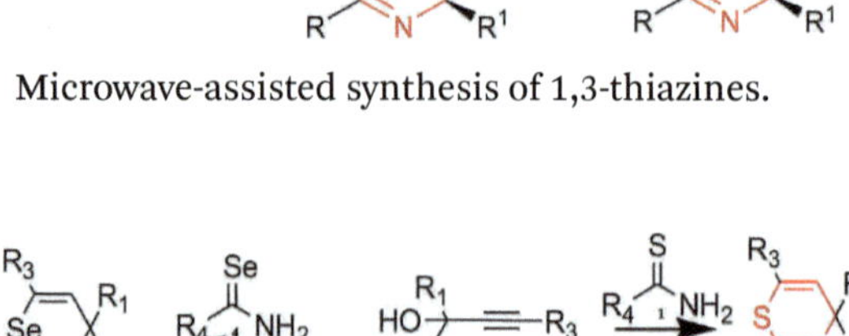

Scheme 9.11 Microwave-assisted synthesis of 1,3-thiazines.

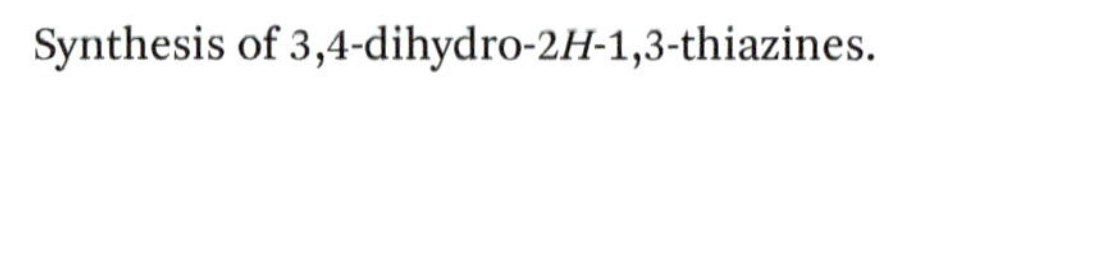

Scheme 9.12 Synthesis of 3,4-dihydro-2*H*-1,3-thiazines.

Scheme 9.13 Synthesis of 5-acyl-4-hydroxy-2*H*-1,3-thiazine-2,6(3*H*)-dione.

activation reaction catalysed by Cu from isocyanides and aniline-benzoyl(acetyl) isothiocyanate adduct (Scheme 9.14).

A series of 2-substituted guanidine-4-(2-amino-5-substituted phenyl)mercapto-6-phenyl-1,3-thiazines were synthesized by the reaction of 1-(2-amino-5-substituted phenyl) mercapto-3-(substituted) phenyl-2-propen-1-one, synthesized from 2-aminothiophenol, with substituted amidinothiocarbamides under reflux in basic medium for 4–5 hours and the synthesized derivatives were evaluated for anti-inflammatory, analgesic and ulcerogenic activity[22] (Scheme 9.15).

Biehl *et al.*[23] have reported the straightforward synthesis of 4*H*-naphtho[2,3-*e*] derivatives of 1,3-thiazines and 1,3-selenazines and naphtha[2′,3′:4,5] derivatives of selenolo[2,3-*b*]pyridines and thieno[2,3-*b*]pyridines from

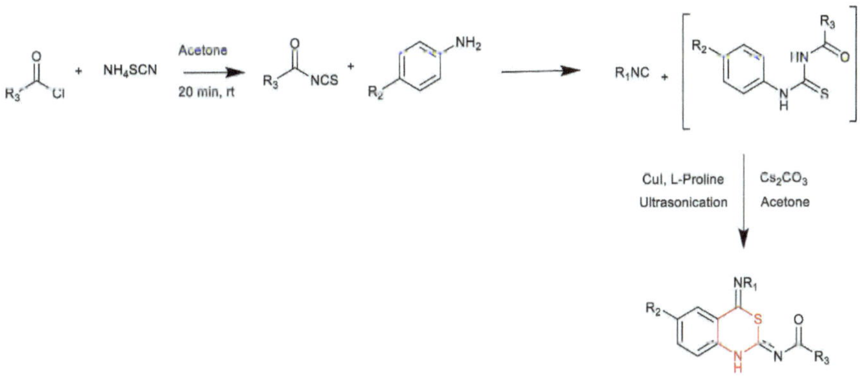

Scheme 9.14 Ultrasound-assisted synthesis of benzo[1,3]thiazine.

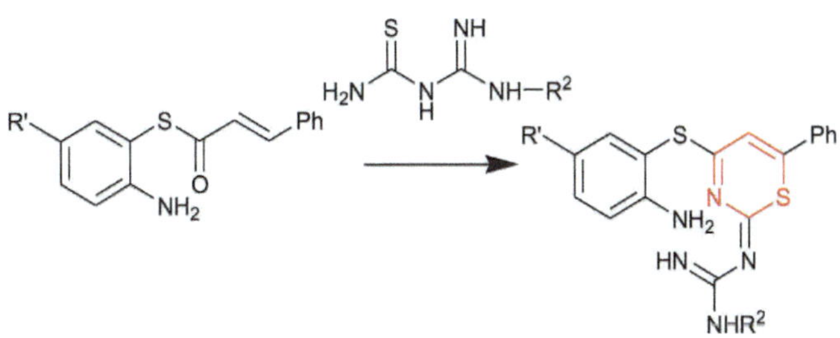

Scheme 9.15 Synthesis of 2-guanidine-4-(2-amino-5-substituted phenyl)1,3-thiazine.

2,3-didehydronaphthalene. By treating (phenyl)[3-(trimethylsilyl)-2-naphthyl]iodonium triflate, 1,3-hydride shift is produced. This is followed by additional aromatization using *N*1,*N*1-disubstituted-*N*2-(dimethylaminomethylidene) thiourea and NMe2 anion displacement (Scheme 9.16).

Thiazine-3,7-dicarbonitrile was synthesized by refluxing (methylthio)methylene malanonitrile with thiourea for 12 hours in the presence of anhydrous potassium carbonate, producing novel 2,6-dihydro-2,6-diimino-4,8-bis(methylthio) pyrimido[2,1-*b*]. Sambhaji and Shivraj[24] found that the incorporation of a 2-methylthio bunch, an initiated nitrogen, and an electron-pulling out cyano bunch builds the reactivity towards the nucleophile to yield subbed 1,3-thiazines (Scheme 9.17).

Fe(III)-montmorillonite was used as a catalyst to produce a series of 1,3-thiazine derivatives, as described by Vinoda *et al.*[25] *In silico* molecular-docking research and antibacterial activity screening were conducted on the synthesized compounds (Scheme 9.18).

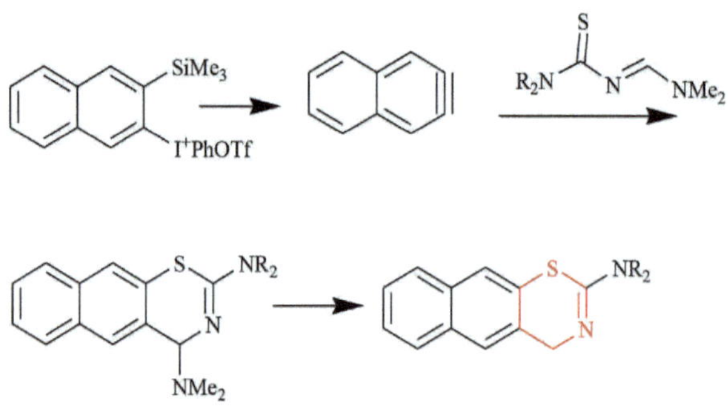

Scheme 9.16 Synthesis of 4*H*-naphtho[2,3-*e*] derivatives of 1,3-thiazines.

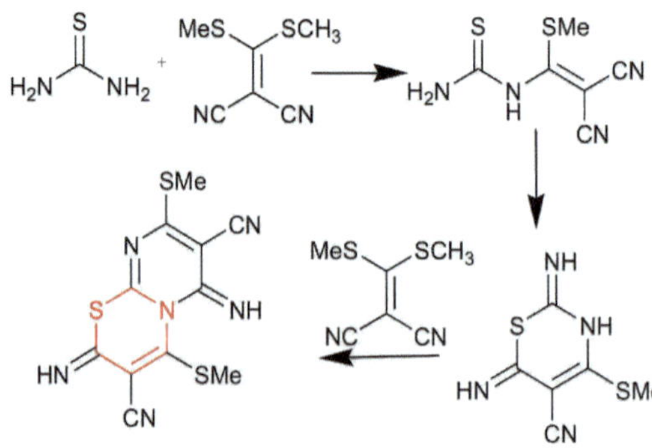

Scheme 9.17 Synthesis of thiazine-3,7-dicarbonitrile.

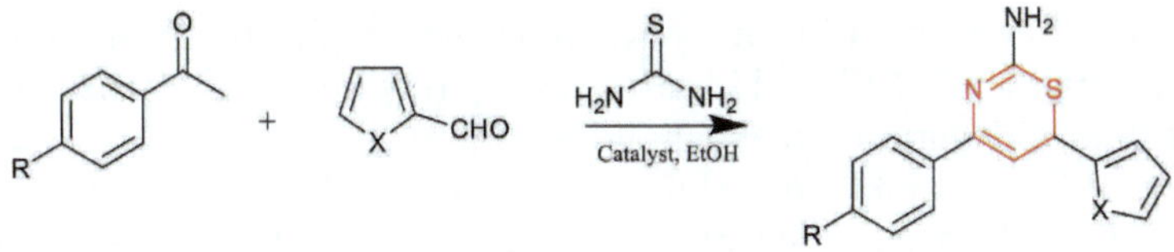

Scheme 9.18 Synthesis of 1,3-thiazine using Fe(III)-montmorillonite as a catalyst.

9.3 Biological Applications

9.3.1 Thiazine as Anticancer Agents

A series of substituted 2-amino-1,3-thiazine derivatives were designed, synthesized, and evaluated for antituberculosis and anticancer activities, as described by Prasad *et al.*[26] (Figure 9.3).

Kalirajan *et al.*[27] reported the design, synthesis, and characterization of novel 9-anilinoacridines substituted with thiazines, as well as their assessment of cytotoxic activity against Daltons lymphoma ascites (DLA) cells (Figure 9.4).

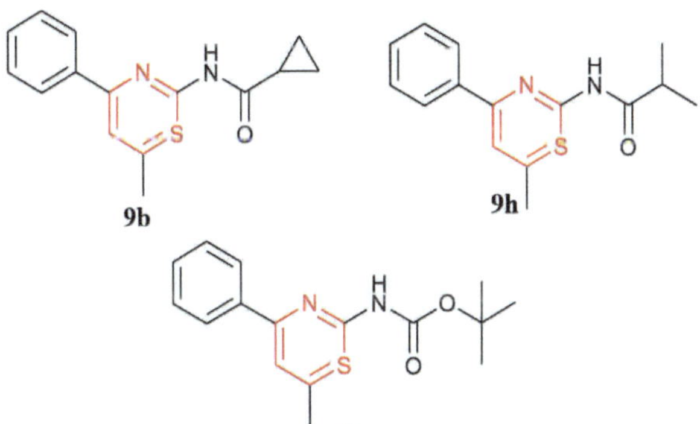

Figure 9.3 Synthesis and anticancer activity of 2-amino-1,3-thiazine derivatives.

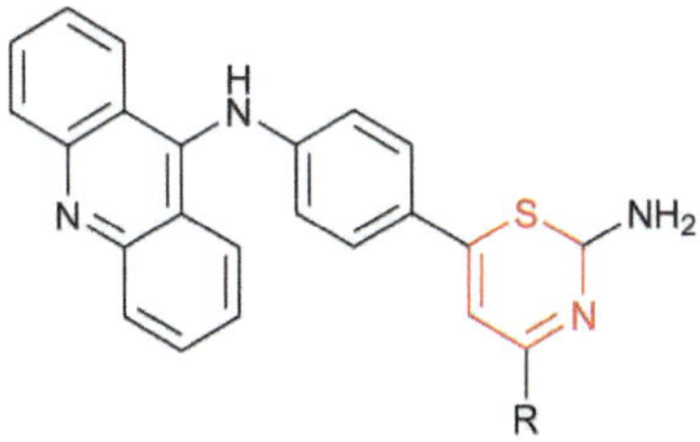

Figure 9.4 Synthesis and cytotoxic activity of 9-anilinoacridines with thiazines.

Ilinykh Elena and Ilinykh[28] have reported on the synthesis and anticancer activity of *S*-alkenyl derivatives of (trifluoromethyl)-4*H*-1,2,4-triazole-3-thiol with various alkenyl halides (Figure 9.5).

New 1*H*-pyrazolo[3,4-*b*]pyridino-1,3-thiazine derivatives have been designed, synthesized, molecularly modelled, and biologically evaluated by Eissa and El-Naggar[29] as possible anticancer drugs with antiproliferative activities against PC3, HCT-116, and HePG-2 cell lines (Figure 9.6).

Bedane and Singh[30] have described the synthesis of substituted 1,3-thiazines from isothiocyanates, which demonstrate anticancer activities in animals treated with chemical carcinogens and impede the metabolic activation of carcinogens (Figure 9.7).

The synthesis and anticancer characteristics of derivatives of thieno[2,3-*d*]pyrimidin-4-one have been reported by Bozorov *et al.*[31] (Figure 9.8).

El-Sayed *et al.*[32] reported on the novel synthesis, ring transformation, and anticancer activity of a new range of heterocyclic compounds, such as 1,3-thiazine, pyrimidine, and triazolo[1,5-*a*]pyrimidine derivatives.

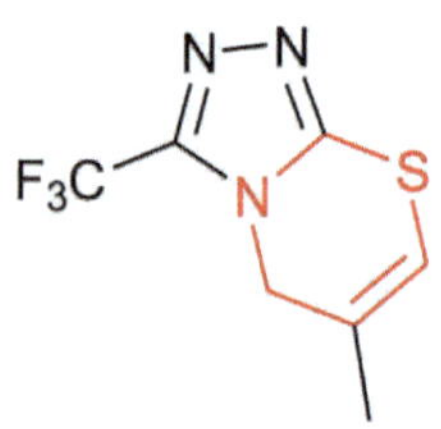

Figure 9.5 Synthesis and anticancer activity of 1,3-thiazine derivatives.

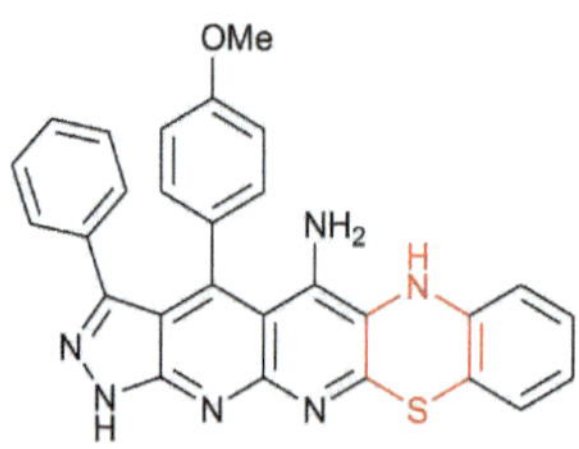

Figure 9.6 Synthesis and anticancer and antiproliferative activities of 1,3-thiazine derivatives.

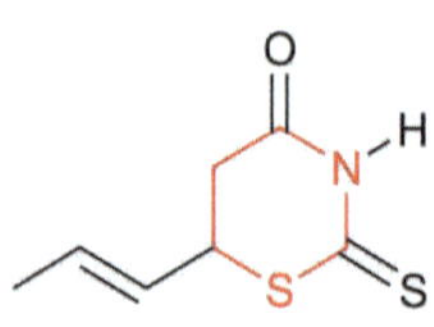

Figure 9.7 Synthesis and anticancer activity of substituted 1,3-thiazine derivatives.

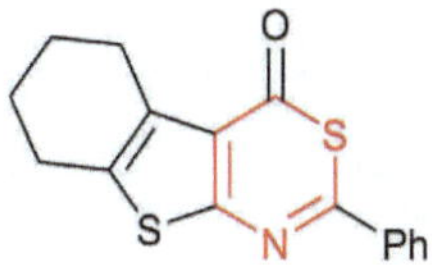

Figure 9.8 Synthesis and anticancer activity of thieno[2,3-*d*]pyrimidin-4-one-1,3-thiazine derivatives.

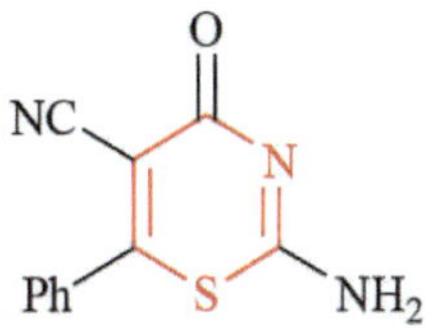

Figure 9.9 Synthesis and anticancer activity of substituted 1,3-thiazine derivatives.

Condensation of thiourea, benzaldehyde, and ethyl cyanoacetate produced aminothiazine. Compounds containing thiazine derivatives reacted with aniline to produce an amino pyrimidine derivative. Oxazine, the acid derivative, was produced by ring opening and recyclization of the *N*-Ph amino pyrimidine derivative during the process.

In the presence of hydroxyl amine hydrochloride, thiazine underwent a ring change, yielding pyrimidine derivatives. Triazolopyrimidines were produced by oxidizing thiazine with I_2/AcOH after thiosemicarbazide was heterocyclized. Thiazine is alkylated through a reaction with ethyl iodide, and when this alkylated thiazine reacts with hydrazine hydrate, it undergoes a ring transformation to produce pyrazole derivatives. The alkylated pyrimidine product was produced when thiazine and ethyl acetate refluxed in the presence of triethyl amine. Pyrimidine derivatives were produced by hydrazinolysis of thiazine with either hydrazine or phenylhydrazine. Triazolopyrimidines were produced by cyclizing the hydrazine–pyrimidine derivative with formaldehyde or carbon disulphide. Novel compounds that were synthesized were tested for their anticancer properties (Figure 9.9).

Using 6-substituted 9-fluoroquinobenzothiazines, Jelen *et al.*[33] produced novel tetracyclic azaphenothiazines, which demonstrated suppression of many blood cancer cell types in addition to anticancer, antiproliferative, and anti-inflammatory activities. The toxicity of 9-fluoroquinobenzothiazine was reduced by substitution at the thiazine nitrogen (Figure 9.10).

9.3.2 Thiazine as Antimicrobial Agents

The morpholine ring was presented by Thanusu *et al.*[34] in a progression of 4-(4-morpholinophenyl)-6-aryl-1,3-thiazin-2-amines, which had

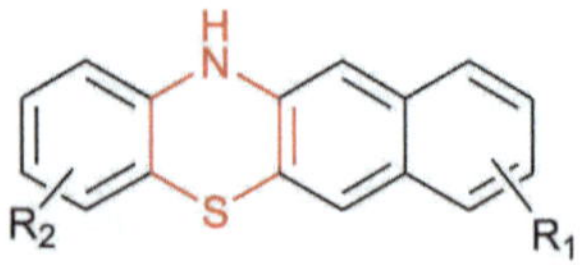

Figure 9.10 Synthesis and anticancer activity of tetracyclic azaphenothiazines.

antibacterial activity against *V. cholera* and different microorganisms as well as antifungal activity against various contagious species, including *M. gypseum* and *M. rhizopus* (Figure 9.11).

The synthesis of some novel bromo-substituted 1,3-thiazine derivatives *via* condensation of 2-hydroxy-3-bromo-5-chlorochalcones with thiourea, phenylthiourea, and diphenyl thiourea in ethanol in the presence of an alkaline solution was reported by Rathore *et al.*[35] The synthesized derivatives demonstrated antibacterial activity against some microbes, such as *B. subtilis*, *S. aureus*, *E. coli*, and *P. aeruginosa* (Figure 9.12).

The α,β-unsaturated aryl ketone was refluxed with phenylthiourea in the presence of aqueous KOH and ethanol to produce novel substituted 1,3-thiazines. Ganorkar *et al.*[36] reported that the antibacterial activity of these compounds was assessed against *Escherichia coli*, *Staphylococcus aureus*, *Bacillus subtilis*, and *Pseudomonas aeruginosa* (Figure 9.13).

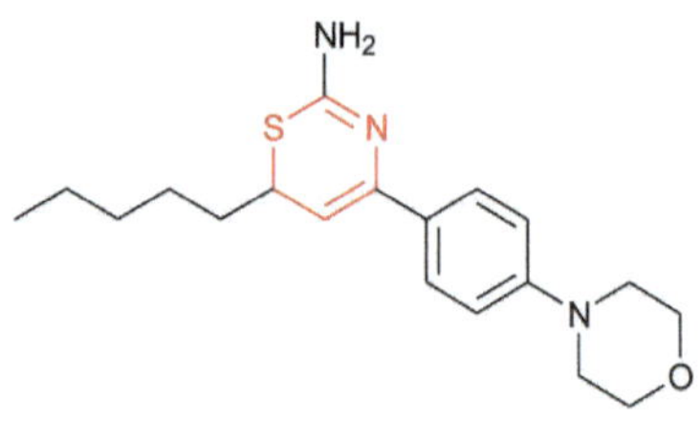

Figure 9.11 Synthesis and antibacterial activity of 4-(4-morpholinophenyl)-6-aryl-1,3-thiazin-2-amines.

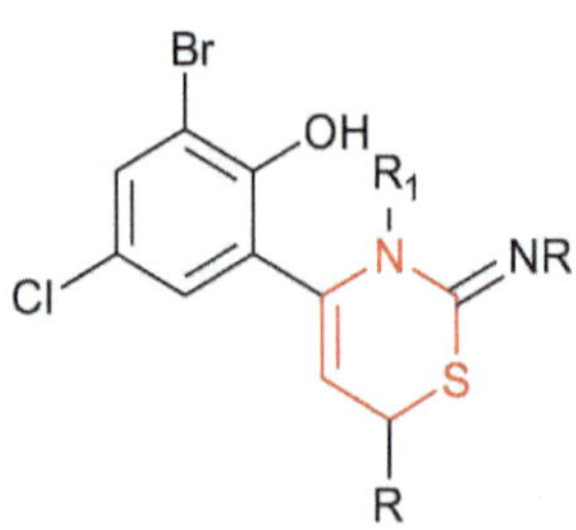

Figure 9.12 Synthesis and antibacterial activity of bromo-substituted 1,3-thiazines.

An array of 4-(2-bromonaphthalen-6-yl) compounds was synthesized, as reported by Prakash *et al.*[37] 1-(2-Bromonaphthalen-6-yl)-3-phenylprop-2-en-1-ones was reacted with thiourea in the presence of alkaline base KOH in alcohol as a solvent to produce 6-(substituted phenyl)-6*H*-1,3-thiazin-2-amines. Using bromo-substituted acetyl naphthalene, a pharmaceutically significant substrate, as a precursor, a series of 1-(2-bromonaphthalen-6-yl)-3-phenylprop-2-en-1-ones were prepared. The antibacterial activity of the produced compounds was assessed *in vitro* (Figure 9.14).

Didwagh *et al.*[38] produced a novel set of potentially potent 3-(4,6-diphenyl-6*H*-1,3-thiazin-2-yl)-2-(4-methoxyphenyl) thiazolidin-4-one derivatives and assessed their antibacterial activity. This was accomplished by cyclizing 4,6-diphenyl-6*H*-1,3-thiazin2-amine, thioglycolic acid, and aromatic aldehyde in polypropylene glycol by condensation. The antimicrobial activity was assessed *in vitro* against two pathogenic fungal strains, *Aspergillus niger* and *Candida albicans*, as well as sensitive organisms, such as Gram-positive and Gram-negative bacteria, *Escherichia coli*, and *Pseudomonas aeruginosa*, using the disc diffusion method (Figure 9.15).

Mitra *et al.*[39] synthesized a library of benzo[*b*][1,4]thiazine-1,1-dioxide derivatives under mild conditions using a peroxide-free reaction method, offering substantial yields and comprehensive functionality. The produced compounds were tested for antibacterial activity and underwent statistical

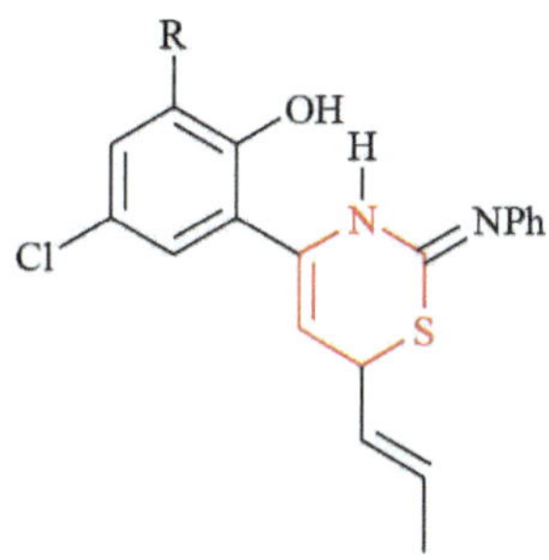

Figure 9.13 Synthesis and antibacterial activity of substituted 1,3-thiazine derivatives.

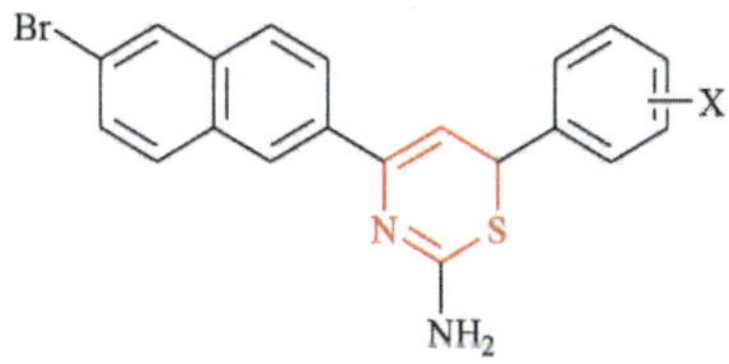

Figure 9.14 Synthesis and antibacterial activity of 6-(substituted phenyl)-6*H*-1,3-thiazin-2-amines.

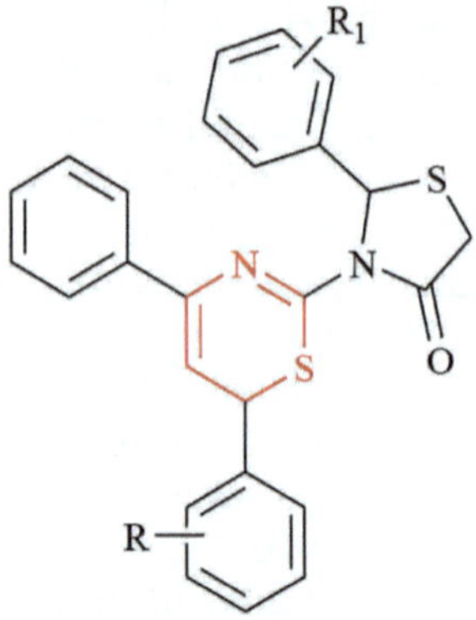

Figure 9.15 Synthesis and antibacterial activity of 3-(4,6-diphenyl-6*H*-1,3-thiazin-2-yl)-2-(4-methoxyphenyl) thiazolidin-4-one derivatives.

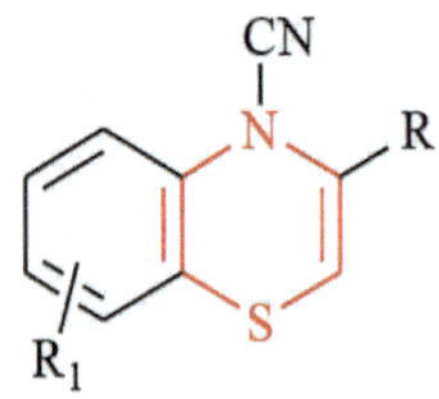

Figure 9.16 Synthesis and antibacterial activity of benzo[*b*][1,4]thiazine-1,1-dioxide.

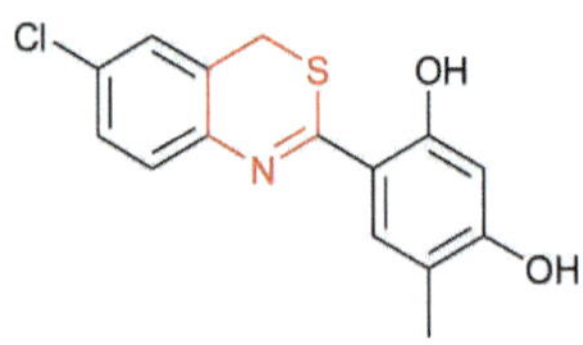

Figure 9.17 Synthesis and antibacterial activity of 1-benzothiazines.

analysis, minimum inhibitory concentration (MIC) analysis, and kinetics analysis of bacterial death (Figure 9.16).

Numerous compounds containing a 2,4-dihydroxyphenyl moiety have been synthesized and shown to have antibacterial activity by Matysiak *et al.*.[40] These compounds include benzo[4*H*-3] thiazoles, 1-benzothiazines, naphtho[2,3-*d*][1,3]thiazole-4,9-diones, 1,3-thiazolo[5,4-*b*]pyridines, and related compounds. These compounds are produced by reacting aryl-modified sulfinyl[bis(2,4-dihydroxyphenylmethanethione)] with suitable different reagents during the endo cyclization processes (Figure 9.17).

9.3.3 Thiazine as Antitubercular Agents

Using *o*-amino thiophenol, maleic anhydride, ethanol, and formaldehyde, Sindhu *et al.*[41] have reported the synthesis of several new sulpha-drug-substituted 1,4-thiazines, whereas 1,3-thiazine was produced from chalcones *via* the Claisen–Schmidt condensation process. The microplate Alamar Blue assay was used to screen the synthesized compounds for antimycobacterial activity (Figure 9.18).

The design, docking research, and synthesis of 1,3-benzothiazinone azide (BTZN3) have been described by Tiwari *et al.*[42] According to computational docking experiments, BTZ-N binds in a pocket that is nearly identical to that of BTZ043. Through our model biochemical reactivity experiments of BTZ043, biological investigations demonstrated that the azide analogue of 1,3-benzothiazinone has cell envelope enzyme fractions of *Mycobacterium smegmatis* for activation of anti-TB activity (Figure 9.19).

9.3.4 Thiazine as Anticonvulsant Agents

The synthesis and anticonvulsant and antimicrobial properties of a couple of new 1,3-thiazines have been accounted for by Bairam *et al.*[43] During the Claisen–Schmidt condensation, 4-*tert*-butyl cyclohexanone was reacted with various aromatic aldehydes (ArCHO) to produce the corresponding

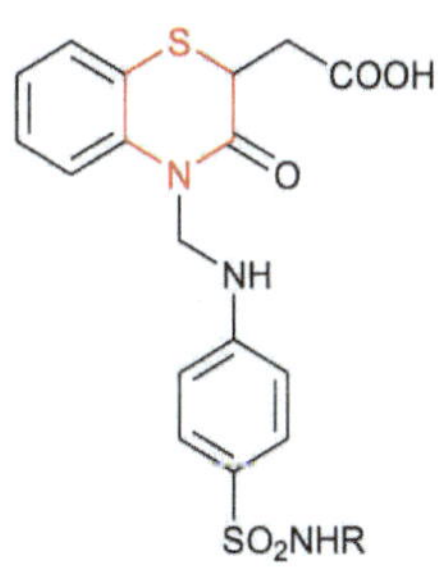

Figure 9.18 Synthesis and antimycobacterial activity of sulpha-drug-substituted 1,4-thiazines.

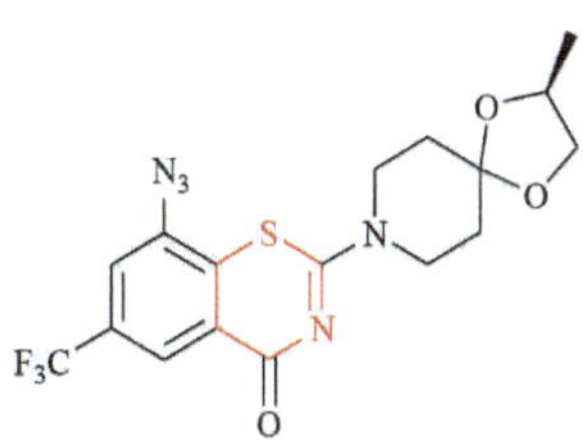

Figure 9.19 Synthesis and anti-TB activity of 1,3-benzothiazinone azide derivatives.

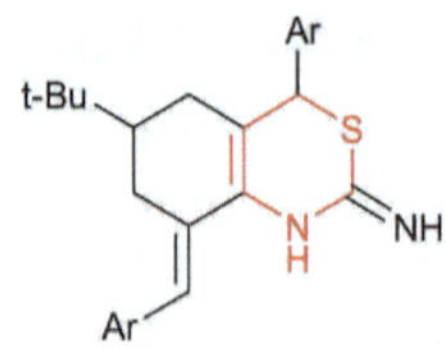

Figure 9.20 Synthesis and anticonvulsant activity of 1,3-thiazine derivatives.

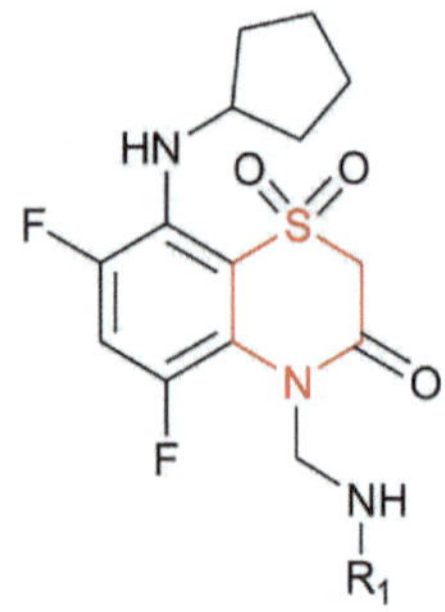

Figure 9.21 Synthesis and antihypertensive activity of thiazines.

chalcones. Furthermore, alkaline potassium hydroxide was used to catalyse the condensation of these compounds with thiourea in the presence of isopropanol, resulting in the formation of 4-aryl-8-arylidene-5,6-dihydro-2-imino-6-methyl-4*H*,7*H*-(1,3)-benzothiazines.

The compounds were assessed for their antimicrobial efficacy utilizing the cup plate agar dispersion strategy and for their antiepileptic action utilizing the pentylenetetrazole (PTZ)-initiated seizure model. Di-Ph hydantoin was utilized as the standard. The mixtures were assessed to forecast their molecular properties, and the Osiris program, Molsoft, Prototox, and ALOGPS 2.1 programming were utilized to decide the harmfulness, drug similarity, lipophilicity, and solvency boundaries. Using docking studies different receptor sites on protein structures have been predicted (Figure 9.20).

9.3.5 Thiazine as Antihypertensive Agents

Kondratev *et al.*[44] have reported the synthesis and assessed the antihypertensive activity of 4-aryl(heteroaryl)methyl-substituted-8-cyclopentylamino-5,7-difluoro-3,4-dihydro-2*H* benzo[1,4] thiazine-1,1-dioxides by treating primary amines with 1-(2-chloroethylsulfonyl)-2,3,5,6-tetrafluorobenzene in ethanol, acetone, or acetonitrile in the presence of triethylamine (Figure 9.21).

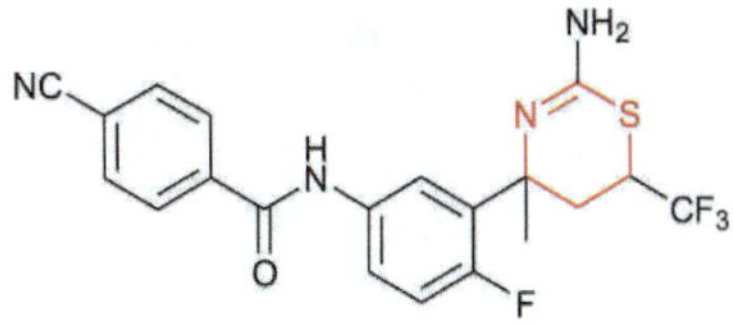

Figure 9.22 Synthesis and secretase inhibitory activity of thiazines.

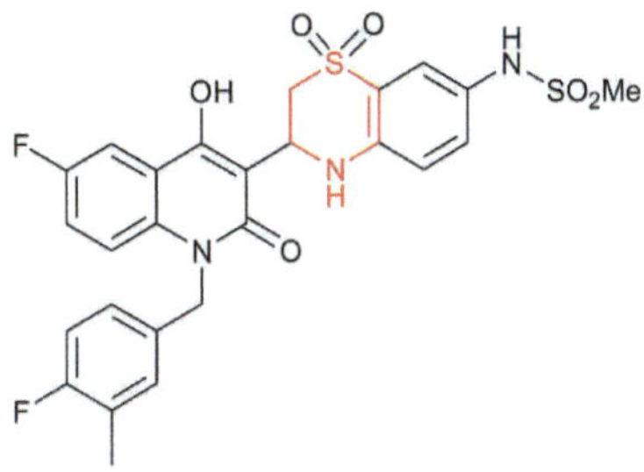

Figure 9.23 Non-nucleoside inhibitors of HCV polymerase activity of benzothiazine-substituted quinolinediones.

9.3.6 Thiazine as Miscellaneous Agents

The combination of 6-CF-substituted-2-amino-dihydro-1,3-thiazines was accounted for by Oguma *et al.*[45] This was accomplished through the cyclization of *N*-hydroxypropyl thiourea by *N*,*N*-diethylaminosulfur trifluoride (DAST). The synthesized 6-CF-substituted-1,3-thiazine derivatives exhibited secretase inhibitory activity (Figure 9.22).

Pharmacokinetic and toxicological studies of a novel series of benzothiazine-substituted quinolinediones have been described by de Vicente *et al.*,[46] and these compounds have been assessed as non-nucleoside inhibitors of HCV polymerase NS5B (Figure 9.23).

The synthesis of 2-aminodihydro[1,3]thiazines and their evaluation as beta secretase 2 (BACE-2) inhibitors in Alzheimer's disease, as well as their preparation and application in the management of diabetes, have been described by Beauchamp *et al.*[47] The derivatives were synthesized by amidating 5-chloropyridine-2-carbonyl chloride with 1-(5-amino-2-fluoro)ethenone, producing 5-chloropyridine-2-carboxylic acid (3-acetyl-4-fluorophenyl) amide, which was then reacted with vinyl magnesium chloride to form 5-chloropyridine-2-carboxylic acid [4-fluoro-3-(1-hydroxy-1-methylallyl)phenyl]amide. This was then added to thiourea to yield the corresponding 3-carbamididoylsulfanyl-1-methylpropenyl derivative, which upon heterocyclization produced the corresponding thiazine derivatives (Figure 9.24).

Novel seven-membered imino sugars coupled with thiazolidin-4-one and benzthiazinan-4-one have been synthesized by Staudinger/aza-Wittig/

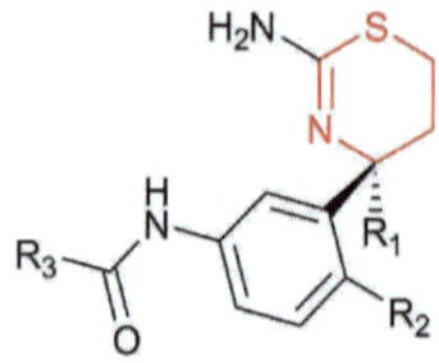

Figure 9.24 Synthesis of 2-aminodihydro[1,3]thiazines and their evaluation as BACE-2 inhibitors.

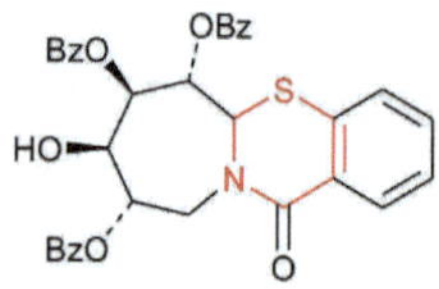

Figure 9.25 Synthesis and anti-HIV activity of benzthiazinan-4-one.

cyclization process under microwave irradiation, as reported by Hou *et al.*[48] The ability of the synthesized chemicals to inhibit HIV reverse transcriptase (RT) was examined (Figure 9.25).

9.4 Conclusions

Literature survey reveals that thiazine-containing derivatives have various biological activities, and hence these derivatives can be considered as good lead compounds for drug design and in medicinal chemistry.

Abbreviations

Ac	Acetyl
Ac_2O	Acetic anhydride
AcOH	Acetic acid
Ar	Aryl
$CDCl_3$	Deuterochloroform
$CHCl_3$	Chloroform
CNS	Central nervous system
Compd	Compound
conc.	Concentrated
CS_2	Carbon disulphide
DCM	Dichloromethane
dil.	Dilute
DMF	Dimethyl formamide
DMSO	Dimethyl sulphoxide

Et	Ethyl
EtOH	Ethanol
g	Grams
h	Hour
H_2SO_4	Sulphuric acid
HCl	Hydrochloric acid
HIV	Human immunodeficiency virus
Hz	Hertz
I_2	Iodine
K_2CO_3	Potassium carbonate
KBr	Potassium bromide
KOH	Potassium hydroxide
M	Molar
Me	Methyl
Mg	Milligrams
MHz	Megahertz
Min	Minutes
mL	Millilitres
mol	Moles
MP	Melting point
MS	Mass spectrum
MWI	Microwave irradiation
NaOAc	Sodium acetate
NaOH	Sodium hydroxide
NMR	Nuclear magnetic resonance
Ph	Phenyl
$POCl_3$	Phosphorous oxychloride
RBF	Round bottom flask
RT	Room temperature
SAR	Structure–activity relationship
THF	Tetrahydrofuran
%	Percentage
))))	Ultrasound
° C	Degree celsius
+ve	Positive;
−ve	Negative
µg	Microgram
µL	Microliter

References

1. Y. Zhao, Y. Wu, J. Jia, D. Zhang and C. Ma, One-Pot Synthesis of Benzo[1,4]thiazin-3(4H)-ones and a Theoretical Study of the S-N Type Smiles Rearrangement Mechanism, *J. Org. Chem.*, 2012, **77**(19), 8501.

2. Z. Hossaini, M. Nematpour and I. Yavari, Ph_3P-mediated one-pot synthesis of functionalized 3,4-dihydro-2H-1,3-thiazines from N,N′-dialkylthioureas and activated acetylenes in water, *Monatsh. Chem.*, 2010, **141**, 229.
3. S. L. Badshah and A. Naeem, Bioactive Thiazine and Benzothiazine Derivatives: Green Synthesis Methods and Their Medicinal Importance, *Molecules*, 2016, **21**, 1054.
4. O. M. I. Adly, H. F. El-Shafiy and M. Shebl, Synthesis, spectroscopic studies, DFT calculations, antimicrobial and antitumor activity of tridentate NNO Schiff base metal complexes based on 5-acetyl-4-hydroxy-2*H*-1,3-thiazine-2,6(3*H*)-dione, *J. Mol. Struct.*, 2019, **1196**, 805.
5. G. Tadano, K. Komano, S. Yoshida, S. Suzuki, K. Nakahara, K. Fuchino, K. Fujimoto, E. Matsuoka, T. Yamamoto, N. Asada, H. Ito, G. Sakaguchi, N. Kanegawa, Y. Kido, S. Ando, T. Fukushima, A. Teisman, V. Urmaliya, D. Dhuyvetter, H. Borghys, A. V. D. Bergh, N. Austin, H. J. M. Gijsen, Y. Yamano, Y. Iso and and K. Kusakabe, Discovery of an Extremely Potent Thiazine-Based β-Secretase Inhibitor with Reduced Cardiovascular and Liver Toxicity at a Low Projected Human Dose, *J. Med. Chem.*, 2019, **62**, 9331.
6. M. Asif, M. Imran and A. Abida, Antimicrobial Activities of Various Thiazine Based Heterocyclic Compounds: A Mini-Review, *Rev. Inorg. Chem.*, 2022, **19**, 166.
7. M. Chandrakanta, T. Shachi and S. Prem Prakash, Comparison of dye (oxazine and thiazine) materials as a photosensitizer for use in photo-galvanic cells based on molecular interaction with sodium dodecyl sulphate by spectral study, *J. Saudi Chem. Soc.*, 2019, **23**, 83.
8. N. Batooie, M. M. Khodaei, K. Bahrami, S. S. Miraghaee, N. Hosseinzadeh and S. Sajadimajd, One-pot synthesis of new benzo[4,5]imidazo[2,1-b]pyrimido[4,5-d][1,3] thiazine-2,4(3H)-dione and benzo[4,5]imidazo[2,1-b][1,3]thiazin-4-one derivatives as new anti-cancer components, *J. Mol. Struct.*, 2023, **1271**, 134037.
9. N. Edayadulla and P. Ramesh, Synthesis of 2,6-dicarbethoxy-3,5-diaryltetrahydro-1,4-thiazine-1,1-dioxide derivatives as potent anticonvulsant agents, *Eur. J. Med. Chem.*, 2015, **106**, 44.
10. B. Tozkoparan, G. Aktay and E. Yeşilada, Synthesis of some 1,2,4-triazolo[3,2-*b*]-1,3-thiazine-7-ones with potential analgesic and antiinflammatory activities, *Il Farmaco*, 2002, **57**, 145.
11. L. D. S. Yadav, V. K. Rai and B. S. Yadav, The first ionic liquid-promoted one-pot diastereoselective synthesis of 2,5-diamino-/2-amino-5-mercapto-1,3-thiazin-4-ones using masked amino/mercapto acids, *Tetrahedron*, 2009, **65**, 1306.
12. T. Utsukihara, M. Matsushita, E. Miyamoto, H. Murukami and C. A. Horiuchi, *Indian J. Chem.*, 2019, **58 B**, 1037.

13. G. Canudo-Barreras, D. Salvador, R. P. Herrera and M. C. Gimeno, Gold-Catalyzed 1,3-Thiazine Formation and Uncommon Tautomer Isolation, *J. Org. Chem.*, 2022, **87**, 10747.
14. Y. Wei, S. Sun and J. Xu, Direct and safe one-pot synthesis of functionalized dimethyl 1-aryl 1,9a-dihydropyrido[2,1-*c*][1,4]thiazine-1-phosphonates, *Tetrahedron Lett.*, 2023, **120**, 154444.
15. T. E. Malah, E. A. Ishak, H. F. Nour, R. M. Shaker, M. M. Ali, A. E. Mahmoud and S. M. Soliman, Synthesis and *In Vitro* Anticancer Evaluation of Symmetrically Bridged 1,3-thiazine Derivatives, *J. Heterocycl. Chem.*, 2018, **55**, 1746.
16. H. A. Al-Difar and M. J. Elarfi, Synthesis of some heterocyclic compounds derived from chalcones, *Sci. Revs. Chem. Commun.*, 2012, **2**(2), 103.
17. X. Wang, K. Rissanen and C. Bolm, A One-Pot Domino Reaction Providing Fluorinated 5,6-Dihydro-1,2-thiazine 1-Oxides from Sulfoximines and 1-Trifluoromethylstyrenes, *Org. Lett.*, 2023, **25**, 1569.
18. R. Legay, J. F. Lohier, F. Peudru and V. Reboul, Facile access to γ-aminothiols from 1,3-thiazines via a microwave assisted three-component reaction, *Tetrahedron*, 2012, **68**, 9016.
19. S. Muthusamy, M. Malarvizhi and E. Suresh, BF3 · OEt2 Catalyzed Synthesis of 1,3-Thiazines/-Selenazines, *Asian J. Org. Chem.*, 2021, **10**, 170.
20. B. A. Ivin, A. V. Moskvin and V. N. Yuskovets, New Method of Synthesis of 5-Acyl 1,3-thiazines, *Russ. J. Gen. Chem.*, 2004, **74**(2), 312.
21. M. Nematpour, E. Rezaee, M. Jahani and S. A. Tabatabai, Ultrasound-assisted synthesis of highly functionalized benzo [1, 3] thiazine via Cu-catalyzed intramolecular CH activation reaction from isocyanides, aniline-benzoyl (acetyl) isothiocyanate adduct, *Ultrason. Sonochem.*, 2019, **50**, 1.
22. V. V. Dabholkar and S. D. Parab, Synthesis of chalcones, 1,3-thiazines and 1,3-pyrimidines derivatives and their biological evaluation for anti-inflammatory, analgesic and ulcerogenic activity, *Heterocycl. Lett.*, 2011, **1**(2), 176.
23. Ed. Biehl and R. Sathunuru, Facile synthesis of 4H-naphtho [2,3-e] derivatives of 1,3- thiazines and 1,3-selenazines and naphtha [2′,3′:4,5] derivatives of selenolo[2,3-b]pyridines and thieno[2,3-b]pyridines via 2,3-didehydronaphthalene, *ARKIVOC*, 2004, **xiv**, 51.
24. P. V. Sambhajiand B. S. Shivraj, Simple and efficient synthesis of novel fused bicyclic heterocycles pyrimido-thiazine and their derivatives, *Org. Chem.: Curr. Res.*, 2012, **1**(5), 1.
25. B. M. Vinoda, Y. D. Bodke, M. Vinuth, R. Kenchappa, S. Telkar, S. Patil and I. Inventi, Synthesis of some novel 1,3- thiazine derivatives using Fe(III)-montmorillonite catalysts as potent antimicrobial agents, *Med. Chem.*, 2016, **3**, 55.

26. T. N. Prasad, K. R. Eeda, V. B. Gudise, S. F. Basha and S. Anwar, Design, synthesis and biological evaluation of substituted 2-amino-1,3-thiazine derivatives as antituberculosis and anti-cancer agents, *Synth. Commun.*, 2019, **49**, 1277.
27. R. Kalirajan, K. Gaurav, A. Pandiselvi, B. Gowramma and S. Sankar, Novel Thiazine Substituted 9-Anilinoacridines: Synthesis, Antitumour Activity and Structure Activity Relationships, *Anti-Cancer Agents Med. Chem.*, 2019, **19**(11), 1350.
28. E. S. Ilinykh Elenaand S. Ilinykh, Synthesis of novel fluorine- and iodine-containin [1,2,4]triazolo [3,4-b][1,3]thiazines based 3-(alkenylthio)-5- (trifluoromethyl)-4H-1,2,4-triazole-3-thiols, *J. Fluorine Chem.*, 2013, **149**, 24.
29. I. H. Eissa and A. M. El-Naggar, *Bioorg. Chem.*, 2016, **67**, 43.
30. K. G. Bedane and G. S. Singh, Reactivity and diverse synthetic applications of acyl isothiocyanates, *ARKIVOC*, 2015, **6**, 206.
31. K. Bozorov, J. Zhao, B. Elmuradov, A. Pataer and H. Aisa, Recent developments regarding the use of thieno[2,3-d]pyrimidin-4-one derivatives in medicinal chemistry, with a focus on their synthesis and anticancer properties, *Eur. J. Med. Chem.*, 2015, **102**, 552.
32. H. A. El-Sayed, M. M. El-Hashash and A. E. Ahmed, Novel synthesis, ring transformation and anticancer activity of 1,3-thiazine, pyrimidine and triazolo[1,5-a]pyrimidine derivatives, *Bull. Chem. Soc. Eth.*, 2018, **32**(3), 513.
33. M. Jelen, K. Pluta, M. Zimecki, B. Morak-Młodawska, J. Artym and M. Kocieba, 6-Substituted 9-fluoroquino[3,2-b]benzo[1,4]thiazines display strong antiproliferative and antitumor properties, *Eur. J. Med. Chem.*, 2015, **89**, 411.
34. J. Thanusu, V. Kanagarajan and M. Gopalakrishnan, Synthesis, spectral characterization, and in vitro antibacterial and antifungal activities of novel 1,3-thiazine-2-amines comprising morpholine nucleus, *J. Enzyme Inhib. Med. Chem.*, 2010, **25**(6), 756.
35. M. M. Rathore, V. V. Parhate and P. R. Rajput, Synthesis and antimicrobial activities of some bromo-substituted-1,3-thiazines, *Int. J. Res. Pharm. Biomed. Sci.*, 2013, **4**(1), 59.
36. R. S. Ganorkar, R. P. Ganorkar and V. V. Parhate, Synthesis, characterisation and antibacterial activities of some new bromo/nitro 1,3-thiazines, *Rasayan J. Chem.*, 2013, **6**(1), 65.
37. N. Prakash, S. Sivagami and N. Ingarsal, A novel bromo naphthyl based 2-amino-1,3-thiazines: synthesis, characterization with in vitro antimicrobial screening, *Res. J. Chem. Sci.*, 2015, **5**(7), 8.
38. S. S. Didwagh, P. B. Piste, A. S. Burungale and A. M. Nalawade, Synthesis and antimicrobial evaluation of novel 3-(4,6-diphenyl-6H-1,3-thiazin-2-yl)-2-(4- methoxyphenyl) thiazolidin-4-one derivatives, *J. Appl. Pharm. Sci.*, 2013, **3**(11), 122.

39. S. Mitra, S. Mukherjee and A. Hajra, Peroxide-free synthesis of benzo[b][1,4]thiazine 1,1-dioxides and their antimicrobial study, *RSC Adv.*, 2016, **6**(1), 201.
40. J. Matysiak, R. Los, A. Malm, M. M. Karpinska, U. Głaszcz, B. Rajtar, M. Polz-Dacewicz, M. Trojanowska-Wesolowska and A. Niewiadomy, Synthesis and antibacterial activity of novel fused 1,3-thiazoles and 1,3-thiazines incorporating a 2,4-dihydroxyphenyl residue, *Arch. Pharm.*, 2012, **345**(4), 302.
41. T. J. Sindhu, M. Chandran, P. David, A. R. Bhat and K. Krishnakumar, Comparitive antitubercular activity of sulfadrug substituted 1,4-thiazines and 1,3- thiazines, *Int. J. Pharm. Res. Scholars.*, 2014, **3**(1), 24.
42. R. Tiwari, P. A. Miller, L. R. Chiarelli, G. Mori, M. Sarkan, I. Centarova, S. Cho, K. Mikusova, S. G. Franzblau and A. G. Oliver, Design, Syntheses, and Anti-TB Activity of 1,3-Benzothiazinone Azide and Click Chemistry Products Inspired by BTZ043, *ACS Med. Chem. Lett.*, 2016, **7**(3), 266.
43. R. Bairam, S. M. Muppavarapu and S. Sreekanth, Synthesis, characterization, biological evaluation and docking of some novel substituted 1, 3- thiazine derivatives, *Int. J. Pharm. Pharm. Sci.*, 2017, **9**(3), 233.
44. A. S. Kondratev, E. V. Tretyakov, A. O. Bryzgalov, T. G. Tolstikova and D. S. Baev, 4-Aryl(heteroaryl)methyl-substituted 8-cyclopentylamine-5,7-difluoro-3,4-dihydro-2H benzo[1,4]thiazine-1,1-dioxides with hypertensive action, *Russian Federation*, 2017, RU2627499 C1.
45. T. Oguma, K. Anan, S. Suzuki, S. Hisakawa, A. Takada, M. Ogawa and K. Kusakabe, Synthesis of a 6-CF-Substituted 2-Amino-dihydro-1,3-thiazine β-Secretase Inhibitor by N,N Diethylaminosulfur Trifluoride-Mediated Chemoselective Cyclization, *J. Org. Chem.*, 2019, **84**(8), 4893.
46. J. de Vicente, R. T. Hendricks, D. B. Smith, J. B. Fell, J. Fischer, S. R. Spencer, P. J. Stengel, P. Mohr, J. E. Robinson and J. F. Blake, Non-nucleoside inhibitors of HCV polymerase NS5B. Part 3: Synthesis and optimization studies of benzothiazine-substituted tetramic acids, *Bioorg. Med. Chem. Lett.*, 2009, **19**(13), 3642.
47. J. Beauchamp, A. Benardeau, H. Hilpert, C. Migliorini, W. Riboulet and H. Wang, 2-Aminodihydro[1,3]thiazines as BACE 2 inhibitors and their preparation and use in the treatment of diabetes, *World Intellectual Property Organization*, WO2011029803 A1, 2011, **03**, 17.
48. Y. Hou, S. Xing, J. Shao, Z. Yin, L. Hao, T. Yang, H. Zhang, M. Zhu, H. Chen and X. Li, Synthesis of seven-membered iminosugars fused thiazolidin-4-one and benzthiazinan-4-one and their HIV-RT inhibitory activity, *Carbohydr. Res.*, 2016, **429**, 105.

CHAPTER 10

Synthesis, Properties, and Biological Applications of Benzothiazepines

PUJA KUMARI, YASHI DWIVEDI AND SANDEEP CHANDRASHEKHARAPPA*

Department of Medicinal Chemistry, National Institute of Pharmaceutical Education and Research Raebareli (NIPER-R), Transit Campus, Bijnor-Sisendi Road, Sarojini Nagar, Near CRPF Base Camp, Lucknow, Uttar Pradesh, 226002, India
*E-mail: c.sandeep@niperraebareli.edu.in; c.sandeep@niperrbl.ac.in

10.1 Introduction

Medicinal chemistry involves the identification, development, interpretation, and discovery of molecular mechanisms underlying the actions of physiologically active molecules. It involves various steps starting from designing the molecule to its final preparation. An important part of this laborious drug development process is the organic synthesis of novel molecules. In particular, the preparation of favored heterocyclic scaffolds is a rapidly expanding field of medicinal chemistry. A "privileged structure" can be demonstrated as a strong and specific ligand for a variety of distinct biological targets due to the flexible binding extensions in its chemical framework.

S-Heterocycles: Retrospect, Prospects, and Biological Applications
Edited by Shrikaant Kulkarni, Hemantkumar Akolkar, Vijay M. Khedkar and A. K. Haghi

Published by the Royal Society of Chemistry, www.rsc.org

Heterocycles containing nitrogen[1–12] and sulfur[13–16] atoms are significant structural motifs found in several compounds with medicinal significance and bioactive natural items.[17] Compounds acquired with imidazole, benzimidazole, and triazole scaffolds have been extensively studied because of their pharmacological applications.[18] Benzothiazepines can be considered one of such scaffolds, which have also shown several biological activities in the central nervous system (CNS) and other therapeutic activities. They have earned the title of "drug prejudice nucleus" because of their versatility in this field.[19] Benzothiazepines are analogs of thiazepines containing a seven-membered ring. Benzothiazepines can be classified into several classes, including 1,2-, 1,3-, 1,4-, 1,5-, and 4,1-benzothiazepines, depending on the location of their heteroatoms—sulfur and nitrogen—on the thiazepine ring. Furthermore, when the thiazepine ring lacks one or both double bonds, benzothiazepines (**i**) can also exist as dihydrobenzothiazepines (**ii**) or tetrahydrobenzothiazepines (**iii**) (see Figure 10.1).[20]

It is interesting to note that a few of these significant seven-membered cores exhibit three annulated rings rather than just two. The seven-membered thiazepine typically switches between an aromatic or a heteroaromatic ring and a benzene ring. Further ring addition on the bicyclic benzothiazepine core helps prevent inversion of the seven-membered ring, enhancing the thermodynamic performance of the scaffold. Benzothiazepine exhibits bicyclic and tricyclic structures, which can exist in a variety of isomeric forms. Ten potential structural benzothiazepine isomers, comprising six dibenzothiazepines,[19] are shown in Figure 10.2.

The journey of benzothiazepines in the therapeutic world began with the introduction of the antidepressant "thiazesim" into the pharmaceutical industry, followed by the cardiovascular medications of this family, such as diltiazem, clentiazem, and siratiazem. Numerous medications, including quetiapine fumarate and thiazesim for the treatment of CNS disorders,

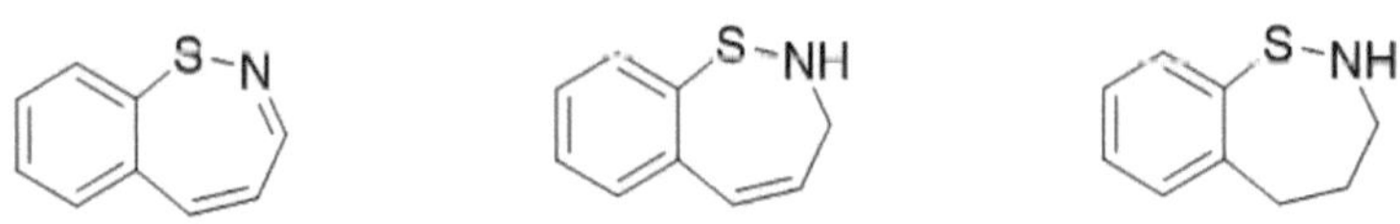

i ii iii

Figure 10.1 Different forms of the benzothiazepine ring.

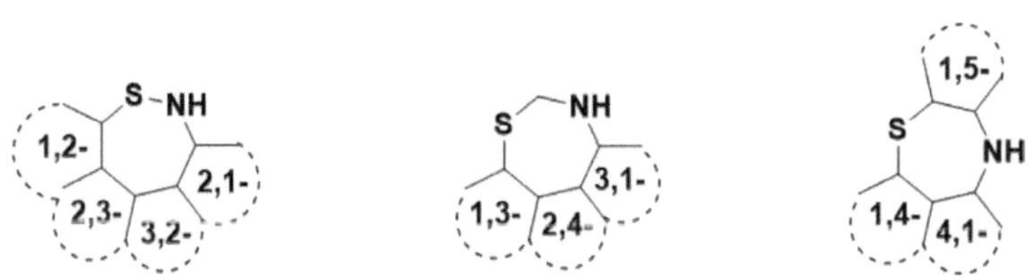

Figure 10.2 Structural isomers of benzothiazepines.

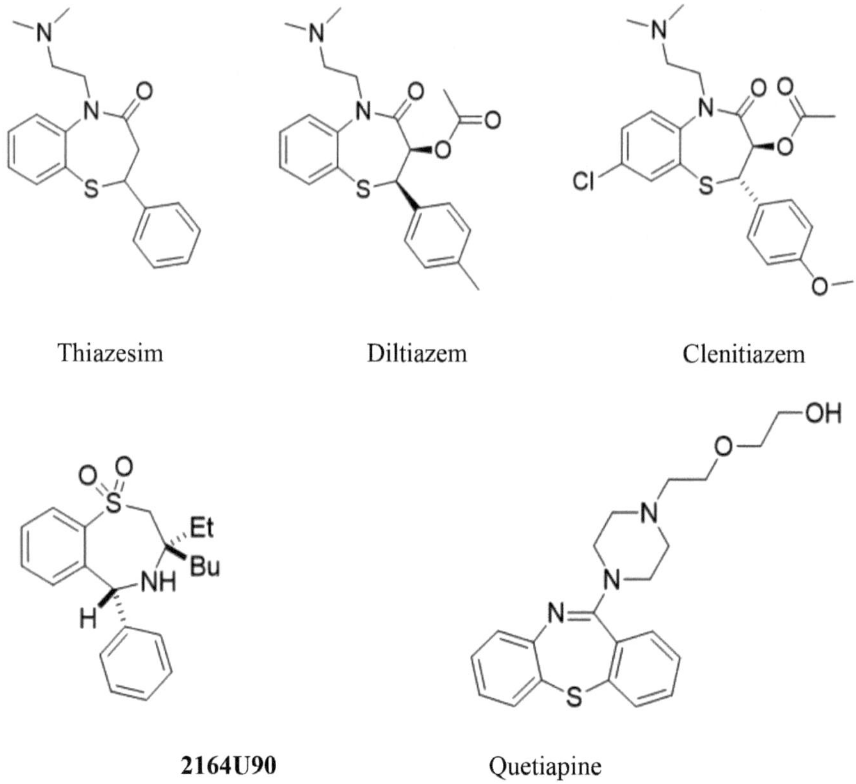

Figure 10.3 Drugs with the benzothiazepine scaffold.

2164U90 as an inhibitor of the bile acid active transport system, and **JMV1645** as a bradykinin receptor antagonist, were developed as a result of further substitution advancements in the benzothiazepine core.[19] They have also been claimed to be renowned cardio and psychotherapeutic moieties (Figure 10.3).

They act as positive allosteric regulators by enhancing GABA-mediated Cl^-channel opening in the neuronal cell membranes *via* binding with gamma-aminobutyric acid (GABA) receptors. This process causes a greater influx of Cl^- ions into the cells, which ultimately leads to the induction of sedative and hypnotic properties.[21] Because of the medicinal properties of benzothiazepine and its derivatives, they have the potential to be used as vasodilators, anticonvulsants, anti-inflammatory agents, antiviral agents, antimalarial agents, CNS depressants, antiplatelet aggregators, anticholinesterase inhibitors, anticancer agents, and Ca^{2+} channel antagonists. In the field of agriculture, they can also be used as antiviral agents and corrosion inhibitors.[22]

Drug-like compounds that include 1,2-thiazepines are a class of preferred skeletons. Among them, the main antidepressants have the fundamental structure of benzo[*f*][1,2]thiazepine dioxide, which demonstrates outstanding biological activity. Tianeptine is one of the well-known antidepressant medications. Benzo[*f*][1,2]thiazepine dioxide-derivatized molecules are explored as farnesyltransferase inhibitors, calcium-sensing receptor agonists, and proteases against the human immunodeficiency virus.[23]

1,3-Thiazepines are known for their cyclic thiourea derivatives with a seven-membered ring. These molecules belong to a significant group that has biological functions such as nitric oxide synthase inhibitors.[24] Omapatrilat, a medication currently undergoing phase IV clinical trials, contains the 1,3-thiazepine core. It reduces blood pressure by blocking the angiotensin-converting enzyme (ACE),[25,26] which narrows blood vessels. The additional benefit of the drug is the suppression of the neutral endopeptidase (NEP) enzyme, which relaxes blood arteries, thereby acting as a vasodilator. Similarly, 1,4-benzothiazepine has been found to be used in various biological approaches such as in the case of neurodegeneration due to its cytoprotective role along with its calcium channel blockade activity. Likewise, this moiety has been used to prepare drugs that exhibit inhibitory activity against the HIV-integrase protein,[27] as well as other targets related to neuropsychiatric and endocrine disorders.

The majority of drugs belong to the 1,5-benzothiazepine class. Currently, some of the most prescribed medications for the treatment of cardiovascular diseases include 1,5-benzothiazepines, such as clentiazem, thiazesim, and diltiazem.

Such adaptability makes benzothiazepines an area of interest in the context of design, discovery, and synthetic preparation of new molecules *via* new methodologies to bring their qualities more in light.

10.2 Synthetic Approaches for Benzothiazepine

This book chapter covers the synthesis of 1,4-benzothiazepines compared to other derivatives due to their less explored biological aspects. The synthesis of benzothiazepine derivatives played a significant role in their therapeutic properties, such as anti-HIV, anti-tuberculosis, anti-diabetic, and anti-microbial activities. For example, JTV5.19 (**10.2.2**)[28,29] and S107 (**10.2.1**)[30] have been found to have antiarrhythmic and cardioprotective effects. The K201 (**10.2.3**)[31] molecule shows the inhibition of annexin V-dependent Ca^{2+} movement in large unilamellar vesicles. Molecule **10.2.4**[20] shows antidiabetic properties, molecule **10.2.5**[29] acts as an apical sodium-dependent bile acid transporter (ASBT), and molecule **10.2.6**[32] depicts *in vitro* ACE inhibitor activity (see Figure 10.4).

A diverse range of 1,4-benzothiazepines have been synthesized by multistep synthesis, heterogeneous catalysis, microwave-aided approach, and ultrasonic synthesis.[20] Szabó *et al.* described the multistep synthesis. The reaction of sodium 3,4-dimethoxythiophenolate **10.2.7** with

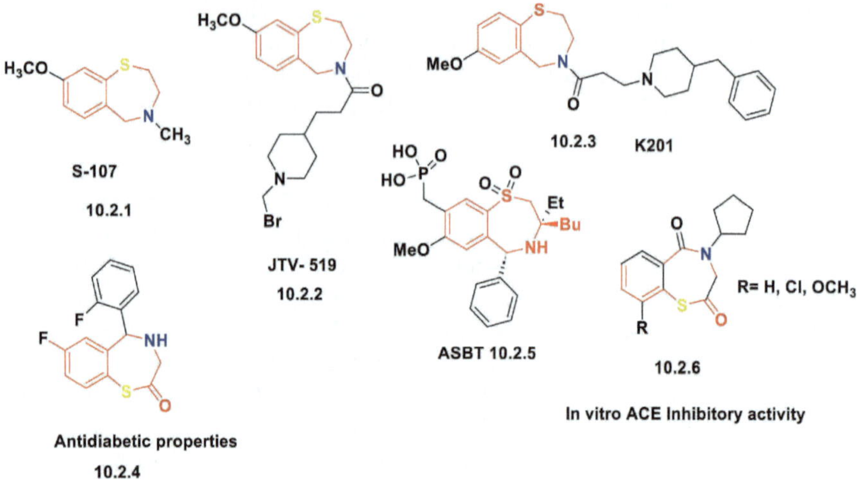

Figure 10.4 Bioactive molecules of 1,4-benzothiazepine.

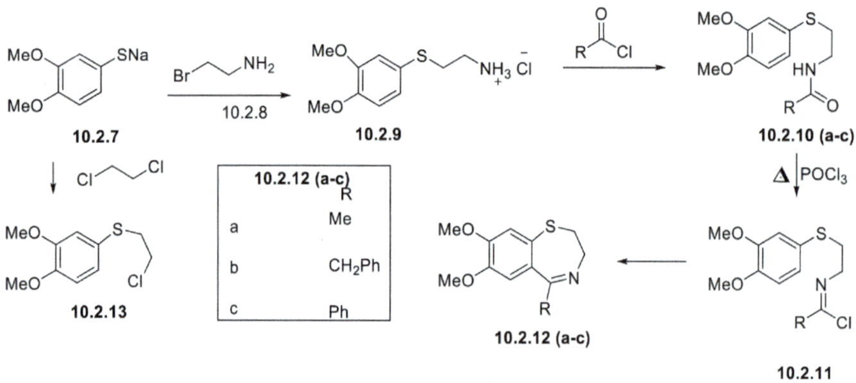

Figure 10.5 Multistep synthesis of 1,4-benzothiazepine.

2-bromoethylamine **10.2.8** generates aminoethyl-3,4-dimethoxyphenyl sulfide **10.2.9** as a salt. The treatment of this salt **10.2.9** with acid chloride produced acid amides **10.2.10 (a–c)**, which, in the presence of POCl3 on heating, furnished benzothiazepine derivatives **10.2.12 (a–c)** with poor yields (Figure 10.5).[33] A microwave-aided approach has also been discussed.

10.3 Synthesis of 1,4-Benzothiazepine

Initially, the esterification of Boc-protected tyrosinol **10.3.1** with methyl thiosalicylate **10.3.2** gives ester derivatives **10.3.3**, which subsequently undergo a series of processes such as Boc deprotection, tosylation, and reduction to give alcohol **10.3.6**. The removal of the tosylate group after intramolecular Mitsunobu cyclization occurs with two functional groups such as hydroxyl and sulfonamide in **10.3.6** to produce an enantiomerically pure benzothiazepine derivative **10.3.8** (Figure 10.6).[34]

Currently, many researchers are utilizing the S-alkylation method for the synthesis of benzothiazepine derivatives. Pie *et al.* used various benzhydrol derivatives as precursors for the synthesis of 1,4-benzothiazepine using the S-alkylation method. They used five methods such as A, B, C, D, and E with different reaction conditions. Method A involves S-alkylation of 2-aminobenzhydrolene with methyl thioglycolate. The thioether derivatives were prepared in pure trifluoroacetic acid (TFA) in just a few hours with good yields. The hydrolysis of thioethers in a basic medium to neutralize the acid results in the synthesis of the desired benzothiazepinones after cyclization with EDC (Figure 10.7).

In Method B, methyl ether generates benzothiazepinones *in situ* in neat TFA by intramolecular aminolysis at a crucial temperature of 80–85°C. In Method C, *N*-acylated benzhydrol also undergoes S-alkylation in TFA with methyl thioglycolate to give the corresponding thioethers in 60% yields. Aqueous NaOH hydrolyzed the ester derivative (**10.3.13**) to the corresponding acid and then EDC induces the lactam formation of an acid to afford **10.3.11** benzothiazepine derivatives at an ambient temperature. Method D involves *ortho*-lithiation of **10.3.14** by *n*-butyl lithium at −78 °C to 0 °C followed by condensation with an aldehyde at 0 °C to room temperature to produce **10.3.15** in good yields. Method E involves the transformation of 2-aminobenzhydrols to **10.3.15** by methyl thioglycolate in the presence of TFA through the microwave-assisted process with the help of basic

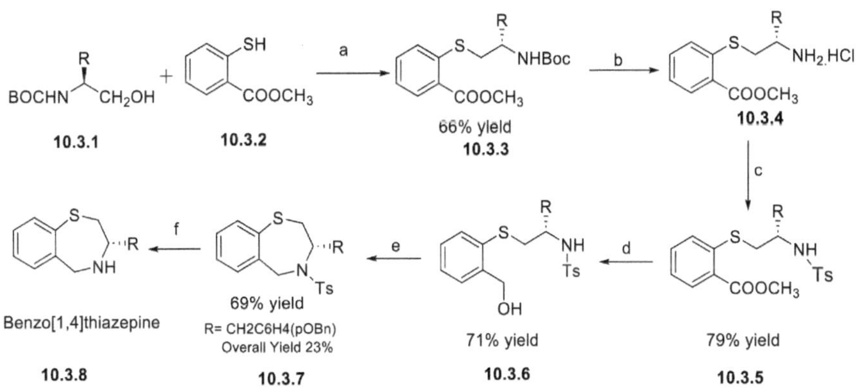

Figure 10.6 Synthesis of benzo[1,4]thiazepine from S-containing amino acids. Reagents and conditions: (a) DEAD, PPh3, THF, 0 °C (1 h) to RT (11 h), N_2, 66% yield; (b) 6 N HCl, MeOH, 0 °C to RT, 45 min, 84% yield; (c) *p*-toluenesulfonyl chloride, DCM, 0 °C to RT, 1 h, N_2, 79% yield; (d) LAH, THF, 0 °C, 1 h, 71% yield; (e) DEAD, PPh_3, THF, 0 °C (1 h) to RT (13 h), N_2, 69% yield; and (f) 4% Na-Hg, Na_2HPO_4, MeOH, 70 °C, 12 h, 64% yield.

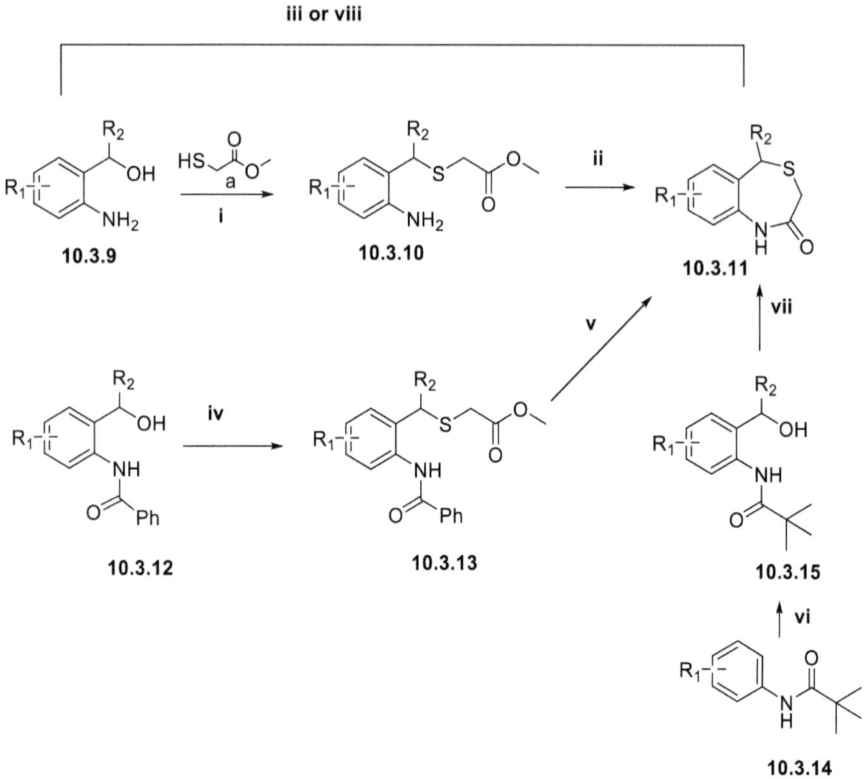

Figure 10.7 Synthesis of benzo[1,4]thiazepine from S-alkylation of benzhydrol derivatives. Reagents and conditions. Method A: (i) TFA/DCM, RT or methyl thioglycolate, TFA (neat); (ii) (a) 1.0 N NaOH aqueous solution, THF/methanol, RT, (b) DIEA, DMAP, EDC, THF, RT. Method B: (iii) methyl thioglycolate, TFA (neat), 80–85 °C. Method C: (iv) TFA, methyl thioglycolate, RT; (v) (a) 1.0 N NaOH aqueous solution, THF/methanol, reflux; (b) DIEA, DMAP, EDC, THF, RT. Method D: (vi) *n*-buLi, THF, −78 to 0 °C, 2 h; aldehydes, 0 °C to RT, 1 h; (vii) methyl thioglycolate, TFA (neat), 80–85 °C, 24 h. Method E: (iix) methyl thioglycolate, TFA, basic $Al2O_3$, MW radiation; saturated aqueous Na_2CO_3, MW radiation at reduced power.

alumina to prevent charring. 1,4-Benzothiazepine derivatives were prepared from **10.3.16** by S-alkylation with mercaptoethanol in TFA to generate thioether alcohol **10.3.17**. Subsequently, bromination and cyclization of compound **10.3.17** furnished **10.3.19** (Figure 10.8). Another benzhydrol derivative **10.3.20** underwent S-alkylation and intramolecular aminolysis in the presence of lithium hexamethyldisilazane (LHMDS) to produce the desired product **10.3.22** (Figure 10.9). Molecule **10.3.25** was obtained by the basic hydrolysis of **10.3.23** and cyclization of **10.3.24** under the given reaction conditions (Figure 10.10).

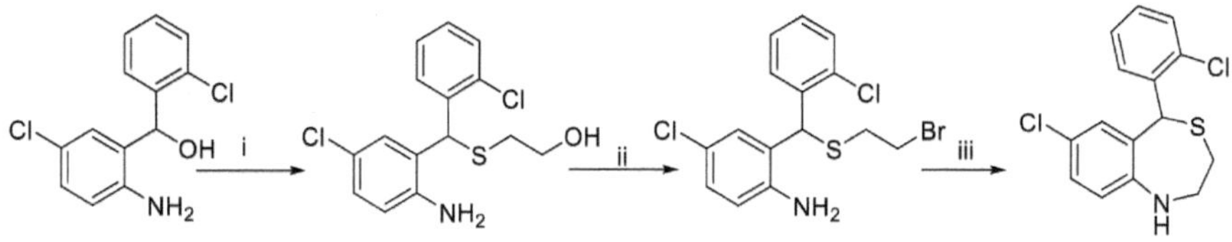

10.3.16 10.3.17 10.3.18 10.3.19

Figure 10.8 Synthesis of aromatic-substituted benzo[1,4]thiazepine using S-alkylation of benzhydrol derivatives. Reagents and conditions: (i) Mercaptoethanol, TFA, RT; (ii) PPh_3, DCM, CBr_4, RT; and (iii) DMF, base K_2CO_3.

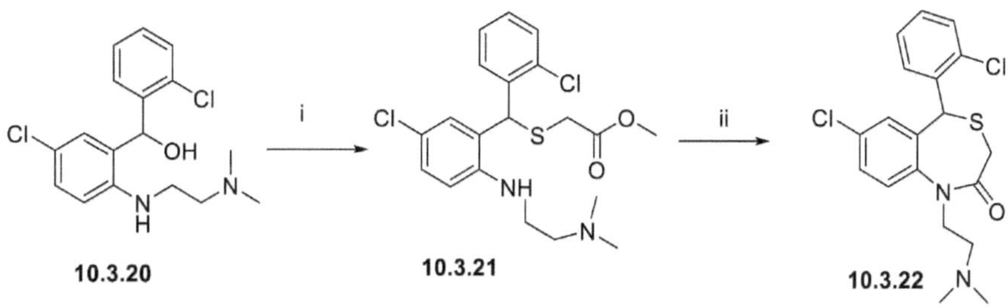

10.3.20 10.3.21 10.3.22

Figure 10.9 Synthesis of *N*-alkyl benzothiazepinones. Reagents and conditions: (i) methyl thioglycolate, TFA, room temperature; (ii) LHMDS, THF, −78 °C.

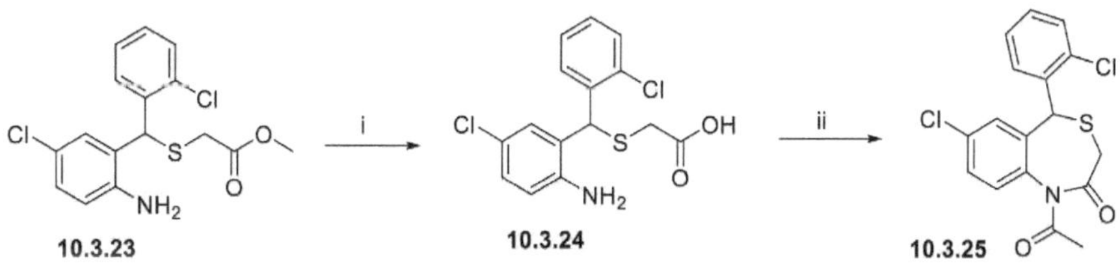

10.3.23 10.3.24 10.3.25

Figure 10.10 Synthesis of *N*-acyl benzothiazepinones.[35] Reagents and conditions: (i) 1.0 N aqueous NaOH, THF/ methanol, RT; (ii) Acyl chloride, DIEA, DMAP, THF, RT, 3 h.

N-(*o*-Chloromethyl)aryl amides (0.10 mmol) **10.3.26** and 1,4-zwitterionic thiolates (0.15 mmol) **10.3.27** react with potassium *tert*-butoxide (0.30 mmol, 34 mg) in dichloromethane (DCM) (2.0 mL) under the basic condition.

The mixture was purified by silica gel column chromatography (petroleum ether/ethyl acetate 3 : 1) at room temperature until the starting material was exhausted (monitored by TLC, 5 min) to provide the desired product **10.3.28** (83–96% yields).[17]

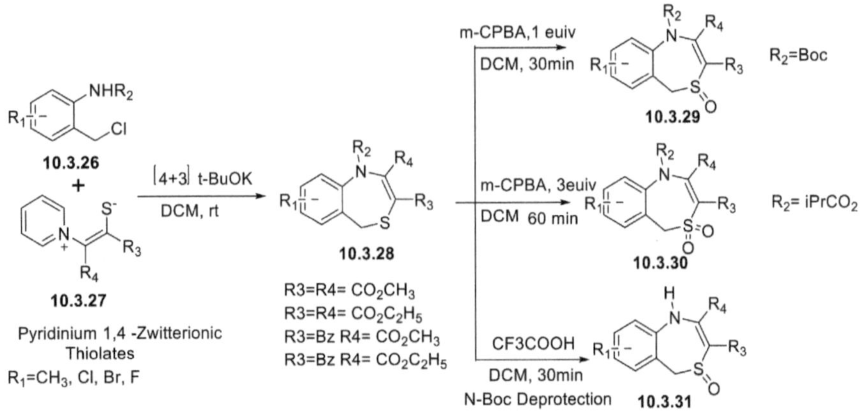

Figure 10.11 The formation of sulfoxide and sulfone derivatives of benzo[*e*][1,4]thiazepine by the [4+3] annulation method.

10.3.1 Synthesis of 4,1-Benzothiazepine-4-oxide 10.3.29

Two reacting materials namely benzo[*e*][1,4]thiazepine (*N*-Boc-protected) derivatives **10.3.28** (0.10 mmol) and *m*-CPBA (0.10 mmol, 1.0) were added to DCM at 0 °C and stirred at room temperature until the starting material was consumed (monitored by TLC, 30 min). The mixture was purified by column chromatography using silica gel (petroleum ether/ethyl acetate 2 : 1) to give sulfoxide **10.3.29** (Figure 10.11).[17]

10.3.2 Synthesis of 4,1-Benzothiazepine-4,4-dioxide 10.3.30

Two reacting materials namely benzo[*e*][1,4]thiazepines (R_2 = $iPrCO_2$) **10.3.28** (0.10 mmol) and *m*-CPBA (0.10 mmol, 1.0) in DCM were added at 0 °C and stirred at room temperature until the starting material was consumed (monitored by TLC, 30 min). The mixture was purified by column chromatography using silica gel (petroleum ether/ethyl acetate 2 : 1) to give sulfoxide **10.3.30** (Figure 10.11).[17]

10.3.3 Synthesis of 1,5-Dihydro-4,1-benzothiazepine Derivatives 10.3.31

2,2,2-Trifluoroacetic acid (0.2 mL) was added at 0 °C to DCM (2 mL) containing 4,1-benzothiazepines **10.3.28** (0.10 mmol). The mixture was continuously stirred at room temperature until the starting material was consumed (monitored by TLC, 60 min). The resulting mixture was washed with water, dried over anhydrous Na_2SO_4, and purified by column chro-

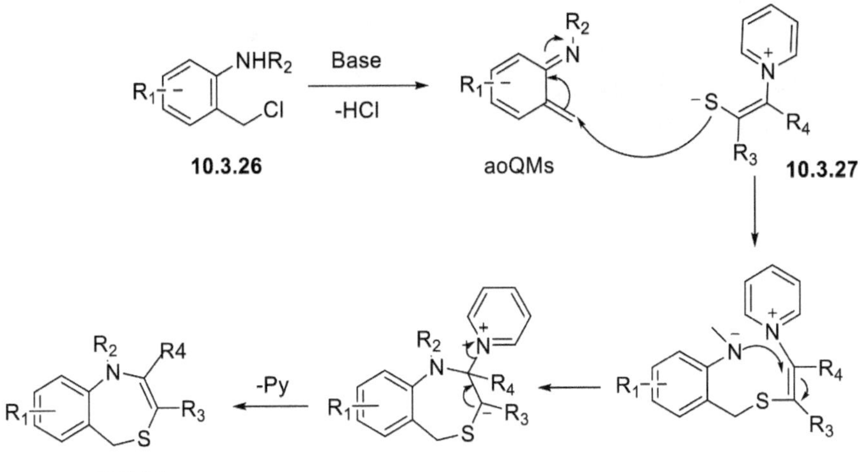

Figure 10.12 Reaction mechanism.

matography on silica gel (petroleum ether/ethyl acetate 1 : 1) to give the 4,1-benzothiazepine derivative **10.3.31** (**Figure 10.11**).

10.3.4 Reaction Mechanism

The base-mediated reduction of *N*-(*o*-chloromethyl)aryl amide **10.3.26** produces aoQM, which reacts with pyridinium thiolate 1,4-zwitterion **10.3.27** *via* the S-Michael addition to form intermediate A. This is followed by an intramolecular Michael pyridine addition/extrusion cascade in a tandem mechanism to provide 4,1-benzothiazepine (Figure 10.12).[17]

10.3.5 Synthesis of 2,3,4,5-Tetrahydrobenzo[1,4]thiazepines *via* *N*-Acyliminium Cyclization[36]

The enhanced reactivity of the *N*-acyliminium ion as a useful reactive intermediate was well demonstrated for the construction of benzo-fused seven-membered rings compared to that of the corresponding iminium ions used in the Mannich and Pictet–Spengler reactions.

In the synthesis of 7-methoxy-2,3,4,5-tetrahydro-1,4-benzothiazepine HBr **10.3.35**; 2-chloroethylamine alkylated the 4-methoxythiophenol and converted the resulting amine derivatives **10.3.31** into the corresponding Cbz derivative **10.3.32**. The treatment of intermediate **10.3.32** with paraformaldehyde in toluene in the presence of *p*-toluenesulfonic acid (PTSA) at 70 °C resulted in the production of the ring-closed *N*-Cbz benzothiazepine **10.3.34** with an 84% yield (**Figure 10.13**).[36]

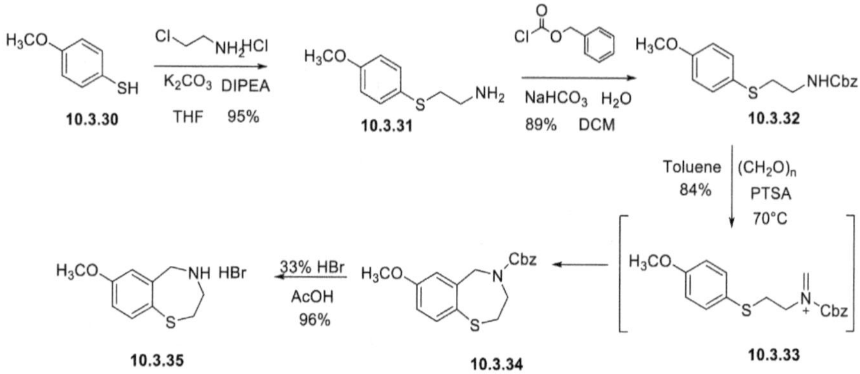

Figure 10.13 Synthesis *via N*-acyliminium cyclization.[36]

10.3.6 Synthesis of Phosphonomethylbenzothiazepine[29]

1,4-Dimethoxy-2-methylbenzene **10.3.36**, which was used for benzophenone synthons, was selected to synthesize phosphonomethylbenzothiazepine **10.3.45**. Benzoyl-1,4-dimethoxy-2-methylbenzene **10.3.36** is suitable for Friedel–Crafts reactions with benzoyl chloride. 1,4-Dimethoxy-2-methylbenzene **10.3.36** and benzoyl chloride were used as starting materials in the reaction in the presence of $AlCl_3$ to afford benzophenone **10.3.37**. The demethylation of the methoxy group in benzophenone and sequential incorporation of phosphorus and sulfur (Pd(*o*)-catalyzed thiol coupling) lead to the synthesis of mercaptan intermediate **10.3.43**. S-Alkylation of **10.3.43** with (*R*)-2-ammonio-2-ethylhexyl sulfate followed by cyclodehydration under Dean–Stark conditions afforded dihydrobenzothiazepine **10.3.45** (Figure 10.14) in 38% overall yield.[29]

10.3.7 Asymmetric Reduction of 10.3.45 and Conversion into ASBT Inhibitor 10.3.48

Figure 10.15 depicts the final three steps of the synthesis of **10.3.48**. As expected, the asymmetric hydrogenation of **10.3.45** with [Ir(COD)$_2$BArF] and ligand iron complex with pressure less than 30 bar (gauge) H_2 predominantly produced *trans*-isomer **10.3.46** (100% conversion; a *trans* : *cis* ratio of 9 : 1). The oxidation of **10.3.46** with H_2O_2/TFA produced sulfone **10.3.47**. The phosphonate hydrolysis of **10.3.47** under 6 N HCl reflux and evaporation of the solvent produced the hydrochloride salt of the target compound as an amorphous solid **10.3.48** in 97% yield (Figure 10.15).

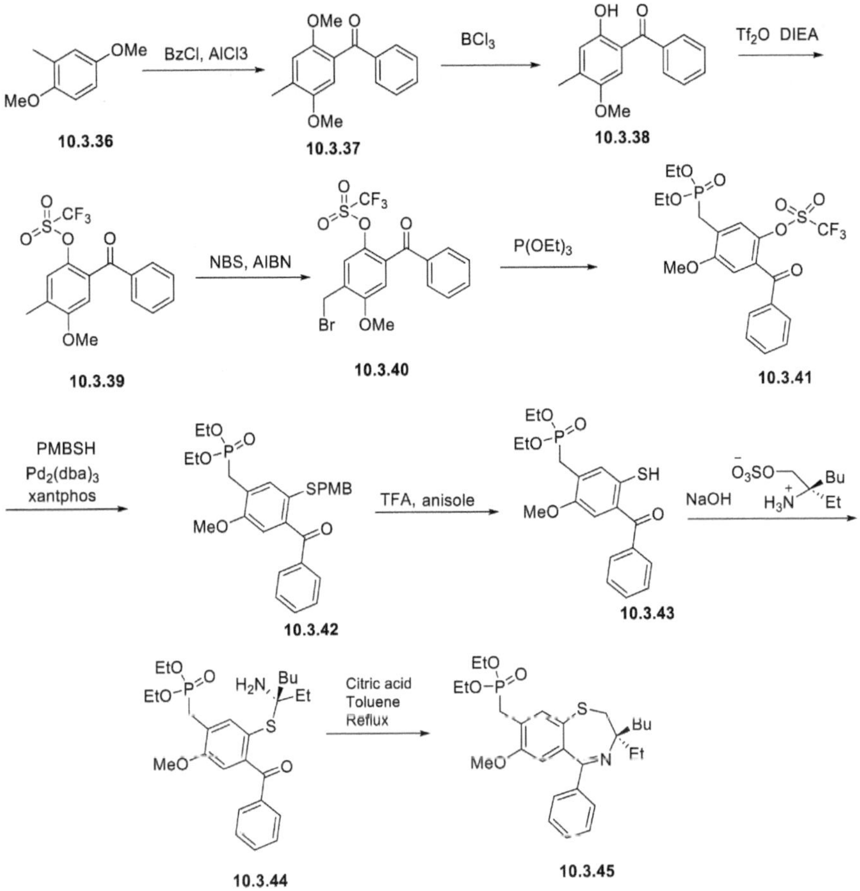

Figure 10.14 Synthesis of phosphonomethylbenzothiazepine.

10.3.8 Synthesis of 1,4-Benzothiazepines from Cyclic Sulfenamides[37]

The chemistry of sulfenamides compared to that of sulfinamides or sulfonamides was reported for the synthesis. The sulfur atom is used as an electrophile due to differences in electronegativity and the nitrogen atom is used as a "supernucleophile" according to the R-effect. The medicinal and industrial applications of sulfenamides include their capacity to generate radicals, which are utilized to expedite rubber vulcanization. Asymmetric sulfenylation of aldehydes and aminosulfenylation of unsaturated aldehydes were also reported. Oxidation processes occur in the presence of *N*-chlorosuccinimide (NCS). Carbonylation or coupling reactions with boronic acids using transition metals were performed. The prodrug omeprazole inhibits the acid-secreting gastric (H^+/K^+)-ATPase,

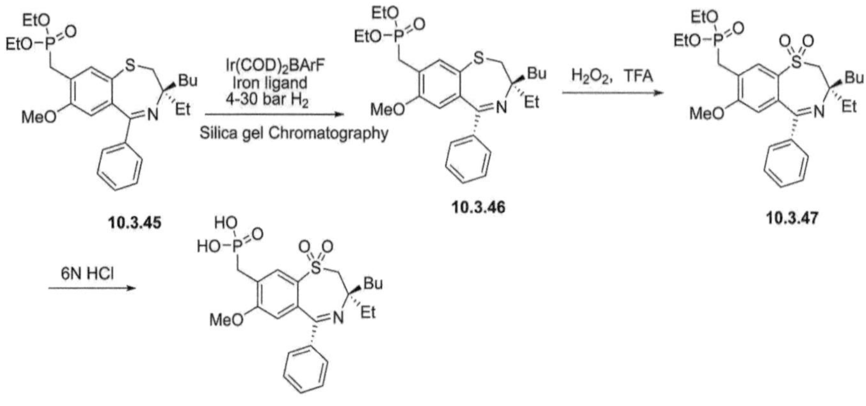

Figure 10.15 Asymmetric reduction of **10.3.45** and its conversion into the ASBT inhibitor **10.3.48**.[29]

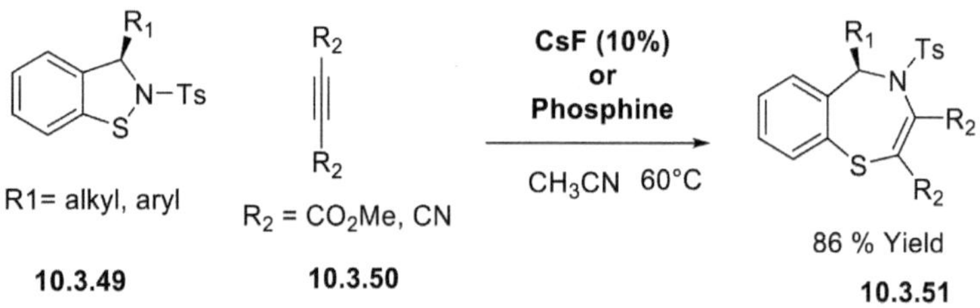

Figure 10.16 Asymmetric synthesis of 1,4-benzothiazepines.

which contains sulfenamide as the main mediator. Carbamazepine is a water-soluble sulfenamide prodrug.

The reactivity of cyclic sulfenamide was determined by asymmetrically synthesizing 1,4-benzothiazepines, which are the precursors of 1,4-benzothiazepine-1,1-dioxides that have a wide range of biological activities. For the synthesis of the desired compound **10.3.51**, sulfenamide **10.3.49** (0.30 mmol) was added to vacuum-dried CsF (0.30 mmol) in acetonitrile in a dried round-bottomed flask. The suspension was heated at 60 °C, and a solution of dimethyl acetylenedicarboxylate **10.3.50** (41 μL, 0.33 mmol) in acetonitrile (3 mL) was added over 30 minutes using a syringe pump. The reaction mixture was then heated to 60 °C for 2 hours. After filtering the suspension and solvent evaporation, the obtained product was refined using silica gel on column chromatography with pentane/Et_2O (8/2) as the eluent, yielding the corresponding benzothiazepine **10.3.51** (122 mg, 86% yield) as yellow crystals (see Figure 10.16).[37] A plausible mechanistic pathway is depicted in Figure 10.17.

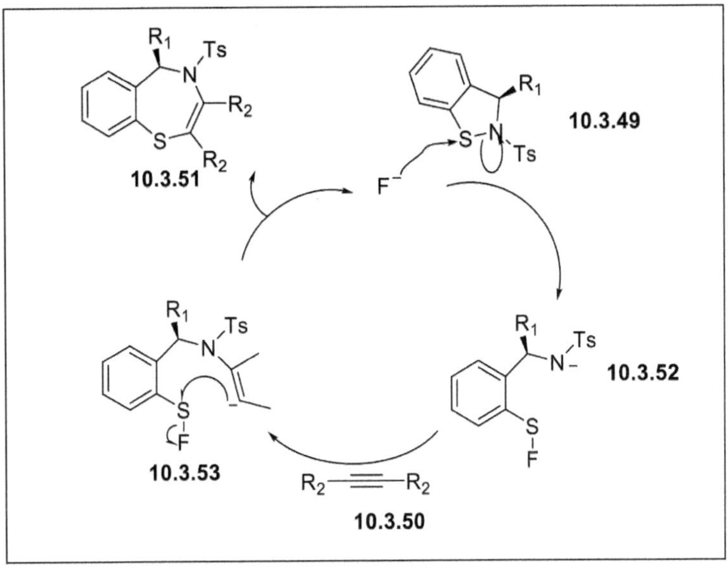

Figure 10.17 A plausible mechanistic pathway.

The reaction mechanism depicts that sulfenamide **10.3.49** reacts with F^- as a nucleophilic catalyst to produce sulfenyl F^- **10.3.52** was reported by Spitz *et al.* Then, the conjugation addition of an amide anion **10.3.52** to electron-deficient acetylene **10.3.50** gave carbanion **10.3.53**, which gets converted into 1,4-benzothiazepine **10.3.51** with the regeneration of F^-. According to the authors, the use of F^- as a nucleophilic catalyst has been demonstrated for the first time. This new method is applied to the synthesis of enantiopure 1,4-benzothiazepine with high efficiency and atomic savings.[37]

10.3.9 Synthesis of 1,4-Benzothiazepine using Cysteine[38]

Manini *et al.* described the synthesis of 1,4-benzothiazepine using cysteine in an overall yield of 70%.

S-(2,3-Dihydroxy-5-methylphenyl)cysteine **10.3.54** was treated with formaldehyde in 0.1 M phosphate buffer, pH 7.4, under reflux. The resulting crude product was collected as a pale brown solid by centrifugation, washed with water, and desiccated under vacuum to give the pure desired product **10.3.55** (Figure 10.18).[38]

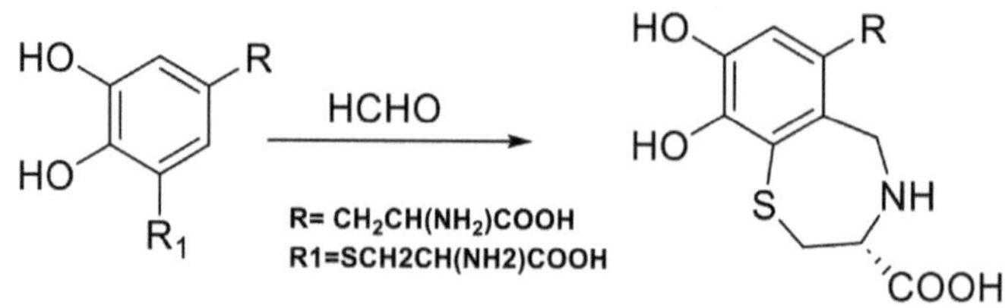

10.3.54

10.3.55

Figure 10.18 Synthesis of 1,4-benzothiazepine using cysteine.[38]

10.3.10 Synthesis of Bicyclic 1,4-Benzothiazepines[39]

Sternbach *et al.* produced 2,3-dihydro-1,4-benzothiazepines **10.3.63** by a multistep synthesis (Figure 10.19) initiated by the diazotization of *o*-amino-substituted aromatics **10.3.56** followed by a potassium- or copper thiocyanate-assisted Sandmeyer-type reaction. Furthermore, the hydrolysis of the base and nucleophilic substitution with 2-bromoethanamine led to substituted *o*-mercapto ketones **10.3.62**, which closed the ring in the presence of pyridine to give 2,3-dihydro-1,4-benzothiazepines **10.3.63** in a moderate yield of 41% (Figure 10.19).[39]

They achieved the synthesis of 1,4-benzothiazepine from 2-chloronicotinic acid in a two-step transformation (Figure 10.20). Mechanistically, 2-chloronicotinic acid **10.3.65** reacts with thionyl chloride to form 2-chloronicotinyl chloride **10.3.69** and with the next starting material 2-(methylamino)acetonitrile **10.3.70**, resulting in **10.3.66**. 2-Chloro-*N*-(cyanomethyl)-*N*-methylnicotinamide subsequently generated a thiolate anion **10.3.72** by removing more acidic protons from an amide derivative, and then the required 1,4-benzothiazepine **10.3.68** is produced by cyclization by intermolecular aromatic nucleophilic substitution after the reaction with CS_2 (Figure 10.21). This process is untrue for nitro-substituted chloronicotinic acid because, in this instance, intermediate **10.3.67** forms the benzothiazole derivative instead of benzothiazepine through failure to complete cyclization.[39]

Fodor *et al.* published two ways for the synthesis of a combination of two diastereomers: **10.3.78** and **10.3.79**. One of these involved the conversion of a methanolic solution of dihydro-1,3-thiazine **10.3.73** into disulfide **10.3.75** when exposed to air. On reaction with 2-bromoacetophenone, the open-ring tautomeric intermediate **10.3.74** formed a mixture of 1,4-benzothiazepine diastereomers, with cyclization of intermediate **10.3.76** yielding somewhat more *cis*-isomer **10.3.78** than the *trans*-isomer **10.3.79** (75%). In contrast, the reaction of 2-benzoylmethylthio-4,5-dimethoxybenzylamine hydrochloride **10.3.77** with benzaldehyde in the presence of NaOMe in MeOH resulted in the diastereomers in a *cis* : *trans* ratio of 6 : 5 (Figure 10.22).[41]

Mironov *et al.* reported Ugi multicomponent reaction (MCR) for the synthesis of 1,4-benzothiazepine by a one-pot process. The synthesis is a

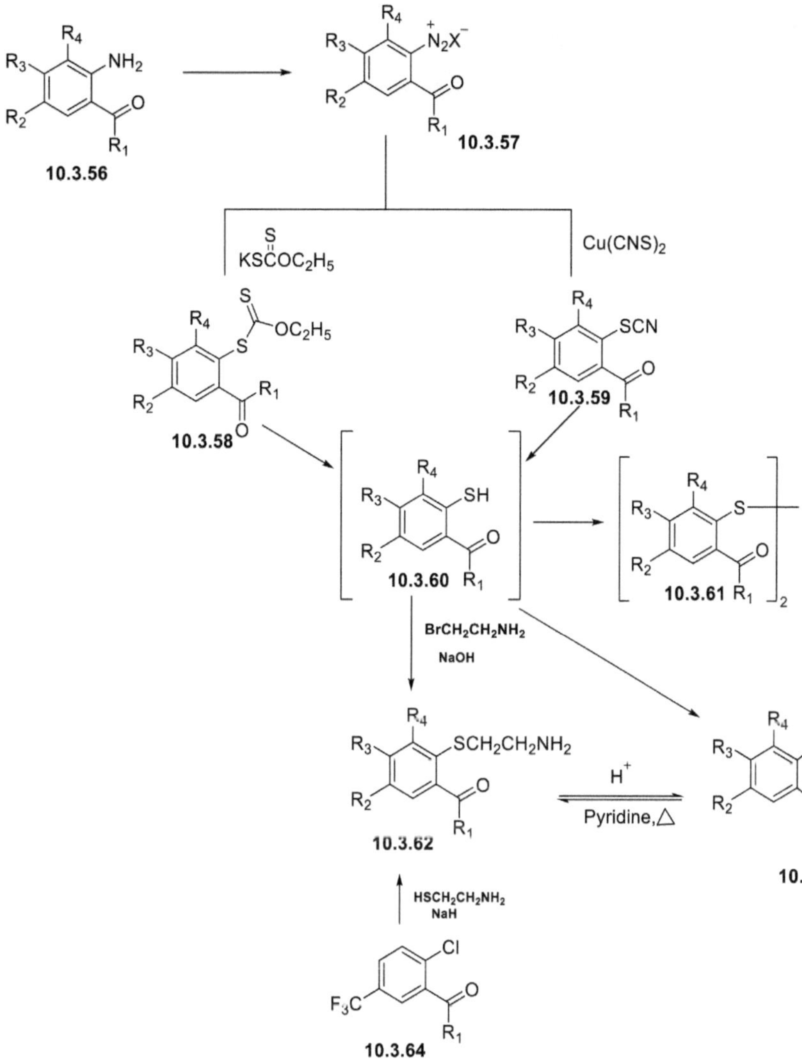

Figure 10.19 2,3-Dihydro-1,4-benzothiazepine synthesis.

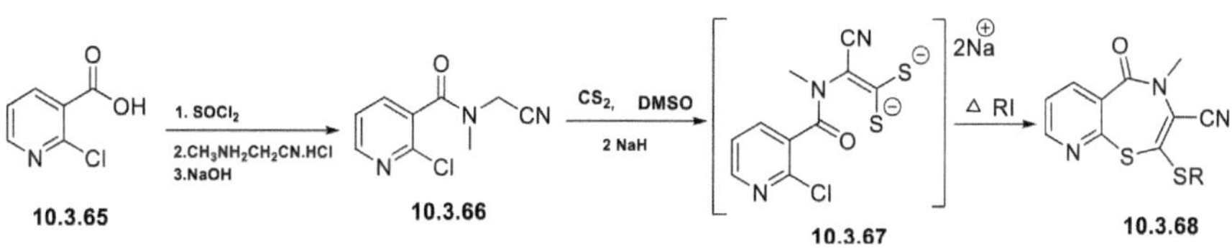

Figure 10.20 Synthesis of 1,4-benzothiazepines from 2-chloronicotinic acid.[40]

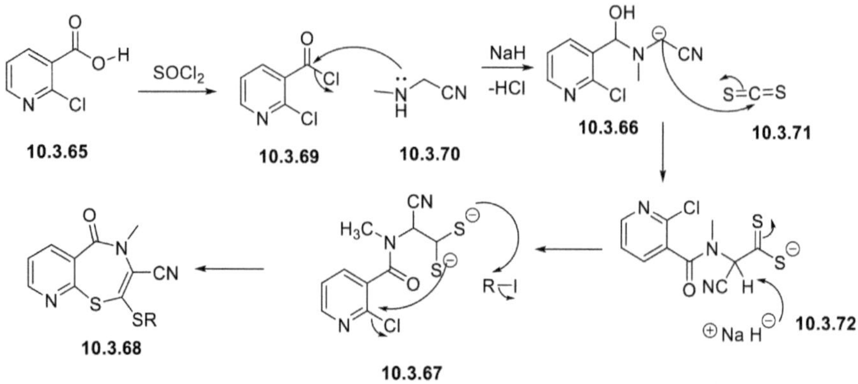

Figure 10.21 Mechanistic pathway from 2-chloronicotinic acid.[40]

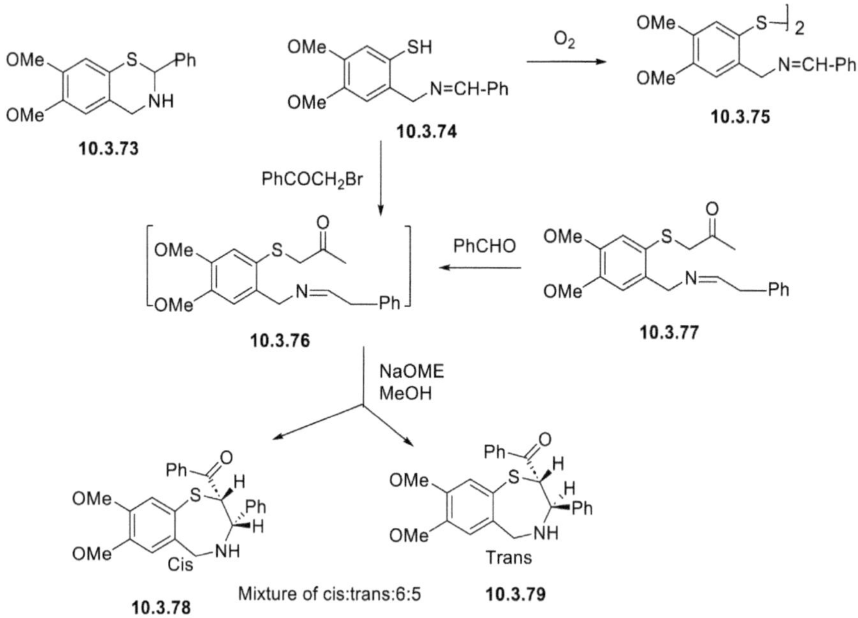

Figure 10.22 Diastereomeric synthesis of 1,4-benzothiazepine derivatives[41].

four-component reaction that involves the condensation of thio-salicylic acid (acid component), chloroacetone (ketone part), *tert*-butyl isonitrile (isonitrile component), and S-alkylated benzylamine (amine component). All four components reacted *via* condensation and nucleophilic addition reactions to produce twelve amide derivatives of 1,4-benzothiazepin-5-ones. Methanol as a polar solvent also participates in this reaction to increase the yield of this reaction. Chloroacetophenone gave better results than

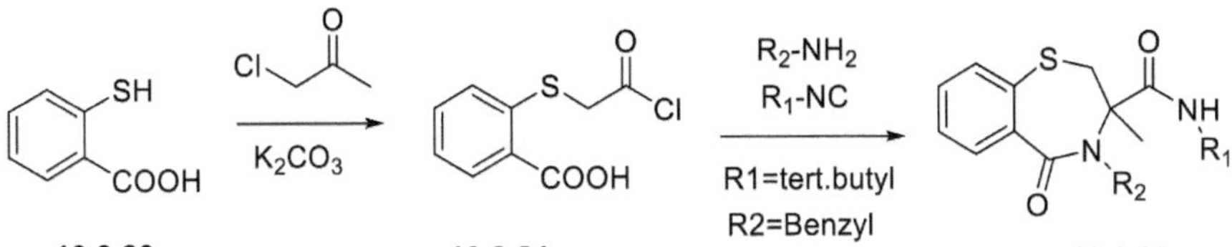

10.3.80 10.3.81 10.3.82

Figure 10.23 Synthesis of 1,4-Benzothiazepin-5-ones by MCR.[42]

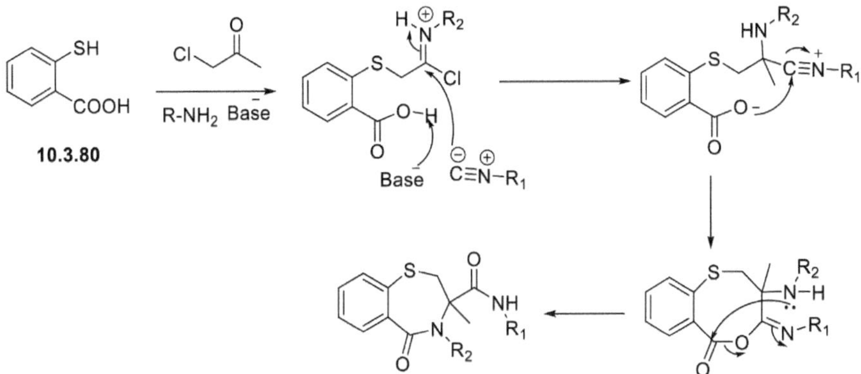

10.3.80

10.3.82

Figure 10.24 The Ugi reaction mechanism to access 1,4-benzothiazepine derivatives.[43]

bromoacetophenone as a source of ketone functional groups in the Ugi reaction (Figure 10.23).[42]

The Ugi reaction mechanism is illustrated in Figure 10.24 to access 1,4-benzothiazepinone derivatives **10.3.82.** K_2CO_3 as a base removed an acidic proton from 2-mercaptobenzoic acid **10.3.80,** followed by nucleophilic substitution with 1-chloropropan-2-one. The iminium ion is formed by the condensation of amine and keto functional groups, which operate as aggressive electrophiles to attack the carbon of isocyanide, forming cyclic intermediate A that undergoes acyl transfer to yield the desired product **10.3.82.**[43]

Spitz *et al.* tried to synthesize 1,4-benzothiazepine *via* two routes. One route is *via* the intermediate allenolate and another route is *via* enamine. An acid-catalyzed condensation reaction between cyclic sulfenamides and enamine gives 1,4-benzothiazepines, where cyclic sulfenamides with an electron withdrawing group (EWG) favored the reaction to form the desired product. The reaction of cyclic sulfenamide with an acetylide C intermediate generated by methyl propionate and allenolate led to the formation of side product 1,3-benzothiazines (Figure 10.25). A mechanistic aspect of the above-mentioned reaction is described in Figure 10.26. They explain that ethyl propionate with allenolate A initially formed a zwitterion

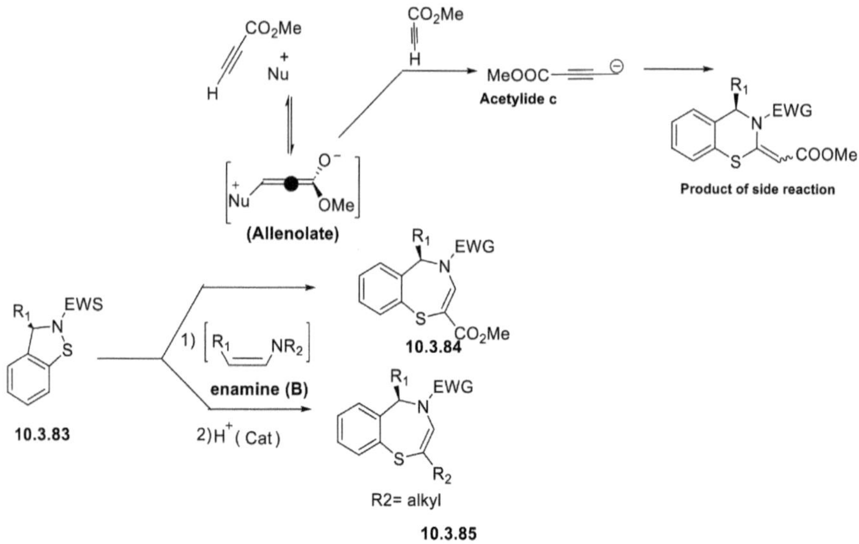

Figure 10.25 The synthesis of 1,4-benzothiazepine from cyclic sulfenamides.

and converted it into the corresponding zwitterion C by reacting with the electrophilic sulfur atom of sulfenamide. After addition–elimination, cyclization occurred, which led to the generation of the 1,4-benzothiazepine product.[43]

Voskressensky *et al.* reported that substituted thiophenols undergo many transformations, such as alkylation, reduction, acylation, and cyclization, with the Bischler–Napieralski reaction to produce 2,3-dihydro-1,4-benzothiazepines **10.3.91** in a low yield (11%), which underwent further reduction leading to the formation of 8-methoxy-4-methyl-2,3,4,5-tetrahydro-1,4-benzothiazepines **10.3.92** in a good yield (92%) and used the previous framework due to the EDG (Figure 10.27).[44]

They described the mechanistic aspects of the synthesis of 2,3,4,5-tetrahydro-1,4-benzothiazepines. In the first step, the electrophilic substitution of substituted thiophenol **10.3.86** with bromoacetonitrile formed thioacetonitrile **10.3.87**, which subsequently underwent reduction and acylation at the nitrile group to form *N*-(2-(phenylthio)ethyl)formamide derivatives **10.3.89**. The cyclization of derivatives of **10.3.89** under Bischler–Napieralski conditions produced dihydrobenzothiazepines **10.3.90**. Dihydrobenzothiazepines **10.3.90** with methyl iodide give quaternary salts **10.3.91** under microwave irradiation, and $NaBH_4$ reduction in methanol produced 2,3,4,5-tetrahydro-1,4-benzothiazepines **10.3.92** (Figure 10.28).

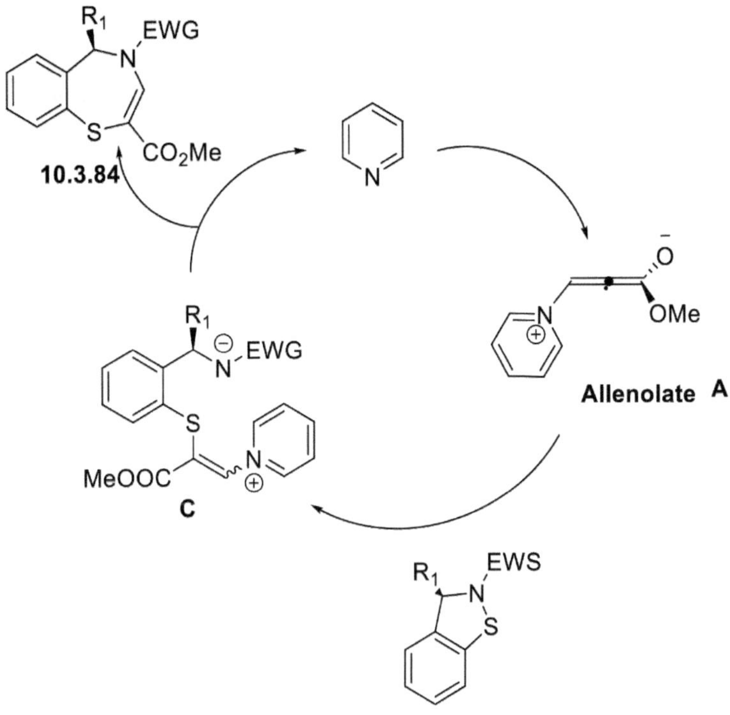

Figure 10.26 Synthesis of 1,4-benzothiazepine from cyclic sulfenamides and an allene: mechanistic details.[43]

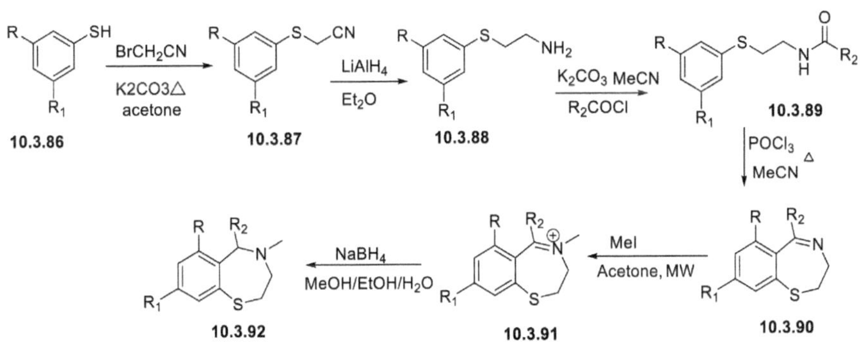

Figure 10.27 Synthesis of 2,3-dihydro-1,4-benzothiazepines using a tandem Bischler–Napieralski process.[44]

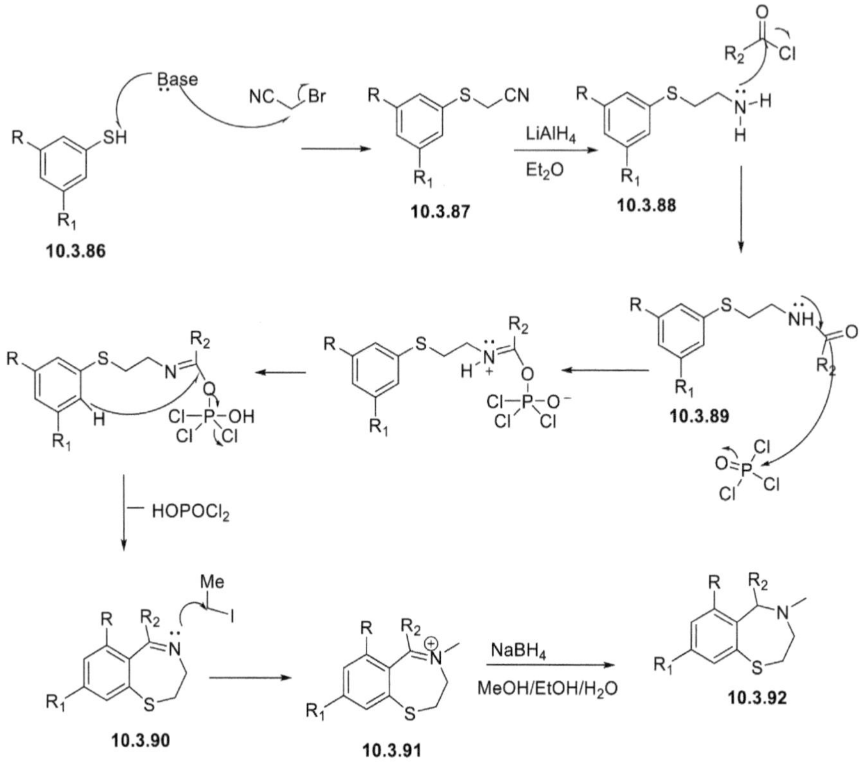

S

Figure 10.28 Mechanistic features of the synthesis of 2,3-dihydro-1,4-benzothiazepines.[44]

10.4 Biological Behavior of 1,4-Benzothiazepines

10.4.1 Anti-tumor potential

When compared to colchicine, a synthetic 1,4-benzothiazepine derivative (**10.4.1**) was shown (Figure 10.29) to exhibit a degree of inhibition on the cancer cell lines at IC_{50} = 1.480 μM for HeLa, 1.940 μM for A549, 1.520 μM for HT29, and 1.470 μM for MCF-7. The low cytotoxicity of **10.4.1** was another factor that made it desirable.[45] Dioxole-conjugated 1,4-benzothiazepines **10.4.2** (Figure 10.29) were shown to have substantial anti-tumor activity against the two human cancer cells, Ec9706 and Eca109, with IC_{50} values of 8.23 and 16.22 mM, respectively. This activity of the compound was found to be superior to that of 5-fluorouracil, which had an IC_{50} value of 23.26 Mm.[46]

New medications for the regimen of cancer have been the focus of research in recent times. Among these discoveries, piperazine-conjugated 1,4-benzothiazepine derivatives (**10.4.3**) are found to have IC_{50} values of

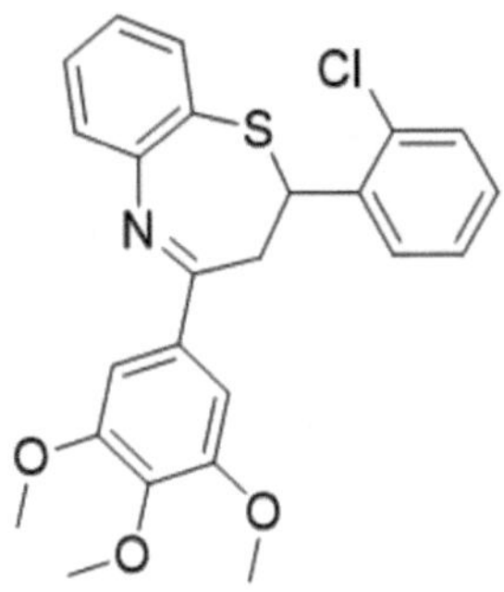

10.4.1

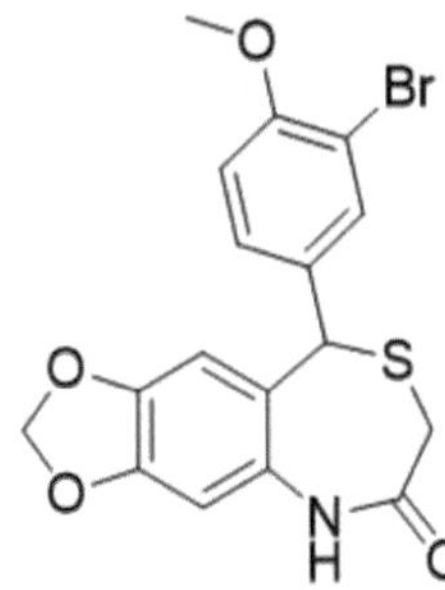

10.4.2

Figure 10.29 Structures of 1,4-benzothiazepine that exhibit anti-tumor activity.

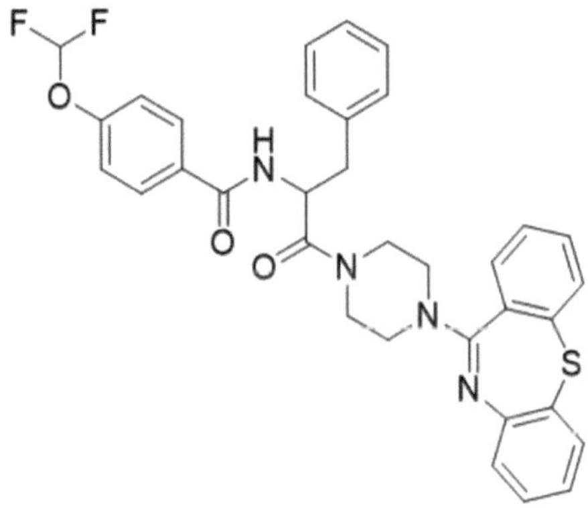

10.4.3

Figure 10.30 Structure of 1,4-benzothiazepine that exhibits anti-cancer activity.

10, 12, and 39 μM for inhibiting the growth of breast cancer, leukemia, and colon carcinoma cells with the estimated mechanism of inhibiting the sirtuin protease (Figure 10.30).[47]

10.4.2 Anti-malarial potential

The concluding results have been published in an effort to find a better treatment for malaria, which is responsible for increased mortality rates in both adults and children. With an IC_{50} value of 0.750 μg mL^{-1}, 1,4-benzothiazepine based on carbazoles (**10.4.4**) exhibited inhibitory activity on the strain *Plasmodium falciparum* (Figure 10.31).[48]

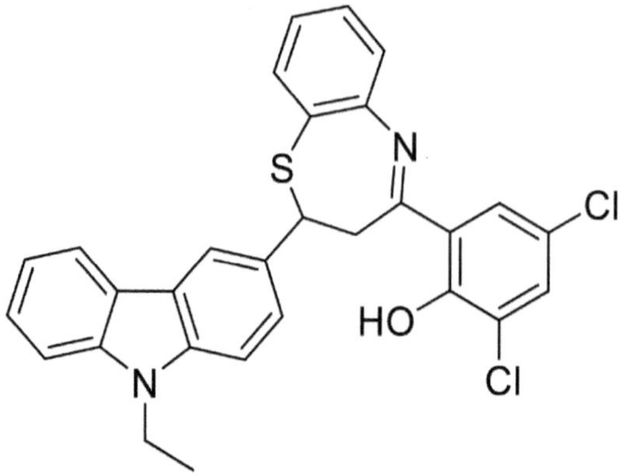

10.4.4

Figure 10.31 Structure of 1,4-benzothiazepine that exhibits anti-malarial activity.

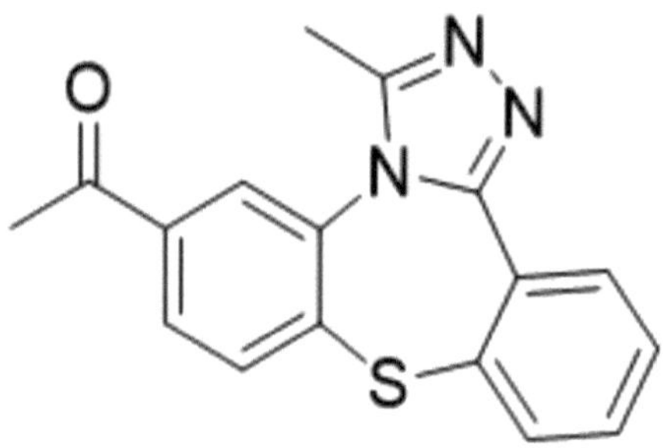

10.4.5

Figure 10.32 Structure of 1,4-benzothiazepine that exhibits anti-bacterial activity.

10.4.3 Anti-bacterial potential

It is no longer implausible that benzothiazepine can inhibit bacterial strains. The effectiveness of benzothiazepine in preventing microbial growth has been documented in a number of studies. Research has revealed that a derivative of triazolo[*b*][1,4]benzothiazepine (**10.4.5**) demonstrated substantial inhibitory impacts on *Staphylococcus aureus*, *Escherichia coli*, and *Enterobacter cloacae*.[49] Some other derivatives were also found active against the strain *Pseudomonas aeruginosa* (Figure 10.32).[50]

10.4.4 Anti-fungal potential

The antifungal properties of 1,4-benzothiazepine derivatives (**10.4.6**) containing an aryl sulfonate moiety were tested against the fungi such as

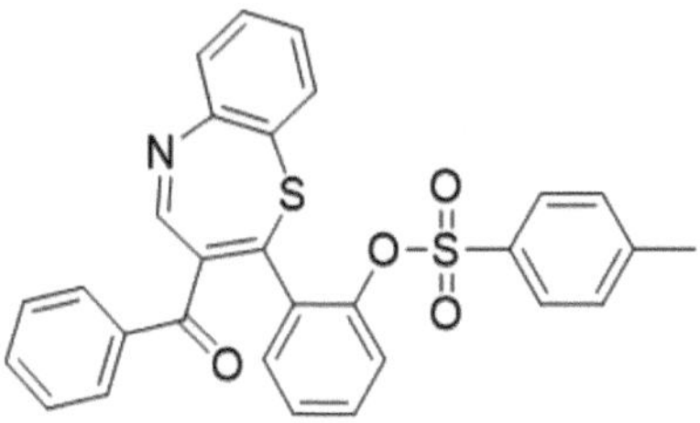

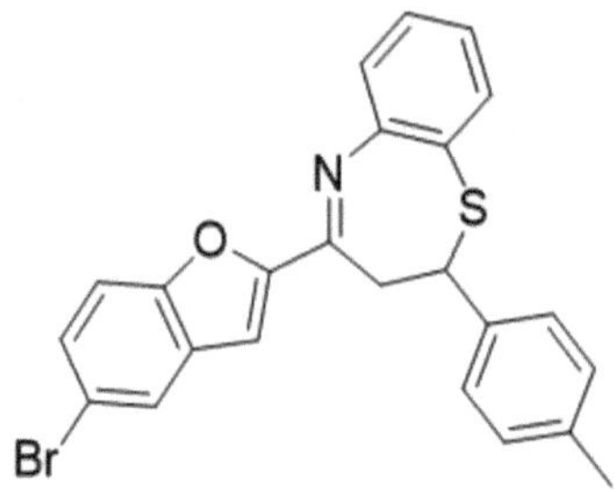

10.4.6 **10.4.7**

Figure 10.33 Structures of 1,4-benzothiazepine that exhibit anti-fungal activity.

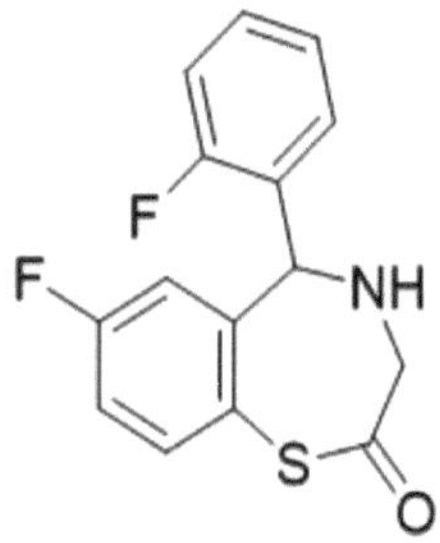

10.4.8

Figure 10.34 Structure of 1,4-benzothiazepine that exhibits anti-cancer activity.

Aspergillus flavus, *Aspergillus niger*, and *Fusarium oxysporum* and discovered that they possess the greatest zone of inhibition measuring, in turn, 18, 22, and 18 mm.[51] 1,5-Benzothiazepine (**10.4.7**) with a benzofuran moiety was found to have significantly more activity than the typical medication against *Aspergillus brasiliensis* and *A. niger*, with inhibition zones of 20.0 and 23.0 mm, respectively (Figure 10.33).[52]

10.4.5 Anti-diabetic Potential

The 1,4-benzothiazepine derivative (**10.4.8**) (Figure 10.34) showed fascinating activity in comparison to metformin and **CGP37197**. The activity was monitored by the electronegative substitution of the benzothiazepine nucleus at position 7′ and the phenyl ring at position 2′.[53]

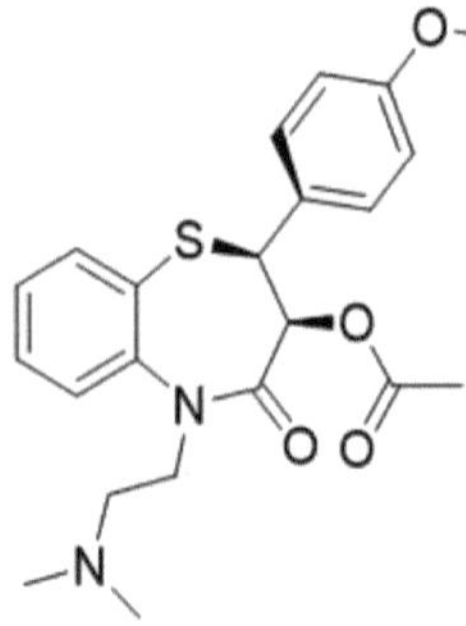

10.4.9

Figure 10.35 Structure of 1,4-benzothiazepine that exhibits antioxidant activity.

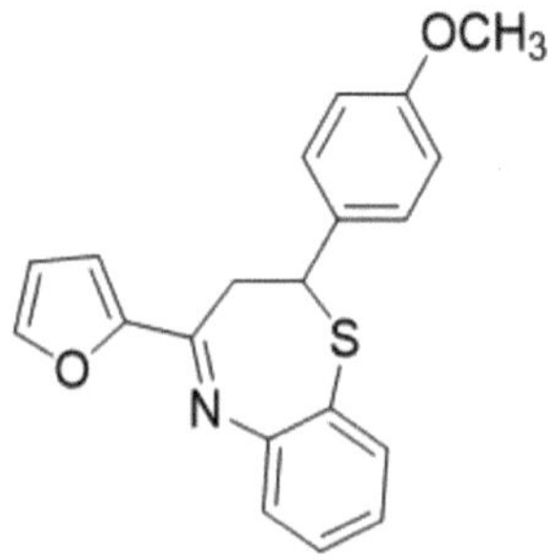

10.4.10

Figure 10.36 Structure of 1,4-benzothiazepine that exhibits anti-inflammatory and analgesic activity.

10.4.6 Antioxidant Potential

Free radicals, such as oxygen, halogen, hydrogen peroxide, and nitric acid that can harm cells, are scavenged by antioxidants or their oxidation is inhibited. The endothelial cells of the human umbilical vein were treated with diltiazem (**10.4.9**) (Figure 10.35) to reduce oxidative stress.[53]

10.4.7 Analgesic and Anti-inflammatory Potential

Anti-inflammatory drugs play a crucial role in the body because they help reduce pain and fight inflammation. The 1,4-benzothiazepine-infused furan analog (**10.4.10**) (Figure 10.36) has demonstrated noteworthy anti-inflammatory and ATPase inhibitory properties, indicating their potential as anti-ulcer medications.[54]

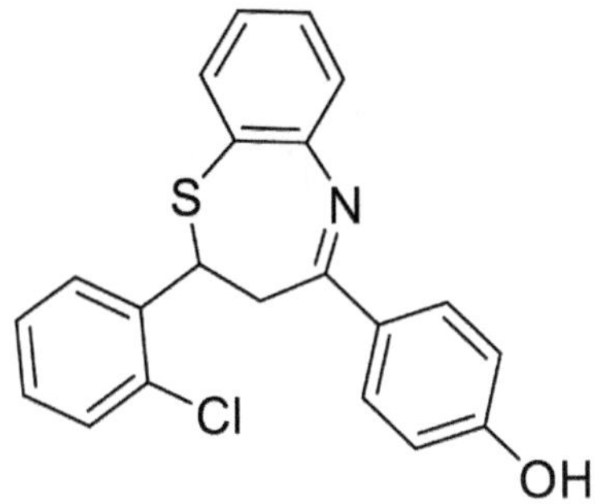

10.4.11

Figure 10.37 Structure of 1,4-benzothiazepine that exhibits anti-convulsant activity.

10.4.8 Anti-convulsant Potential

Epilepsy is a major health concern worldwide. The benzothiazepine derivative (**10.4.11**) has about 80% inhibition of seizures in the supramaximal electroshock seizure pattern test compared to phenytoin sodium as the standard (Figure 10.37).

10.4.9 Other Biological Potentials

Apart from these areas, the scope of 1,4-benzothiazepine efficacy can be seen in other fields as well. D2DAR, dopamine receptor D2, is a drug target in which some 1,4-benzothiazepine scaffolds effectively fit in and treat neuropsychiatric and endocrine disorders.[55] Likewise, **CGP37157** (Figure 10.38) and its isosteric analog show neuroprotective activity by stabilizing the entry of neuronal and mitochondrial calcium ions and thereby

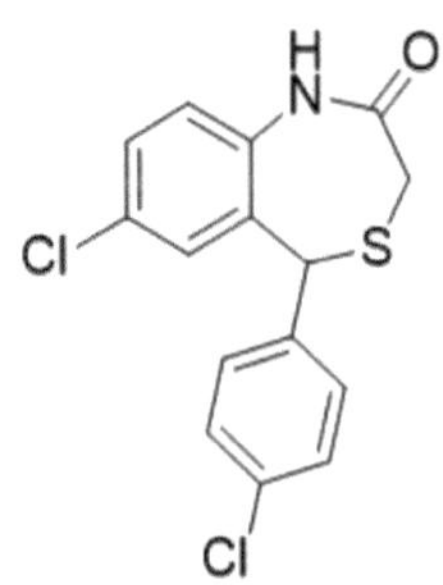

CGP37157

Figure 10.38 Structure of 1,4-benzothiazepine that exhibits neuroprotective activity.

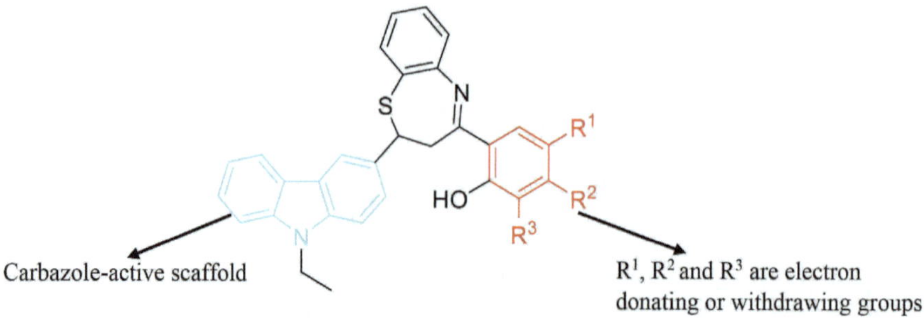

Figure 10.39 Structure–activity relationship (SAR) for anti-microbial and anti-malarial activity.

providing protection against neurodegenerative diseases.[56–58] 1,4-Benzothiazepine-2,5-diones have also proved to be anti-hypertensive agents *via* the ACE inhibitor pathway.[32] Anti-viral,[27] anti-protozoal,[50] and anti-leishmanial[50] properties are some other domains where 1,4-benzothiazepine scaffolds bag the reward as well.

10.5 Structure–Activity Relationship Study

A comparable finding regarding the anti-bacterial and anti-malarial properties of the benzothiazepine derivatives (Figure 10.39) was observed. The various functions increased with the existence of electron-withdrawing groups such as Cl and Br substituents on the aromatic ring. Furthermore, the antimicrobial activity of the compounds is enhanced by the addition of the carbazole group. It has been noted in various investigations that substituting benzothiazepine derivatives with electron-withdrawing groups results in greater activity than substituting them with electron-donating groups.

Similarly, anti-convulsant derivatives (Figure 10.40) were also found to show good activities with electron-withdrawing groups at 2- or 4-phenyl

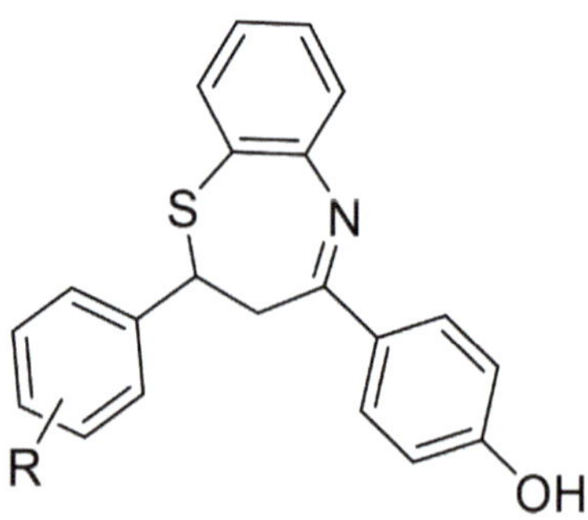

Figure 10.40 SAR for anti-convulsant activity.

substitutions. It was majorly observed in various other activities that electron-deficient groups favor their activity to a greater extent.

10.6 Conclusion

In summary, several reaction mechanisms were used for the development of the skeleton of 1,4-benzothiazepines, such as the Mitsunobu reaction (Figure 10.12), Mannich and Pictet–Spengler reaction (Figure 10.13), Friedel–Crafts reaction (Figure 10.14), Ugi reaction (Figure 10.24), and Bischler–Napieralski reaction (Figure 10.27). Several methods, such as S-alkylation, annulation, and others, are utilized. Zhang *et al.* demonstrated that the [4+3] cyclization reaction, in which *in situ* generated aza-*o*-quinone methides (aoQMs) with 1,3-zwitterionic intermediates are used for the synthesis of seven-membered nitrogen-containing heterocycles, has drawn much attention due to high efficiency and step economy. Researchers have used different types of precursors, such as S-amino acids, cysteine, various benzhydrol derivatives, thiophenols, 2-chloronicotinic acid, and dimethoxy methyl benzene, for synthesizing 1,4-benzothiazepine derivatives. Furthermore, the most recent information on the essential pharmacological effects of 1,4-benzothiazepine derivatives has been reviewed to comprehend their therapeutic potential in drug development.

References

1. K. Lakshmikanth, S. M. Saini, S. T. Dorai and S. Chandrashekharappa, Tandem-Michael-cyclization cascade to make pyridines: Use of electron-deficient acetylenes for the synthesis of indolizines in aqueous media, *Tetrahedron*, 2023, **142**(2), 133516.
2. K. N. Venugopala, S. Chandrashekharappa, S. Bhandary, D. Chopra, M. A. Khedr and B. E. Aldhubiab, *et al.*, Efficient Synthesis and Characterization of Novel Substituted 3-Benzoylindolizine Analogues via the Cyclization of Aromatic Cycloimmoniumylides with Electrondeficient Alkenes, *Curr. Org. Synth.*, 2018, **15**(3), 388–395.
3. K. N. Venugopala, O. H. A. Al-Attraqchi, C. Tratrat, S. K. Nayak, M. A. Morsy and B. E. Aldhubiab, *et al.*, Novel Series of Methyl 3-(Substituted Benzoyl)-7-Substituted-2-Phenylindolizine-1-Carboxylates as Promising Anti-Inflammatory Agents: Molecular Modeling Studies, *Biomolecules*, 2019, **9**(11), 661.
4. R. Ningegowda, S. Chandrashekharappa, V. Singh, V. Mohanlall and K. N. Venugopala, Design, synthesis and characterization of novel 2-(2, 3-dichlorophenyl)-5-aryl-1, 3, 4-oxadiazole derivatives for their anti-tubercular activity against Mycobacterium tuberculosis, *Chem. Data Collect.*, 2020, **28**, 100431.

5. C. Sandeep, B. Padmashali, R. S. Kulkarni, S. M. Mallikarjuna, M. B. Siddesh, H. K. Nagesh and K. S. Thriveni, Synthesis of Substituted 5-acetyl-3-benzoylindolizine-1-carboxylate from substituted 2-acetyl pyridinium bromides, *Heterocycl. Lett.*, 2014, **04**(03), 371–376.
6. C. Sandeep, P. Basavaraj and R. S. Kulkarni, Synthesis of Isomeric Subtituted 6-acetyl-3-benzoylindolizine-1-carboxylate and 8-acetyl-3-benzoylindolizine-1-carboxylate from subtituteded 3-acetyl pyridinium bromides and their antimicrobial activity, *J. Appl. Chem.*, 2013, **2**(5), 1049.
7. K. N. V. Sandeep Chandrashekharappa and B. P. Rashmi Venugopala, Qualitative anti-tubercular activity of synthetic ethyl 7-acetyl2-substituted-3-(4-substituted benzoyl) indolizine-1-carboxylate analogues, *J. App. Pharm. Sci.*, 2019, **9**(2), 124–128.
8. S. C. Vijayakumar Uppar, A. I. Basarikatti, G. Banuprakash, M. K. Mohan, M. Chougala, K. K. Mudnakudu-Nagaraju, R. Ningegowda and B. Padmashali, Synthesis, antibacterial, and antioxidant studies of 7-amino-3-(4-fluorobenzoyl)indolizine-1-carboxylate derivatives, *J. Appl. Pharm. Sci.*, 2020, **10**(2), 077–085.
9. K. N. Venugopala, S. Chandrashekharappa, P. K. Deb, C. Tratrat, M. Pillay and D. Chopra, *et al.*, Anti-tubercular activity and molecular docking studies of indolizine derivatives targeting mycobacterial InhA enzyme, *J. Enzyme Inhib. Med. Chem.*, 2021, **36**(1), 1472–1487.
10. V. Uppar, S. Chandrashekharappa, C. Shivamallu, P. Sushma, S. P. Kollur and J. Ortega-Castro, *et al.*, Investigation of Antifungal Properties of Synthetic Dimethyl-4-Bromo-1-(Substituted Benzoyl) Pyrrolo[1,2-a] Quinoline-2,3-Dicarboxylates Analogues: Molecular Docking Studies and Conceptual DFT-Based Chemical Reactivity Descriptors and Pharmacokinetics Evaluation, *Molecules*, 2021, **26**(9), 2722.
11. P. Mundhe, S. Kidwai, S. M. Saini, H. R. Singh, R. Singh and S. Chandrashekharappa, Design, synthesis, characterization, and anti-tubercular activity of novel ethyl-3-benzoyl-6, 8-difluoroindolizine-1-carboxylate analogues: Molecular target identification and molecular docking studies, *J. Mol. Struct.*, 2023, **1284**(147), 135359.
12. K. Shivaprasad, S. Kidwai, S. Gopavaram, S. M. Saini, K. Reddy and S. Chugh, *et al.*, Design, synthesis and in vitro antitubercular evaluation of novel 7-methoxy pyrrolo[1,2-a]quinoline analogues as CYP 121 inhibitors, *J. Mol. Struct.*, 2023, **1284**(5), 135439.
13. K. N. Venugopala, M. A. Khedr, M. Pillay, S. K. Nayak, S. Chandrashekharappa and B. E. Aldhubiab, *et al.*, Benzothiazole analogs as potential anti-TB agents: computational input and molecular dynamics, *J. Biomol. Struct. Dyn.*, 2019, **37**(7), 1830–1842.
14. K. N. Venugopala, S. Chandrashekharappa, M. Pillay, S. Bhandary, M. Kandeel and F. M. Mahomoodally, *et al.*, Synthesis and structural elucidation of novel benzothiazole derivatives as anti-tubercular

agents: In-silico screening for possible target identification, *Med. Chem.*, 2019, **15**(3), 311–326.
15. H. Nagesh, B. Padmashali, C. Sandeep, T. Yuvaraj, M. Siddesh and S. Mallikarjuna, Synthesis and antimicrobial activity of benzothiophene substituted coumarins, pyrimidines and pyrazole as new scaffold, *Int. J. Pharm. Sci. Rev. Res.*, 2014, **28**(2), 6–10.
16. M. Siddesh, B. Padmashali, K. Thriveni and C. Sandeep, Synthesis of thiophene-linked pyrimidopyrimidines as pharmaceutical leads, *J. Chem. Sci.*, 2014, **126**, 821–826.
17. L. Zhang, L. Fang, H. Huang, C. Wang, F. Gaoand Z. Wang, Synthesis of Benzo [e][1, 4] thiazepines by Base-Induced Formal [4+ 3] Annulation Reaction of Aza-o-quinone Methides and Pyridinium 1, 4-Zwitterionic Thiolates, *J. Org. Chem.*, 2021, **86**(24), 18156–18163.
18. G. D. Mahale, Y. M. Kolekar, A. Kumar, D. Singh, K. M. Kodam and S. B. Waghmode, Synthesis and Antibacterial Activities of New Dibenzothiazepine Derivatives, *Indian J. Chem.*, 2011, **50B**, 1196–1201.
19. D. Saha, G. Jain and A. Sharma, Benzothiazepines: Chemistry of a privileged scaffold, *RSC Adv.*, 2015, **5**(86), 70619–70639.
20. T. A. Ogunnupebi, O. O. Ajani, G. O. Oduselu, O. F. Elebiju and E. Adebiyi, Chemistry and pharmacological diversity of benzothiazepine—Excellent pathway to drug discovery, *J. Mol. Struct.*, 2023, **1280**, 135071.
21. F. Zamania and E. Doustkhahb, An introduction to benzodiazepines and benzothiazepines, *Benzodiazepine-Based Drug Discovery*, 2022, 1.
22. T. Sasikala, K. Parameswari and S. Chitra, Synthesis and corrosion inhibition study of benzothiazepine derivatives on mild steel in acid medium, *Orient. J. Chem.*, 2016, **32**(2), 1215.
23. S. Kasper and B. S. McEwen, Neurobiological and clinical effects of the antidepressant tianeptine, *CNS Drugs*, 2008, **22**, 15–26.
24. M. Struga, J. Kossakowski, A. E. Koziol, E. Kedzierska, S. Fidecka and P. La Colla, *et al.*, Synthesis, pharmacological and antiviral activity of 1, 3-thiazepine derivatives, *Eur. J. Med. Chem.*, 2009, **44**(12), 4960–4969.
25. N. G. Delaney, J. C. Barrish, R. Neubeck, S. Natarajan, M. Cohen and G. C. Rovnyak, *et al.*, Mercaptoacyl dipeptides as dual inhibitors of angiotensin-converting enzyme and neutral endopeptidase. Preliminary structure-activity studies, *Bioorg. Med. Chem. Lett.*, 1994, **4**(15), 1783–1788.
26. J. A. Robl, C.-Q. Sun, J. Stevenson, D. E. Ryono, L. M. Simpkins and M. P. Cimarusti, *et al.*, Dual metalloprotease inhibitors: mercaptoacetyl-based fused heterocyclic dipeptide mimetics as inhibitors of angiotensin-converting enzyme and neutral endopeptidase, *J. Med. Chem.*, 1997, **40**(11), 1570–1577.
27. N. Neamati, J. A. Turpin, H. E. Winslow, J. L. Christensen, K. Williamson and A. Orr, *et al.*, Thiazolothiazepine inhibitors of HIV-1 integrase, *J. Med. Chem.*, 1999, **42**(17), 3334–3341.

28. K. Inagaki, Y. Kihara, T. Izumi and S. Sasayama, The cardioprotective effects of a new 1,4-benzothiazepine derivative, JTV519, on ischemia/reperfusion-induced Ca2+ overload in isolated rat hearts, *Cardiovasc. Drugs Ther.*, 2000, **14**, 489–495.
29. D. J. Cowan, J. L. Collins, M. B. Mitchell, J. A. Ray, P. W. Sutton and A. A. Sarjeant, *et al.*, Enzymatic- and iridium-catalyzed asymmetric synthesis of a benzothiazepinylphosphonate bile acid transporter inhibitor, *J. Org. Chem.*, 2013, **78**(24), 12726–12734.
30. Y. Mei, L. Xu, H. H. Kramer, G. H. Tomberlin, C. Townsend and G. Meissner, Stabilization of the skeletal muscle ryanodine receptor ion channel-FKBP12 complex by the 1, 4-benzothiazepine derivative S107, *Biophys. J.*, 2013, **104**(2), 443a.
31. N. Kaneko, R. Matsuda, M. Toda and K. Shimamoto, Inhibition of annexin V-dependent Ca2+ movement in large unilamellar vesicles by K201, a new 1, 4-benzothiazepine derivative, *Biochim. Biophys. Acta, Biomembr.*, 1997, **1330**(1), 1–7.
32. J. W. Skiles, J. T. Suh, B. E. Williams, P. R. Menard, J. N. Barton and B. Loev, *et al.*, Angiotensin-converting enzyme inhibitors: new orally active 1, 4-thiazepine-2, 5-diones, 1, 4-thiazine-2, 5-diones, and 1, 4-benzothiazepine-2, 5-diones possessing antihypertensive activity. Journal of medicinal chemistry, *J. Med. Chem.*, 1986, **29**(5), 784–796.
33. J. Szabó, G. Bernáth, Á. Katócs, L. Fodor and P. Sohár, Synthesis and spectroscopic investigations of 1,4-benzothiazepine derivatives, *Can. J. Chem.*, 1987, **65**(1), 175–181.
34. J. K. Mishra and G. Panda, Diversity-oriented synthetic approach to naturally abundant S-amino acid based benzannulated enantiomerically pure medium ring heterocyclic scaffolds employing inter- and intramolecular mitsunobu reactions, *J. Comb. Chem.*, 2007, **9**(2), 321–338.
35. Y. Pei, M. J. Lilly, D. J. Owen, L. J. D'Souza, X.-Q. Tang and J. Yu, *et al.*, Efficient Syntheses of Benzothiazepines as Antagonists for the Mitochondrial Sodium– Calcium Exchanger: Potential Therapeutics for Type II Diabetes, *J. Org. Chem.*, 2003, **68**(1), 92–103.
36. S. Deng, Z. Cheng, C. Ingalls, F. Kontes, J. Yan and S. Belvedere, Synthesis of 2, 3, 4, 5-Tetrahydrobenzo [1, 4] thiazepines via N-Acyliminium Cyclization, *Org. Process Res. Dev.*, 2017, **21**(11), 1801–1805.
37. C. Spitz, J.-F. Lohier, J. S.-D. O. Santos, V. Reboul and P. Metzner, Fluoride ion and phosphines as nucleophilic catalysts: synthesis of 1,4-benzothiazepines from cyclic sulfenamides, *J. Org. Chem.*, 2009, **74**(10), 3936–3939.
38. P. Manini, M. d'Ischia and G. Prota, A novel octahydropyridobenzothiazepine metabolite in human urine: Biomimetic formation from the melanogen 5-S-cysteinyldopa and formaldehyde via a peculiar sulfur-controlled double Pictet– Spengler condensation, *J. Org. Chem.*, 2000, **65**(14), 4269–4273.

39. L. Sternbach, H. Lehr, E. Reeder, T. Hayes and N. Steiger, A new type of 1, 4-benzothiazepine derivatives, *J. Org. Chem.*, 1965, **30**(8), 2812–2818.
40. W. Dölling, M. Biedermann and H. Hartung, A Facile Synthesis of Substituted 1, 4 - Benzothiazepin - 5 (4H) - ones and Pyrido [3, 2 - f][1, 4] thiazepin - 5 (4H) - ones– Crystal and Molecular Structure of 2 - Ethylthio - 4 - methyl - 5 (4H) - oxopyrido [3, 2 - f][1, 4] thiazepine - 3 - carbonitrile, *European. J. Org. Chem.*, 1998, **1998**(6), 1237–1242.
41. L. Fodor, J. Szabó, G. Bernáth and P. Sohár, New convenient synthesis of 1, 4-benzothiazepines, *Tetrahedron Lett.*, 1995, **36**(5), 753–756.
42. M. Mironov, M. Ivantsova, M. Tokareva and V. Mokrushin, New synthesis of 1, 4-benzothiazepin-5-ones, *Russ. Chem. Bull.*, 2004, **53**(6), 1232–1236.
43. C. Spitz, V. Reboul and P. Metzner, Cyclic sulfenamide: versatile template for the synthesis of 1,4-benzothiazepines, *Tetrahedron Lett.*, 2011, **52**(48), 6321–6324.
44. L. Voskressensky, S. Akbulatov, T. Borisova, L. Kulikova, A. Listratova and E. Sorokina, *et al.*, Transformations of tetrahydro-1, 4-benzoxazepines and tetrahydro-1, 4-benzothiazepines under the action of alkynes. First example of the synthesis of tetrahydro-1, 4-benzothiazonine-6-carboxylate, *Chem. Heterocycl. Comp.*, 2013, **49**, 331–340.
45. B. Wang, L.-R. Wang, L.-L. Liu, W. Wang, R.-J. Man and D.-J. Zheng, *et al.*, A novel series of benzothiazepine derivatives as tubulin polymerization inhibitors with anti-tumor potency, *Bioorg. Chem.*, 2021, **108**, 104585.
46. L. Wu, X. Yang, Q. Peng and G. Sun, Synthesis and anti-proliferative activity evaluation of novel benzo[d][1, 3] dioxoles-fused 1, 4-thiazepines, *Eur. J. Med. Chem.*, 2017, **127**, 599–605.
47. J. J. S. García, Y. Toledano-Magaña, M. Flores-Alamo, E. Martínez-Klimova, R. Galindo-Murillo and L. F. Hernández-Ayala, *et al.*, Polycyclic ferrocenyl (dihydro) thiazepine derivatives: Diastereo-selective synthesis, characterization, electrochemical behavior, theoretical and biological investigation, *J. Inorg. Biochem.*, 2017, **166**, 141–149.
48. V. Kadnor and S. Shelke, Synthesis, antimicrobial and antimalarial activity of 1, 4-benzothiazepine and pyrazolinederivatives incorporating carbazolemoiety, *Bulg. Chem. Commun.*, 2019, 234–241.
49. A. Sharma, D. Kishore and B. Singh, An Expedient Method for the Synthesis of 1, 2, 4 - Triazolo - fused 1, 5 - Benzodiazepine, 1, 5 - Benzoxazepine, and 1, 5 - Benzothiazepine Scaffolds: A Novel Seven - membered Ring System of Biological Interest, *J. Heterocycl. Chem.*, 2018, **55**(3), 586–592.
50. P. Navin, P. Sarvil, P. Amit, P. Divyesh, R. Dhansukh and R. Moo-Puc, *et al.*, Synthesis and biological evaluation of newer 1, 3, 4-oxadiazoles incorporated with benzothiazepine and benzodiazepine moieties, *Zeitschrift für Naturforschung C*, 2017, **72**(3–4), 133–146.

51. B. V. Kendre, M. G. Landge and S. R. Bhusare, Synthesis and biological evaluation of some novel pyrazole, isoxazole, benzoxazepine, benzothiazepine and benzodiazepine derivatives bearing an aryl sulfonate moiety as antimicrobial and anti-inflammatory agents, *Arabian J. Chem.*, 2019, **12**(8), 2091–2097.
52. M. Mostofi, G. M. Ziarani and N. Lashgari, Design, synthesis and biological evaluation of benzofuran appended benzothiazepine derivatives as inhibitors of butyrylcholinesterase and antimicrobial agents, *Bioorg. Med. Chem.*, 2018, **26**(12), 3076–3095.
53. V. Srivastav and M. Tiwari, Synthesis and Evaluation of Antihyperglycemic Activity of 1, 4-Benzothiazepine-2-One Derivatives in Alloxan Induced Diabetic Rats, *Indian J. Pharm. Sci.*, 2021, **83**(5), 1033–1043.
54. D. M. Lokeshwari, N. D. Rekha, B. Srinivasan, H. K. Vivek and A. K. Kariyappa, Design, synthesis of novel furan appended benzothiazepine derivatives and in vitro biological evaluation as potent VRV-PL-8a and H^+/K^+ ATPase inhibitors, *Bioorg. Med. Chem. Lett.*, 2017, **27**(14), 3048–3054.
55. J. Xiao, R. B. Free, E. Barnaeva, J. L. Conroy, T. Doyle and B. Miller, *et al.*, Discovery, optimization, and characterization of novel D2 dopamine receptor selective antagonists, *J. Med. Chem.*, 2014, **57**(8), 3450–3463.
56. J.-C. Fernández-Morales, J.-A. Arranz-Tagarro, E. Calvo-Gallardo, M. Maroto, J.-F. Padín and A. G. García, Stabilizers of neuronal and mitochondrial calcium cycling as a strategy for developing a medicine for Alzheimer's disease, *ACS Chem. Neurosci.*, 2012, **3**(11), 873–883.
57. F. J. Martínez-Sanz, R. Lajarín-Cuesta, A. J. Moreno-Ortega, L. González-Lafuente, J. C. Fernández-Morales and R. López-Arribas, *et al.*, Benzothiazepine CGP37157 analogues exert cytoprotection in various in vitro models of neurodegeneration, *ACS Chem. Neurosci.*, 2015, **6**(9), 1626–1636.
58. L. González-Lafuente, J. Egea, R. León, F. J. Martínez-Sanz, L. Monjas and Cn. Perez, *et al.*, Benzothiazepine CGP37157 and its isosteric 2′-methyl analogue provide neuroprotection and block cell calcium entry, *ACS Chem. Neurosci.*, 2012, **3**(7), 519–529.

Part IV: Thiophenes

CHAPTER 11

Synthesis, Properties, and Biological Applications of Thiophene

DIPANKAR DAS*, ARPITA DAS, SOURAV SUTRADHAR, PARNASHABARI SARKAR AND BISWA NATH GHOSH

National Institute of Technology Silchar, Silchar, Assam, 788010, India
*E-mail: ddipankardas01@gmail.com

11.1 Introduction

Heterocyclic compounds are extensively dispersed in nature and have widespread synthetic applications. They are extremely significant in organic chemistry and medicinal chemistry because of their profound pharmacological activities, which attracted medicinal chemists to design, organize, and apply innovative strategies for the development of novel medications. Heterocyclic compounds are organic compounds that possess a ring structure consisting of at least one carbon atom, along with one or more other elements, such as N, O, or S. Thiophene, a monocyclic five-membered ring with the chemical formula C_4H_4S and the chemical name thiacyclopentadiene, is an example of such heterocyclic compounds, containing a sulfur heteroatom. The term 'thiophene' was derived from the Greek words 'theion' and 'phaino', which mean sulfur and shining, respectively.[1–4]

Thiophene has a molecular mass of 84.14 g mol^{-1}, a density of 1.051 g mL^{-1}, and a melting point of −38 °C. At room temperature, thiophene is

S-Heterocycles: Retrospect, Prospects, and Biological Applications
Edited by Shrikaant Kulkarni, Hemantkumar Akolkar, Vijay M. Khedkar and A. K. Haghi

Published by the Royal Society of Chemistry, www.rsc.org

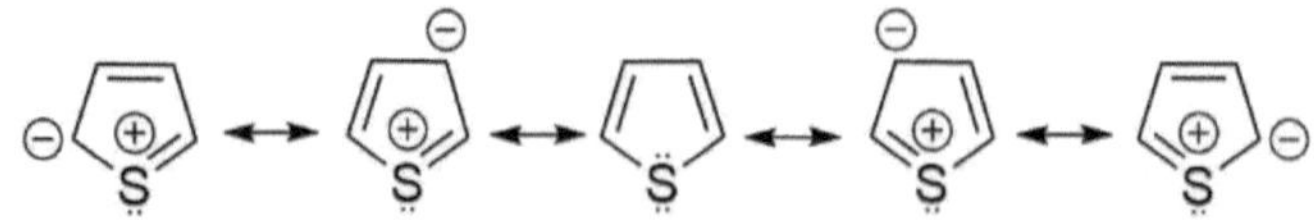

Scheme 11.1 Schematic representation of delocalization of electron pairs in thiophene.

a colorless, hazardous, and flammable liquid. Thiophene is soluble in the majority of organic solvents, including alcohol and ether, but insoluble in water.[5] The presence of π-electron clouds in thiophene makes it reactive similarly to benzene derivatives. Out of the two electron pairs, one electron pair of the sulfur atom is highly delocalized within the five-membered ring, making it obey Hückel's rule of aromaticity, and the other pair is located in an sp^2-hybrid orbital (Scheme 11.1). Although thiophene is resonance-stabilized and aromatic in nature, theoretical calculations showed that its degree of aromaticity is less than that of benzene. Thiophene has an ionization potential of 8.9 eV (and benzene has 9.3 eV), making it slightly more nucleophilic than benzene.[6] Furthermore, thiophene can easily undergo nitration, sulfonation, acylation, and halogenation but fails to undergo alkylation and oxidization reactions.[7]

Among sulfur-containing heterocyclic receptor systems, the synthesis of thiophene and its derivatives has drawn significant attention in the pharmaceutical chemistry and materials science field as inhibitors of metal corrosion, and in the industrial chemistry field, they play crucial roles in the advancement of organic semiconductors and fabrication of organic field-effect transistors and organic light-emitting diodes.[7,8] Thiophene derivatives and thiophene-fused heterocyclic moieties display a wide range of biological activities, including antibacterial,[9,10] anticancer,[11,12] antileishmanial,[13,14] anticonvulsant,[15] enzyme inhibitor,[16] antiurease,[17] antithrombotic,[18] anti-influenza,[19,20] antioxidant,[21,22] anti-inflammatory,[23,24] antifungal,[25,26] and antitubercular activities. Thiophene has demonstrated itself to be an alluring isostere that enhances the efficacy of various drugs, enabling them to be incorporated as an integral component of numerous commercially available drugs, such as dorzolamide, suprofen, tiaprofenic acid, methapyrilene, OSI-930, clopidogrel, tienilic acid, ticlopidine, olanzapine, etizolam, raloxifene, morantel, tigabine, tipepidine, timepidium bromide, tioconazole, citizolam, sertaconazole, cefoxitin, penthiopyrad, raltitrex, thiophenfurin, tiamenidine, articaine, and dimethenamide. Schemes 11.2 and 11.3 show the structures of some important commercially available thiophene-based drugs.

Therefore, considering the importance of thiophene-based derivatives in pharmaceutical chemistry for the development of novel medications, the synthesis of thiophene-based derivatives has drawn significant attention. The condensation process is a common and important synthetic route to thiophene derivatives, including the Gewald,[27] Paal–Knorr,[28] Fiesselmann,[29] and Hinsberg[30] synthesis. Aminothiophene derivatives are produced *via*

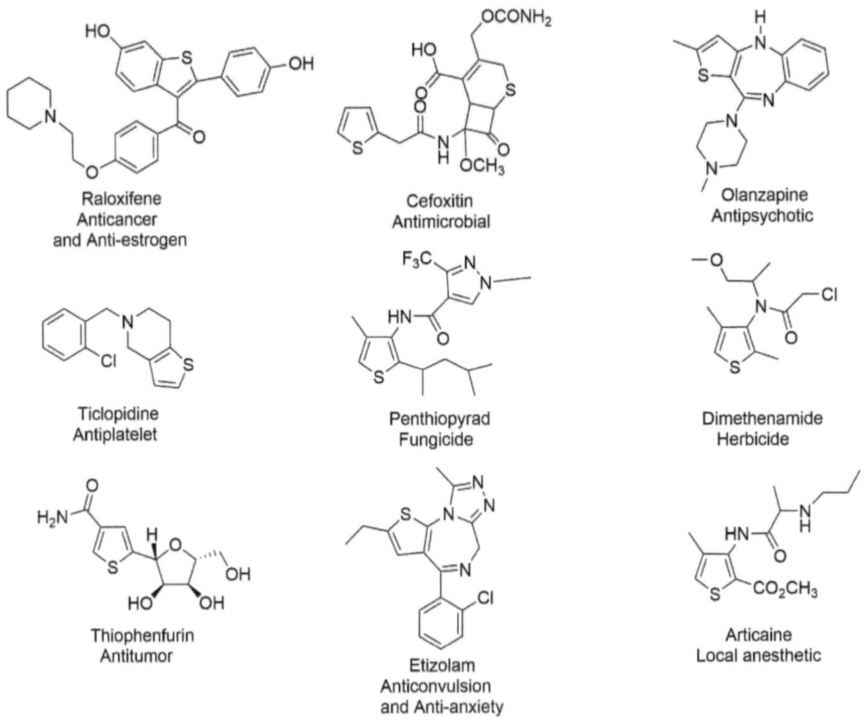

Scheme 11.2 Structures of some important commercially available thiophene-based drugs.

the Gewald reaction, which involves an α-cyano ester, a sulfur atom, and an α-methylene carbonyl molecule. The Paal–Knorr reaction involves the condensation of 1,4-dicarbonyl moieties in the presence of phosphorus pentasulfide (P_4S_{10}). 3-Hydroxy-2-thiophene carboxylic derivatives are produced *via* the Fiesselmann reaction, which involves condensation between α,β-acetylenic ester and thioglycolic acid derivatives under the basic condition. Thiophene carboxylic acid derivatives are produced *via* the Hinsberg reaction, which involves the formation of the δ-lactone intermediate from dialkyl thiodiacetate and 1,2-diketones in the presence of potassium or sodium alkoxide (see Scheme 11.4 for synthetic schemes of Gewald, Paal–Knorr, Fiesselmann, and Hinsberg synthesis of substituted thiophene).

In recent years, various innovative synthetic methods have been reported to improve the yield, efficiency, regioselectivity, and tolerability of thiophene derivatives through metal-catalyzed reaction, iodocyclization reaction or iodine-promoted heterocyclization, one-pot synthetic methodology in the absence or presence of catalysts, and multicomponent reaction procedures.

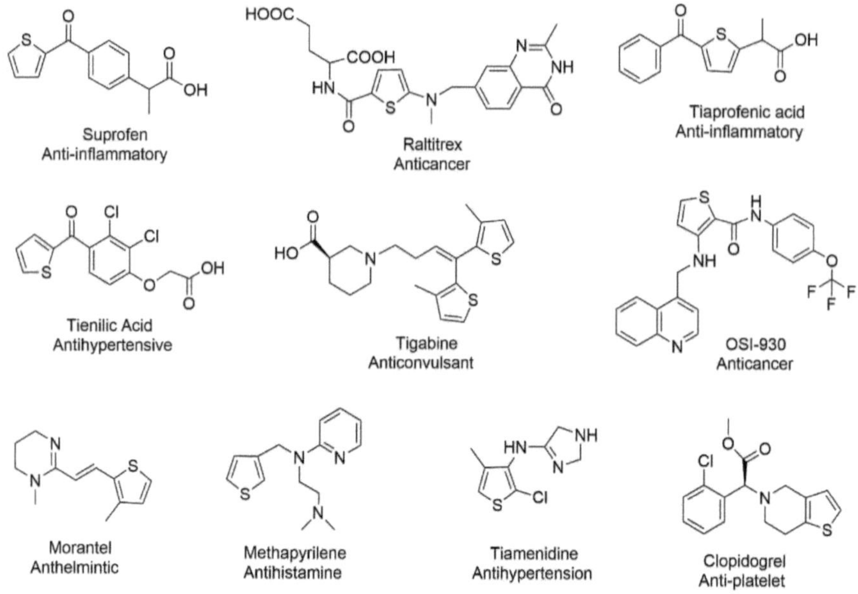

Scheme 11.3 Structures of some important commercially available thiophene-based drugs.

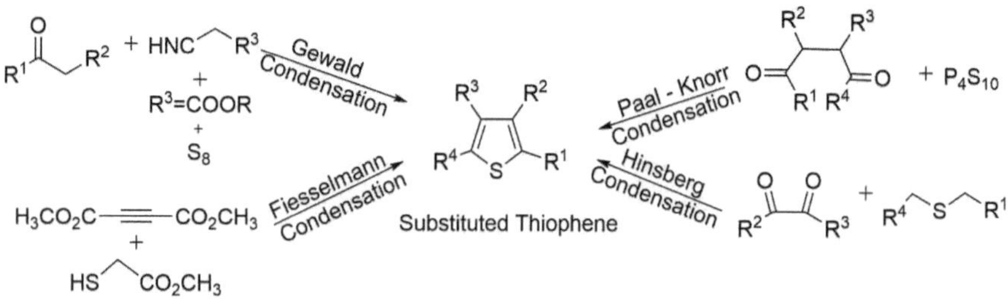

Scheme 11.4 Synthetic schemes of substituted thiophene through Gewald, Paal–Knorr, Fiesselmann, and Hinsberg reactions.

11.2 Recent Developments in the Synthesis of Thiophene Derivatives

11.2.1 Synthesis of Thiophene Derivatives Through Metal-catalyzed Reaction

Among all the synthetic procedures, the most effective method for the synthesis of substituted thiophene derivatives from easily accessible acyclic substrates in a regioselective and cost-effective manner is metal-catalyzed heterocyclization. In 2019, Yu *et al.*[31] proposed a synthetic procedure for a series of substituted thiophene derivatives (**3**) (for synthetic protocol, see

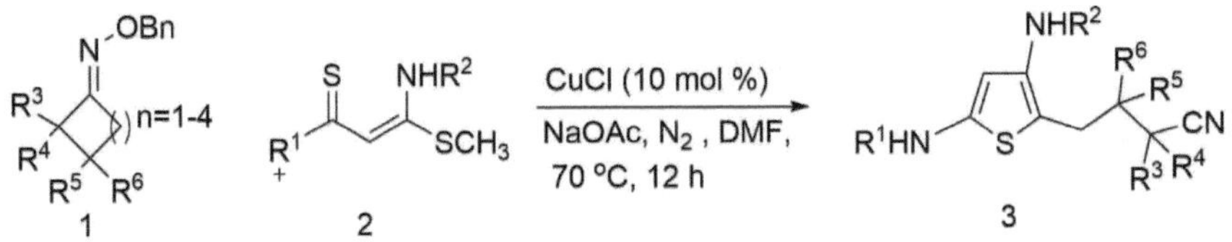

R^1 =Ph, 2-OMeC_6H_4, 2-FC_6H_4, 3-MeC_6H_4, 3-$CF_3C_6H_4$, 4-$CF_3C_6H_4$, 4-OMeC_6H_4, 4-BrC_6H_4
R^2 =Me, Et, Bu, Allyl, Ph, Bn, $CH_2CH_2C_6H_5$, 2-Cl-C_7H_6, 3-OMeC_7H_6 , 4-OMeC_7H_6, 4-FC_7H_6, 4-ClC_7H_6, furan-2-yl, 3-ethyl-indoyl
R^3= H, Me, Allyl, Ph, Bn R^4=H, Me
R^5= H, Ph, $(CH_3)_2CH_2CO_2Me$, 4-tBuC_6H_4, CO_2tBu R^6= H, Me, Ph

Scheme 11.5

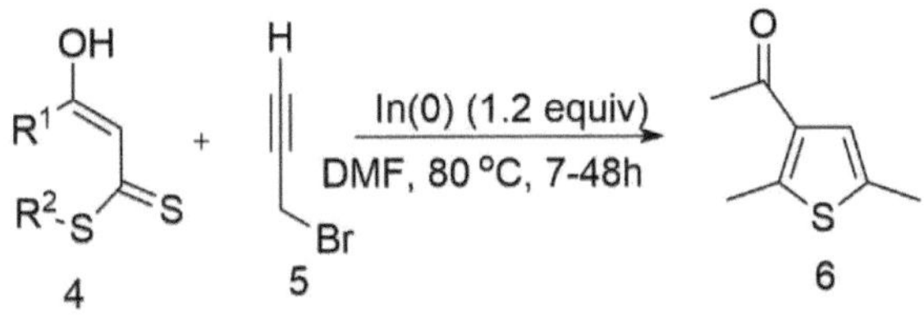

R^1 =Me, iPr, Ph, 2-ClC_6H_4, 4-MeC_6H_4, 4-OMeC_6H_4, 3-BrC_6H_4, 4-ClC_6H_4,4-$CF_3C_6H_4$, biphenyl, furan-2yl, thiophene--yl
R^2= Me, Et, tBu, CH_2Ph,

Scheme 11.6

Scheme 11.5) through a radical ring-opening reaction between 2-substituted cyclobutanone oxime esters (**1**) and enaminothiones (**2**) *via* C–C bond cleavage (a thermodynamically less favorable reaction).

In 2015, Singh *et al.*[32] proposed a synthetic procedure for the synthesis of 2-(alkylthio)-5-alkyl thiophenes (**6**) through a Csp^3–S cross-coupling reaction between enolic dithioester (**4**) and propargyl bromide (**5**) with a high yield (for synthetic protocol, see Scheme 11.6).

In 2014, Deng *et al.*[33] proposed a synthetic method for poly-substituted 2-aminothiophene derivative (**9**) by a reaction between thioamides (**7**) and alkynoates (**8**) in the presence of Cu(II) as a catalyst (see Scheme 11.7).

In 2018, Deng *et al.*[34] proposed a Cu(II)-catalyzed synthetic procedure for dihydrothiophene derivative (**12**) in ethanolamine at 70 °C *via* a reaction between β-ketothioamides (**10**) and enynones (**11**) through the Michael addition in the presence of Cu(II) as a catalyst, followed by intramolecular cyclization and the protonation cascade reaction (see Scheme 11.8).

In 2018, Singh *et al.*[35] utilized substituted naphthalene-2-ol (**13**) to react with α-enolic dithioesters (**14**) in the presence of Cu(II) as a catalyst, producing a fused tricyclic thiophene derivative (**15**) with a yield of 65–90% (see Scheme 11.9).

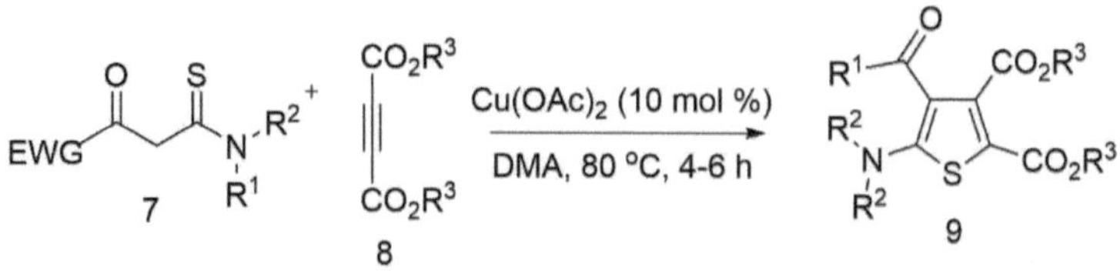

EWG =-COMe,-COEt, -COt-Bu, -CON(Me)$_2$, -CSN(Me)$_2$, -CO$_2$Me,OBn
R^1=R^2= Me, Et, Ph, Morpholine R^3= Me, Et

Scheme 11.7

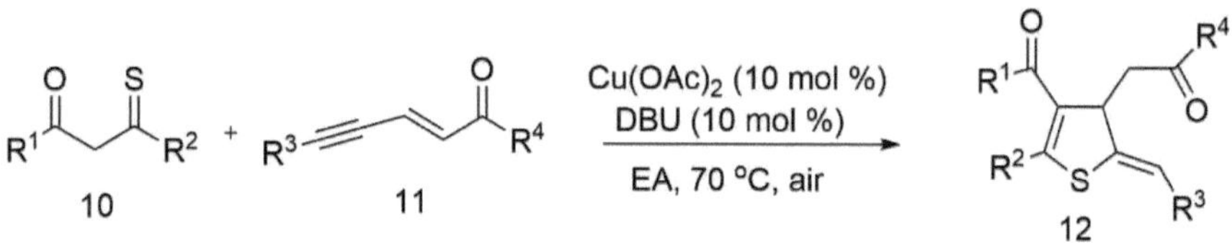

R^1= -OMe, iPr, Me, Et, OBn, R^2= -N(Me)$_2$, morpholine, -NMePh
R^3= Cyclopropyl, Ph, 4-OMeC$_6$H$_4$, 4-BrC$_6$H$_4$ R^4= 4-OMeC$_6$H$_4$, 4-ClC$_6$H$_4$, Ph

Scheme 11.8

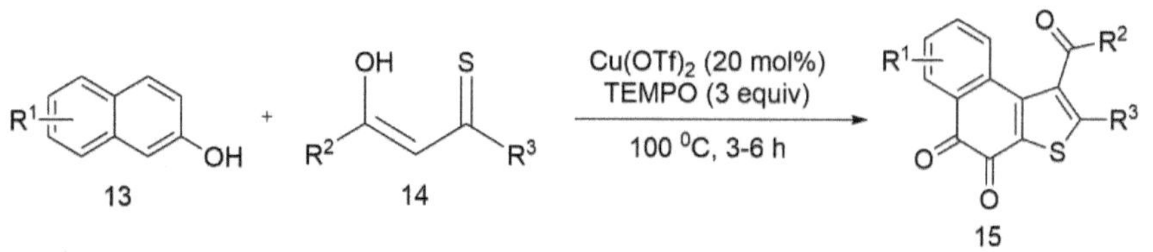

R^1 = H, CN, Br, CO$_2$Me, p-MeOC$_6$H$_4$—≡
R^2 = Me, cyclopropyl, Ph, 4-PhC$_6$H$_4$, 3-BrC$_6$H$_4$, 4-MeC$_6$H$_4$, 4-OMeC$_6$H$_4$, 4-CF$_3$C$_6$H$_4$, 4-ClC$_6$H$_4$, 2-Furyl, 2-thiophenyl
R^3= Me, Et, Pr, Bu, Bn

Scheme 11.9

In 2016, Kurandina and Gevorgyan[36] reported a simple one-step synthesis of a series of tetrasubstituted thiophene derivatives (**18**) with an excellent yield of 57–99% by a reaction between 1,2,3-thiadiazole (**16**) and a substituted alkyne (**17**) moiety. In this reaction, rhodium was used as a catalyst (see Scheme 11.10).

In 2016, Jiang *et al.*[37] proposed a synthetic method for arylated indeno[1,2-*c*]thiophenes (**20**) with a good yield of 45–79% by Cu(II)-catalyzed Michael addition followed by cyclization of 1,6-enynes (**19**) in the presence of potassium sulfide (see Scheme 11.11).

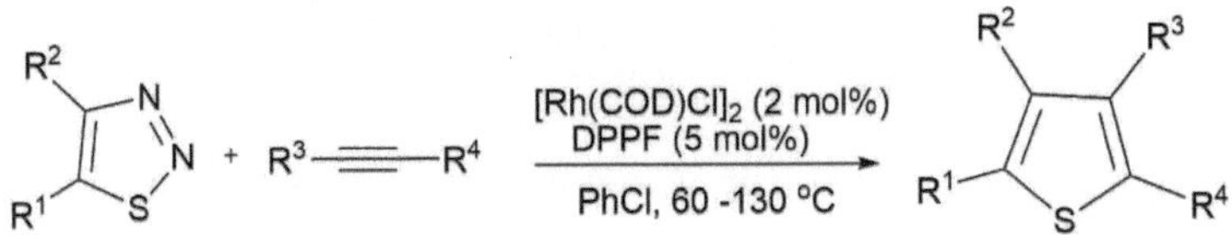

16 17 18

R^1 = H, CO_2Et, Ph R^2 = Ph, CO_2Et R^3 = H, C_3H_7
R^4 = C_3H_5, C_5H_{11}, cyclopentyl, tBu, C_3H_6Cl, C_3H_7, CH_2OPiV, $SiEt_3$, C_8H_{17}, Ph, 2-OMeC_6H_4, 4-OMeC_6H_4, 4-FC_6H_4, 4-$CF_3C_6H_4$, thiophene-2-yl

Scheme 11.10

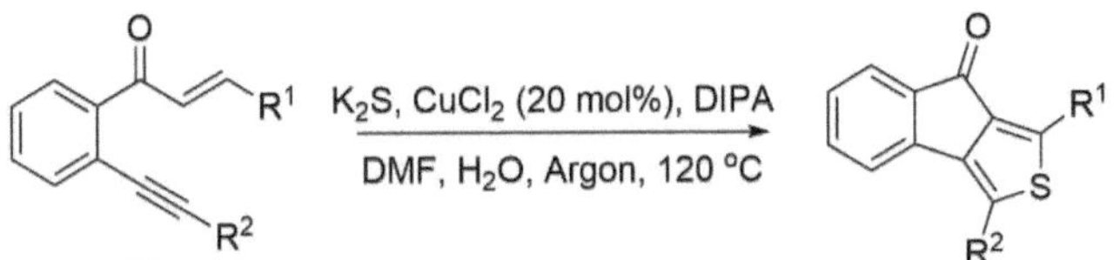

19 20

R^1 = Ph, 4-MeC_6H_4, 4-OMeC_6H_4, 4-ClC_6H_4,

R^2 = *n*Bu, Ph, 4-MeC_6H_4, 4-OMeC_6H_4, 4-ClC_6H_4, 4-FC_6H_4, 4-tBuC_6H_4, 4-EtC_6H_4,

Scheme 11.11

11.2.2 Synthesis of Thiophene Derivatives Through Iodocyclization Reaction

In recent years, the electrophilic iodocyclization process has attracted a lot of interest as an adaptable method for the synthesis of new highly functionalized heterocyclic derivatives. Iodocyclization reaction takes place through a nucleophilic attack on alkyne moiety in the presence of iodine. Mancuso *et al.*[38] proposed the one-step synthesis of substituted thiophene derivatives (**22**) through the iodocyclization reaction by utilizing 1-mercapto-3-alkyne-2-ol (**21**) in the presence of iodine and EmimEtSO_4 as a solvent (see Scheme 11.12).

Maity and Ranu[39] proposed a synthetic method for 2,3-disubstituted thiophene derivatives (**24**) from substituted benzene having a 1,3-dienyl bromide moiety (**23**) on reaction with potassium sulfide in the presence of iodine in argon gaseous medium (see Scheme 11.13).

Deng *et al.*[40] proposed a synthetic method for poly-substituted thiophene (**27**) with a good yield of 47–91% from thioamide derivatives (**25**) on reaction with allene (**26**). In this reaction, tetrabutylammonium iodide (TBAI) and *tert*-butyl hydroperoxide (TBHP) catalysts were utilized (see Scheme 11.14).

Zali-Boeini *et al.*[41] proposed the synthesis of tetra-arylated thiophenes with moderate yields of 58–80%. They utilized an α-halo-acetophenone

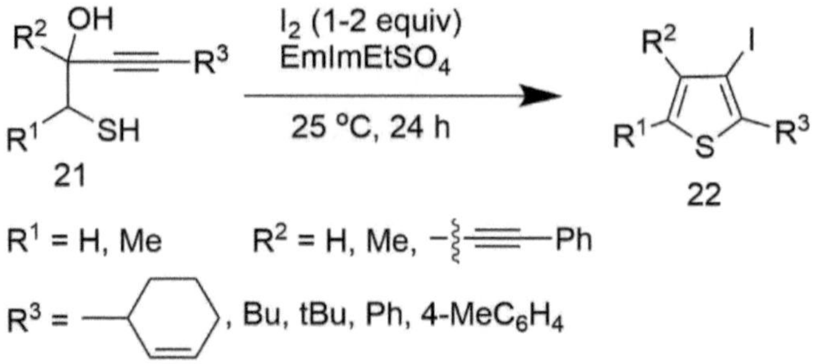

Scheme 11.12

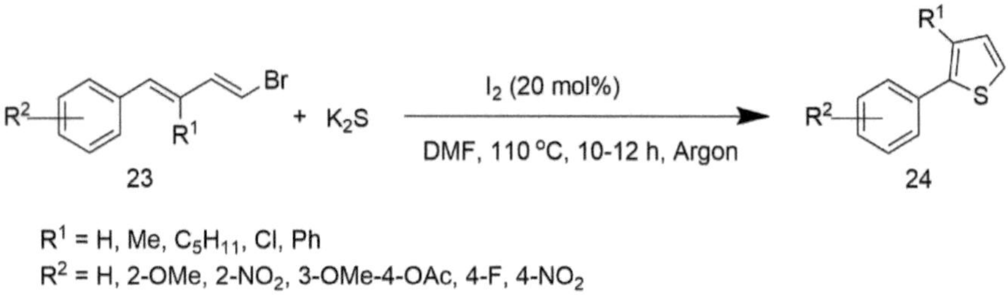

Scheme 11.13

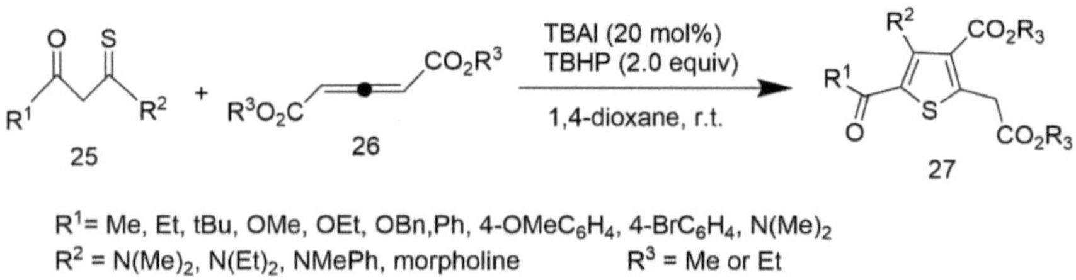

Scheme 11.14

derivative (**28**) to react with sodium sulfide in an ethanol medium, yielding a sulfide analogue, which on further reaction with another α-halo-acetophenone (**29**) derivative in the presence of 1,8-diazabicyclo[5.4.0] undec-7-ene (DBU), NaI, and phase-transfer *N*,*N*,*N*-trimethylhexadecylammonium bromide (HTAB) catalyst resulted in tetra-arylated thiophene derivatives (**30**) with a good yield (see Scheme 11.15).

Zeni *et al.*[42] proposed the synthesis of tetrasubstituted thiophene derivatives (**33**) with a high yield of 39–81% from substituted 1,3-diynes (**31**) on reaction with diorganyl disulfide (**32**) as cyclizing agents in the presence of iron salts, such as $FeCl_3$, in dry DCM (see Scheme 11.16).

Zali-Boeini and Fadaei[43] proposed a synthetic method for tetrasubstituted thiophene derivatives (**35**) having morpholine moieties substituted

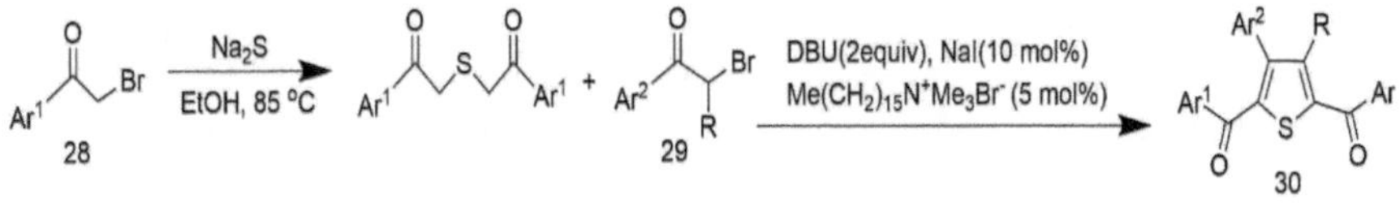

R = H, C_6H_4 Ar[1] = 4-MeC_6H_4, 4-ClC_6H_4, 4-BrC_6H_4, 4-pyridyl
Ar[2] = 4-MeC_6H_4, 4-ClC_6H_4, 4-BrC_6H_4, 4-$NO_2C_6H_4$, 4-$OMeC_6H_4$, 4-$AcNHC_6H_4$, 4-pyridyl

Scheme 11.15

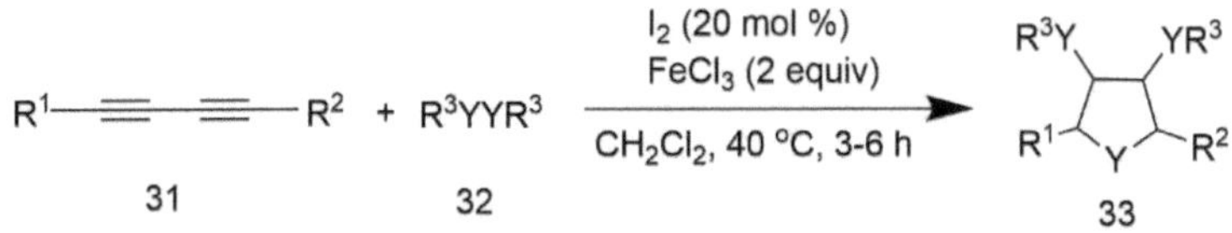

Y = S or Se R^1 = Bu, C_8H_{17}, Ph, 2-MeC_6H_4, 4-FC_6H_4
R^2 = Bu, C_8H_{17}, Ph, 2-MeC_6H_4, 4-MeC_6H_4, 4-FC_6H_4 R^3 = Me, Bu

Scheme 11.16

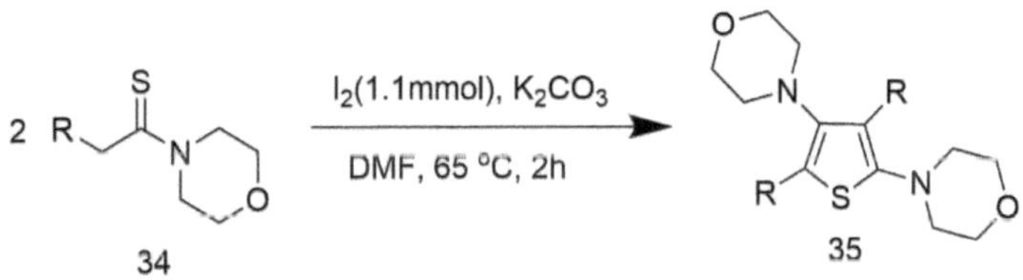

R = 4-MeC_6H_4, 4-ClC_6H_4, 4-BrC_6H_4, 4-$OMeC_6H_4$, Ph, 4-biphenyl, 2-naphthyl

Scheme 11.17

at positions 2 and 4 through a self-condensation reaction of substituted 1-morpholinoethan-1-thione (**34**) by utilizing K_2CO_3 and iodine as catalysts (see Scheme 11.17).

11.2.3 Synthesis of Thiophene Derivatives Through Metal-free Approaches

Tetrahydrothiophene derivatives were synthesized by utilizing a 1,4-dithiane-2,5-diol (**37**) reagent under basic conditions through the formation of mercaptoacetaldehyde, which later on reaction with alkyne derivative through [3+2] annulation gives the thiophene derivative. Punniyamurthy *et al.*[44] proposed the synthesis of tetrasubstituted thiophene derivatives (**38**) with a good yield (29–81%) by utilizing an alkyne derivative (**36**) with 1,4-dithiane-2,5-diol (37) and 1,4-diazabicyclo[2,2,2]octane (DABCO) as catalysts (see Scheme 11.18).

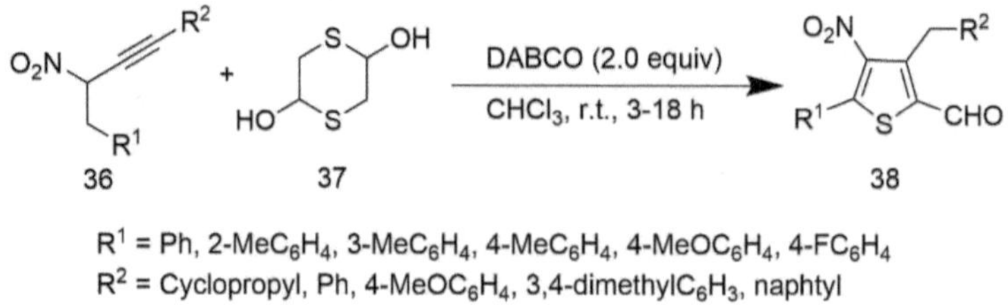

Scheme 11.18

Peng *et al.*[45] proposed a synthetic procedure for 2-aminothiophenes derivatives (**40**) and thieno[2,3-*c*]isothiazoles (**41**) by utilizing linear alkynyl nitriles (**39**) and sulfur-transferring reagents, such as P_4S_{10} or Lawesson's reagent (LR), with an excellent yield up to 95% and 73%, respectively (see Scheme 11.19).

Reddy *et al.*[46] utilized substituted 4-en-1-yn-3-yl acetates (**42**) through the nucleophilic addition reaction by potassium thioacetate, followed by annulation/aromatization, resulting in 2,4-disubstituted thiophene derivatives (**43**) with a good yield of 62–95% (see Scheme 11.20).

Harrity *et al.*[47] proposed the synthesis of thiophene derivatives (**46**) through a condensation reaction between alkyl thiols (**45**) and trifluoroborate salts of the alkyn-one substrate (**44**) in the basic condition of potassium carbonate, with a good yield of 55–95% (see Scheme 11.21).

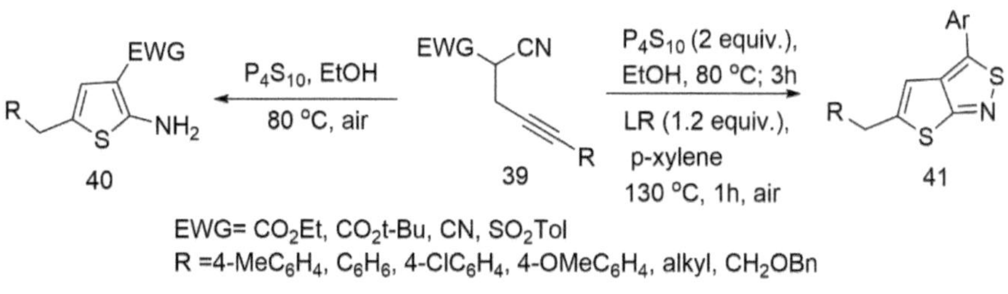

Scheme 11.19

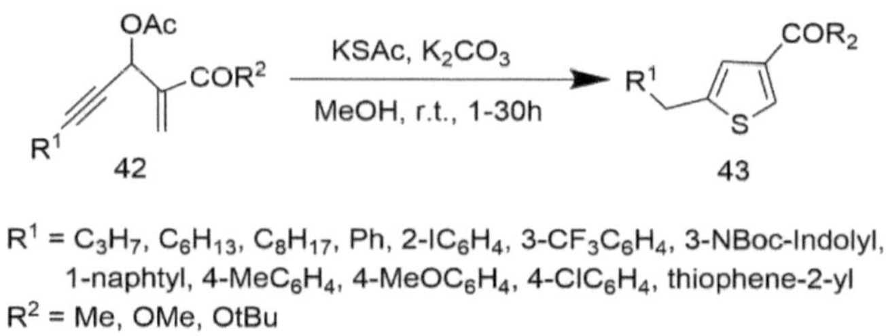

Scheme 11.20

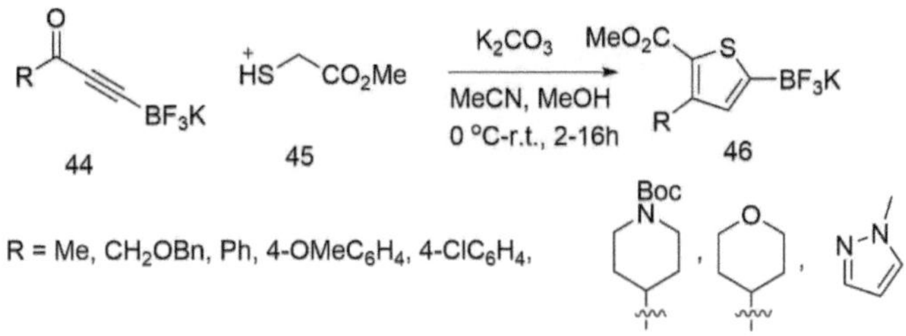

Scheme 11.21

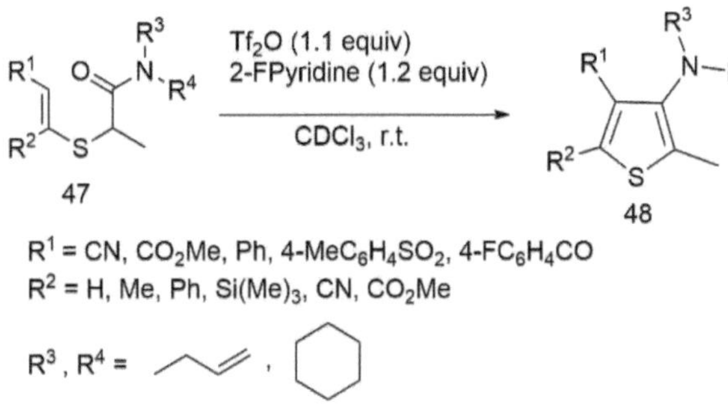

Scheme 11.22

Mesmaeker *et al.*[48] proposed a simple synthetic procedure for highly substituted 3-aminothiophene (**48**) derivatives through a keteniminium salt formation between vinylthioacetamide moieties (**47**) and triflic anhydride, along with 2-fluoropyridine (see Scheme 11.22).

Li *et al.*[49] utilized 1,4-zwitterionic thiolates having an acyl moiety (or trifluoromethyl) (**49**) to react with activated alkynes (**50**) for synthesizing poly-substituted thiophenes (**51**) through a [3+2] S-annulation process with an excellent yield up to 40–99% (see Scheme 11.23).

11.2.4 Synthesis of Thiophene Derivatives Through Multicomponent Reaction Approaches

Gong *et al.*[50] proposed a multicomponent reaction approach for the synthesis of 2-nitrothiophenes (**53**) from 3-chloroacrylaldehyde (**52**) on reaction with potassium thiocyanate in an acidic medium, producing 3-thiocyanatoacrylaldehyde and then 3-thiocyanatoenal, which on further reaction with nitromethane in a basic medium produced a requisite product (**53**) with a good yield of 33–65% (see Scheme 11.24).

Mantellini *et al.*[51] proposed a novel multicomponent synthetic procedure for tetrasubstituted 2,5-dihydrothiophene derivatives (**59**) by utilizing

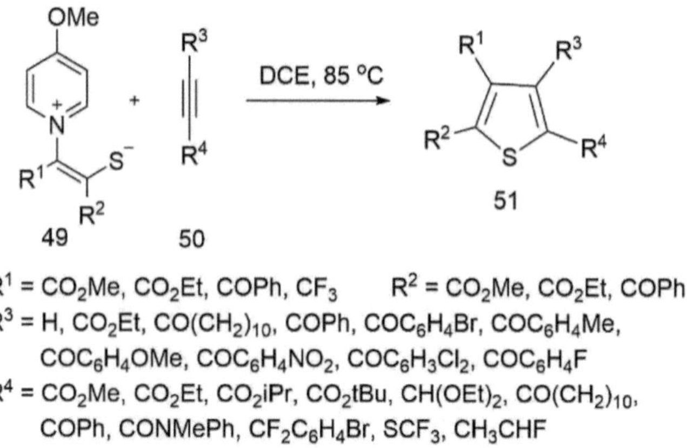

Scheme 11.23

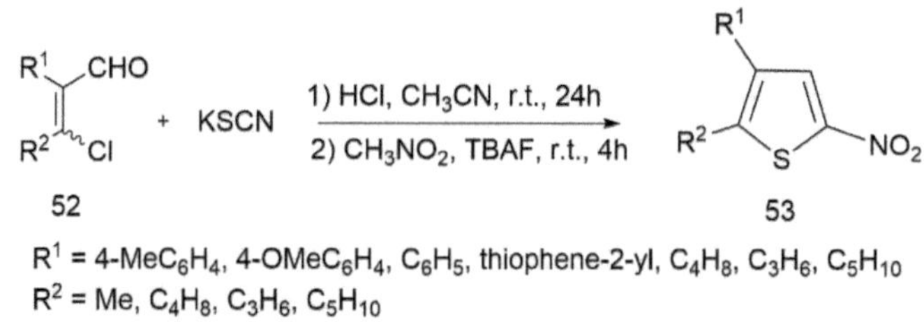

Scheme 11.24

four-component reactants, *viz.* β-ketoesters (**54**), aryl isothiocyanates (**55**), primary alkyl amines (**57**), and 1,2-diaza-1,3-dienes (**56**). The hydrazone moiety of 5-2,5-dihydrothiophenes (**58**) can be hydrolyzed to 5-aminothiophene-2,4-dicarboxylates (**59**) with a good yield (44–69%) with the help of Amberlyst 15*H* catalyst (see Scheme 11.25).

Mantellini *et al.*[52] proposed a synthetic procedure for tetrasubstituted thiophene derivative (**61**) by the hydrolysis of 2-arylamino-5-hydrazono-thiophene (**60**) under the basic condition (see Scheme 11.26).

Luo *et al.*[53] proposed a metal-free synthetic procedure for 2-aminothienylether derivatives (**64**) by utilizing a reaction between thioamides (**62**) and 2-ynals (**63**) in an alcoholic medium (see Scheme 11.27).

Lee *et al.*[54] proposed a solid-phase synthetic procedure for substituted thiophene derivatives (**69**) by utilizing a Merrifield resin solid (**65**). The reaction proceeded through the generation of a polymer-supported thiolate moiety (**67**) from the dipotassium salt of thiolate (**66**). The synthesized compound (**67**) is further reacted with α-haloaryl derivatives in acetone to produce 2-amino-thiophene resin (**68**), followed by a reaction with

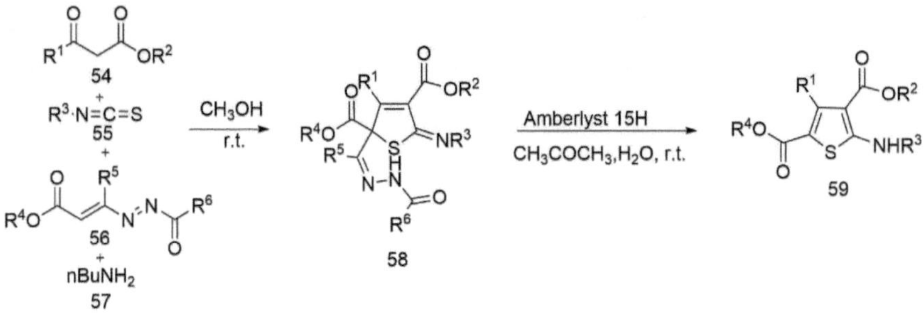

R_1= Me, Et, Pr R^2= Me, Et, iPr, tBu, allyl, Bn R^3= Ph, 4-ClC_6H_4, 4-$OMeC_6H_4$
R^4= Me, Et, tBu R^5= Me, Et R^6= Me, tBu, NH_2, NHC_6H_4

Scheme 11.25

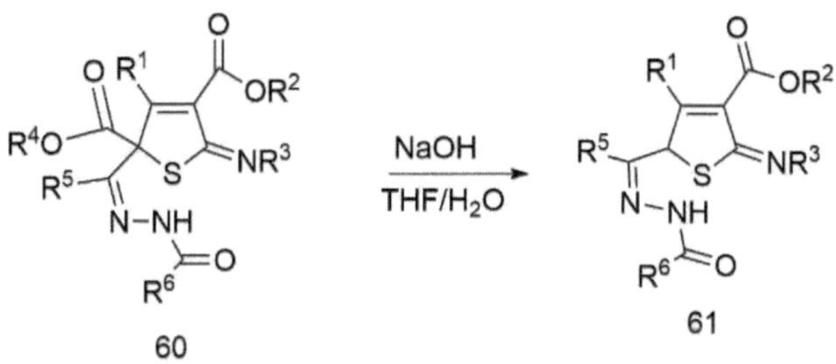

R^1= Me, Et R^2= Me, Et R^3= Ph, 4-ClC_6H_4, 4-$OMeC_6H_4$
R^4= Ph, 4-BrC_6H_4, 4-$NO_2C_6H_4$ R^5= OMe, OEt, OtBu, 4-pyridyl
R^6= OMe, Ot-Bu, OEt, Ph, py

Scheme 11.26

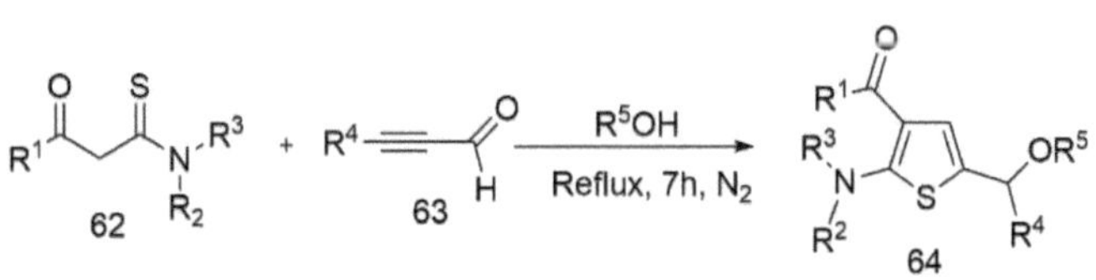

R_1= Me, Et, tBu, $N(Me)_2$, OMe, OEt, O-tBu, OBn, Ph, 4-$CF_3C_6H_4$, 4-BrC_6H_4, 4-$OMeC_6H_4$, 4-$OC_8H_{17}C_6H_4$, thiophene-2-yl
R^2= H, Me, Et, Ph, morpholine R_3= Me, Et, Ph, morpholine
R_4= Ph, 4-BrC_6H_4, 4-$OMeC_6H_4$ R_5= Me, Et, iPr

Scheme 11.27

m-chloroperoxybenzoic acid (*m*-CPBA) and then R^2H yielding tetrasubstituted thiophene (**69**) (see Scheme 11.28).

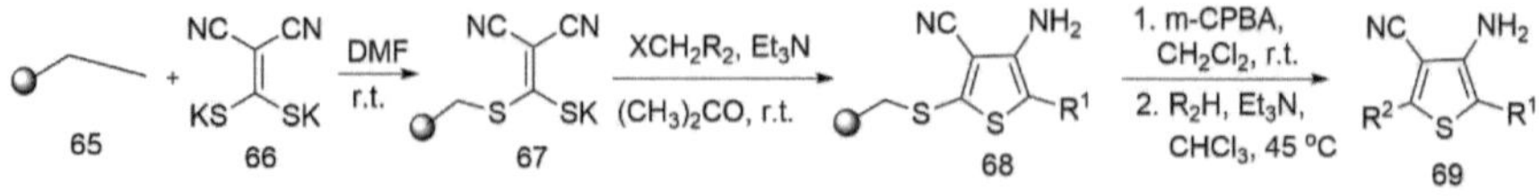

R^1= COPh, CO_2Et, COMe X=Cl, Br
R^2= BnNH, 4-OMeBnNH, n-BuNH, i-PrNH, PhS, n-PrS, pyrrolidene, piperidine, morpholine

Scheme 11.28

11.3 Recent Advances in Biological Applications of Thiophene Derivatives

11.3.1 Antimicrobial Activity of Thiophene Derivatives

Among microbial infections, fungal and bacterial infections have not drawn much attention in recent decades, but the prevalence of both infections and their associated morbidity and mortality rates are rapidly increasing.

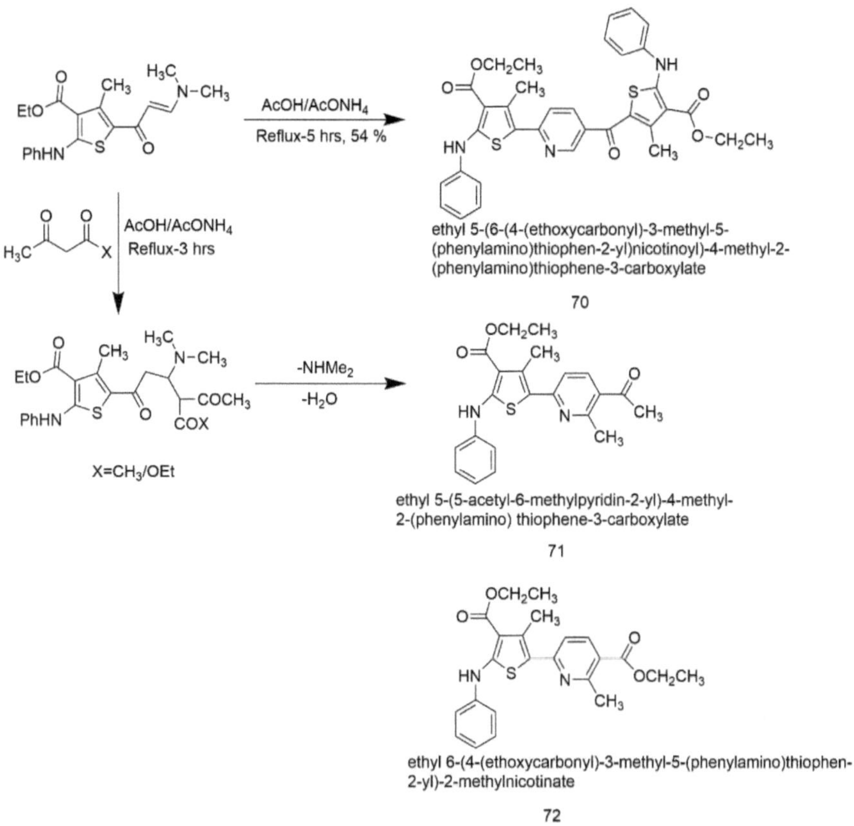

Scheme 11.29 Schematic representation of the proposed compounds having antimicrobial activity.

However, pathogenic fungi—especially *Aspergillus* and *Candida* species, *Bacillus subtilis*, and *Streptococcus pneumoniae*—are the most often identified pathogens causing illness and mortality in people having weakened immune systems.[55,56] Among the numerous heterocyclic compounds, thiophene-based derivatives were found to possess excellent antimicrobial properties.

Mabkhot *et al.*[10] proposed the synthesis of three heterocyclic compounds (**70–72**) by incorporating thiophene derivatives (Scheme 11.29) and assessed their antimicrobial activities. The synthesized compounds were tested for their effectiveness against various fungal strains, such as *Candida albicans*, *Geotrichum candidum*, *Syncephalastrum racemosum*, *Geotrichum candidum*, and *Aspergillus fumigatus*; Gram-negative bacterial strains, such as *Pseudomonas aeruginosa* and *Escherichia coli*; and Gram-positive bacterial strains, such as *Streptococcus pneumoniae* and *Bacillus subtilis*, using ampicillin, gentamicin, and amphotericin B as standard drugs. The screening results of these three compounds revealed excellent antimicrobial activity and promising antibacterial activity.

Ajdačić *et al.*[55] proposed the synthesis of three guanylhydrazone derivatives (**73–75**) containing thiophene moieties (see Scheme 11.30) and assessed their antifungal activity (*in vitro*) against *Fusarium oxysporum* AB18, *Microsporum canis* DMT2, *C. albicans* CA5, *A. fumigatus* ATCC 13073, *Candida parapsilosis* C27, *Trichophyton mentagrophytes* DMT4, and *C. albicans* ATCC10231, taking voriconazole as a standard drug. All these compounds expressed excellent antimicrobial activity against *C. albicans* CA5.

Rizk *et al.*[9] proposed the synthesis of two thiophene-2-carboxamide derivatives (**76** and **77**) (see Scheme 11.31) and assessed their *in vitro* antibacterial activity against Gram-positive (*S. aureus* and *B. subtilis*) and Gram-negative (*E. coli* and *P. aeruginosa*) bacterial strains using ampicillin as a standard drug. Furthermore, the antifungal activity was investigated against *C. albicans* using clotrimazole as a standard drug, and anti-hepatitis C virus (HCV) activity was also assessed. Both the compounds exhibited good antimicrobial activity and anti-HCV activity.

Mabkhot *et al.*[26] proposed the synthesis of three tetrasubstituted thiophene derivatives (**78–80**) (see Scheme 11.32) and assessed their *in vitro* antibacterial and antifungal activities. They considered *G. candidum*, *C. albicans*, *A. fumigatus*, and *S. racemosum* for testing antifungal activity and two each of Gram-negative (*P. aeruginosa* and *E. coli*) and Gram-positive (*S. pneumoniae* and *B. subtilis*) bacterial strains for testing antibacterial activity. Amphotericin B, ampicillin, and gentamicin were used in this study as standard antifungal and antibacterial drugs. The screening results showed that all these compounds exhibited excellent antifungal and antibacterial activities.

Mabkhot *et al.*[25] also proposed the synthesis of another tetrasubstituted thiophene derivative (**81**) (see Scheme 11.33) and screened it for *in vitro*

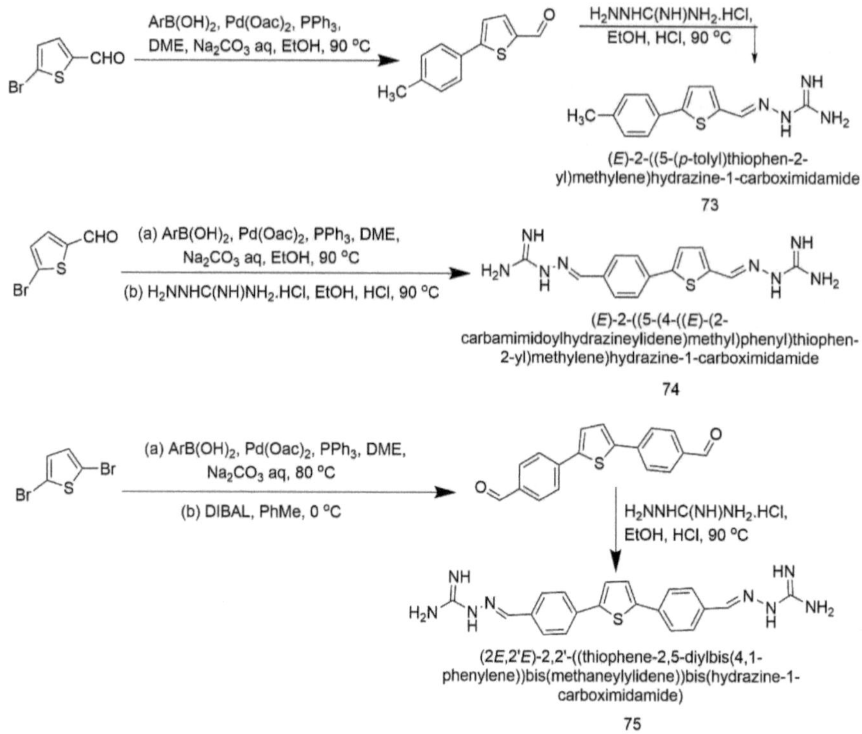

Scheme 11.30 Schematic representation of the proposed compounds having antifungal activity.

antifungal and antibacterial activities. In this study, they considered the same four fungal species and two each of Gram-negative and Gram-positive bacterial strains. The results revealed excellent antimicrobial activity.

11.3.2 Antileishmanial Activity of Thiophene Derivatives

Protozoans belonging to the genus *Leishmania* are responsible for a variety of illnesses, including leishmaniasis. Félix *et al.*[14] synthesized cycloalkyl-fused thiophene derivatives (**82–84**) containing an indole moiety (see Scheme 11.34) and assessed their *in vitro* antileishmanial activity. In this study, they have utilized *Leishmania amazonensis* promastigotes parasite and, as reference drugs, pentavalent and trivalent antimonates. All three compounds exhibited excellent antileishmanial activity with IC_{50} values of 2.1–3.2 μg mL^{-1}.

Jacomini *et al.*[13] proposed the synthesis of a series of thiophene derivatives (**85–87**) (see Scheme 11.35) and assessed their antileishmanial activity. All three compounds exhibited excellent antileishmanial activity against both axenic amastigote and promastigote forms of *L. amazonensis* parasite.

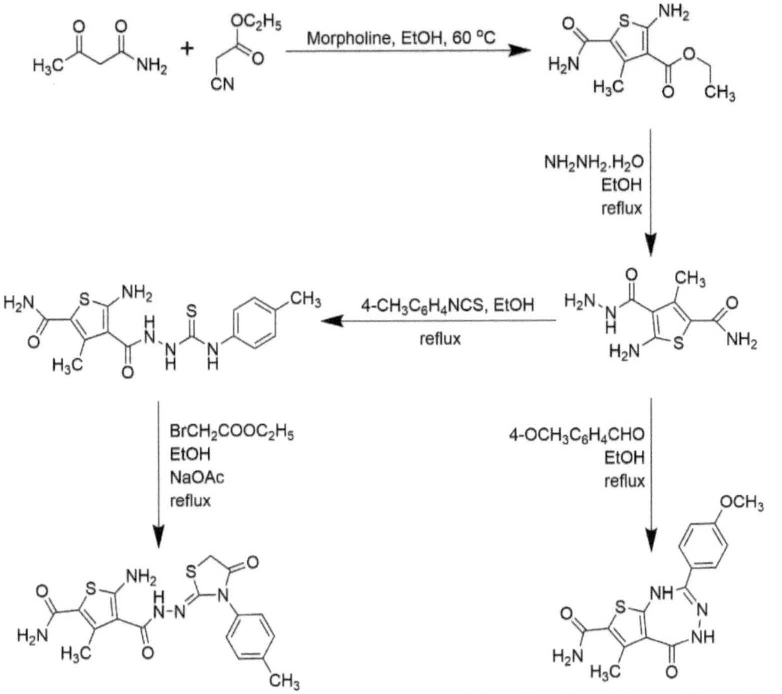

(*Z*)-5-amino-3-methyl-4-(2-(4-oxo-3-(*p*-tolyl)thiazolidin-2-ylidene)hydrazine-1-carbonyl)thiophene-2-carboxamide

76

2-(4-methoxyphenyl)-6-methyl-5-oxo-4,5-dihydro-1*H*-thieno[2,3-*e*][1,2,4]triazepine-7-carboxamide

77

Scheme 11.31 Schematic representation of the proposed compounds having antibacterial activity.

Rodrigues *et al.*[14] proposed the synthesis of cycloalkyl-fused 2-amino-thiophene derivatives (**88** and **89**) (see Scheme 11.36) and assessed their *in vitro* antileishmanial activity against axenic amastigote and promastigote forms of *L. amazonensis* using meglumine antimoniate as a reference drug. Both of the synthesized derivatives expressed potent antileishmanial activity.

Gonzalez *et al.*[57] proposed a synthetic procedure for thiophene derivatives (**90a–d**) utilizing palladium catalyst (for synthetic protocols, see Scheme 11.37) and investigated their *in vivo* and *in vitro* antiparasitic activity. Compounds **a**, **b**, **c**, and **d** were found to be very effective against *Leishmania donovani*, *Plasmodium falciparum*, *Trypanosoma cruzi*, and *Trypanosoma brucei* rhodesiense, and showed *in vivo* antiparasitic activity against *Trypanosoma brucei* rhodesiense. In these experimental studies, pentamidine and furamidine were used as standard drugs.

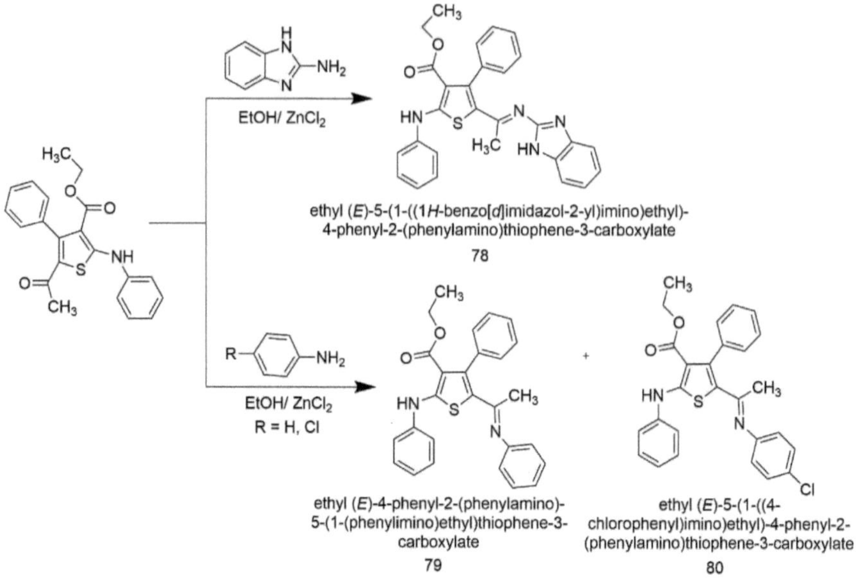

Scheme 11.32 Schematic representation of the proposed compounds having antimicrobial activity.

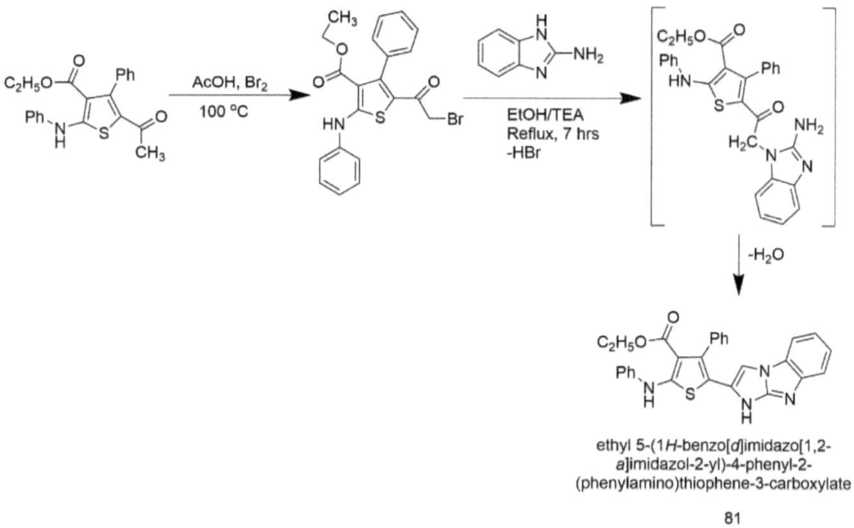

Scheme 11.33 Schematic representation of the proposed compounds having antimicrobial activity.

11.3.3 Antiviral Activity of Thiophene Derivatives

One of the major pathogens contributing to the high morbidity and fatality rates of respiratory infectious illnesses in humans is the influenza virus,

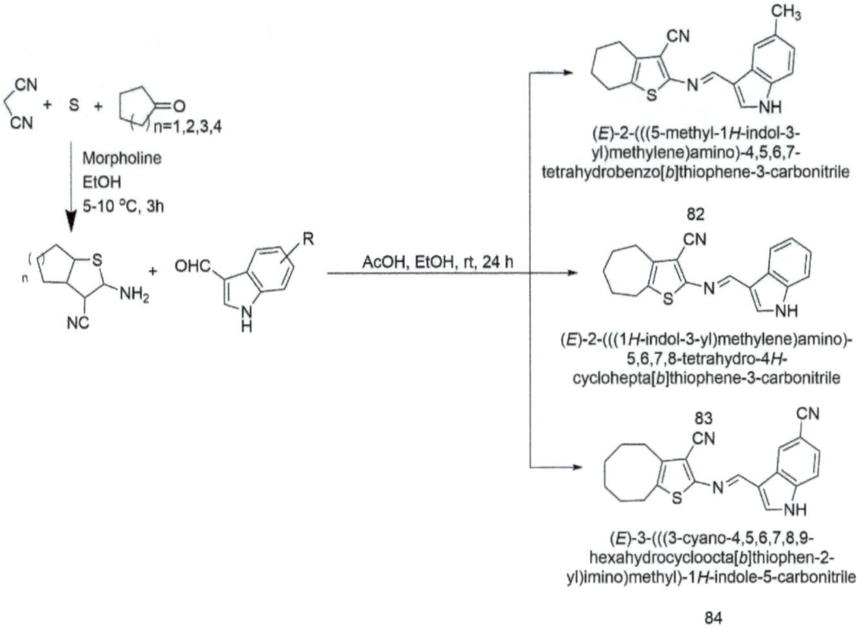

Scheme 11.34 Schematic representation of the proposed compounds having antileishmanial activity.

which causes flu. High rates of mutation and genomic reassortment across the strains cause the virus to evade vaccination-induced immunity and become resistant to the available antiviral medications.[57,58] Therefore, the synthesis of new drugs having anti-influenza/anti-flu activities attained a lot of attention. Desantis *et al.*[19] synthesized four thiophene derivatives, in which the thiophene ring is fused with the cycloheptane ring and connected to 2-hydroxybenzamide and other substituted benzamide groups to produce cycloheptathiophene-3-carboxamide derivatives (**91–94**) (see Scheme 11.38). These thiophene derivatives exhibited anti-influenza/anti-flu activities.

Rashad *et al.*[20] synthesized various thiophene-containing compounds (**95–97**) (see Scheme 11.39) and assessed their antiviral activity against the H5N1 virus through the 3-(4,5-dimethylthiazol-2-yl)-2,5-diphenyl tetrazolium bromide (MTT) assay and plaque reduction assay using zanamivir as a standard drug. The results of this study revealed that these compounds exhibited high antiviral activity against the H5N1 virus.

11.3.4 Anticancer Activity of Thiophene Derivatives

As cancer is one of the worst diseases, there is always a need for the development of anticancer drugs. To alleviate the drug resistance and

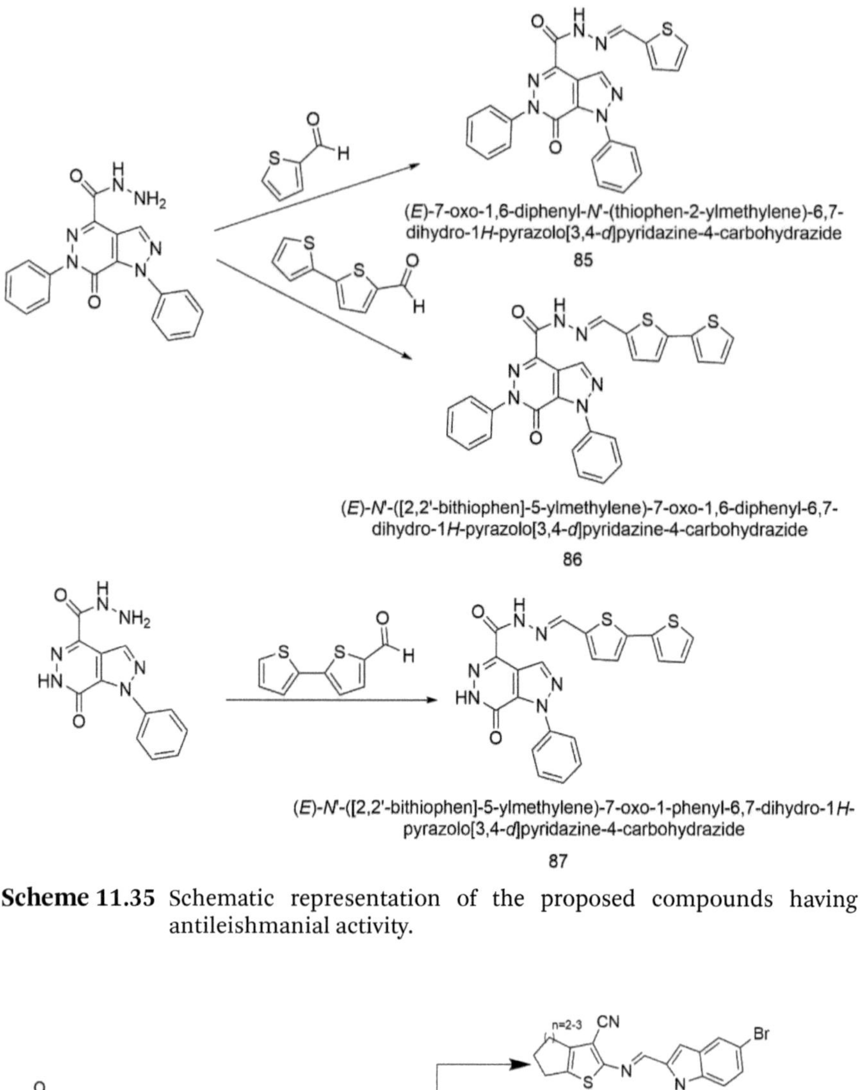

Scheme 11.35 Schematic representation of the proposed compounds having antileishmanial activity.

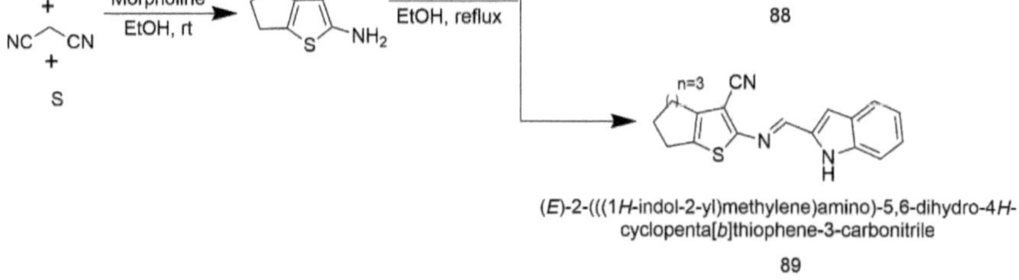

Scheme 11.36 Schematic representation of the proposed compounds having antileishmanial activity.

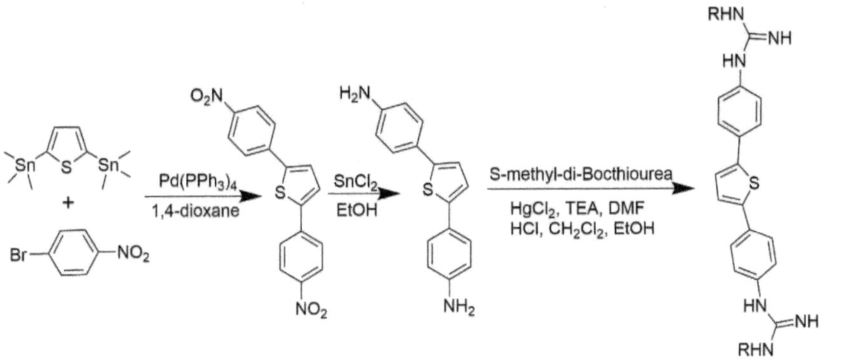

a) R = H b) R = Me
c) R = c-pent d) R = c-hex

1,1'-(thiophene-2,5-diylbis(4,1-phenylene))diguanidine
90

Scheme 11.37 Schematic representation of the proposed compounds having antiparasitic activity.

toxicity issues with the anticancer medications that are already available on the market and to meet the urgent demand for the development of more potent chemotherapeutic agents, research is still undergoing at full pace. Zhang *et al.*[11] synthesized and proposed various aminobenzo[*b*]thiophene-1,1-dioxides derivatives (see Scheme 11.40) and investigated their antitumor activity against MDAMB-231, MCF-7 (breast carcinoma), HCT116 (colon carcinoma), and A549 (lung carcinoma) cancer cell lines through the MTT assay experiment. The results of this study revealed that compound **98** has exhibited the most potent antitumor activity.

Mabkhot *et al.*[12] synthesized a tetrasubstituted thiophene derivative (**99**) (see Scheme 11.41) and assessed its *in vitro* cytotoxic activity. The antiproliferative study results revealed that the synthesized drug was potent against HeLa and PC3 cancer cell lines (human cervical and prostate cancer cells) but less effective compared to reference drugs soxorubicin and doxorubicin.

Ikram *et al.*[59] proposed the synthesis of a series of trisubstituted thiophene derivatives (**100a–g**) (see Scheme 11.42) with a good yield by utilizing the Suzuki cross-coupling reaction. The synthesized compounds were assessed against human breast cancer cell lines MCF-7, MDAMB-231, and 4 T1 by the MTT assay utilizing cisplatin as a standard drug. The synthesized compounds (**100a–g**) were found to be highly effective against all these cancer cell lines.

Perin *et al.*[60] proposed a synthetic procedure for fused tricyclic thiophene derivative (**101**) (see Scheme 11.43) and studied its efficacy against CFPAC-1, SW620, HeLa, A549, and HepG2 cancer cell lines using 5-fluorouracil as a standard drug. The compound was found to be most effective against HeLa with an IC_{50} value of 0.21–0.45 μM mol^{-1}.

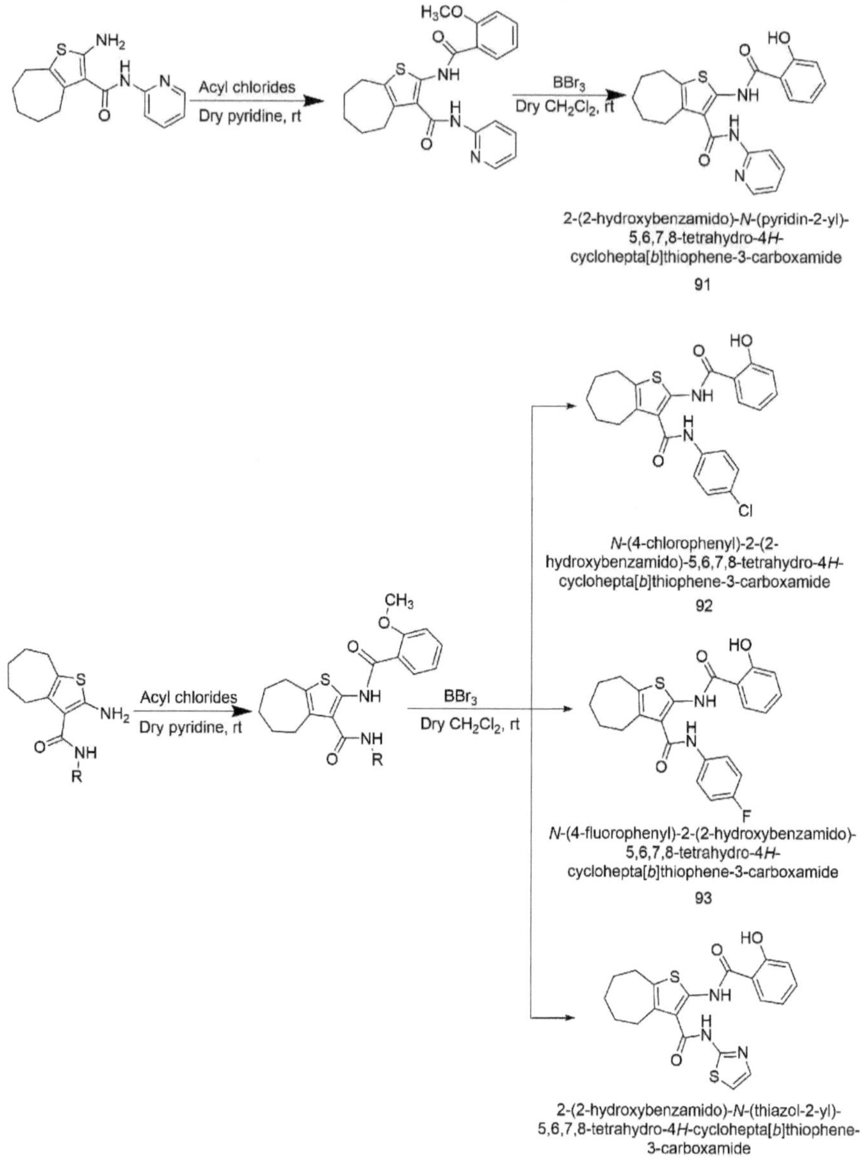

Scheme 11.38 Schematic representation of the proposed compounds having antiviral activity.

Fouad *et al.*[61] synthesized substituted thiophene derivatives (see Scheme 11.44) and studied their anticancer activity against MCF-7 (breast), Hep-2 (epidermoid), HepG-2 (liver), and PC-3 (prostate) cancer cell lines using doxorubicin as a reference drug. The results of the MTT assay experiment

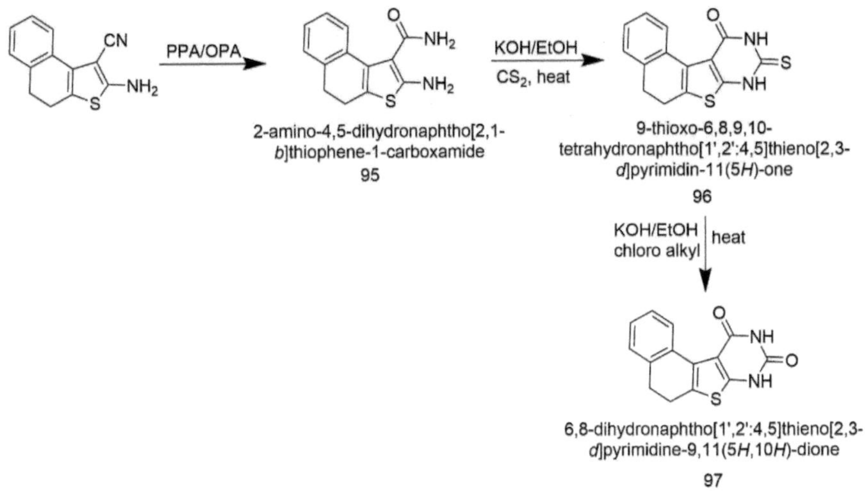

Scheme 11.39 Schematic representation of the proposed compounds having antiviral activity.

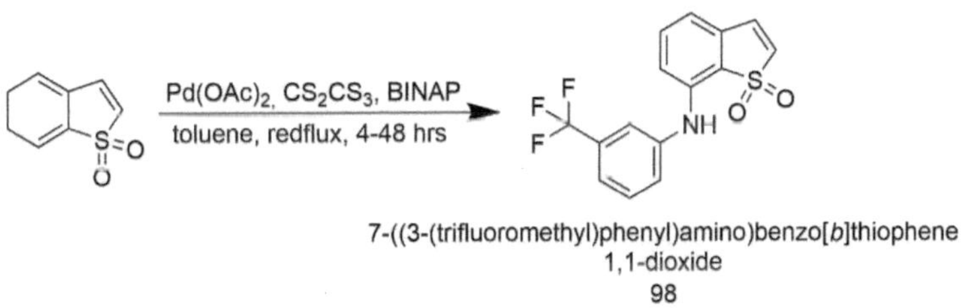

Scheme 11.40 Schematic representation of the proposed compound having antitumor activity.

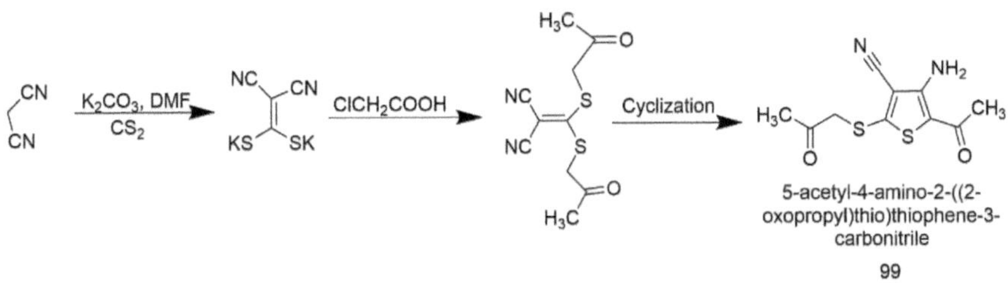

Scheme 11.41 Schematic representation of the proposed compound having anticancer activity.

revealed that compound **102** possessed a high efficacy similar to the reference drug used.

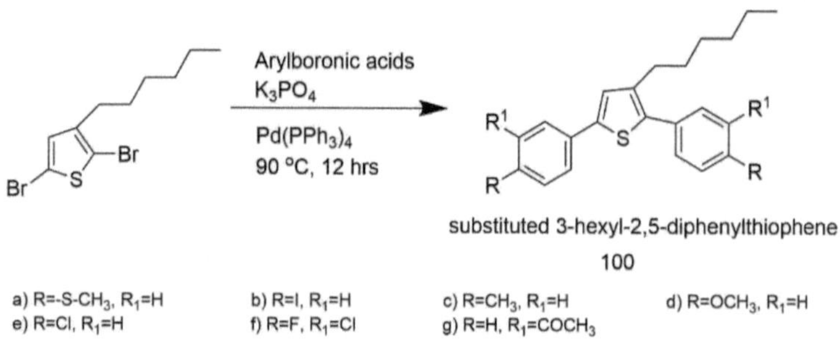

Scheme 11.42 Schematic representation of the proposed compounds having anticancer activity.

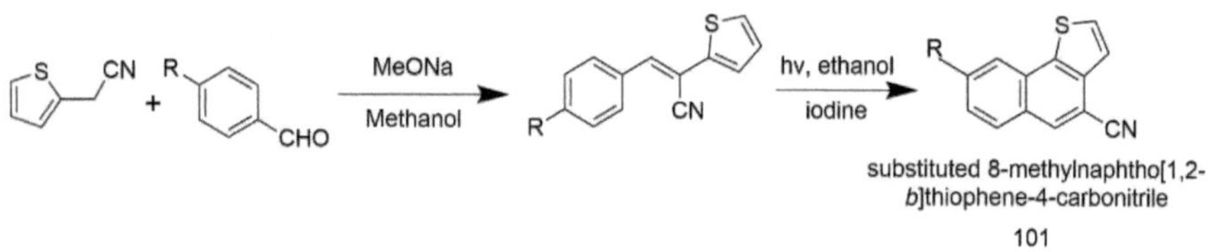

Scheme 11.43 Schematic representation of the proposed compound having anticancer activity.

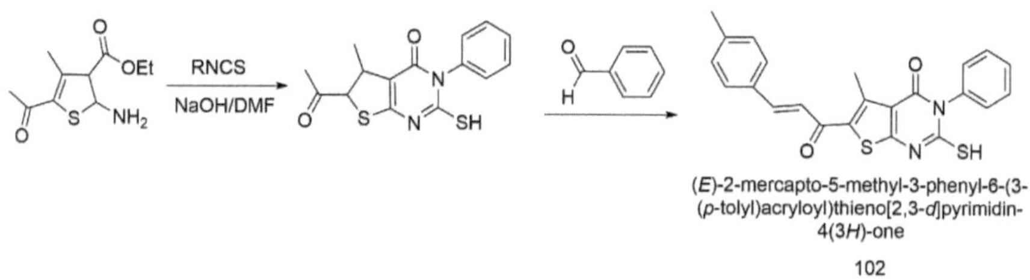

Scheme 11.44 Schematic representation of the proposed compound showing anticancer activity against Hep-2, HepG-2, PC-3, and MCF-7 cancer cell lines.

Ghorab *et al.*[62] proposed the synthesis of fused aminothiophene derivatives containing quinoline and isoquinoline moieties (see Scheme 11.45). The synthesized compounds were tested against MCF-7 breast cancer cell lines through the sulforhodamine B (SRB) stain assay. The experimental results revealed a good anticancer activity (IC_{50} = 47.9 μmol L^{-1}).

Hafez *et al.*[63] proposed a synthetic procedure for a trisubstituted thiophene derivative containing pyrrolopyrimidine moieties (**104**) (see Scheme 11.46) and assessed its anticancer activity against NCI-H460 (lung cancer), HT-29 (colon cancer), and HepG-2 (liver cancer) cell lines. The

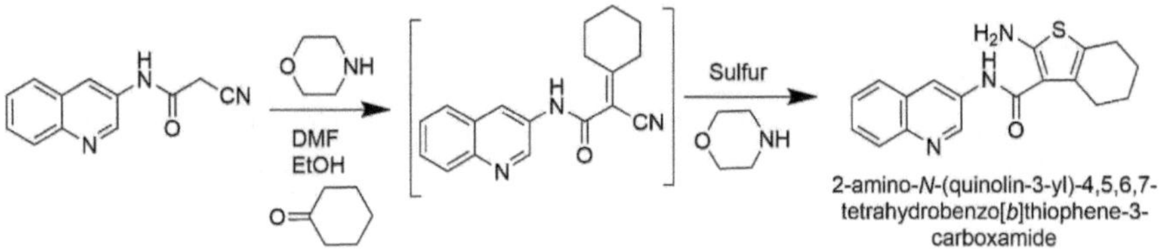

2-amino-*N*-(quinolin-3-yl)-4,5,6,7-tetrahydrobenzo[*b*]thiophene-3-carboxamide

103

Scheme 11.45 Schematic representation of the proposed compound having potent anticancer activity.

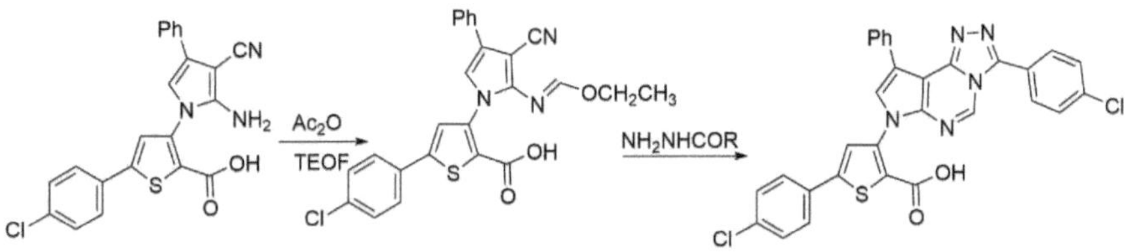

5-(4-chlorophenyl)-3-(3-(4-chlorophenyl)-9-phenyl-7*H*-pyrrolo[3,2-*e*][1,2,4]triazolo[4,3-*c*]pyrimidin-7-yl)thiophene-2-carboxylic acid

104

Scheme 11.46 Schematic representation of the proposed compound having anticancer activity against HepG-2, HT-29, and NCI-H460 cancer cell lines.

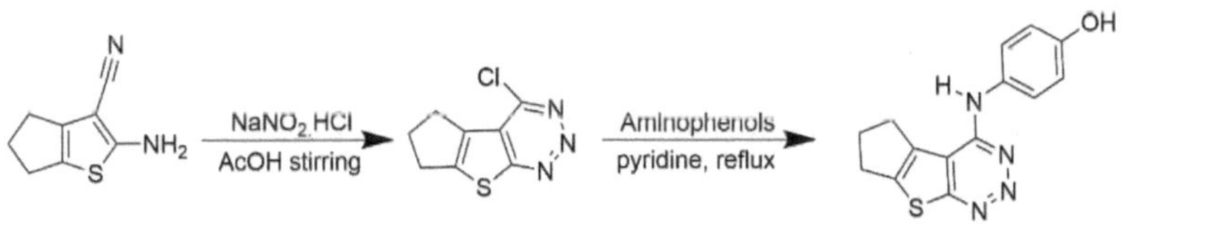

4-((6,7-dihydro-5*H*-cyclopenta[4,5]thieno[2,3-*d*][1,2,3]triazin-4-yl)amino)phenol

105

Scheme 11.47 Schematic representation of the proposed compounds having anticancer activity against HCT116, MCF-7, MIAPaCa-2, HUVEC, and LS180 cancer cell lines.

results of the MTT assay revealed a remarkable anticancer activity against all these cancer cell lines.

Mudududdla *et al.*[64] synthesized disubstituted thiophene derivatives containing an arylamino moiety and a carboxyamide moiety (**105**) (see Scheme 11.47) and assessed their activity against HCT116, MCF-7, MIA-PaCa-2, HUVEC, and LS180 cancer cell lines. The synthesized compounds exhibited good anticancer activity against HUVEC cancer cell lines.

Bozorov *et al.*[65] proposed a synthetic procedure for tetrasubstituted thiophene (**106**) (see Scheme 11.48) and assessed its anticancer activity

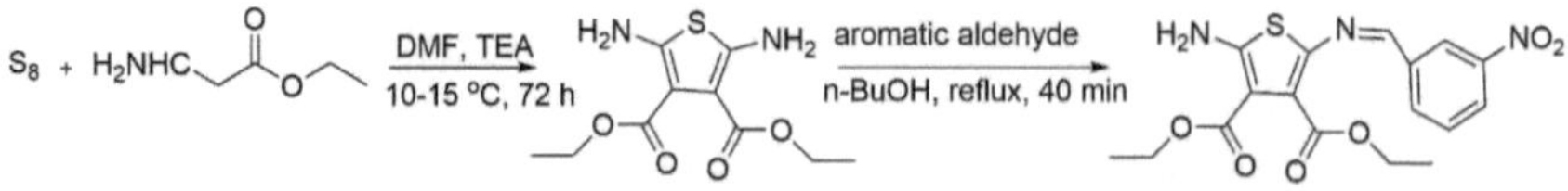

diethyl (Z)-2-amino-5-((3-nitrobenzylidene)amino)thiophene-3,4-dicarboxylate

106

Scheme 11.48 Schematic representation of the proposed compound having anticancer activity.

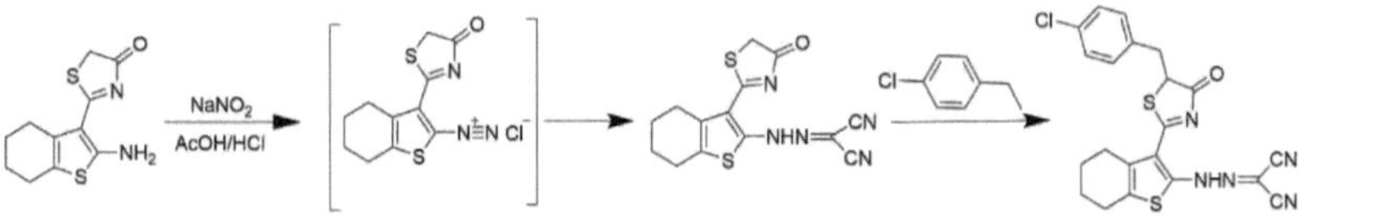

N-(3-(5-(4-chlorobenzyl)-4-oxo-4,5-dihydrothiazol-2-yl)-4,5,6,7-tetrahydrobenzo[b]thiophen-2-yl)carbonohydrazonoyl dicyanide

107

Scheme 11.49 Schematic representation of the proposed compound having anticancer activity.

against HeLa, MCF-7, and T47D (breast cancer) and Ishikawa (endometrial cancer) cell lines. The synthesized compound was found to be most effective against T47D (IC_{50} = 15.5 μmol L^{-1}) and MCF-7 (IC_{50} = 6.75 μmol L^{-1}) cancer cell lines with great efficacy.

Mohareb *et al.*[66] synthesized a substituted thiophene derivative containing thiazole and cyano moieties (**107**) (see Scheme 11.49) and studied its anticancer activity against WI-38 (fibroblast cells), MCF-7, DLD-1 (colon cancer), NUGC (gastric cancer), HONE-1 (nasopharyngeal carcinoma), HEPG-2, and HA22T (liver cancer) cell lines using CHS-828 as a reference compound. The results of these anticancer studies revealed that the compound possessed a remarkable efficacy against all these cancer cell lines.

11.3.5 Anti-inflammatory Activity of Thiophene Derivatives

Anti-inflammatory drugs, especially nonsteroidal drugs, are widely prescribed to treat swelling and inflammatory diseases. Bahashwan *et al.*[24] proposed the synthesis of a series of fused thiophene moieties containing triazolo-pyrimidine and tetrazolopyrimidine derivatives (see Scheme 11.50). They synthesized these compounds (**108–112**) and investigated their anti-inflammatory activity using indomethacin as a standard drug. All these compounds were found to possess high anti-inflammatory activity, and the electron-donating moieties attached to the main body were supposed to be the reason behind this enhanced anti-inflammatory activity.

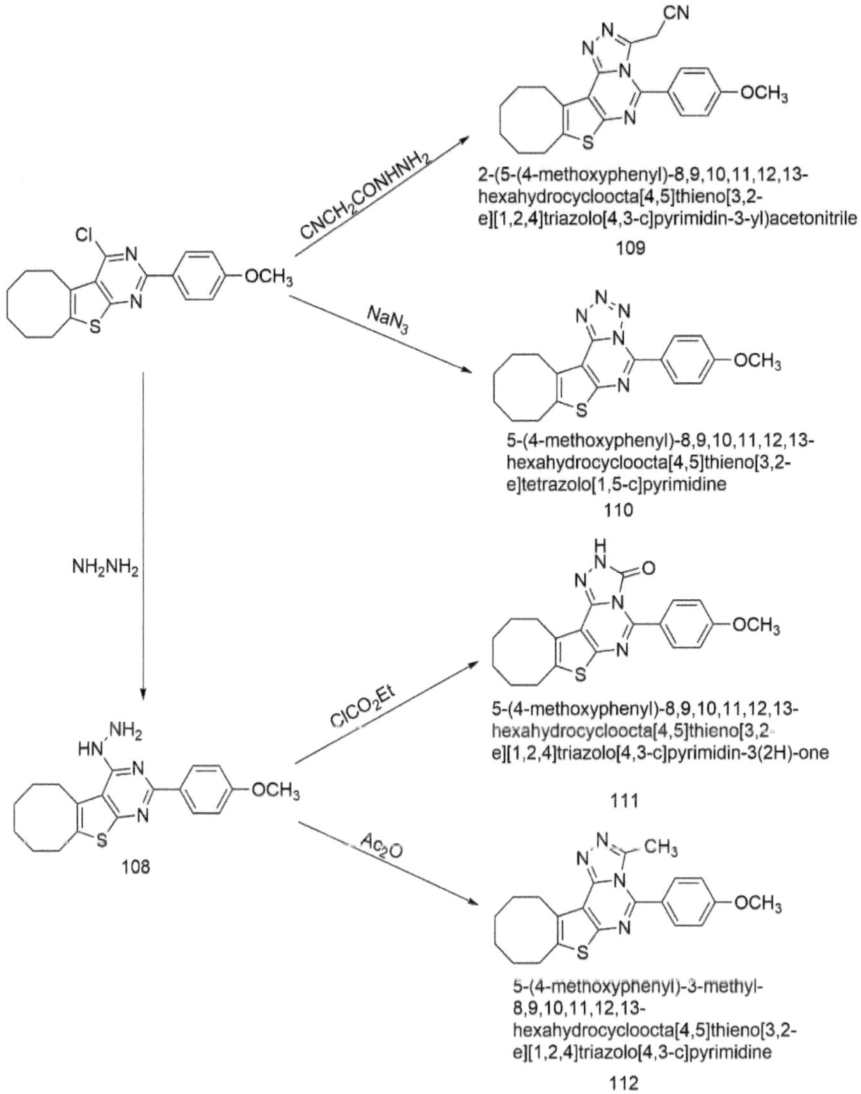

Scheme 11.50 Schematic representation of the proposed thiophene derivative having anti-inflammatory activity.

Beers *et al.*[67] synthesized a disubstituted thiophene containing phenoxide and sulphonyl moieties (**113**) (see Scheme 11.51) and assessed their anti-inflammatory and inhibition activities in adjuvant arthritic rat and human peripheral blood leukocyte (PBL) cells. The results revealed that the synthesized compound possessed good anti-inflammatory activity.

Ouf and Amr[23] proposed the synthesis of a fused thiophene moiety containing hydrazone derivatives (**114**) (see Scheme 11.52), and these

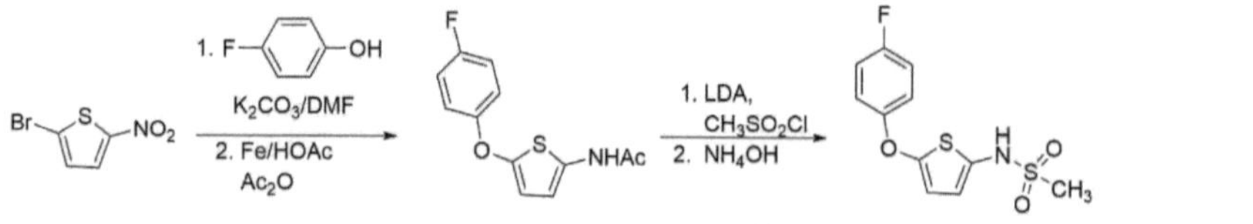

N-(5-(4-fluorophenoxy)thiophen-2-yl)methanesulfonamide

113

Scheme 11.51 Schematic representation of the proposed thiophene derivative.

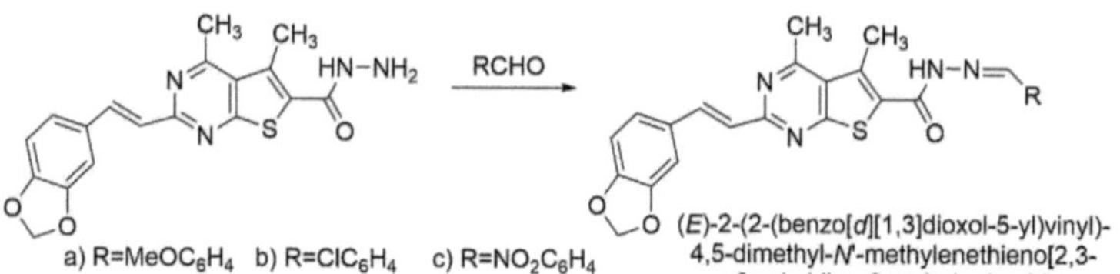

(E)-2-(2-(benzo[d][1,3]dioxol-5-yl)vinyl)-4,5-dimethyl-N′-methylenethieno[2,3-d]pyrimidine-6-carbohydrazide

114

Scheme 11.52 Schematic representation of the proposed thiophene derivative.

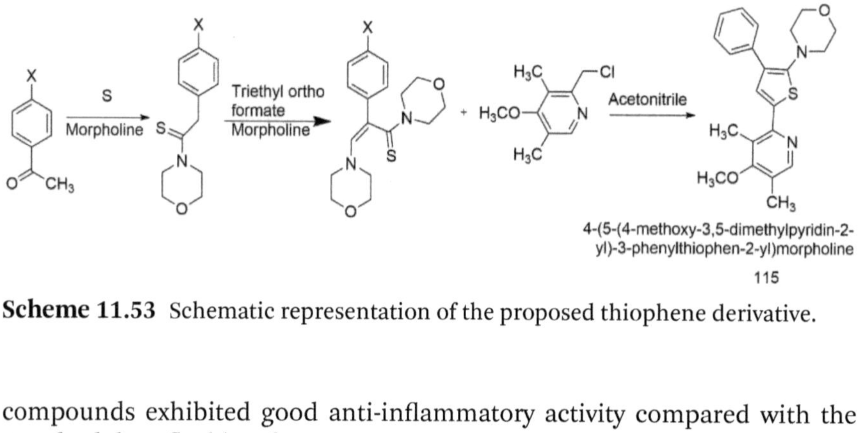

4-(5-(4-methoxy-3,5-dimethylpyridin-2-yl)-3-phenylthiophen-2-yl)morpholine

115

Scheme 11.53 Schematic representation of the proposed thiophene derivative.

compounds exhibited good anti-inflammatory activity compared with the standard drug flurbiprofen.

Pillai *et al.*[68] proposed a synthetic procedure for trisubstituted thiophene derivative containing morpholine and pyridine moieties (**115**) (see Scheme 11.53) and assessed their *in vivo* anti-inflammatory activity using ibuprofen as a standard drug. The synthesized compound exhibited a good anti-inflammatory activity.

Pillai *et al.*[69] also proposed a synthetic procedure for tetrasubstituted thiophene derivatives (see Scheme 11.54) showing anti-inflammatory activity in a rat paw edema model using ibuprofen as a standard drug. Among all the synthesized compounds, compound **116** was found to possess the most effective anti-inflammatory activity.

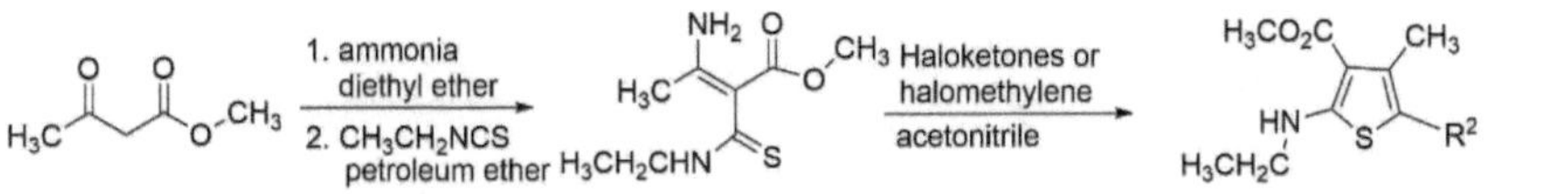

substituted 2-(ethylamino)-4,5-dimethylthiophene-3-carboxylate

116

Scheme 11.54 Schematic representation of the proposed compound having anti-inflammatory activity.

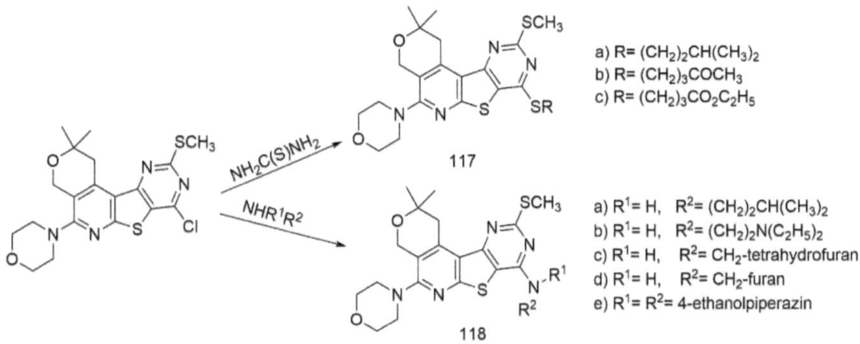

Scheme 11.55 Schematic representation of the proposed compound having anticonvulsant activity.

11.3.6 Anticonvulsant and Antiurease Activity of Thiophene Derivatives

Anticonvulsants are a class of medications primarily used to treat epilepsy, which are intended to prevent and stop seizures. Epilepsy is a long-term, non infectious neurological condition marked by recurrent convulsions. Paronikyan *et al.*[15] proposed a synthetic method for thiophene-fused heterocyclic compounds containing the morpholine moiety (**117a–c** and **118a–e**) (see Scheme 11.55) and assessed their anticonvulsant activity for preventing clonic twitches and clonic convulsions in mice. The synthesized compounds possessed an anticonvulsive effect.

Helicobacter pylori, a human pathogenic bacterium, causes diseases such as stomach cancer, gastritis, and ulcerative reflux disease, by generating an excessive amount of urease and breaking down the urea to live in the acidic environment of the stomach.[70] The synthesis of new compounds having urease-inhibiting properties is very important. Rasool *et al.*[17] synthesized a disubstituted thiophene containing sulphonylacetamide moiety (**119**) (see Scheme 11.56) and assessed its antiurease activities using thiourea as a standard drug. The results of the studies revealed that the synthesized compound possessed excellent urease-inhibiting activity.

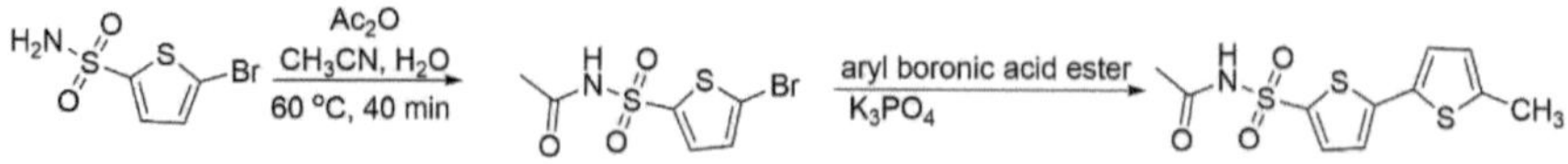

N-((5'-methyl-[2,2'-bithiophen]-5-yl)sulfonyl)acetamide

119

Scheme 11.56 Schematic representation of the proposed compound having antiurease activity.

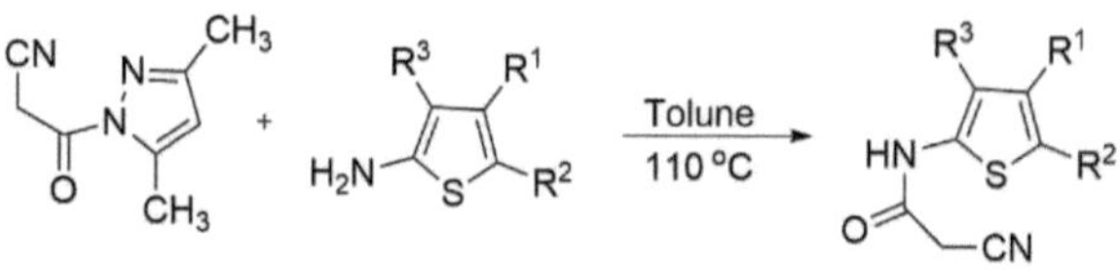

substituted 2-cyano-N-(thiophen-2-yl)acetamide

120

Scheme 11.57 Schematic representation of the proposed compound having antioxidant activity.

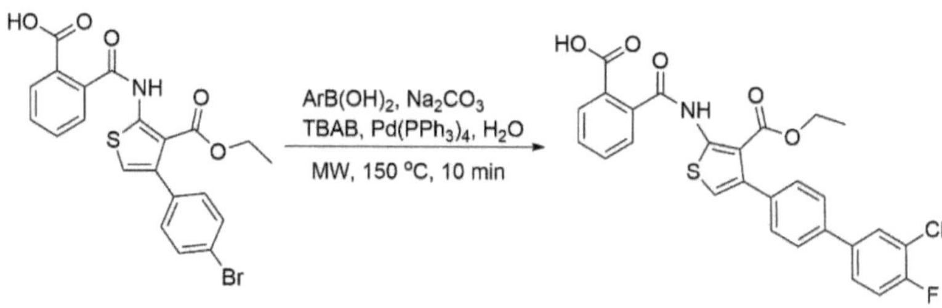

2-((4-(3'-chloro-4'-fluoro-[1,1'-biphenyl]-4-yl)-3-(ethoxycarbonyl)thiophen-2-yl)carbamoyl)benzoic acid

121

Scheme 11.58 Schematic representation of the proposed compound having enzyme inhibitory activity.

11.3.7 Antioxidant, Enzyme Inhibition, and Antithrombotic Activity of Thiophene Derivatives

Madhavi and Sree Ramya[21] synthesized substituted thiophene derivatives containing a cyanoacetamide group (**120**) (see Scheme 11.57) and assessed their *in vitro* antioxidant activity. These compounds possessed a good antioxidant activity in 2,2-diphenyl-1-picrylhydrazyl (DPPH)-scavenging and NO free radical assay experiments.

Ye *et al.*[16] synthesized trisubstituted thiophene (see Scheme 11.58), and the synthesized compound (**121**) was found to possess enzyme-inhibitory activity through a PTP-1B inhibition study.

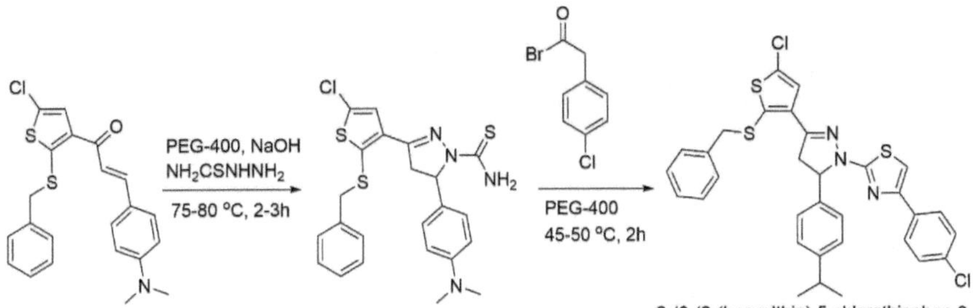

2-(3-(2-(benzylthio)-5-chlorothiophen-3-yl)-5-(4-isopropylphenyl)-4,5-dihydro-1*H*-pyrazol-1-yl)-4-(4-chlorophenyl)thiazole

122

Scheme 11.59 Schematic representation of the proposed compound having antioxidant activity and enzyme inhibitory activity.

2-(2-(4-(4-methoxyphenyl)piperazin-1-yl)acetamido)-4,5,6,7-tetrahydrobenzo[*b*]thiophene-3-carboxamide

123

Scheme 11.60 Schematic representation of the proposed compound having acetylcholinesterase-inhibitory activity.

Mandawad *et al.*[22] synthesized many trisubstituted thiophene derivatives containing pyrazoline moieties (see Scheme 11.59) and assessed their antioxidant and xanthine oxidase-inhibitory activities. For antioxidant activity, ascorbic acid was used as a standard against DPPH-scavenging assay, and for inhibitory activities, allopurinol was used as a standard drug. The results of these experiments showed that the synthesized thiophene moiety (**122**) possessed both antioxidant and inhibitory activities.

Ismail *et al.*[71] synthesized fused thiophene derivatives containing amide moieties (see Scheme 11.60) following the Gewald protocol. The acetylcholinesterase-inhibitory activity study results using donepezil as a standard revealed that the synthesized compound (**123**) possessed a significant inhibitory activity.

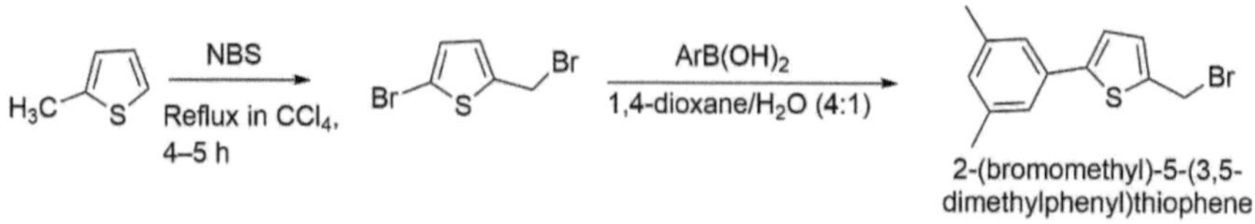

124

Scheme 11.61 Schematic representation of the proposed compound having antithrombotic activity.

Zubair *et al.*[18] synthesized disubstituted thiophene derivatives (see Scheme 11.61) and assessed their antithrombotic activity in venous blood using streptokinase as a standard. The synthesized compound (**124**) exhibited good clot lysis activity results.

References

1. R. Mancuso and B. Gabriele, *Molecules*, 2014, **19**, 15687–15719.
2. R. Mishra, K. K. Jha, S. Kumar and I. Tomer, *Der. Pharma. Chem..*, 2011, **3**, 38–54.
3. R. Mishra, P. C. Neetu Sachan, N. Kumar and I. Mishra, *J. Heterocycl. Chem.*, 2019, **55**, 2019–2034.
4. S. Pathania, R. Kumar and R. K. Rawal, *Eur. J. Med. Chem.*, 2019, **180**, 486–508.
5. A. Chaudhary, K. K. Jha and S. Kumar, *J. Adv. Sci. Res.*, 2012, **3**, 3–10.
6. A. Singh, G. Singh and P. M. S. Bedi, *J. Heterocycl. Chem.*, 2020, **57**, 2658–2703.
7. R. Shah and P. K. Verma, *Chem. Cent. J.*, 2018, **12**, 1–22.
8. S. Anandan, S. Manoharan, N. K. S. Narendran, T. C. S. Girisun and A. M. Asiri, *Opt. Mater.*, 2018, **85**, 18–25.
9. O. H. Rizk, O. G. Shaaban and A. E. Abdel Wahab, *Open Med. Chem. J.*, 2017, **11**, 38–53.
10. Y. N. Mabkhot, N. A. Kaal, S. Alterary, S. S. Al-Showiman, T. A. Farghaly and M. S. Mubarak, *Chem. Cent. J.*, 2017, **11**, 75.
11. W. Zhang, T. Ma, S. Li, Y. Yang, J. Guo, W. Yu and L. Kong, *Eur. J. Med. Chem.*, 2017, **125**, 538–550.
12. Y. N. Mabkhot, F. D. Aldawsari, S. S. Al-Showiman, A. Barakat, S. M. Soliman, M. I. Choudhary, S. Yousuf, T. Ben Hadda and M. Mubarak, *Molecules*, 2016, **21**, 214.
13. A. P. Jacomini, M. J. V. Silva, R. G. M. Silva, D. S. Gonçalves, H. Volpato, E. A. Basso, F. R. Paula, C. V. Nakamura, M. H. Sarragiotto and F. A. Rosa, *Eur. J. Med. Chem.*, 2016, **124**, 340–349.
14. M. B. Félix, E. R. de Souza, M. do C. A. de Lima, D. K. G. Frade, V. de L. Serafim, K. A. da F. Rodrigues, P. L. do N. Néris, F. F. Ribeiro, L. Scotti,

M. T. Scotti, T. M. de Aquino, F. J. B. Mendonça Junior and M. R. de Oliveira, *Bioorg. Med. Chem.*, 2016, **24**, 3972–3977.
15. E. G. Paronikyan, S. S. Dashyan, A. S. Noravyan, I. A. Dzhagatspanyan, R. G. Paronikyan, I. M. Nazaryan and A. G. Akopyan, *Russ. J. Bioorg. Chem.*, 2015, **41**, 663–669.
16. D. Ye, Y. Zhang, F. Wang, M. Zheng, X. Zhang, X. Luo, X. Shen, H. Jiang and H. Liu, *Bioorg. Med. Chem.*, 2010, **18**, 1773–1782.
17. M. Noreen, N. Rasool, Y. Gull, M. Zubair, T. Mahmood, K. Ayub, F. U. H. Nasim, A. Yaqoob, M. Zia-Ul-Haq and V. De Feo, *Molecules*, 2015, **20**, 19914–19928.
18. K. Rizwan, M. Zubair, N. Rasool, S. Ali, A. F. Zahoor, U. A. Rana, S. U. D. Khan, M. Shahid, M. Zia-Ul-Haq and H. Z. E. Jaafar, *Chem. Cent. J.*, 2014, **8**, 74.
19. J. Desantis, G. Nannetti, S. Massari, M. L. Barreca, G. Manfroni, V. Cecchetti, G. Palù, L. Goracci, A. Loregian and O. Tabarrini, *Eur. J. Med. Chem.*, 2017, **138**, 128–139.
20. A. E. Rashad, A. H. Shamroukh, R. E. Abdel-Megeid, A. Mostafa, R. El-Shesheny, A. Kandeil, M. A. Ali and K. Banert, *Eur. J. Med. Chem.*, 2010, **45**, 5251–5257.
21. K. Madhavi and G. Sree Ramya, *Asian J. Pharm. Clin. Res.*, 2017, **10**, 95–100.
22. G. G. Mandawad, B. S. Dawane, S. D. Beedkar, C. N. Khobragade and O. S. Yemul, *Bioorg. Med. Chem.*, 2013, **21**, 365–372.
23. N. H. Ouf and A. E. G. E. Amr, *Monatsh. Chem.*, 2008, **139**, 579–585.
24. S. A. Bahashwan, A. A. Fayed, A. E. G. E. Amr, E. M. Flefel and A. Kalmouch, *Molecules*, 2013, **18**, 15051–15063.
25. Y. N. Mabkhot, S. S. Al-Showiman, S. M. Soliman, H. A. Ghabbour, M. A. AlDamen and M. S. Mubarak, *Chem. Cent. J.*, 2017, **11**, 51.
26. Y. N. Mabkhot, F. Alatibi, N. N. E. El-Sayed, S. Al-Showiman, N. A. Kheder, A. Wadood, A. Rauf, S. Bawazeer and T. Ben Hadda, *Molecules*, 2016, **21**, 222.
27. Z. Puterová, A. Krutošíková and D. Végh, *ARKIVOC*, 2010, 209–246.
28. S. Khaghaninejad and M. M. Heravi, in *Advances in Heterocyclic Chemistry*, Elsevier, 2014, vol. **111**, pp. 95–146.
29. H. Fiesselmann and S. Uber, *Chem. Ber.*, 1954, **87**, 835–841.
30. H. Wynberg and H. J. Kooreman, *J. Am. Chem. Soc.*, 1965, **87**, 1739–1742.
31. Y. He, J. Lou, K. Wu, H. Wang and Z. Yu, *J. Org. Chem.*, 2019, **84**, 2178–2190.
32. S. Chowdhury, T. Chanda, S. Koley, B. J. Ramulu, R. C. F. Jones and M. S. Singh, *Tetrahedron*, 2015, **71**, 1844–1850.
33. L. S. Ge, Z. L. Wang, X. L. An, X. Luo and W. P. Deng, *Org. Biomol. Chem.*, 2014, **12**, 8473–8479.
34. M. Wen, P. P. Sun, X. Luo and P. Deng, *Tetrahedron*, 2018, **74**, 4168–4173.

35. G. Shukla, A. Srivastava, D. Yadav and M. S. Singh, *J. Org. Chem.*, 2018, **83**, 2173–2181.
36. D. Kurandina and V. Gevorgyan, *Org. Lett.*, 2016, **18**, 1804–1807.
37. Y. N. Wu, R. Fu, N. N. Wang, W. J. Hao, G. Li, S. J. Tu and B. Jiang, *J. Org. Chem.*, 2016, **81**, 4762–4770.
38. R. Mancuso, C. S. Pomelli, C. Chiappe, R. C. Larock and B. Gabriele, *Org. Biomol. Chem.*, 2014, **12**, 651–659.
39. P. Maity and B. C. Ranu, *Adv. Synth. Catal.*, 2017, **359**, 4369–4378.
40. T. Han, Y. Wang, H. L. Li, X. Luo and W. P. Deng, *J. Org. Chem.*, 2018, **83**, 1538–1542.
41. H. Zali-Boeini and M. Ghani, *Synth.*, 2013, **45**, 913–918.
42. F. N. Bilheri, A. L. Stein and G. Zeni, *Adv. Synth. Catal.*, 2015, **357**, 1221–1228.
43. H. Zali-Boeini and N. Fadaei, *Synlett*, 2015, **26**, 1819–1822.
44. G. Bharathiraja, G. Sathishkannan and T. Punniyamurthy, *J. Org. Chem.*, 2016, **81**, 2670–2674.
45. T. T. Kao, B. K. Peng, M. C. Liang, C. J. Lee, I. C. Chen, K. S. Shia and Y. K. Wu, *J. Org. Chem.*, 2018, **83**, 14688–14697.
46. C. R. Reddy, R. R. Valleti and M. D. Reddy, *J. Org. Chem.*, 2013, **78**, 6495–6502.
47. P. Fricero, L. Bialy, W. Czechtizky, M. Méndez and J. P. A. Harrity, *Org. Lett.*, 2018, **20**, 198–200.
48. D. Dagoneau, A. Kolleth, A. Lumbroso, G. Tanriver, S. Catak and S. Sulzer-Mossé, De MesmaekerA., *Helv. Chim. Acta*, 2019, **102**, 1900031.
49. B. Cheng, X. Duan, Y. Li, X. Zhang, H. Li, F. Wu, Y. Li, T. Wang and H. Zhai, *Eur. J. Org. Chem.*, 2020, **2020**, 1896–1906.
50. L. Rong, Y. Shen, G. Xiong and Y. Gong, *J. Heterocycl. Chem.*, 2019, **56**, 670–675.
51. G. Mari, M. Verboni, L. De Crescentini, G. Favi, S. Santeusanio and F. Mantellini, *Org. Chem. Front.*, 2018, **5**, 2108–2114.
52. G. Mari, L. De Crescentini, G. Favi, S. Santeusanio and F. Mantellini, *Eur. J. Org. Chem.*, 2018, **2018**, 6548–6556.
53. X. Luo, L. S. Ge, X. L. An, J. H. Jin, Y. Wang, P. P. Sun and W. P. Deng, *J. Org. Chem.*, 2015, **80**, 4611–4617.
54. S. Lee, D. Lee, K. S. Song, K. H. Liu, Y. D. Gong and T. Lee, *Tetrahedron*, 2014, **70**, 9183–9190.
55. V. Ajdačić, L. Senerovic, M. Vranić, M. Pekmezovic, V. Arsic-Arsnijevic, A. Veselinovic, J. Veselinovic, B. A. Šolaja, J. Nikodinovic-Runic and I. M. Opsenica, *Bioorg. Med. Chem.*, 2016, **24**, 1277–1291.
56. B. Shan, Y. Z. Cai, J. D. Brooks and H. Corke, *J. Agric. Food Chem.*, 2007, **55**, 5484–5490.
57. J. L. Gonzalez, C. E. Stephens, T. Wenzler, R. Brun, F. A. Tanious, W. D. Wilson, T. Barszcz, K. A. Werbovetz and D. W. Boykin, *Eur. J. Med. Chem.*, 2007, **42**, 552–557.

58. L. Naesens, A. Stevaert and E. Vanderlinden, *Curr. Opin. Pharmacol.*, 2016, **30**, 106–115.
59. H. M. Ikram, N. Rasool, M. Zubair, K. M. Khan, G. A. Chotana, M. N. Akhtar, N. Abu, N. B. Alitheen, A. M. Elgorban and U. A. Rana, *Molecules*, 2016, **21**, 977.
60. N. Perin, V. Rep, I. Sović, S. Juričić, D. Selgrad, M. Klobučar, N. Pržulj, C. L. Gupta, N. Malod-Dognin, S. K. Pavelić and M. Hranjec, *Eur. J. Med. Chem*, 2020, **185**, 111833.
61. M. M. Fouad, E. R. El-Bendary, G. M. Suddek, I. A. Shehata and M. M. El-Kerdawy, *Bioorg. Chem.*, 2018, **81**, 587–598.
62. M. M. Ghorab, M. S. Alsaid, M. S. Al-Dosari, F. A. Ragab, A. A. Al-Mishari and A. N. Almoqbil, *Acta Pharm.*, 2016, **66**, 155–171.
63. H. N. Hafez, S. A. Alsalamah and A. R. B. A. El-Gazzar, *Acta Pharm.*, 2017, **67**, 275–292.
64. R. Mudududdla, S. K. Guru, A. Wani, S. Sharma, P. Joshi, R. A. Vishwakarma, A. Kumar, S. Bhushan and S. B. Bharate, *Org. Biomol. Chem.*, 2015, **13**, 4296–4309.
65. K. Bozorov, H. R. Ma, J. Y. Zhao, H. Q. Zhao, H. Chen, K. Bobakulov, X. L. Xin, B. Elmuradov, K. Shakhidoyatov and H. A. Aisa, *Eur. J. Med. Chem.*, 2014, **84**, 739–745.
66. R. M. Mohareb, A. E. M. Abdallah and E. A. Ahmed, *Acta Pharm.*, 2017, **67**, 495–510.
67. S. A. Beers, E. A. Malloy, W. Wu, M. Wachter, J. Ansell, M. Singer, M. Steber, A. Barbone, T. Kirchner, D. Ritchie and D. Argentieri, N-(5-substituted) thiophene-2-alkylsulfonamides as potent inhibitors of 5-lipoxygenase, *Bioorg. Med. Chem.*, 1997, **5**, 779–786.
68. A. D. Pillai, P. D. Rathod, F. P. Xavier, K. K. Vasu, H. Padh and V. Sudarsanam, *Bioorg. Med. Chem.*, 2004, **12**, 4667–4671.
69. A. D. Pillai, P. D. Rathod, F. P. Xavier, H. Padh, V. Sudarsanam and K. K. Vasu, *Bioorg. Med. Chem.*, 2005, **13**, 6685–6692.
70. S. Svane, J. J. Sigurdarson, F. Finkenwirth, T. Eitinger and H. Karring, *Sci. Rep.*, 2020, **10**, 8503.
71. M. M. Ismail, M. M. Kamel, L. W. Mohamed, S. I. Faggal and M. A. Galal, *Molecules*, 2012, **17**, 7217–7231.

CHAPTER 12

Synthesis, Properties, and Biological Applications of Benzothiophene

JAIDIP B. WABLE[a], CHERYL MASCARENHAS[a], HEMANTKUMAR N. AKOLKAR[b], NIRMALA R. DAREKAR[c] AND PRADNYA PRABHU*[a]

[a]K. J. Somaiya College of Science and Commerce, Vidyavihar (Autonomous College), Mumbai, 400077, India; [b]Department of Chemistry, Abasaheb Marathe Arts and New Commerce, Science College, Rajapur, Ratnagiri, Maharashtra, 416702, India; [c]P. G. and Research, Department of Chemistry, Radhabai Kale Mahila Mahavidyalaya, Ahmednagar, Maharashtra, 414001, India
*E-mail: pradnya.prabhu@somaiya.edu

12.1 Introduction

Benzothiophene is a type of heterocyclic aromatic compound. The thiophene ring is a five-membered ring containing one sulfur atom. The benzene nucleus and thiophene fuse to form the well-known heterocycle benzo[*b*]thiophene (Figure 12.1). In the bicyclic system known as benzo[*b*]thiophene, the thiophene ring and the benzene ring are fused at positions 4 and 5.[1]

Benzo[*b*]thiophene is a special heterocyclic core among the sulfur heterocycles that has piqued curiosity as it has been conceptualized

S-Heterocycles: Retrospect, Prospects, and Biological Applications
Edited by Shrikaant Kulkarni, Hemantkumar Akolkar, Vijay M. Khedkar and A. K. Haghi

Published by the Royal Society of Chemistry, www.rsc.org

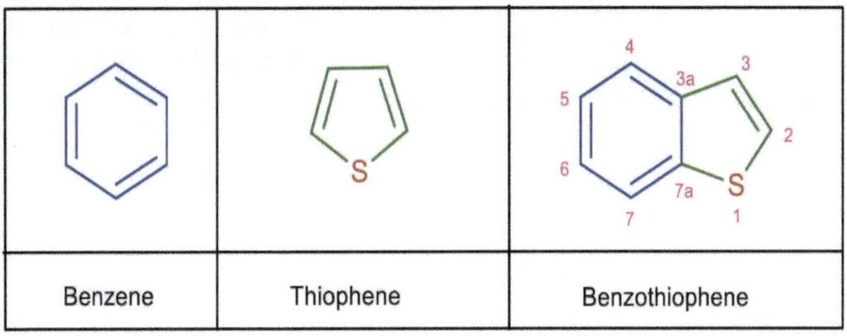

Figure 12.1 Benzene nucleus and thiophene fuse to form benzo[*b*]thiophene.

as a significant pharmacophore of several bioactive compounds.[2] Naturally occurring deposits associated with petroleum, such as lignite tar,[3,4] and some natural products, such as bryoanthrathiophene, which have beneficial biological properties, *viz.* anti-angiogenic properties, contain benzo[*b*]thiophene.[5,6] Coffee beans contain parent benzothiophene,[7] and its analogs have a broad range of applications in medicinal chemistry and are of great interest to both industry and academia.

They exhibit numerous biological and pharmacological properties, including anti-inflammatory, antifungal,[8] antimitotic,[9] anti-diabetic,[10] enzyme-inhibitory,[11] analgesic,[12] anti-hyperglycemic,[13] antidepressant,[14,15] anthelmintic,[16] anti-angiogenic,[17,18] anti-convulsant,[19] anti-malarial,[20] estrogen receptor-modulating,[21,22] pesticidal,[23] anti-tubercular,[24] anti-cancer,[25,26] and kinase-inhibiting[27] activities. They demonstrate a total of twenty-seven activities.

12.1.1 Physical Properties

Benzothiophene demonstrates the following physical properties:

1. Chemical formula: C_8H_6S and molar mass per mole: 134.20 g.
2. The solid benzothiophene is white in color.
3. The boiling and melting points of benzothiophene are 221–222 °C and 28–32 °C, respectively.
4. At 25 °C, benzothiazole has a density of 1.149 g ml^{-1}.
5. The smell of benzothiophene is comparable to that of naphthalene (mothballs).
6. The moiety is thermally stable.

Benzothiophene is soluble in typical organic solvents, such as acetone, ether, and benzene, but it is insoluble in water. In addition, it is soluble in alcohol and concentrated sulfuric acid.[28,29]

12.2 Recent Synthesis

12.2.1 Cyclization Reactions by Lewis Acid

Yu *et al.* proposed a unique technique for the synthesis of benzo[*b*]thiophenes and benzothiazoles.[30,31] They used thioacids, 1,10-phenanthroline, copper iodide, and Pr_3N as bases for their reaction with 2-iodophenyliminotriphenylphosphorane. Scheme 12.1 presents the synthesis route.

12.2.2 Cyclization Reactions by Halogen Catalysis

Yan *et al.*[32] proposed an additional synthesis route in which they generated benzo[*b*]thiophenes by a ring-closure reaction of substituted thiophenols with alkynylcarboxylates at 110 °C in a solvent-free reaction environment. This reaction (Scheme 12.2) is catalyzed by iodine.

12.2.3 Cyclization Reactions by Transition Metal Catalysis

The reaction of 2-bromo alkynyl benzenes in the presence of sodium sulfide *via* the thiolation annulation process furnishes 2-substituted benzothiophenes, as reported by Sun *et al.*[33] (Scheme 12.3).

Another method to synthesize 2,3-disubstituted benzothiophenes was described by Nakamura *et al.*[34] This approach involved a gold-catalyzed

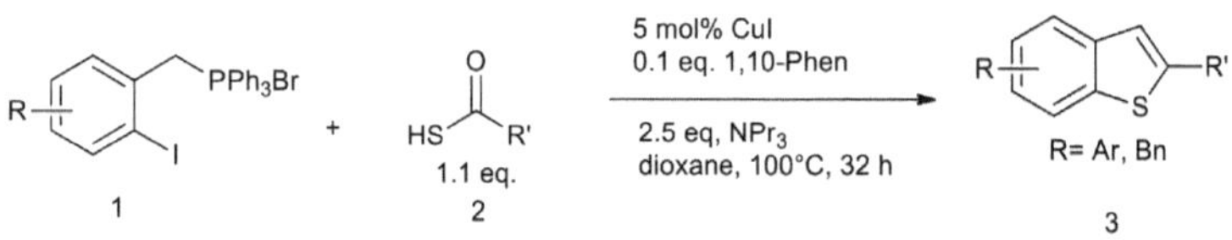

Scheme 12.1 Synthesis of benzo[*b*]thiophenes.

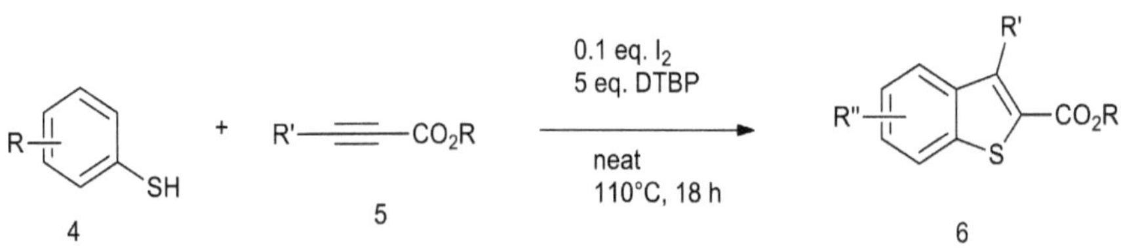

Scheme 12.2 Synthesis of substituted benzo[*b*]thiophenes *via* an iodine-catalyzed reaction.

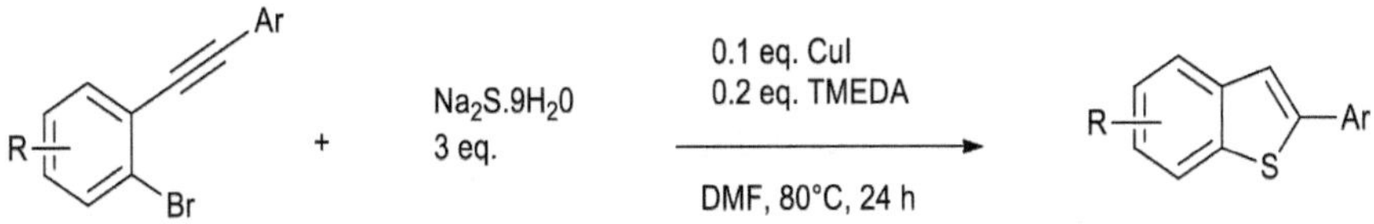

7 8

Scheme 12.3 Synthesis of 2-aryl benzo[*b*]thiophenes *via* the thiolation annulation reaction.

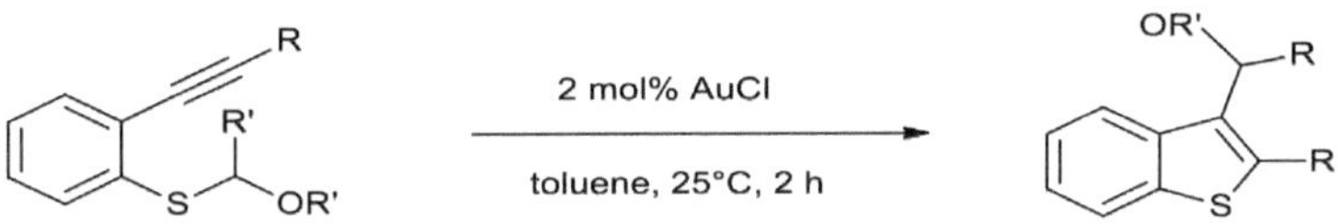

9 10

Scheme 12.4 Synthesis of disubstituted benzo[*b*]thiophenes by gold-catalyzed carbothiolation.

carbothiolation reaction of α-alkoxy alkylortho-alkynyl phenyl sulfides. The equivalent synthesis process is depicted in Scheme 12.4, and this method allowed for the measurement of the synthesized heterocycles containing sulfur. Willis *et al.*[35] reported a Pd-catalyzed reaction between an aryl halide and thioketone for an intramolecular C–S coupling. This method converted analogs of thioketone ketone into benzofurans. They found that the sulfur reacted favorably under the same conditions.

12.2.4 Cyclization Reactions by Base Catalysis

According to Nguyen and Retailleau,[36] elemental sulfur and 2-nitrochalcones (**20**) can be reacted to form 2-benzoylbenzo[*b*]thiophenes in the presence of *N*,*N*-diisopropylethylamine (DIPEA), as shown in Scheme 12.5.

Kuhn *et al.* have demonstrated further techniques for the preparation of 2-arylbenzothiophene using benzyl thiol and *ortho*-fluoro ketone[37] followed by Knoevenagel condensation (Scheme 12.6).

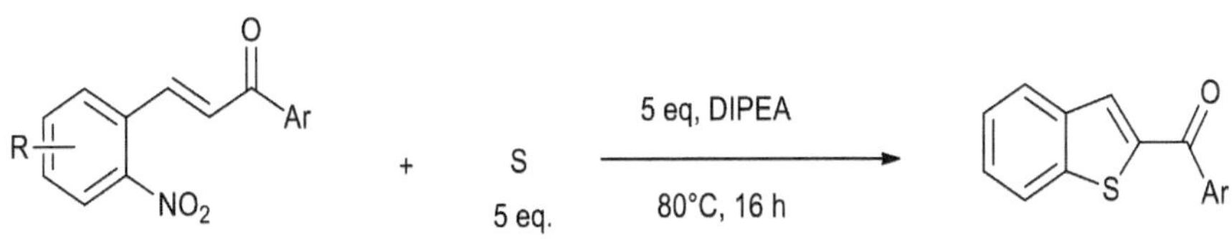

15 16

Scheme 12.5 Synthesis of 2-benzoylbenzo[*b*]thiophenes from 2-nitrochalcones.

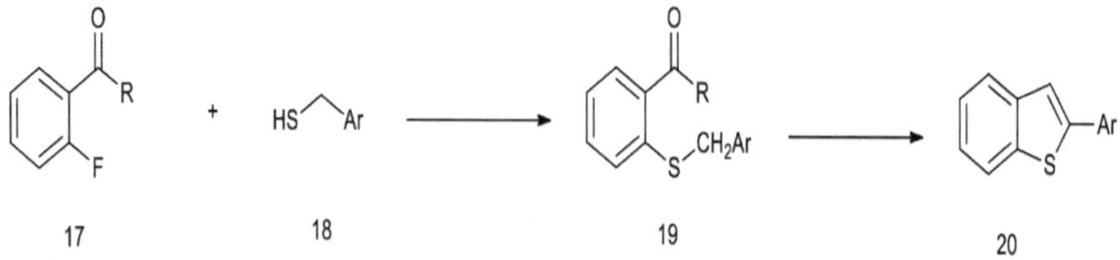

Scheme 12.6 Synthesis of benzo[*b*]thiophenes.

12.2.5 Electrophilic Cyclization Reactions

Kong *et al.* described the production of 3-halobenzo[*b*]thiophene analogs from the acetylenic compounds[38] in the presence of I_2 or *N*-bromosuccinimide (NBS) (Scheme 12.7).

Singh *et al.*[39] devised a new synthesis method based on current reports *via* one-pot benzannulation reactions. They synthesized good yields of di-, tri-, or higher substituted benzofurans and benzo[*b*]thiophenes. In the presence of a base, nitromethane was added to alkyl- and aryl-substituted furans and thiophene-3-ynones to produce benzofurans and benzo[*b*]thiophenes, which react similarly well. Trisubstituted benzo[*b*]thiophenes and benzofurans were produced as a result of the reaction with additional nitroalkanes, increasing the production of a variety of products (Scheme 12.8).

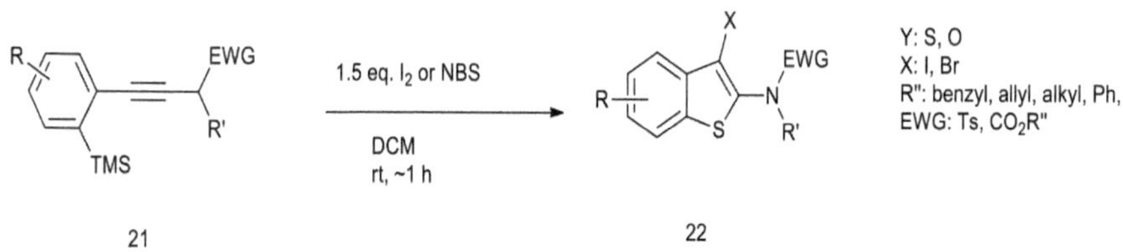

Scheme 12.7 Synthesis of benzo[*b*]thiophenes *via* the electrophilic cyclization reaction.

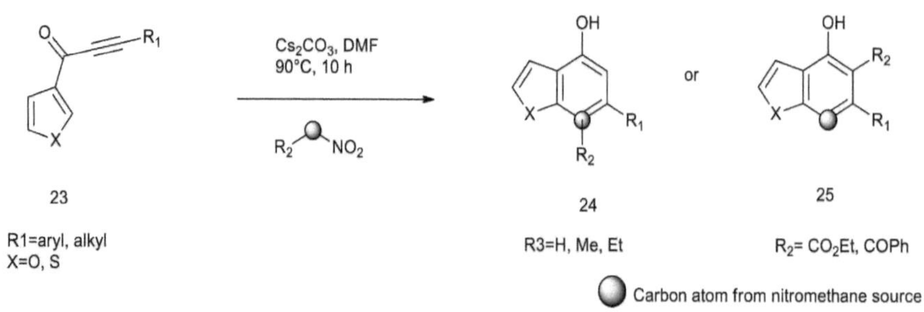

Scheme 12.8 Benzoannulation reaction between furan-3- and thiophene-3-ynones.

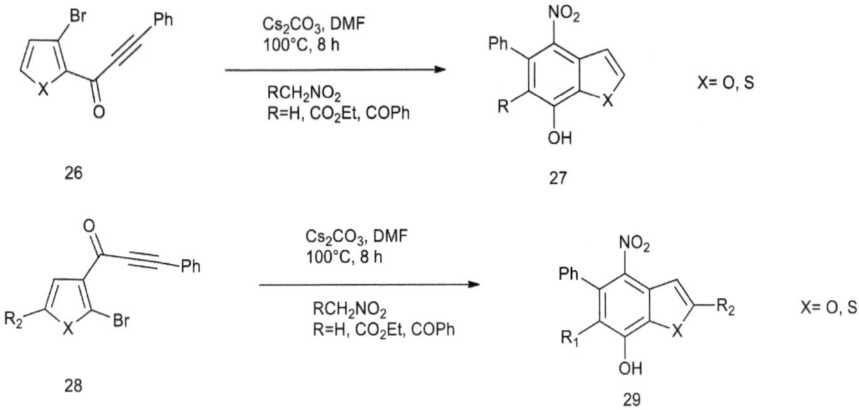

Scheme 12.9 Synthesis of benzofurans and benzo[*b*]thiophenes from bromofuran and bromothiophene-2-yones.

Nitroalkanes were added to 5-substituted thiophene-3-ynones and furan to broaden the application of this benzannulation process. As a result, there are high yields of benzo[*b*]-thiophenes (Scheme 12.9). Derivatives of benzofuran and benzo[*b*]thiophene are produced when 3-bromothiophene-2-yones and 3-bromofuran-2-ynones combine with nitromethane.

Michael addition and 6p electrocyclization, involving the process of elimination, promote aromaticity and overall benzannulation, in which an $-NO_2$ group serves as a repeating carbanion activator. Michael addition is a standard step in the benzannulation of *ortho*-halo furanynones and *ortho*-halo thiopheneynones. This step begins with an aromatic nucleophilic substitution (SNAr) reaction. To enhance the synthesis of nitrobenzofurans and nitrobenzo[*b*]-thiophenes, nitromethanes also undergo positional changes. One of the intriguing aspects of the one-pot benzannulation process of furan/thiophene-derived ynones is that it promoted aromatization, in which the $-NO_2$ group functions as a precursor for the *ortho*-haloynones and as a suicidal promotor, which eventually disappears as the normal ynones participate in the reaction. A crucial point to note is that the one-pot benzannulation process was compatible with the incorporation of nitroalkanes, efficiently producing benzo[*b*]thiophenes and benzofurans from both 2- and 3-substituted furan- and thio-pheneynones.

12.3 Biological Activities

12.3.1 Benzothiophenes as Anticancer Agents

Cancer, a terrible illness, is the major cause of mortality in society. It is expected to surpass the worldwide death toll by over 10 million by 2020 and become the world's biggest cause of death by 2030. Anticancer drug[40,41] treatments[42,43] exacerbate the issue by killing the bulk of the cells rather than targeting only the damaged portion. The subclassification of malignant cell proliferation in specific tumors differs based on the characteristics that lead to the recurrence of the disease. Researchers must overcome a number of obstacles to create new medications that are safe and effective and have minimal adverse effects, and these medications are also needed to improve resistance.

Raloxifene (**2**) and arzoxifene (**1**), two cancer medications, are produced by modifications of the benzothiophene nucleus (Figure 12.2). Being more potent than raloxifene, arzoxifene is an extremely effective medication for preventing breast cancer in rats caused by the carcinogen nitrosomethylurea.[44] In postmenopausal women who have osteoporosis and who are at risk of developing aggressive breast cancer, raloxifene is used to prevent osteoporosis and reduce the likelihood of breast cancer spreading.[45]

Penthala *et al.* reported novel benzothiophene acrylonitrile derivatives (Figure 12.3) as a promising anticancer agent. Three novel benzothiophene acrylonitrile derivatives showed a growth inhibition 50% (GI_{50}) value in the range of 10–100 nM.[46]

According to Bai *et al.*, the primary target of pharmacologically active therapeutic medicines used to treat breast cancer is the newly generated estrogen receptor.[47] Their investigation focused on the anti-estrogen receptor alpha (anti-ERa) activity of six different hydroxy benzo[*b*]thiophenes, which were isolated from raloxifene. The results showed that compound (**4**) exhibited significant antagonistic activity in estrogen receptor protein (ERP) tumor cells. The discovery of a new family of chemicals exhibiting superior covalent antagonistic activity was a significant advancement in the field's push for high performance (Figure 12.4).

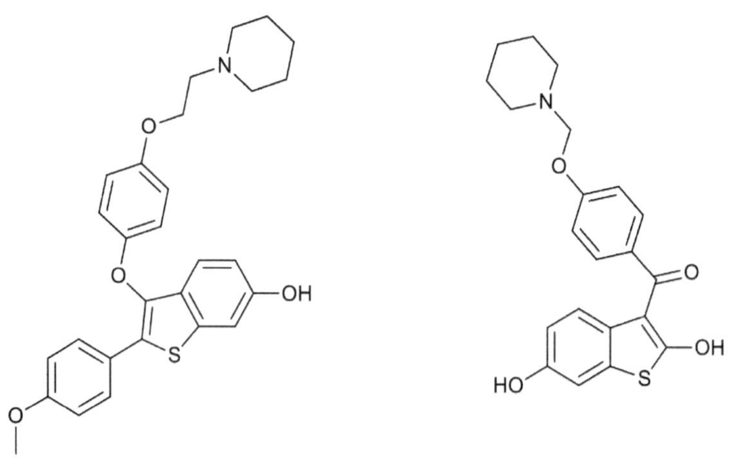

arzoxifene (1) raloxifene (2)

Figure 12.2 Drugs used in the treatment of cancer.

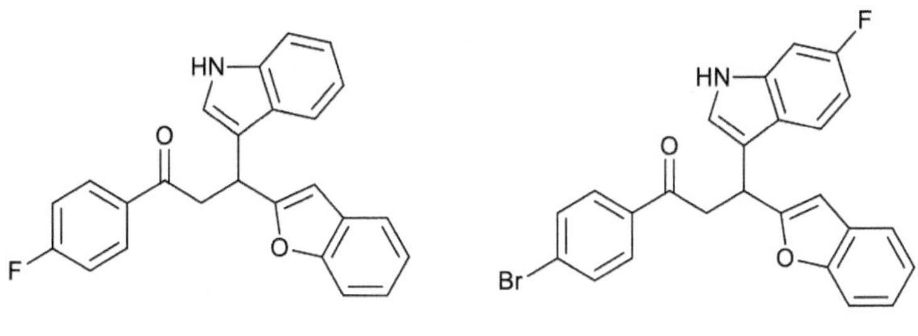

3 4

Figure 12.3 Benzo[*b*]thiophene acrylonitrile with promising anticancer activity.

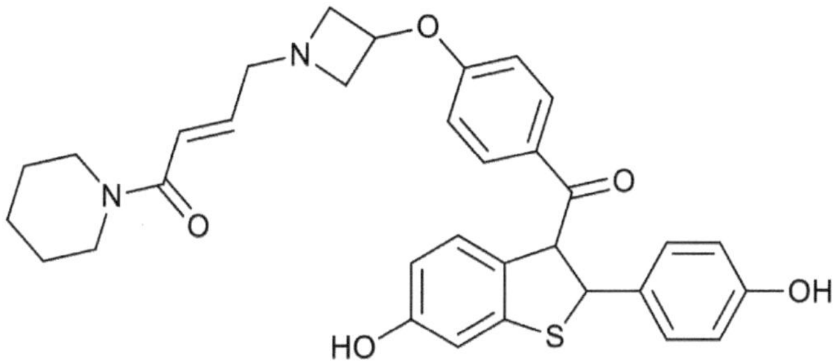

4

Figure 12.4 The covalently bonded cysteine molecule attached to the helix H11.

Benzothiophene compounds synthesized by Chen *et al.* are an appealing therapeutic target for cancer because they inhibit nicotinamide phosphoribosyltransferase (Nampt).[48] The majority of benzothiophene compounds showed moderate-to-good anti-Nampt efficacy. The Nampt inhibitory action was significantly reduced when the benzofuran scaffold was substituted by the benzothiophene scaffold. The substituted thiophene derivatives were synthesized by removing a benzene nucleus from a benzothiophene core. Adding a methyl substituent to the C-5 position of the thiophene produced numerous effective Nampt inhibitors. The half-maximal inhibitory concentration (IC_{50}) values of compounds (**5–8**) (Figure 12.5) were found to be 1.1, 0.86, 0.95, and 0.17 μM, respectively. Furthermore, compound (**8**) had the most anti-proliferative activity (IC_{50} = 3.9 μM) against the HepG2 cell line.

Davis *et al.* synthesized novel benzothiophene chemicals that function as inhibitors of Rho-associated protein kinase (ROCK).[49] In a skin cancer model on mice, it has been shown that they contributed to the thickness and stiffness of the tissue surrounding tumors, mostly through an increase in the amount of collagen in the surrounding tissue. Then, in comparison to a phenyl group with hydrogen, the propyl-substituted molecule (**9**) has shown outstanding activity in enzymatic potency of around 5-, 15-, and 100-fold. These chemicals dramatically reduced intraocular pressure *in*

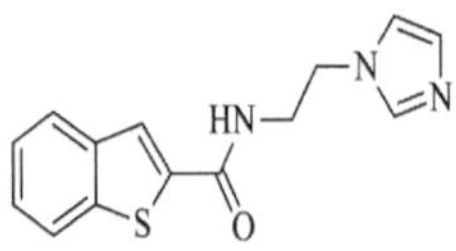

5

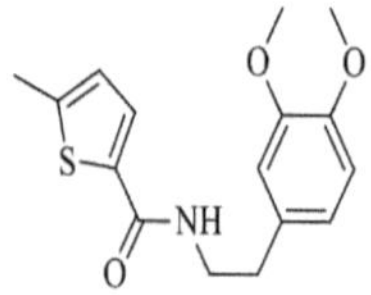

6

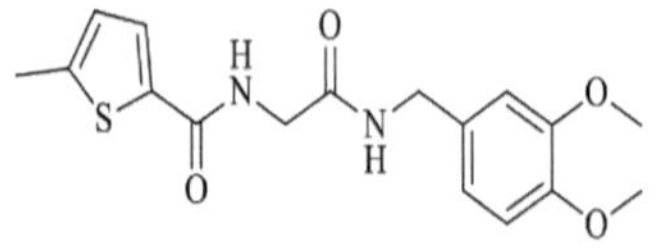

7

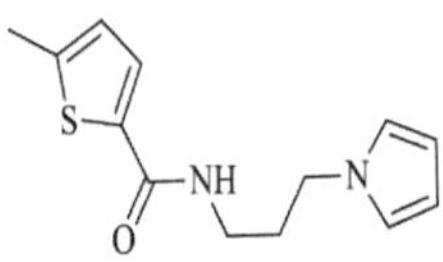

8

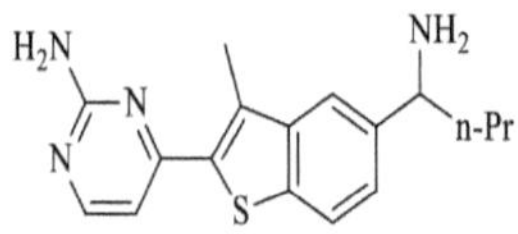

9

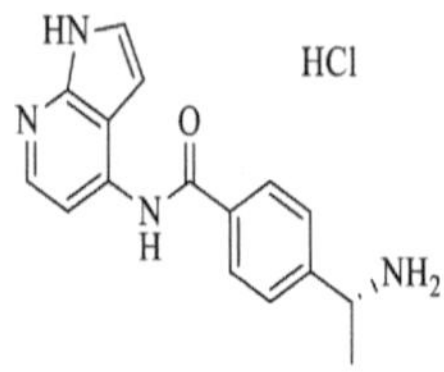

10

Figure 12.5 Compounds showing anticancer activity.

vivo and had an *in vitro* action similar to that of the clinical candidate Y-39983 (**10**). According to structure–activity relationship (SAR) research, the addition of an amino methylene linker between the tail group and the benzothiophene core significantly affected the enzymatic and cellular SAR, producing molecules that might be applied topically to the eye.[50]

12.3.2 Benzothiophenes as Anti-inflammatory and Analgesic Agents

Many diseases have inflammations as their primary cause. Inflammation is an essential part of the immune system's reaction in the human body toward pathogens, wounds, or trauma. Even though certain conditions are not directly caused by inflammation, inflammatory processes frequently exacerbate the associated pain and suffering.[51] Recent research has revealed the crucial part inflammation plays in autoimmune illnesses, such as retinitis, multiple sclerosis, Crohn's disease, sclerosis, psoriasis, rheumatoid arthritis, intestinal disorders, type I diabetes, and retinitis, in addition to infections and cancer.[52] Nonsteroidal anti-inflammatory drugs (NSAIDs) have demonstrated significant efficacy in the management of a variety of inflammatory conditions. For treating inflammatory illnesses, NSAIDs targeting the cyclooxygenase (COX) pathway are the very first line of clinical treatments. The production of several prostanoids, including prostacyclin, thromboxane, and prostaglandins, which are implicated in pathological and physiological mechanisms of inflammation, depends on COX enzymes. Traditional NSAIDs possess anti-inflammatory properties through the selective inhibition of either COX-1, COX-2, or both. COX-2 inhibition is responsible for the beneficial effects of analgesia and reduced edema, whereas COX-1 inhibition is responsible for the harmful effects of NSAIDs. It is challenging to find a treatment for chronic inflammatory diseases since the critical etiology and aggravating processes are poorly known. Thus, it is crucial to develop novel, effective, and selective NSAIDs by medicinal chemistry techniques.

The discussion of the anti-inflammatory and analgesic properties of benzothiophene derivatives is given as follows.

Fakhr *et al.* described a series of 2-substituted benzothiophene analogs. Rats were used for the acetic acid-induced writhing test, hot-plate testing, and carrageenan-induced paw edema assay to evaluate the analgesic and anti-inflammatory properties of the molecules. The outcomes were contrasted with those of piroxicam. Additionally, a hot-plate test and an acetic acid-induced writhing test have been utilized to find their analgesic effect. Anti-inflammatory and anti-nociceptive properties of compounds (**11–13**) (Figure 12.6) were shown to be more powerful than those of piroxicam (30% at 1 hour, 25 mg kg^{-1}) at doses of 25 or 50 mg kg^{-1}. According to SAR research the potency of the system is higher for carbohydrazide (**11**) compared to semicarbazide (**12**), and phenylthiosemicarbazide (**13**).[53]

According to Isloor *et al.*, in the rat hind paw edema method, hybrids of benzothiophenediaryl pyrazoles have shown good anti-inflammatory efficacy. Compound (**14**), exhibiting substitutions of *p*-anisyl and *p*-chlorophenyl in pyrazole at the third and fifth positions, has demonstrated inhibition (51.25%), which is nearly identical to that of the reference medication, diclofenac sodium (51.88%).[54]

The anti-inflammatory properties of a series of 4,5,6,7-tetrahydrobenzothiophene derivatives were investigated by Mohammed Khan *et al.*,[55] and

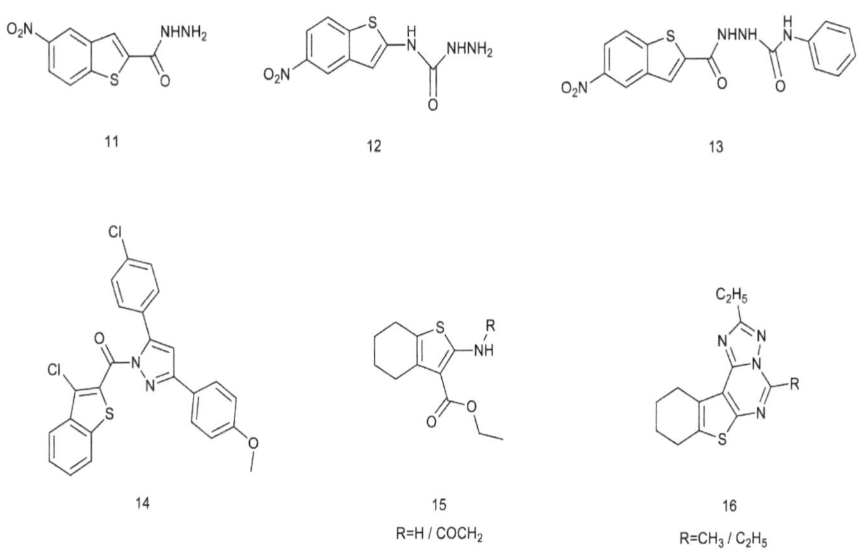

Figure 12.6 Compounds showing anti-inflammatory and analgesic activity.

the outcomes were compared to those of indomethacin. Compound (**15**, R = H) showed 61% anti-inflammatory activity with an IC_{50} value of 121 μM, whereas compound (**105**, R = COCH2) showed 81% anti-inflammatory activity with an IC_{50} value of 396 μM.[56] Compound **16** had better analgesic and anti-inflammatory properties than the benchmark medication, diclofenac sodium (55% and 40% at 1 hour, 10 mg kg^{-1}).[57]

12.3.3 Benzothiophenes as Anti-tubercular Agents

One of the fatal, infectious diseases that has added to humanity's problems is tuberculosis (TB), a lung infection. The majority of tuberculosis cases are caused by a pathogenic bacterial species *Mycobacterium tuberculosis* (Mtb), which belongs to the genus *Mycobacterium*. Currently, one-third of the world's population is infected with Mtb, and 8.9–9.9 million cases are newly added tuberculosis and relapses are reported annually. The current course of treatment involves a multidrug regimen that must be followed for at least six months, but there is no assurance that the pathogen will completely be eradicated. Thus, there is a pressing necessity to create novel anti-TB medications with a quick mode of action against mycobactericidal organisms in the intracellular milieu, reduced toxicity, and an efficient means of eliminating multiple drug-resistant (MDR) strains.[58]

Under aerobic growth conditions in liquid culture, Chandrasekera *et al.* investigated the efficacy of a heteroaryl or aryl thioether substituent at the C-3 position of the benzothiophene-1,1-dioxide (BTD) analog against a virulent strain of Mtb. Tetrazole (**17**), oxadiazole, and oxazole **18(a–e)** analogs are found to be the most efficient compounds; ineffective derivatives included aromatic 6-membered substitution at the C-3 position. The potency of oxadiazole compounds was not affected by the phenyl ring substitution by any group. When electron-donating groups (such as methoxy, methyl, and **18(b,e)**) or electron-withdrawing groups (such as the chloro group and **18d**) are added to the para position, the anti-TB effect remains unchanged. According to SAR research, benzothiophene compounds linked *via* a thioether and featuring a heterocyclic substituent with five members at position C-3 had superior efficacy against *M. tuberculosis* when compared to non-heterocyclic compounds having a pyridyl group and naphthyl, pyrimidyl, quinolinyl, or isoquinolinyl groups.[59]

Quinazolinone–benzothiophene hybrids were found by Rao and Subramaniam to be effective anti-TB drugs against the Mtb strain. The compounds (**19a**) and (**19b**), containing 2 Cl or 2 NO_2 attached to the phenyl ring at the second position of the quinazolinone scaffold, demonstrated noteworthy activity.[60]

After studying the thiazolidinones (**20**) and azetidinones (**21**) with benzothiophene nuclei using quantitatively structure–activity relationship (QSAR), Narute *et al.* found that the molecules with bulky substitution and strong nucleophilicity nature exhibited favorable anti-TB action.[61]

Thaker *et al.* reported on 1,2,4-triazoles and imidazolones with a benzothiophene nucleus. Additionally, these drugs were examined for their ability to inhibit MTB, and the results were contrasted with those of traditional rifampin. Maximum activity is shown by the 3-bromo (**22a**, 67%) group of imidazolone and 1,2,4-triazole nucleus.[62] Imidazolones and 1,2,4-triazoles took the place of thiosemicarbazide and 1,3,4-thiadiazole in the same array. Higher activity was demonstrated by the compounds containing 2-(3′-chloro-5-phenoxybenzothiophen-2′-yl)-5-(*p*-methoxyphenyl)amino-1,3,4-thiadiazole (**24c**, 90%), 2-methyl (**24a**, 60%), and 2-methoxy (**24b**, 60%) analogs[63–65] (Figure 12.7).

12.3.4 Benzothiophenes as Antioxidant Agents

Foods rich in antioxidant components are significant health-promoting agents. Scientific study suggests that antioxidants may lower the danger of chronic conditions including heart diseases and cancer. Cellular oxidizable

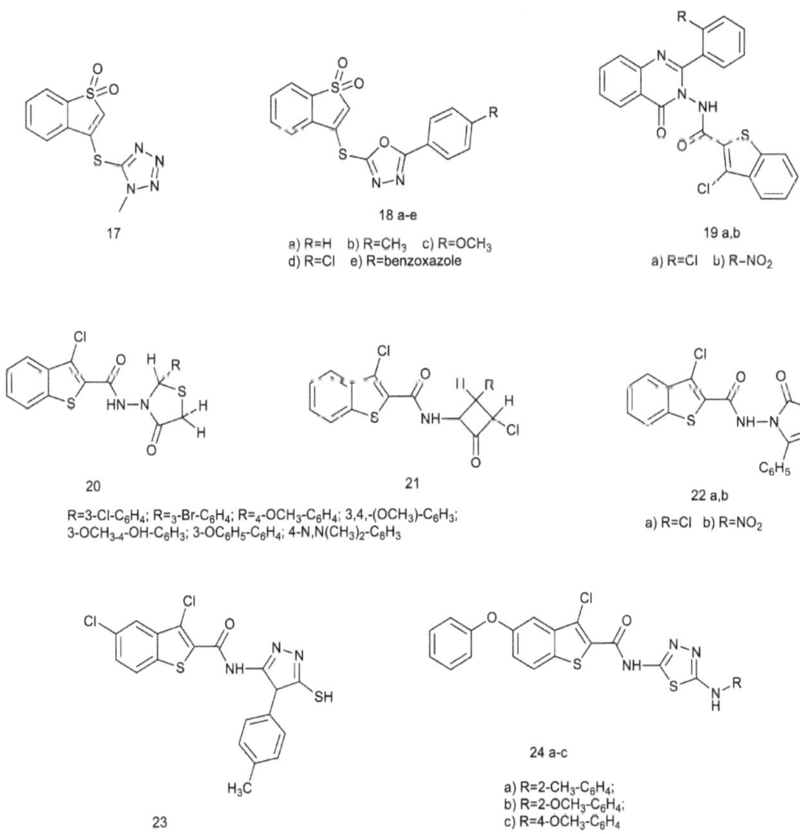

Figure 12.7 Compounds showing anti-tubercular activity.

compounds may be prevented/delayed from oxidizing by antioxidants. Synthetic antioxidants, such as butylated hydroxytoluene (BHT) and butylated hydroxyanisole (BHA), were widely used in industrial processing to reduce health risks and increase food shelf life.[66] Flavonoids and carotenoids are two of the numerous naturally occurring antioxidants.[67] The goal of the research in recent years has been to produce natural, safe, and effective antioxidants that can withstand oxidative stressors.

Tetrahydrobenzothiophene analogs containing the naphthoquinone nucleus were synthesized by Gouda *et al.*,[68] and such derivatives were tested for their antioxidant potential. The antioxidant activity of compounds (**25**) (89.87%), (**26a**) (89.93%), and (**26b**) (95.97%) (Figure 12.7) was promising.

Compounds (**25–28**) have the capacity to shield DNA from bleomycin-induced damage. Naphthoquinones (**25**) and (**26a**) are more effective than 5-fluorouracil (5-Fu) based on SAR research. Compound (**25**) may be less effective than compound (**26**), which could be due to the replacement of the $-CONH_2$, –N=N, and –OH moieties with –COOEt, –NH, and –OMe, respectively.

Gouda *et al.* described benzothiophen-2-amine carboxamide derivatives as antioxidant agents in the same array. By using the superoxide anion radical-scavenging assay, they revealed that compounds (**29–32**) demonstrated strong antioxidant properties with percentages of inhibition of 58.6%, 64.1%, 58.9%, and 57.1%, respectively. The presence of the thiophene moiety, amino group, and sulfanyl core enables the SAR investigation to demonstrate the antioxidant action[69] (Figure 12.8).

12.3.5 Benzothiophenes as Anti-HIV Agents

Human immunodeficiency virus (HIV) infection causes the deadly illness known as acquired immune deficiency syndrome (AIDS). HIV infection compromises the immune system, making a person more vulnerable to other illnesses.[70] Over 30 million people are thought to be infected with HIV globally, and the treatment for the virus continues to pose a significant challenge due to the advent of drug-resistant varieties as well as harmful intricacy. Therefore, it is still crucial to find and develop newer anti-HIV drug candidates to address the issues surrounding this illness. To determine the potency of the chalcones of 3-keto salicylic acid, Sharma *et al.* synthesized them using benzothiophene and evaluated their integrase-inhibitory activity. Substitution of the phenyl ring with heteroaromatic and polycyclic rings is revealed by SAR research; exchange of the phenyl ring with thiophene or benzothiophene (**33**) showed enhanced activity.[71]

In an effort to better understand how resveratrol derivatives can boost the antiviral effects of decitabine, which is a nucleoside analog that inhibits viral replication by increasing the frequency of HIV-1 mutation, resveratrol compounds with a variety of substituents, including benzoxazole and benzothiophene, were synthesized by Clouser *et al.* Compounds (**34**) and

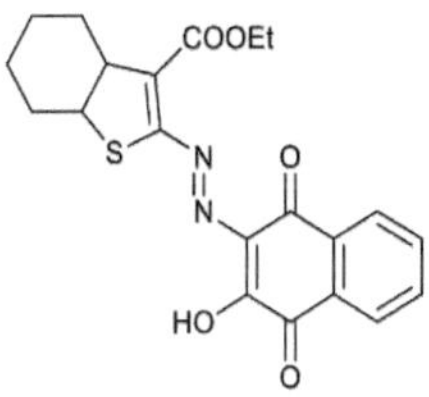

25

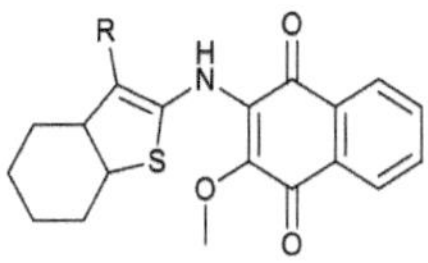

26 a,b

a) R=COOEt b) R=CONH$_2$

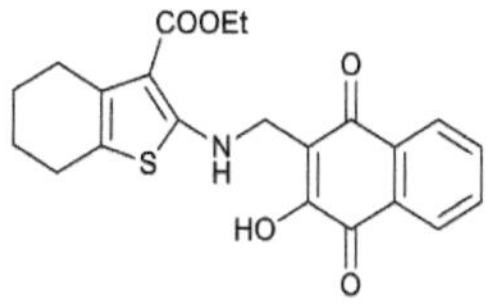

27

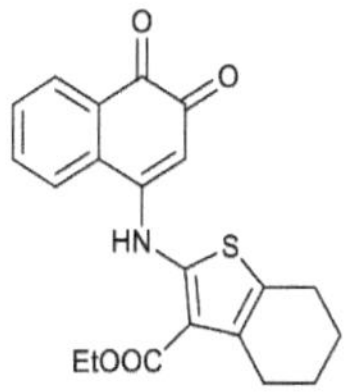

28

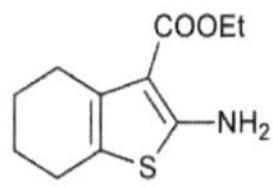

29

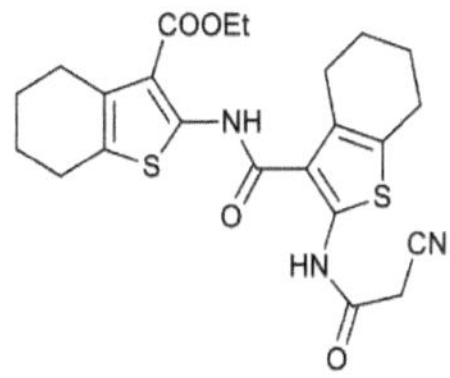

30

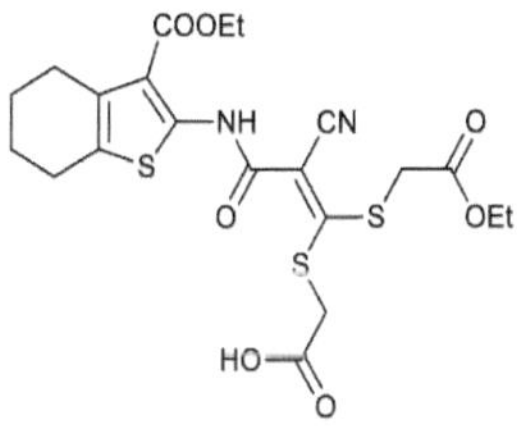

31

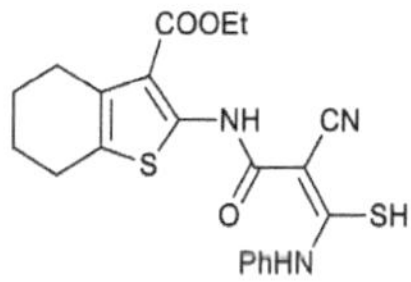

32

Figure 12.8 Compounds showing antioxidant activity.

(**35**), in particular, exhibit enhanced potency for cytotoxicity. The results of the SAR investigation showed that decitabine and resveratrol work together to suppress HIV-1 infectivity without causing additional cellular damage.[72]

To synthesize possible HIV-1 protease inhibitors, Chiummiento *et al.* reported synthesizing compounds with heteroarene moieties, such as benzofuran and benzothiophene. Compound (**36**), which contained benzothiophene, demonstrated a 16-times improvement with an IC_{50} value of 60 nM when compared to benzofuran (130 μM).[73]

Analogs of dimethyl-1-(1piperidynyl)cyclobuta[1]benzothiophene-2,2*a* (7*H*)-dicarboxylate (NSC-380292) were described by Krajewski *et al.* as HIV-1 reverse-transcriptase inhibitors. A unique reverse-transcriptase inhibitor, the lead molecule (**37**) has limited leniency for pyrrolidinyl modifications[74] and is methyl-sensitive (Figure 12.9).

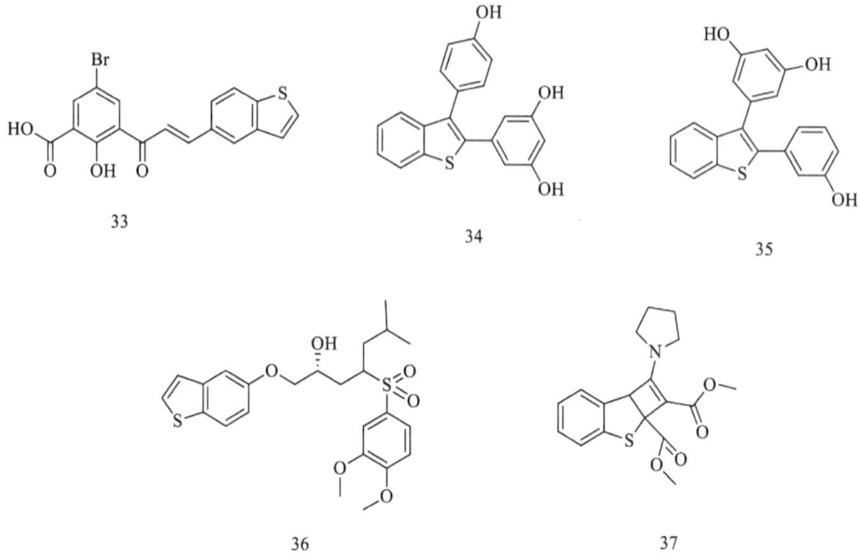

Figure 12.9 Compounds as anti-HIV agents.

12.3.6 Benzothiophenes as Anti-Alzheimer's Disease Agents

Alzheimer's disease (AD) is a central nervous system (CNS) neurodegenerative disease that causes early mental decline.[75] In India and throughout the world, AD is growing in importance as a public health concern for the elderly. As per the World Health Organization's projection, the population of those aged 65 years or older is expected to rise from 524 million in 2010 to approximately 1.5 billion by 2050, primarily in developing nations.[76] The more dominant course of action is to detect Alzheimer's disease in its preclinical stage, as this may assist manage symptoms, start therapies as soon as feasible, and enhance the quality of life.

Finding novel, fast-acting medications that effectively treat AD with few to no adverse effects is necessary due to the intricacy of the disease and the available pharmacological options. Thus, research and development of novel medications to treat AD are ongoing.

According to Lee *et al.*, benzothiophenes with a carbonyl group and a hydrophobic group are anti-cholinesterases. The inhibitory effects of benzothiophenone (**38**) and -nitro-substituted one (**39**) on cholinesterase (ChE) are significantly weaker. In comparison to (**38**), the $-NO_2$ group in (**39**) boosted an inhibitory effect toward AChE by a factor of five. With an inhibition constant value of 4.9 mM, compound (**38**) is the least effective BuChE inhibitor. This is because the nitro group has no electronic effect on BuChE, and the hydrophobic site of the protein has expanded.[77]

Cerminara *et al.* produced benzothiophenes with substituents at 4th and 5th positions and evaluated their beta-secretase (BACE) activity.

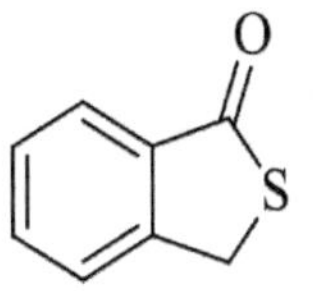

38

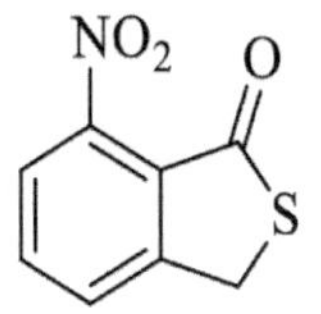

39

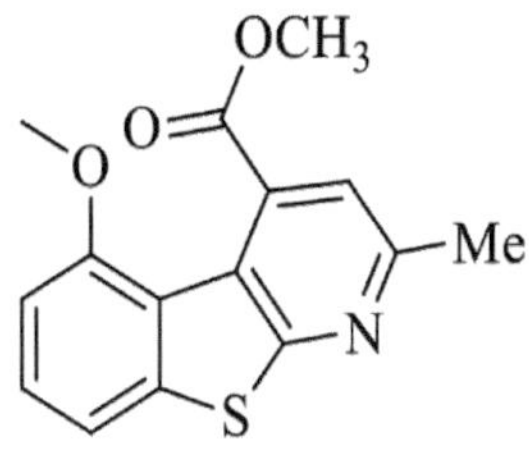

40

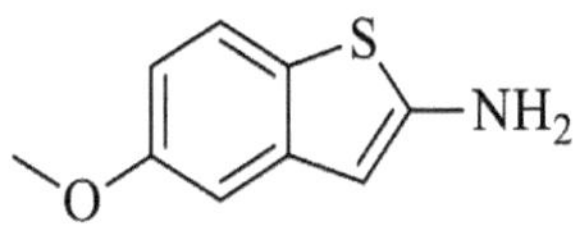

41

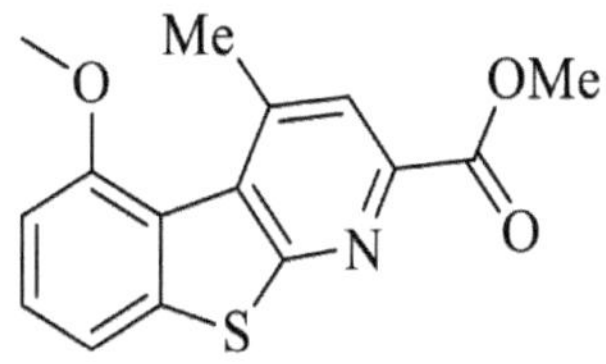

42

Figure 12.10 Benzothiophenes as Alzheimer's disease (AD) agents.

Compounds (**40–42**) (Figure 12.10), in particular, demonstrated modest BACE-1 activity; compounds with methoxy and hydroxyl substituents on a benzene nucleus will improve BACE activity based on an SAR analysis.[78]

12.3.7 Benzothiophenes as Anti-diabetic Agents

The most prevalent chronic illness in the modern world, diabetes mellitus (DM) is characterized by increased blood glucose levels brought on by either an absolute or a relative inefficiency of circulating insulin levels. DM is a serious health concern.[79] Diabetes may eventually result in retinopathy, nephropathy, and neuropathy due to microvascular damage in important tissues. There is still a great demand for better anti-diabetes medications even though there are currently around a dozen anti-diabetes medications on the market. Ipragliflozin, ASP1941, **43** (Figure 12.10), is a benzothiophene derivative that is currently being investigated to treat type-2 diabetes mellitus.[80]

Imamura *et al.* synthesized various C-glucosides containing benzothiophene and then assessed their inhibitory effects. These compounds function differently and lower blood glucose levels without affecting insulin secretion. It was discovered that compound (**43a**), a derivative of benzothiophene, is an effective and specific sodium–glucose co-transporter 2 (SGLT2) inhibitor, and compound (**43b**) is an extremely potent and selective SGLT2 inhibitor. The ortho position of these compounds has oxygen functional groups, which enhance the inhibitory activity against SGLT2.

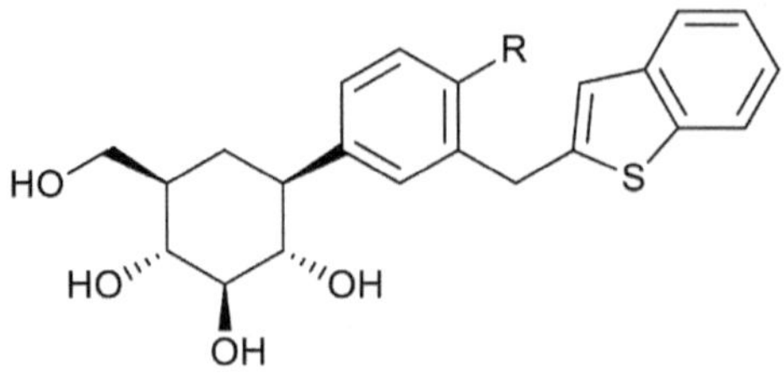

43 a.b

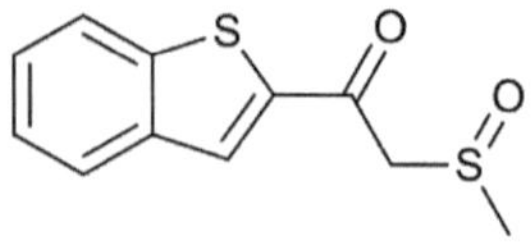

44

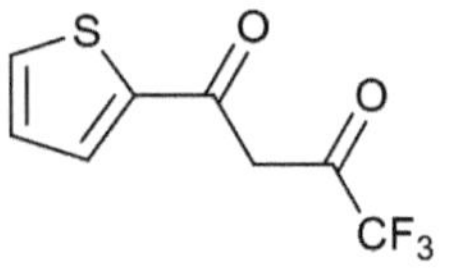

45

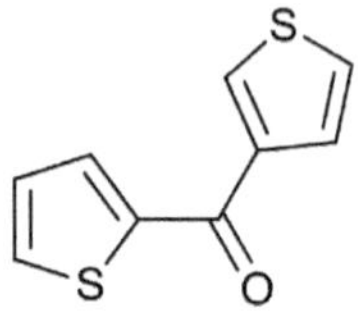

46

Figure 12.11 Benzothiophenes as anti-diabetic agents.

At this position (**43b**), substitution with a fluorine atom produced a very strong *in vitro* profile with a 254-fold selectivity *vs.* SGLT1[81,82] and an IC_{50} value of 7.4 nM.

Shao *et al.* produced benzothiophene and thiophene derivatives and evaluated their *in vitro* peroxisome proliferator-activated receptor γ (PPARγ) activity in HepG2 cells. Compound (**44**) transactivated PPARγ by 120.97%, while three compounds (**44–46**) transactivated PPARγ by >100%. Research conducted *in vivo* revealed that chemicals (**45**) and (**46**), when tested on animals, dramatically ameliorated the diabetes condition. Based on SAR research, substances containing 2-(β-carbonyl) or 2-(β-sulfonyl)butyryl-thiophene (**45, 46**) have been shown to increase PPARγ activity[83] (Figure 12.11).

12.3.8 Benzothiophenes as Anti-convulsant Agents

A wide class of drugs known as anti-convulsants is prescribed to treat tremors. The anti-convulsant agents are used more often to treat neuropathic pain and bipolar illness as many of them appear to function as mood stabilizers. The production of piperidines with varying substitutions as inhibitors of μ-calpain was elucidated by Lubisch *et al.* These inhibitors have shown efficacy in experimental models of muscular dystrophy, stroke, and myocardial infarction. The most potent inhibitor, a piperidine-tethered benzothiophene produced from ketoamides, has a Ki value of 9 nM. Moreover, it shows a strong calpain-inhibitory activity.[84]

Ling *et al.* synthesized new 2,3-benzodiazepines and associated isoquinoline compounds by substituting a 2nd-benzothiophenyl moiety at the

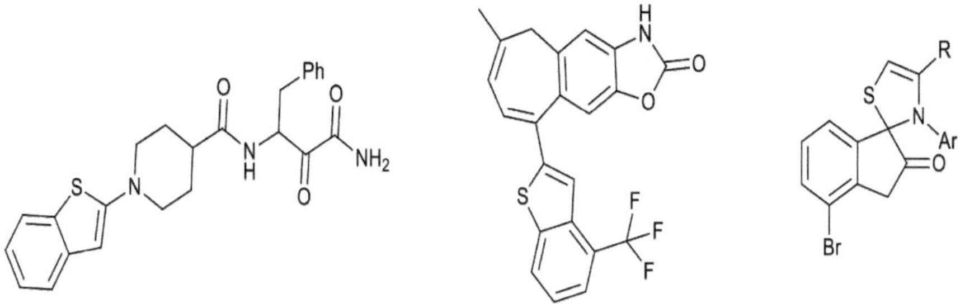

47 48 49,50

49: Ar=1-$C_{10}H_7$, R=OH
50: Ar=1-$C_{10}H_7$, R=CH_3

Figure 12.12 Benzothiophenes as anti-convulsant agents.

1st position. These substances were further examined for the gamma-aminobutyric acid (GABA)-alpha-5 receptor inhibitory activity. Compound (**47**) improved the performance of rats in the thing recognition test and exhibited good GABA-alpha-5 receptor inhibitory activity without sedative, proconvulsive, or motor adverse effects.[85]

Zaher *et al.* have elucidated the process of synthesizing spiro[1]benzothiophene-thiazolidine compounds and ascertained their anti-convulsant properties by assessing the degree of sedation influenced by phenobarbitone and a test compound's capacity to avert strychnine-induced seizures in mice. Numerous substances that were produced have shown noteworthy and encouraging anti-convulsant properties. The most efficient molcecules for both protecting against strychnine-induced seizures and extending the duration of phenobarbitone-induced sleep were determined to be the naphthyl derivatives (**48**) and (**49**)[86] (Figure 12.12).

12.3.9 Benzothiophenes as Anti-Parkinson's Disease Agents

Neurodegenerative illnesses are now a significant age-related wellness issue, particularly in developed nations where the proportion of the elderly population is growing. There are medications on the market with significant adverse effects. Thus, scientists are trying to create novel compounds to treat Parkinson's disease. *In vitro*, Rautio *et al.* produced a benzothiophene derivative (**51**) that inhibited the catechol-*O*-methyltransferase inhibitor (COMT) enzyme with an outstanding inhibition constant (Ki < 1 nM). Levodopa's primary enzyme of metabolism is now COMT. Compound (**51**) was discovered to have a high systemic clearance and a poor oral bioavailability.[87]

N-Heterocyclic-substituted piperazines containing various heterocycles were synthesized by Brown *et al.* The drug was then analyzed for

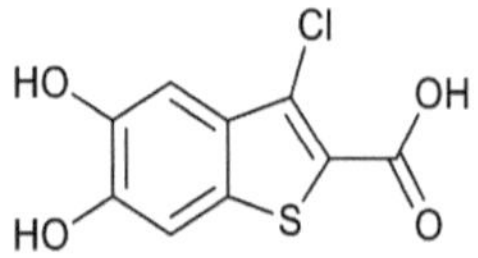

51

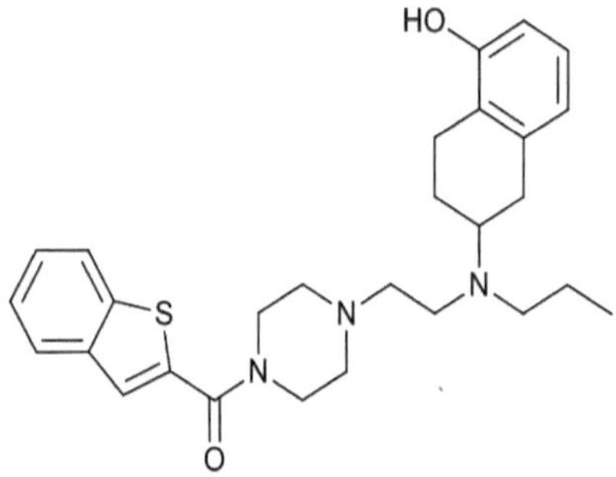

52

Figure 12.13 Benzothiophenes as anti-Parkinson's disease agents.

[3*H*]spiperone-binding tests using HEK-293 cells expressing D2 or D3 dopamine receptors. The benzothiophene derivative (**52**) showed a high affinity for D2 and D3 receptors (Ki = 76.9 and 1.69 nM for D2 and D3 receptors, respectively)[88] (Figure 12.13).

12.3.10 Benzothiophenes as Antimicrobial Agents

Given the current elevated incidence of bacterial infections, there has been a proliferation of approaches to investigate novel avenues for combating multidrug-resistant pathogenic microorganisms and transmissible diseases. Scientific evolution has allowed for the integration of these additional elements, which are used as a source for treating many bacterial infections.[89,90] Given this, a number of benzothiophene derivatives are receiving a greater focus, and significant investigative studies have been carried out in such fields. Antibacterial activities of benzo[*b*]thiophenes are primarily derived from derivatives containing nitro, halogen, methoxy, and phenyl residues.[91] Benzothiophene derivatives and semicarbazide, 2-urea, and heteroaryl/-acyl-substituted compounds are produced by a substitution reaction involving the main ingredients having a halogen substituent in either the third or seventh position.[92,93] Heterocyclic compounds, such as thiazolidinediones, have numerous biological uses. Glitazones are the form in which they are found and are used in medicine. This approach is also used to treat diabetes.

Meena *et al.* suggested synthesizing a thiazolidine derivative-tethered benzo[*b*]thiophene moiety using the cup or well method to assess its fungal and bacterial activities against different strains of bacteria and fungi, such as *A. fumigatus*, *P. aeruginosa*, *S. aureus*, *P. aeruginosa*, *E. Coli*, and *C. albicans*.[94]

The range of inhibition for various bacterial strains was 10–16 mm, whereas that for fungal strains was 14–23 mm. These values are considerably closer to the regions of inhibition of reference substances

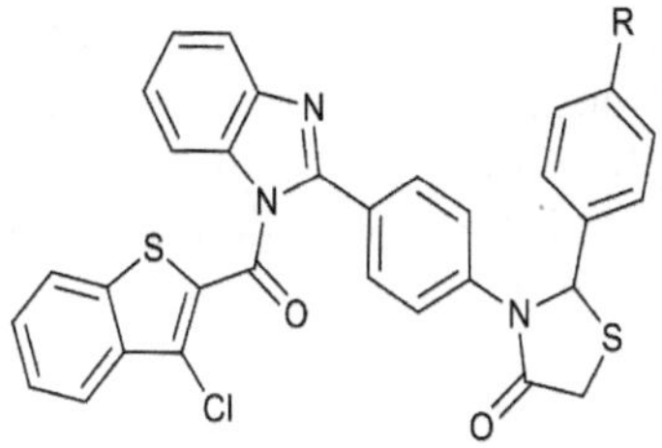

Figure 12.14 Benzo[*b*]thiophenes for screening of bacterial and fungal activities against bacterial and fungal strains.

griseofulvin and ciprofloxacin, respectively. Strong inhibition was shown by the derivatives (**53a–d**) against *P. aeruginosa* and *S. aureus*. When it comes to *E. coli*, the chemicals **53a** and **53b** (Figure 12.13) work well. A modest degree of antibacterial activity was shown by the synthesized compounds in comparison to a standard medication, ciprofloxacin. In a similar vein, (**53b**) has the maximum amount of fungal activity against *Candida albicans*, whereas (**53c**) and (**53b**) show a broad range of inhibition against *A. niger. A. fumigatus,* and *A. niger,* which were susceptible to the antibacterial properties of the dimethylamino derivative (**53d**) (Figure 12.14).

According to Patil *et al.*,[95] a series of physiologically active dimethyl formamide (DMF) derivatives were synthesized using the benzothiophene moiety of 2-amino-4,5,6,7-tetrahydrobenzothiophene-3-carbonitrile. The spectral and analytical data of the products verified their structure. These produced compounds were examined to determine their antibacterial properties against *A. niger*, *C. albicans*, *E. coli*, and *S. aureus*. The synthesis of *N*-aryl-tetrahydrobenzothiophen-2-yl-formamidines was the main goal (**54**). In contrast to gentamicin and fluconazole, the antimicrobial activities of (**54a–d**) were good, those of (**54a** and **54b**) were moderate, and those of (**54d**) were excellent. Additionally, as Figure 12.14 illustrates, compounds **54a**, **54b**, and **54c** had outstanding antifungal activity (Figure 12.15).

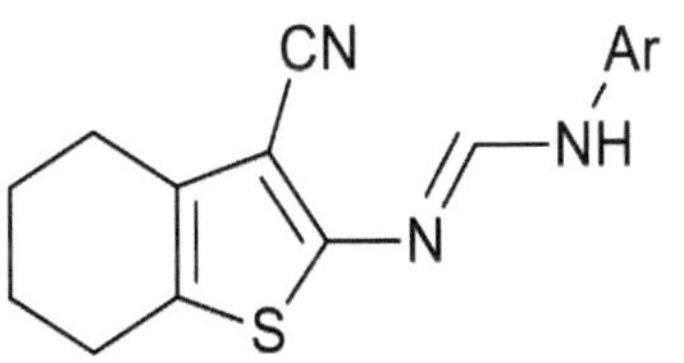

54a-d

Figure 12.15 Biologically active *N*,*N*-dimethylformamide derivatives containing benzo[*b*]thiophene.

The need for newly developed medications and the problem of antibiotic resistance are growing. Hooda *et al.* synthesized 1-(2,4-dinitrophenyl)-2-(3-methylbenzo[*b*]thiophen-6-yl) to address this issue. Figure 12.15 shows the structure of 1*H*-benzo[*d*]imidazole-5-carboxamides (**55a**). They focused on tRNA synthetase. The inhibition leads to an antibacterial effect, and the generated compounds were assessed using the binding energy derived from the docking screening data.[96] In comparison to the reference, the compounds (**55a1**, **55a2**, and **55a3**) (Figure 12.16) exhibited good minimum inhibitory concentrations against the six selected microorganisms. Amide 55a1, which has the 2,4-dinitrophenyl substituent that draws electrons, exhibited a favorable high level of antibacterial activity. The 2- and 5-chloro derivatives

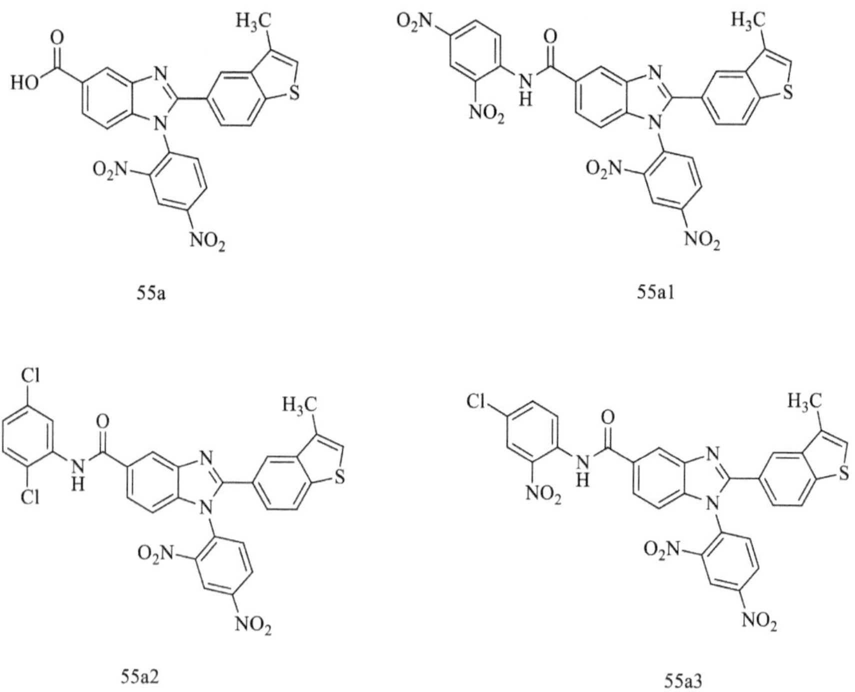

Figure 12.16 *N*-Arylbenzimidazole-5-carboxamides with 3-methylbenzo[*b*] thiophene substituents.

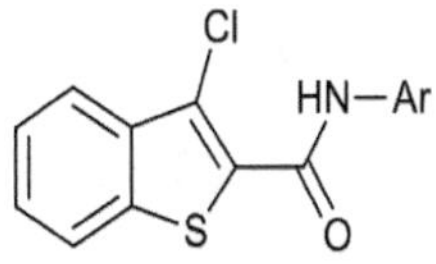

56a-c

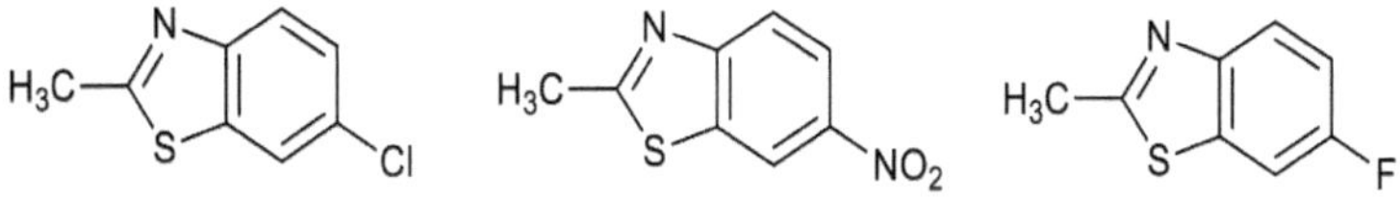

56a 56b 56c

Figure 12.17 Substituted benzo[*b*]thiophene carboxyamides as antimicrobial agents.

exhibited remarkable antibacterial activity as well, while the antifungal activity of (**55a2**) was diminished. Based on the *in vitro* results, the novel compounds show good antibacterial activity and may hold potential for *in vivo* study.

The modified benzo[*b*]thiophenecarboxamides (**56a–c**) (Figure 12.16) were synthesized by Kallur *et al.* using carbonyl chloride, aminobenzothiazoles, and aminopyridines by refluxing the mixture for ten to fifteen hours.[97] The yields of pure products, compared with those of the commonly sold miraculous drug, were astonishingly high (50–58%). To determine the anthelmintic activity of the investigated substances, further research was conducted. Earthworms were used in the tests, and a sample solution (50 mg ml^{-1}) prepared from a 1% gum acacia solution (derived from common saline) was used as a control. In comparison to other compound series, the results demonstrate that **56a**, **56b**, and **56c** possess a wide range of anthelmintic activity. As a result, the substituted benzo[*b*]-thiophenes (**56**) may present a lucrative business opportunity (Figure 12.17).

Abbreviations (Solvents and Reagents)

1,10-Phen	1,10-Phenanthroline
5-Fu	5-Florouracil
AChE	Anti-cholinesterase
AD	Alzheimer's disease
AIDS	Acquired immune deficiency syndrome
Ar	Aryl
AuCl	Gold(I) chloride
BACE	Beta-secretase
BHA	Butylated hydroxyanisole

BHT	Butylated hydroxytoluene
Bn	Benzyl
BTD	Benzo[*b*]thiophene-1,1-dioxide
BuChE	Butyrylcholinesterase
ChE	Cholinesterase
CNS	Central nervous system
COMT	Catechol-*O*-methyltransferase inhibitor
COX	Cyclooxygenase
Cs_2CO_3	Cesium carbonate
CuI	Copper iodide
D2, D3	Dopamine receptor
DCM	Dichloromethane
DIPEA	*N*,*N*-Diisopropylethylamine
DM	Diabetes mellitus
DMF	Dimethyl formamide
DPE Phos	Bis-[2-(diphenylphosphino)phenyl]ether
DTBP	Di-*tert*-butyl peroxide
EC50	Half-maximal effective concentration
ERP	Estrogen receptor protein
FAAH	Fatty acid amide hydrolase
GABA	Gama-aminobutyric acid
HIV	Human immunodeficiency virus
I_2	Iodine
IC_{50}	Half-maximal inhibitory concentration
kg	Kilogram
MDR	Multiple drug resistant
mg	Milligram
ml	Milliliter
Mtb	*Mycobacterium tuberculosis*
Nampt	Nicotinamide phosphoribosyl transferase
NBS	*N*-Bromosuccinimide
NK2	Neurokinin 2 receptor
nM	Nanomolar
NPr3	*n*-Isopropyl amine
NSAID	Nonsteroidal anti-inflammatory drug
PD	Parkinson's disease
Pd_2dba_2	Tris(dibenzylideneacetone)palladium(0)
PPAR	Peroxisome proliferator-activated receptor
QSAR	Quantitative structure–activity relationship
R	Alkyl
ROCK	Rho-associated protein kinase
SAR	Structure–activity relationship
SGLT	Sodium–glucose co-transporters
TMEDA	Tetramethylenediamine
WHO	World Health Organization

μg	Microgram
μM	Micromolar

References

1. National Center for Biotechnology Information, PubChem Compound Summary for CID 7221, Benzothiophene, 2023, https://pubchem.ncbi.nlm.nih.gov/compound/Benzothiophene. Accessed 25 Aug 2023.
2. J. Joule, in *Heterocyclic chemistry*, 3rd edn, 1995.
3. E. Campaigne, D. R. Knapp, E. S. Neiss and T. R. Bosin, Biologically active benzo[b]thiophene derivatives Part I, *Adv. Drug. Res.*, 1970, **5**, 1–54.
4. T. R. Bosin and E. Campaigne, biologically active benzo[b]thiophene derivatives, *Adv. Drug Res.*, 1977, **11**, 191–232.
5. F. Boberg, W. Bruns, R. Kaller and D. Musshoff, Schwefelverbindungen des erdöls xvii.1 dibenzothiophene mit einem verzweigten C3-C5-alkylrest, *Phosphorus, Sulfur Silicon Relat. Elem.*, 1992, **72**, 33–47.
6. S. J. Jeong, R. Higuchi, T. Miyamoto, M. Ono, M. Kuwano and S. F. Mawatari, Bryoanthrathiophene, a new antiangiogenic constituent from the bryozoan Watersipora subtorquata, *J. Nat. Prod.*, 2002, **65**, 1344–1345.
7. T. R. Kelly, Y. Fu, J. T. Sieglen and H. De Silva, Synthesis of an orange anthrathiophene pigment isolated from a Japanese Bryozoan, *Org. Lett.*, 2000, **2**, 2351–2352.
8. L. Jianqi, G. Kai and L. V. Na, *Benzothiophene alkanol piperazine derivatives and their use as antidepressant*, 2010, WO2010000198 A1.
9. M. A. Spiller, in *Caffeine*, ed. G. A. Spiller, CRC Press, Boca Raton, 1998.
10. Z. Qin, I. Kastrati, R. E. Chandrasena, H. Liu, P. Yao, P. A. Petukhov, J. L. Bolton and G. R. Thatcher, Benzothiophene selective estrogen receptor modulators with modulated oxidative activity and receptor affinity, *J. Med. Chem.*, 2007, **50**, 2682–2692.
11. R. Romagnoli, P. G. Baraldi, M. D. Carrion, C. L. Cara, D. Preti, F. Fruttarolo, M. G. Pavani, M. A. Tabrizi, M. Tolomeo, S. Grimaudo, A. Di Cristina, J. Balzarini, J. A. Hadfield, A. Brancale and E. Hamel, Synthesis and biological evaluation of 2- and 3-aminobenzo[b]thiophene derivatives as antimitotic agents and inhibitors of tubulin polymerization, *J. Med. Chem.*, 2007, **50**, 2273–2277.
12. K. Sweidan, J. Engelmann, W. A. Rayyan, D. Sabbah, M. A. Zarga, T. Al-Qirim, Y. Al-Hiari, G. A. Sheikha and G. Shattat, Synthesis and preliminary biological evaluation of new heterocyclic carboxamide models, *Lett. Drug Design Dis.*, 2015, **12**, 417–429.
13. A. Martorana, C. Gentile, U. Perricone, A. P. Piccionello, R. Bartolotta, A. Terenzi, A. Pacc, F. Mingoia, A. M. Almerico and A. Lauria, Synthesis, antiproliferative activity, and in silico insights of new

3-benzoylamino-benzo[b]thiophene derivatives, *Eur. J. Med. Chem.*, 2015, **90**, 537–546.
14. G. K. Rao and R. Subramaniam, Synthesis, antitubercular and antibacterial activities of Some quinazolinone analogs substituted with benzothiophene, *Chem. Sci. J.*, 2015, **6**, 92–96.
15. T. Banerjee, N. Kapoor, N. Surolia and A. Surolia, Benzothiophene carboxamide derivatives as novel antimalarials, *IUBMB Life*, 2011, **63**, 1111–1115.
16. I. M. I. Fakhr, M. A. A. Radwan, S. El-Batran, O. M. E. Abd El-Salam and S. M. ElShenawy, Synthesis and pharmacological evaluation of 2-substituted benzo[b]thiophenes as anti-inflammatory and analgesic agents, *Eur. J. Med. Chem.*, 2009, **44**, 1718–1725.
17. L. Berrade, B. Aisa, M. J. Ramirez, S. Galiano, S. Guccione, L. R. Moltzau, F. O. Levy, F. Nicoletti, G. Battaglia, G. Molinaro, I. Aldana, A. Monge and S. Perez-Silanes, Novel benzo[b]thiophene derivatives as new potential antidepressants with rapid onset of action, *J. Med. Chem.*, 2011, **54**, 3086–3090.
18. G. Moinet, C. Leriche and M. Kergoat, *Antidiabetic compounds comprising benzofuran and benzothiophene derivatives*, EP1685121 A1. PCT/EP2004/012075.
19. M. S. Malamas, J. Sredy, C. Moxham, A. Katz, W. Xu, R. McDevitt, F. O. Adebayo, D. R. Sawicki, L. Seestaller, D. Sullivan and J. R. Taylor, Novel benzofuran and benzothiophene biphenyls as inhibitors of protein tyrosine phosphatase 1B with antihyperglycemic properties, *J. Med. Chem.*, 2000, **43**, 1293–1310.
20. D. A. Kennedy and L. A. Summers, Chemical constitution and activity of Herbicides, *J. Heterocycl. Chem.*, 1981, **39**, 409–410.
21. G. Naganagowda and B. Padmashali, Synthesis, antimicrobial, and anthelmintic activities of some new 3-chlorobenzothiophene-2-carbonylchloride derivatives, *Phosphorus, Sulfur Silicon Relat. Elem.*, 2010, **8**, 1691–1700.
22. Y. Loidreau, E. Deau, P. Marchand, M. R. Nourrisson, C. Logé, G. Coadou, N. Loaëc, L. Meijer and T. Besson, Synthesis and molecular modelling studies of 8-arylpyrido[3′,2′:4,5]thieno[3,2-d]pyrimidin-4-amines as multitarget Ser/Thr kinases inhibitors, *Eur. J. Med. Chem.*, 2015, **92**, 124–134.
23. M. D. Rackham, J. A. Brannigan, D. K. Moss, Z. Yu, A. J. Wilkinson, A. A. Holder, E. W. Tate and R. J. Leatherbarrow, Discovery of novel and ligand-efficient inhibitors of plasmodium falciparum and plasmodium vivax N-myristoyltransferase, *J. Med. Chem.*, 2013, **56**, 371–375.
24. R. J. Mourey, B. L. Burnette, S. J. Brustkern, J. S. Daniels, J. L. Hirsch, W. F. Hood, M. J. Meyers, S. J. Mnich, B. S. Pierce, M. J. Saabye, J. F. Schindler, S. A. South, E. G. Webb, J. Zhang and D. R. Anderson, A benzothiophene inhibitor of mitogen-activated protein kinase-activated protein kinase 2 inhibits tumor necrosis factor alpha production and has oral

anti-inflammatory efficacy in acute and chronic models of inflammation, *J. Pharmacol. Exp. Ther.*, 2010, **333**, 797–807.

25. V. A. Jagtap and Y. S. Agasimundin, Synthesis and preliminary evaluation of some 2-amino-n′- [substituted]-4,5,6,7-tetrahydro-1-benzothiophene-3-carbohydrazide as antimicrobial agents, *J. Pharm. Res.*, 2015, **9**, 10–14.
26. H. U. Bryant and W. H. Dere, Selective estrogen receptor modulators: An alternative to hormone replacement therapy, *Proc. Soc. Exp. Biol. Med.*, 1998, **217**, 45–52.
27. M. D. Rackham, J. A. Brannigan, D. K. Moss, Z. Yu, A. J. Wilkinson, A. A. Holder, E. W. Tate and R. J. Leatherbarrow, Discovery of novel and ligand-efficient inhibitors of Plasmodium falciparum and plasmodium vivax N-myristoyltransferase, *J. Med. Chem.*, 2013, **56**, 371–375.
28. National Center for Biotechnology Information, PubChem Compound Summary for CID 7221, Benzothiophene, 2023, https://pubchem.ncbi.nlm.nih.gov/compound/Benzothiophene. Accessed 25 Aug 2023.
29. J. Joule, in *Heterocyclic chemistry*, 5th edn, 2010.
30. T. Kondo and T.-a. Mitsudo, Metal-Catalyzed Carbon-Sulfur Bond Formation, *Chem. Rev.*, 2000, **100**, 3205–3220.
31. H. Yu, M. Zhang and Y. Li, Copper-Catalyzed Synthesis of Benzo[b]Thiophenes and Benzothiazoles Using Thiocarboxylic Acids as a Coupling Partner, *J. Org. Chem.*, 2013, **78**, 8898–8903.
32. K. Yan, D. Yang, M. Zhang, W. Wei, Y. Liu, L. Tian and H. Wang, Facile Access to Benzothiophenes through Metal-Free Iodine-Catalyzed Intermolecular Cyclization of Thiophenols and Alkynes, *Synlett*, 2015, **26**, 1890–1894.
33. L.-L. Sun, C.-L. Deng, R.-Y. Tang and X.-G. Zhang, CuI/ TMEDA-Catalyzed Annulation of 2-Bromo Alkynyl Benzenes with Na2S: Synthesis of Benzo[b]Thiophenes, *J. Org. Chem.*, 2011, **76**, 7546–7550.
34. I. Nakamura, T. Sato and Y. Yamamoto, Gold-Catalyzed Intramolecular Carbothiolation of Alkynes: Synthesis of 2,3-Disubstituted Benzothiophenes from (a-Alkoxy Alkyl) (ortho-Alkynyl Phenyl) Sulfides, *Angew. Chem., Int. Ed. Engl.*, 2006, **45**, 4473–4475.
35. M. C. Willis, D. Taylor and A. T. Gillmore, Palladium-Catalyzed Intramolecular O-Arylation of Enolates: Application to Benzo[b]Furan Synthesis, *Org. Lett.*, 2004, **6**, 4755–4757.
36. T. B. Nguyen and P. Retailleau, DIPEA-Promoted Reaction of 2-Nitrochalcones with Elemental Sulfur: An Unusual Approach to 2-Benzoylbenzothiophenes, *Org. Lett.*, 2017, **19**, 4858–4860.
37. M. Kuhn, F. C. Falk and J. Paradies, Palladium-Catalyzed C–S Coupling: Access to Thioethers, *Org. Lett.*, 2011, **13**, 4100–4103.
38. Y. Kong, L. Yu, L. Fu, J. Cao, G. Lai, Y. Cui, Z. Hu and G. Wang, Electrophilic Cyclization of o-Anisole- and o-Thioanisole-Substituted Ynamides: Synthesis of 2-Amidobenzofurans and 2-Amidobenzothiophenes, *Synthesis*, 2013, **45**, 1975–1982.

39. S. Singh, S. Nerella, S. Pabbaraja and G. Mehta, Stitching Ynones with Nitromethanes: Domino Synthesis of Functionally Enriched Benzofurans and Benzothiophenes, *J. Org. Chem.*, 2021, **86**, 12093–12106.
40. H. Sung, J. Ferlay, R. L. Siegel, M. Laversanne, I. Soerjomataram, A. Jemal and F. Bray, Global Cancer Statistics 2020: GLOBOCAN Estimates of Incidence and Mortality Worldwide for 36 Cancers in 185 Countries, *Ca-Cancer J. Clin.*, 2021, **71**, 209–249.
41. WHO. Cancer. World Health Organization, American Cancer Society, 2006, https://www.cancer.org/research/cancer-facts-statistics/global. Accessed 1 Jun 2020.
42. M. Garcia, A. Jemal, E. M. Ward, Y. Hao, R. L. Siegel and M. J. Thun, *Global Cancer Facts& Figures*, American Cancer Society, Atlanta, GA, 2007, https://www.cancer.org/research/cancer-facts-statistics/ global. Accessed 1 Jun 2020.
43. World Health Organization, *Preventing Chronic Diseases: A Vital Investment: WHO Global Report*, World Health Organization, Iris, 2005, https://apps.who.int/iris/handle/10665/43314.
44. C. R. Overk, K. W. Peng, R. T. Asghodom, D. D. Lantvit, Z. Qin, J. Frasor, J. L. Bolton and G. R. Thatcher, Structure-activity relationships for a family of benzothiophene selective estrogen receptor modulators including raloxifene and arzoxifene, *ChemMedChem*, 2007, **2**, 1520–1526.
45. *FDA approves new uses for evista, U.S.*, Food and Drug Administration, 2007.
46. N. R. Penthala, V. N. Sonar, J. Horn, M. Leggas, J. S. K. B. Yadlapalli and P. A. Crooks, Synthesis and evaluation of a series of benzothiophene acrylonitrile analogs as anticancer agents, *MedChemComm*, 2013, **4**(7), 1073.
47. C. Bai, S. Wu, S. Ren, M. Zhu, G. Luo and H. Xiang, Benzothiophene Derivatives as Selective Estrogen Receptor Covalent Antagonists: Design, Synthesis and anti- Era Activities, *Bioorg. Med. Chem.*, 2021, **47**, 116395.
48. W. Chen, G. Dong, S. He, T. Xu, X. Wang, N. Liu, W. Zhang, C. Miao and C. Sheng, Identification of benzothiophene amides as potent inhibitors of human nicotinamide phosphoribosyl transferase, *Bioorg. Med. Chem. Lett.*, 2016, **26**, 765–768.
49. R. L. Davis, M. Kahraman, T. J. Prins, Y. Beaver, T. G. Cook, J. Cramp, C. S. Cayanan, E. M. M. Gardiner, M. A. McLaughlin, A. F. Clark, M. R. Hellberg, A. K. Shiau, S. A. Noble and A. J. Borchardt, Benzothiophene containing Rho kinase inhibitors: Efficacy in an animal model of glaucoma, *Bioorg. Med. Chem. Lett.*, 2010, **20**, 3361–3366.
50. P. L. Bosquesi, T. R. F. Melo, E. O. Vizioli, J. L. D. Santos and M. C. Chung, Anti-inflammatory drug design using a molecular hybridization approach, *Pharmaceuticals*, 2011, **4**, 1450–1474.

51. M. B. Goldring and M. Otero, Inflammation in osteoarthritis, *Curr. Opin. Rheumatol.*, 2011, **23**, 471–478.
52. N. Herrmann, S. A. Chau, I. Kircanski and K. L. Lanctôt, Current and emerging drug treatment options for Alzheimer's disease: A systematic review, *Drugs*, 2011, **71**, 2031–2065.
53. I. M. I. Fakhr, M. A. A. Radwan, S. El-Batran, O. M. E. Abd el-Salam and S. M. El-Shenawy, Synthesis and pharmacological evaluation of 2-substituted benzo[b]thiophenes as anti-inflammatory and analgesic agents, *Eur. J. Med. Chem.*, 2009, **44**, 1718–1725.
54. A. M. Isloor, B. Kalluraya and K. S. Pai, Synthesis, characterization and biological activities of some new benzo[b]thiophene derivatives, *Eur. J. Med. Chem.*, 2010, **45**, 825–830.
55. K. Mohammed Khan, Z. Nullah, M. Arif lodhi, S. Jalil, M. I. Choudhary and A. U. Rahman, Synthesis and anti-inflammatory activity of some selected aminothiophene analogs, *J. Enzy. Inhib. Med. Chem*, 2006, **21**, 139–143.
56. J. A. S. Mulla, M. I. A. Khazi, S. I. Panchamukhi, Y. Gong and I. A. M. Khazi, Synthesis and pharmacological evaluation of novel thienopyrimidine and triazolothienopyrimidine derivatives, *Med. Chem. Res.*, 2014, **23**, 3235–3243.
57. World Health Organization WHO, http://whqlibdoc.who.int/publication/2010/9789241564069_eng.pdf.
58. T. R. Frieden, T. R. Sterling, S. S. Munsiff, C. J. Watt and C. Dye, Tuberculosis, *Lancet*, 2008, **362**, 887–899.
59. N. S. Chandrasekera, M. A. Bailey, M. Files, T. Alling, S. K. Florio, J. Ollinger, J. O. Odingo and T. Parish, Synthesis and anti-tubercular activity of 3-substituted benzo*[b]*thiophene-1,1-dioxides, *PeerJ.*, 2014, **2**, 1–10.
60. G. K. Rao and R. Subramaniam, Synthesis, antitubercular and antibacterial activities of some quinazolinone analogs substituted with benzothiophene, *Chem. Sci. J.*, 2015, **6**, 92–96.
61. A. S. Narute, P. B. Khedekar and K. P. Bhusari, QSAR studies on 4-thiazolifdinones and 2-azetidinones bearing benzothiohene nucleus as potential anti-tubercular agents, *Ind. J. Chem.*, 2008, **47B**, 586–591.
62. K. Thaker, P. Zalavadiya and H. S. Joshi, Synthesis of some new imidazolones and 1,2,4-triazoles bearing benzo*[b]*thiophene nucleus as antimicrobial agents, *J. Sci. Islamic Repub. Iran.*, 2005, **16**, 139–144.
63. S. L. Vasoya, D. J. Paghdar, P. T. Chovatia and H. S. Joshi, Synthesis of some new thiosemicarbazide and 1,3,4-thiadiazole heterocycles bearing benzo[b]thiophene nucleus as a potent antitubercular and antimicrobial agents, *J. Sci., Islamic Repub. Iran.*, 2005, **16**, 33–36.
64. S. L. Vasoya, P. T. Chovatia, D. J. Paghdar and H. S. Joshi, Synthesis of some new thiosemicarbazide and 1,2,4-triazoles heterocycles bearing thiophene nucleus as a potent antitubercular and antimicrobial agents, *J. Ind. Chem. Soc.*, 2007, **84**, 709–710.

65. K. M. Thaker and H. S. Joshi, Synthesis and pharmacological evaluation of 2-(3′,5′-dichlorobenzo[b]thiophen-2′-yl)-5-arylamino-1,3,4-thiadiazoles, *Ind. J. Chem. Sec B*, 2005, **44B**, 410–412.
66. Y. Tanuja, S. Mishra, S. Das, S. Aggarwal and V. Rani, Anticedants and natural prevention of environmental toxicants induced accelerated aging of skin, *Environ. Toxicol. Pharmacol.*, 2014, **39**, 384–391.
67. N. Getoff, Anti-aging and aging factors in life. The role of free radicals, *Radiat. Phys. Chem.*, 2007, **76**, 1577–1586.
68. M. A. Gouda, H. F. Eldien, M. M. Girges and M. A. Berghot, Synthesis and antioxidant activity of novel series of naphthoquinone derivatives attached to benzothiophene moiety, *Med. Chem.*, 2013, **3**, 228–232.
69. M. A. Gouda and S. A. A. El Bialy, Synthesis and antioxidant evaluation of novel sophisticated carboxamides based on 3-(ethoxycarbonyl)-4, 5, 6, 7-tetrahydro-1-benzothiophen-2-amine, *Eur. J. Chem.*, 2014, **5**, 644–651.
70. K. Sepkowitz, AIDS-the first 20 years, *N. Engl. J. Med.*, 2001, **344**, 1764–1772.
71. H. Sharma, S. Patil, T. W. Sanchez, N. Neamati, R. F. Schinazi and J. K. Buolamwini, Synthesis, biological evaluation and 3D-QSAR studies of 3-keto salicylic acid chalcones and related amides as novel HIV-1 integrase inhibitors, *Bioorg. Med. Chem.*, 2011, **19**, 2030–2045.
72. C. L. Clouser, J. Chauhan, M. A. Bess, J. L. van Oploo, D. Zhou, S. Dimick-Gray, L. M. Mansky and S. E. Patterson, Anti-HIV-1 activity of resveratrol derivatives and synergistic inhibition of HIV-1 by the combination of resveratrol and decitabine, *Bioorg. Med. Chem. Lett.*, 2012, **22**, 6642–6646.
73. L. Chiummiento, M. Funicello, P. Lupattelli, F. Tramutola, F. Berti and F. Marino-Merlo, Synthesis and biological evaluation of novel small non-peptidic HIV-1 PIs: The benzothiophene ring as an effective moiety, *Bioorg. Med. Chem. Lett.*, 2012, **22**, 2948–2950.
74. K. Krajewski, Y. Zhang, D. Parrish, J. Deschamps, P. P. Roller and V. K. Pathak, New HIV-1 reverse transcriptase inhibitors based on a tricyclic benzothiophene scaffold: Synthesis, resolution, and inhibitory activity, *Bioorg. Med. Chem. Lett.*, 2006, **16**, 3034–3038.
75. C. L. Masters, R. Cappai, K. J. Barnham and V. L. Villemagne, Molecular mechanisms for Alzheimer's disease: implications for neuroimaging and therapeutics, *J. Neurochem.*, 2006, **97**, 1700–1725.
76. M. Prince, M. Prina and M. Guerchet, *World Alzheimer Report 2013: Journey of Caring – an Analysis of Long-term Care for Dementia*, September 2013.
77. E. S. Lee, B. W. Choi, D. Jung, H. J. Hwang, J. T. Hahn and B. H. Lee, Design and synthesis of phenyl boronic acids and benzothiophenones as anticholinesterases, *Bull. Korean Chem. Soc.*, 2003, **24**, 243–245.
78. F. M. I. Cerminara and L. Chiummiento, Heterocycles for alzheimer disease: 4- and 5-substituted benzothiophenes as starting scaffold in

the construction of potential new inhibitors of BACE 1, *Med. chem.*, 2016, **6**, 377–384.
79. S. Wild, G. Roglic, A. Green, R. Sicree and H. King, Global prevalence of diabetes: Estimates for the year 2000 and projections for 2030, *Diabetes Care*, 2004, **27**, 1047–1053.
80. L. S. Rotenstein, B. M. Kozak, J. P. Shivers, M. Yarchoan, J. Close and K. L. Close, The ideal diabetes therapy: what will it look like? how close are we?, *Clin. Diabetes*, 2012, **30**, 44–53.
81. R. M. Poole and R. T. Dungo, Ipragliflozin: First global approval, *Drugs*, 2014, **74**, 611–617.
82. M. Imamura, K. Nakanishi, T. Suzuki, K. Ikegai, R. Shiraki, T. Ogiyama, T. Murakami, E. Kurosaki, A. Noda, Y. Kobayashi, M. Yokota, T. Koide, K. Kosakai, Y. Ohkura, M. Takeuchi, H. Tomiyama and M. Ohta, Discovery of ipragliflozin (ASP1941): A novel C-glucoside with benzothiophene structure as a potent and selective sodium glucose co-transporter 2 (SGLT2) inhibitor for the treatment of type 2 diabetes mellitus, *Bioorg. Med. Chem.*, 2012, **20**, 3263–3279.
83. H. Shao, D. Li, Y. Yang, H. Guo, Z. Liu, S. Si, Z. Yang and Z. Li, The effect of substituted thiophene and benzothiophene derivates on PPARγ expression and glucose metabolism, *J. Enzyme Inhib. Med. Chem.*, 2010, **25**, 282–289.
84. W. Lubisch, H. P. Hofmann, H. J. Treiber and A. Möller, Synthesis and biological evaluation of novel piperidine carboxamide derived calpain inhibitors, *Bioorg. Med. Chem. Lett.*, 2000, **10**, 2187–2191.
85. I. Ling, B. Mihalik, L. Etherington, G. Kapus, A. Pálvölgyi, G. Gigler, S. Kertész, A. Gaál, K. Pallagi, P. Kiricsi, É. Szabó, G. Szénási, L. Papp, L. G. Hársing, G. Lévay, M. Spedding, J. J. Lambert, D. Belelli, J. Barkóczy, B. Volk, G. Simig, I. Gacsályi and F. A. Antoni, A novel GABA(A) alpha 5 receptor inhibitor with therapeutic potential, *Eur. J. Pharmacol.*, 2015, 497–507.
86. A. F. Zaher, N. A. Khalil and E. M. Ahmed, Synthesis and anticonvulsant activity of new 3′-aryl-7-bromo-spiro[[1]benzothiophene-3,2′-[1,3]thiazolidine]-2,4′-dione derivatives, *Orient. J. Chem.*, 2010, **26**, 1241–1248.
87. J. Rautio, J. Leppänen, M. Lehtonen, K. Laine, M. Koskinen, J. Pystynen, J. Savolainen and M. Sairanen, Design, synthesis and *in vitro/in vivo* evaluation of orally bioavailable prodrugs of a catechol-*O*-methyltransferase inhibitor, *Bioorg. Med. Chem. Lett.*, 2010, **20**, 2614–2616.
88. D. A. Brown, M. Mishra, S. Zhang, S. Biswas, I. Parrington, T. Antonio, M. E. A. Reith and A. K. Dutta, Investigation of various *N*-heterocyclic substituted piperazine versions of 5/7-{[2-(4-aryl-piperazin-1-yl)-ethyl]-propyl-amino}-5,6,7,8-tetrahydro-naphthalen-2-ol: Effect on affinity and selectivity for dopamine D3 receptor, *Bioorg. Med. Chem.*, 2009, **17**, 3923–3933.

89. Q. Jiang, J. Chen, C. Yang, Y. Yin and K. Yao, Quorum Sensing: A Prospective Therapeutic Target for Bacterial Diseases, *Biomed. Res. Int.*, 2019, **2019**, 2015978.
90. S. Y. Priyanka and G. P. Senthilkumar, Synthesis and Evaluation of Some Benzothiazole Derivatives as Antidiabetic Agents, *Int. J. Pharmtech. Res.*, 2011, **3**, 1–7.
91. R. S. Keri, K. Chand, S. Budagumpi, S. B. Somappa, S. A. Patil and B. M. Nagaraja, An Overview of Benzo[b]thiophene-Based Medicinal Chemistry, *Eur. J. Med. Chem.*, 2017, **138**, 1002–1033.
92. H. K. Nagesh,B. Padmashali, C. Sandeep, T. C. M. Yuvaraj, M. B. Siddesh and S. M. Mallikarjuna, Synthesis and Antimicrobial Activity of Benzothiophene Substituted Coumarins, Pyrimidines and Pyrazole as New Scaffold, *Int. J. Pharm. Sci. Rev. Res.*, 2014, **28**, 6–10.
93. R. Chawla, A. Arora, M. K. Parameswaran, P. C. Sharma, S. Michael and T. Ravi, Synthesis of Novel 1,3,4-Oxadiazole Derivatives as Potential Antimicrobial Agents, *Acta Poloniae Pharmaceutica-Drug Res.*, 2010, **67**, 247–253.
94. V. K. Meena, A. K. Patidar and M. Bhatnagar, Antimicrobial Activity of New Series of Benzothiophene Containing Thiazolidinone Derivatives, *J. Adv Sci. Res.*, 2020, **11**, 228–231.
95. P. S. Patil, S. M. Chavan, R. B. Toche, A. D. Bhole, V. M. Patil and P. B. Aware, Synthesis of (3-Cyano-4,5,6,7-Tetrahydrobenzo[b]Thiophen-2-yl)Formamidine Derivatives as Effective Antimicrobial Agents, *IOSR J. Appl. Chem.*, 2017, **10**, 96–100.
96. T. Hooda, S. Sharma and N. Goyal, Synthesis, in Silico Designing, Microbiological Evaluation and Structure Activity Relationship of Novel Amide Derivatives of 1-(2,4-Dinitrophenyl)-2-(3-Methylbenzo[b]Thiophen-6yl)-1H-Benzo[d]Imidazole-5-Carboxylic Acid, *Polycyclic Aromat. Compd.*, 2021, **42**, 3361–3376.
97. H. J. Kallur, P. S. Mathapati, C. K. Singh and A. Malpani, Synthesis, Characterization and Anthelmintic Screening of Some New Benzothiophene Derivatives, *World J. Pharm. Res.*, 2020, **9**, 1838–1849.

Subject Index